Lecture Notes
in Control and Information Sciences 364

Editors: M. Thoma, M. Morari

Alessandro Chiuso, Augusto Ferrante,
Stefano Pinzoni (Eds.)

Modeling, Estimation and Control

Festschrift in Honor of Giorgio Picci on the
Occasion of his Sixty-Fifth Birthday

 Springer

Editors

Prof. Alessandro Chiuso
Dipartimento di Tecnica e Gestione dei
Sistemi Industriali
Università di Padova
sede di Vicenza
stradella San Nicola, 3
I-36100 Vicenza, Italy
E-mail: chiuso@dei.unipd.it

Prof. Augusto Ferrante
Dipartimento di Ingegneria
dell'Informazione
Università di Padova
via Gradenigo, 6/B
I-35131 Padova, Italy
E-mail: augusto@dei.unipd.it

Prof. Stefano Pinzoni
Dipartimento di Ingegneria
dell'Informazione
Università di Padova
via Gradenigo, 6/B
I-35131 Padova, Italy
E-mail: pinzoni@dei.unipd.it

Library of Congress Control Number: 2007930050

ISSN print edition: 0170-8643
ISSN electronic edition: 1610-7411
ISBN-10   3-540-73569-0 Springer Berlin Heidelberg New York
ISBN-13   978-3-540-73569-4 Springer Berlin Heidelberg New York

Springer is a part of Springer Science+Business Media
springer.com

© Springer-Verlag Berlin Heidelberg 2007
Printed in Germany

Typesetting: By the authors and SPS using a Springer LaTeX macro package

Printed on acid-free paper   SPIN: 11777557   89/SPS   5 4 3 2 1 0

To Giorgio Picci

on the occasion of his sixty-fifth birthday

Preface

This *Festschrift* is intended as a homage to our esteemed colleague, friend and *maestro* Giorgio Picci on the occasion of his sixty-fifth birthday.

We have known Giorgio since our undergraduate studies at the University of Padova, where we first experienced his fascinating teaching in the class of System Identification. While progressing through the PhD program, then continuing to collaborate with him and eventually becoming colleagues, we have had many opportunities to appreciate the value of Giorgio as a professor and a scientist, and chiefly as a person. We learned a lot from him and we feel indebted for his scientific guidance, his constant support, encouragement and enthusiasm. For these reasons we are proud to dedicate this book to Giorgio.

The articles in the volume will be presented by prominent researchers at the "International Conference on Modeling, Estimation and Control: A Symposium in Honor of Giorgio Picci on the Occasion of his Sixty-Fifth Birthday", to be held in Venice on October 4-5, 2007.

The material covers a broad range of topics in mathematical systems theory, estimation, identification and control, reflecting the wide network of scientific relationships established during the last thirty years between the authors and Giorgio. Critical discussion of fundamental concepts, close collaboration on specific topics, joint research programs in this group of talented people have nourished the development of the field, where Giorgio has contributed to establishing several cornerstones.

We are happy and honored that these distinguished contributors have joined us in paying tribute to Giorgio as a token of esteem and friendship. We heartily thank them, with a special acknowledgment to Chris Byrnes, Anders Lindquist and Sanjoy Mitter for writing Giorgio's Laudatio.

With these feelings of gratitude and recognition, we all together wish the best to Giorgio for a happy birthday and many more to come.

Padova, May 2007

Alessandro Chiuso
Augusto Ferrante
Stefano Pinzoni

The book editors would like to gratefully thank the Department of Information Engineering, University of Padova for generously sponsoring the Conference and Springer-Verlag for publishing this book as a volume in the prestigious "Lecture Notes in Control and Information Sciences" Series.

Laudatio

Giorgio Picci has made profound contributions to several important topics in systems and control, notably stochastic realization theory, statistical theory of identification, image processing and dynamic vision.

The seminal paper [1] by Giorgio on splitting subspaces became the impetus for a whole field of stochastic systems theory. One of us (AL) had the privilege of long and fruitful collaboration with Giorgio in this direction, leading to contributions to stochastic realization theory [2], geometric theory of linear stochastic systems [3, 4, 5] and the geometric structure of matrix Riccati inequalities [6]. This research turned out to be timely and important as the system identification community turned their interest in the 90's toward a new class of identification procedures known as subspace methods, which turned out to be based on the same principles as those in the geometric theory of Markovian representations [7, 8].

Giorgio also applied the geometric theory of Markovian representations to statistical physics [9] and, together with Pinzoni, to factor analysis models [10, 11]. In a completely different direction, he has also studied different aspects of stochastic aggregation [12, 13] as well as positive Markov chains [14, 15]. In [16] he investigated the connections between the theory of sufficient statistics and the identifiability problem.

Among some of Giorgio's more recent contributions to subspace identification, we would like to mention [17], the follow-up papers [18, 19] co-authored with Katayama, and [20], which introduced oblique splitting subspaces as a tool for modeling systems with inputs. This is a key idea for understanding the geometry of subspace identification with inputs. The paper [21] together with Chiuso provides an in-depth analysis of the plethora of various subspace identification methods, showing that they are essentially all equivalent. This analysis, based on fundamental principles of stochastic realization theory, also provides simple expressions for the asymptotic variance of subspace identification estimates [22] and tools for understanding closed-loop subspace identification [23, 24]. These problems had remained open for a long time. Giorgio has also made important contributions to smoothing together with Ferrante [25] and the structure of stochastic systems together with Ferrante and Pinzoni [26].

In a quite different direction, Giorgio made a significant contribution to Dynamic Vision, especially through his students. The first problem is in "Structure from Motion," that is the reconstruction of the three-dimensional motion of the camera as well as the three-dimensional structure of the scene. Although there were prior attempts to use Kalman filters and their extended versions for this problem, Giorgio was the driving force behind its proper deployment, which led to a series of papers starting from the CDC 1994 paper "motion estimation via dynamic vision" [27]. The problem can be cast as a filtering problem where the state space is the product of a quotient space (the shape space of points in Euclidean space, in the sense of Kendall) and the Lie group of rigid motions. It is, however, a non-standard filtering problem, because the model is not observable (there is an ambiguous Gauge transformation), the state-space is variable (points can appear and disappear due to occlusions), and non-linear. Giorgio was the first to point out that there were observability issues to be studied, and that a proper model had to take into account the geometry of the state-space.

The second problem is in the study of "Dynamic Textures" [28] that are essentially stochastic realizations of video signals. The idea is to think of a video sequence as a realization of a linear system driven by white noise. The model colors the noise, and with a simple identification algorithm one can identify some 20-dimensional models that can be used to (a) simulate novel sequences, for instance of moving foliage, smoke, steam, fog etc. by just feeding Gaussian white noise to a linear system, and (b) to recognize these processes from video, for instance to detect smoke, fire etc. using distances between observability spaces. These ideas are straightforward, and in principle one would not need a background in identification theory or stochastic realization to have them. However, it is only because of Giorgio's work that his former students were able to apply these ideas to a different context. Although Giorgio was not an author in the first paper that appeared at ICCV 2001, his influence is direct and immediately visible in that paper, as well as in the many papers that followed from research groups in the US, Europe and Asia.

In image processing, as in the stochastic realization of signals with only finitely many known covariance lags, one needs to deal with finite sequences of data. In this case, however, the data are pixels and there are nontrivial boundary effects. As Giorgio has recently observed, this produces a problem far more complex than the traditional stochastic modeling problem, but one which Giorgio is currently researching using his wonderful ability to develop novel insights into elegant formalisms.

The University of Padova has today one of the strongest groups in Systems and Control in the world. There is no doubt whatsoever that this is primarily due to the scientific leadership of Giorgio Picci, a great researcher and teacher.

We congratulate Giorgio Picci – a great friend and a great scholar – on the occasion of his 65th birthday.

St. Louis, Stockholm, and Cambridge, May 2007

Christopher I. Byrnes
Anders Lindquist
Sanjoy K. Mitter

References

1. Picci G. (1976), Stochastic realization of Gaussian processes. Proc. IEEE 64: 112–122.
2. Lindquist A., Picci G. (1979), On the stochastic realization problem. SIAM J. Control Optim. 17: 365–389.
3. Lindquist A., Picci G. (1985), Realization theory for multivariate stationary Gaussian processes. SIAM J. Control Optim. 23: 809–857.
4. Lindquist A., Picci G. (1985), Forward and backward semimartingale models for Gaussian processes with stationary increments. Stochastics 15: 1–50.
5. Lindquist A., Picci G. (1991), A geometric approach to modelling and estimation of linear stochastic systems. J. Math. Systems Estim. Control 1: 241–333.
6. Lindquist A., Michaletzky Gy., Picci G. (1995), Zeros of spectral factors, the geometry of splitting subspaces and the algebraic Riccati inequality. SIAM J. Control Optim. 33: 365–401.
7. Lindquist A., Picci G. (1996), Canonical correlation analysis, approximate covariance extension, and identification of stationary time series. Automatica 32: 709–733.
8. Lindquist A., Picci G. (1996), Geometric methods for state space identification. In: S. Bittanti and G. Picci (eds), Identification, Adaptation, Learning: The Science of Learning Models from Data, Nato ASI Series (Series F, Vol. 153): 1–69, Springer.
9. Picci G. (1986), Application of stochastic realization theory to a fundamental problem of statistical physics. In: C.I. Byrnes and A. Lindquist (eds), Modelling, Identification and Robust Control: 211–258, North-Holland, Amsterdam.
10. Picci G., Pinzoni S. (1986), Factor analysis models for stationary stochastic processes. Analysis and optimization of systems, Lecture Notes in Control and Inform. Sci., 83: 411–424, Springer, Berlin.
11. Picci G. (1989), Parametrization of factor analysis models. J. Econometrics 41: 17–38.
12. Picci G. (1988), Stochastic aggregation. Linear Circuits, Systems and Signal Processing: Theory and Application: 493–501, North-Holland, Amsterdam.
13. Picci G., Taylor T.J. (1990), Stochastic aggregation of linear Hamiltonian systems with microcanonical distribution. Realization and modelling in system theory (Amsterdam, 1989), Progr. Systems Control Theory 3: 513–520, Birkhäuser, Boston.
14. Picci G., van Schuppen J. H. (1984), On the weak finite stochastic realization problem. Filtering and control of random processes (Paris, 1983), Lecture Notes in Control and Inform. Sci., 61: 237–242, Springer, Berlin.
15. Picci G., van den Hof J. M., van Schuppen J. H. (1998), Primes in several classes of the positive matrices. Linear Algebra Appl. 277: 149–185.
16. Picci G. (1977), Some connections between the theory of sufficient statistics and the identifiability problem. SIAM J. Appl. Math. 33: 383–398.
17. Picci G. (1997), Oblique splitting subspaces and stochastic realization with inputs. In: U. Helmke, D. Prätzel-Wolters and E. Zerz (eds), Operators, Systems and Linear Algebra: 157–174, Teubner, Stuttgart.
18. Picci G., Katayama T. (1996), Stochastic realization with exogenous inputs and subspace identification methods. Signal Processing 52: 145–160.
19. Katayama T., Picci G. (1999), Realization of stochastic systems with exogenous inputs and subspace identification methods. Automatica 35: 1635–1652.
20. Picci G. (1997) Stochastic realization and system identification. In: T. Katayama and S. Sugimoto (eds), Statistical Methods in Control and Signal Processing: 1–63, M. Dekker.
21. Chiuso A., Picci G. (2004), On the ill-conditioning of subspace identification with inputs. Automatica 40: 575–589.

22. Chiuso A., Picci G. (2004), Asymptotic variance of subspace estimates. Journal of Econometrics 118: 257–291.
23. Chiuso A., Picci G. (2005), Prediction error vs. subspace methods in closed loop identification. Proc. of the 16th IFAC World Congress, Prague.
24. Chiuso A., Picci G. (2005), Consistency analysis of some closed-loop subspace identification methods. Automatica 41: 377-391.
25. Ferrante A., Picci G. (2000), Minimal realization and dynamic properties of optimal smoothers. IEEE Trans. Automatic Control 45: 2028–2046.
26. Ferrante A., Picci G., Pinzoni S. (2002), Silverman algorithm and the structure of discrete-time stochastic systems. Linear Algebra Appl. 351-352: 219–242.
27. Soatto S., Perona P., Frezza R. and Picci G. (1994), Motion estimation via dynamic vision. Proc. CDC94, Orlando, pp. 3253–3258.
28. Soatto S., Doretto G. and Wu Y. (2001), Dynamic textures. Proc. of the Intl. Conf. on Computer Vision, 2001, pp. 439–446.

Contents

Differential Forms and Dynamical Systems

An Algebraic Framework for Bayes Nets of Time Series

A Birds Eye View on System Identification

Global Identifiability of Complex Models, Constructed from Simple Submodels

Identification of Hidden Markov Models - Uniform LLN-s

Identifiability and Informative Experiments in Open and Closed-Loop Identification

Prediction-Error Approximation by Convex Optimization

Patchy Solutions of Hamilton-Jacobi-Bellman Partial Differential Equations

A Geometric Assignment Problem for Robotic Networks

New Development of Digital Signal Processing Via Sampled-Data Control Theory

List of Contributors

Masanao Aoki
Department of Economics, University of California, Los Angeles,
Los Angeles, CA 90095-1477, USA.
`aoki@econ.ucla.edu`

Alexandre Sanfelice Bazanella
Electrical Engineering Department, Universidade Federal do Rio Grande do Sul,
Av. Osvaldo Aranha 103, 90035-190 Porto Alegre-RS, Brazil.
`bazanela@ece.ufrgs.br`

Sergio Bittanti
Dipartimento di Elettronica e Informazione, Politecnico di Milano,
Piazza Leonardo da Vinci 32, I-20133 Milano, Italy.
`bittanti@elet.polimi.it`

Jeroen Boets
Department of Electrical Engineering (ESAT-SCD), Katholieke Universiteit Leuven,
Kasteelpark Arenberg 10, B-3001 Leuven, Belgium.
`jeroen.boets@esat.kuleuven.be`

Francesco Bullo
Department of Mechanical Engineering, Center for Control, Dynamical Systems and
Computation, University of California, Santa Barbara, CA 93106-5070, USA.
`bullo@engineering.ucsb.edu`

Christopher I. Byrnes
Department of Electrical and Systems Engineering, Washington University in St. Louis,
One Brookings Drive, St. Louis, MO 63130, USA.
`chrisbyrnes@wustl.edu`

Peter E. Caines
Department of Electrical and Computer Engineering, McGill University,
3480 University Street, Montreal, QC H3A 2A7, Canada.
`peterc@cim.mcgill.ca`

Marco C. Campi
Dipartimento di Elettronica per l'Automazione, Università di Brescia,
Via Branze 38, I-25123 Brescia, Italy.
marco.campi@ing.unibs.it

Katrien De Cock
Department of Electrical Engineering (ESAT-SCD), Katholieke Universiteit Leuven,
Kasteelpark Arenberg 10, B-3001 Leuven, Belgium.
katrien.decock@esat.kuleuven.be

Manfred Deistler
Institut für Wirtschaftsmathematik, Forschungsgruppe Ökonometrie und System-
theorie, Technische Universität Wien, Argentinierstraße 8, A-1040 Wien, Austria.
Manfred.Deistler@tuwien.ac.at

Bart De Moor
Department of Electrical Engineering (ESAT-SCD), Katholieke Universiteit Leuven,
Kasteelpark Arenberg 10, B-3001 Leuven, Belgium.
bart.demoor@esat.kuleuven.be

Augusto Ferrante
Dipartimento di Ingegneria dell'Informazione, Università di Padova,
Via Gradenigo 6/B, I-35131 Padova, Italy.
augusto@dei.unipd.it

Lorenzo Finesso
Institute of Biomedical Engineering, CNR-ISIB, Padova,
Corso Stati Uniti 4, I-35127 Padova, Italy.
lorenzo.finesso@isib.cnr.it

Paul A. Fuhrmann
Department of Mathematics, Ben-Gurion University of the Negev,
P.O.B. 653, Beer Sheva 84105, Israel.
fuhrmannbgu@gmail.com

Tryphon T. Georgiou
Department of Electrical and Computer Engineering, University of Minnesota,
200 Union Street S.E., Minneapolis, MN 55455, USA.
tryphon@umn.edu

Markus Gerdin
NIRA Dynamics AB, Gothenburg, Sweden.
markus.gerdin@gmail.com

László Gerencsér
MTA SZTAKI (Computer and Automation Institute, Hungarian Academy of Sciences),
Kende u. 13-17, H-1111 Budapest, Hungary.
gerencser@sztaki.hu

Michel Gevers
CESAME, Université Catholique de Louvain,
Avenue Georges Lemaître 4, B-1348 Louvain-la-Neuve, Belgium.
gevers@csam.ucl.ac.be

Torkel Glad
Automatic Control, Department of Electrical Engineering, Linköping University,
SE-58183 Linköping, Sweden.
torkel@isy.liu.se

Andrea Gombani
Institute of Biomedical Engineering, CNR-ISIB, Padova,
Corso Stati Uniti 4, I-35127 Padova, Italy.
gombani@isib.cnr.it

Uwe Helmke
Institut für Mathematik, Universität Würzburg,
D-97074 Würzburg, Germany.
helmke@mathematik.uni-wuerzburg.de

Daniel Holder
Department of Mathematics and Statistics, Texas Tech University,
2500 Broadway, Lubbock, TX 79409-1042, USA.
daniel.holder@ttu.edu

Xiaoming Hu
Optimization and Systems Theory, Department of Mathematics, Royal Institute of
Technology, SE-10044 Stockholm, Sweden.
hu@kth.se

Lin Huo
Department of Mathematics and Statistics, Texas Tech University,
2500 Broadway, Lubbock, TX 79409-1042, USA.
lin.huo@ttu.edu

Magnus Jansson
Signal Processing, School of Electrical Engineering, Royal Institute of Technology,
SE-10044 Stockholm, Sweden.
magnus.jansson@ee.kth.se

Maja Karasalo
Optimization and Systems Theory, Department of Mathematics, Royal Institute of
Technology, SE-10044 Stockholm, Sweden.
karasalo@math.kth.se

Tohru Katayama
Faculty of Culture and Information Science, Doshisha University,
KyoTanabe, Kyoto 610-0394, Japan.
tokataya@mail.doshisha.ac.jp

Matthias Kawski
Department of Mathematics and Statistics, Arizona State University,
Tempe, AZ 85287-1804, USA.
kawski@asu.edu

Arthur J. Krener
Department of Applied Mathematics, Naval Postgraduate School,
Monterey, CA 93943-5216, USA.
ajkrener@nps.edu

Anders Lindquist
Optimization and Systems Theory, Department of Mathematics, Royal Institute of
Technology, SE-10044 Stockholm, Sweden.
alq@math.kth.se

Lennart Ljung
Automatic Control, Department of Electrical Engineering, Linköping University,
SE-58183 Linköping, Sweden.
ljung@isy.liu.se

Clyde F. Martin
Department of Mathematics and Statistics, Texas Tech University,
2500 Broadway, Lubbock, TX 79409-1042, USA.
clyde.f.martin@ttu.edu

Ted Matsko
ABB USA.
ted.matsko@us.abb.com

György Michaletzky
Department of Probability Theory and Statistics, Eötvös Loránd University,
Pázmány Péter sétány 1/C, H-1117 Budapest, Hungary.
michgy@ludens.elte.hu

Ljubiša Mišković
CESAME, Université Catholique de Louvain,
Avenue Georges Lemaître 4, B-1348 Louvain-la-Neuve, Belgium.
miskovic@csam.ucl.ac.be

Sanjoy K. Mitter
LIDS, Massachusetts Institute of Technology,
77 Massachusetts Avenue, Cambridge, MA 02139-4307, USA.
mitter@mit.edu

Mats A. Molander
ABB Corporate Research, Västerås, Sweden.
`mats.a.molander@se.abb.com`

Gábor Molnár-Sáska
Morgan Stanley Hungary Analytics, Budapest,
Deák Ferenc u. 15, H-1052 Budapest, Hungary.
`gabor.molnar-saska@morganstanley.com`

Carmeliza Navasca
ETIS Lab - UMR CNRS 8051,
Avenue du Ponceau 6, F-95014 Cergy-Pontoise, France.
`cnavasca@gmail.com`

Michele Pavon
Dipartimento di Matematica Pura ed Applicata, Università di Padova,
Via Trieste 63, I-35131 Padova, Italy.
`pavon@math.unipd.it`

Maria Prandini
Dipartimento di Elettronica e Informazione, Politecnico di Milano,
Piazza Leonardo da Vinci 32, I-20133 Milano, Italy.
`prandini@elet.polimi.it`

Federico Ramponi
Dipartimento di Ingegneria dell'Informazione, Università di Padova,
Via Gradenigo 6/B, I-35131, Padova, Italy.
`rampo@dei.unipd.it`

Stephen L. Smith
Department of Mechanical Engineering, Center for Control, Dynamical Systems and
Computation, University of California, Santa Barbara, CA 93106-5070, USA.
`stephen@engineering.ucsb.edu`

Stefano Soatto
Computer Science Department, University of California, Los Angeles,
3531 Boelter Hall, Los Angeles, CA 90095-1596, USA.
`soatto@ucla.edu`

Peter Spreij
Korteweg-de Vries Institute for Mathematics, Universiteit van Amsterdam,
Plantage Muidergracht 24, 1018 TV Amsterdam, The Netherlands.
`spreij@science.uva.nl`

Thomas J. Taylor
Department of Mathematics and Statistics, Arizona State University,
Tempe, AZ 85287-1804, USA.
`tom.taylor@asu.edu`

Jan H. van Schuppen
CWI, P.O. Box 94079, 1090 GB Amsterdam, The Netherlands.
`J.H.van.Schuppen@cwi.nl`

Bo Wahlberg
Automatic Control, School of Electrical Engineering, Royal Institute of Technology,
SE-10044 Stockholm, Sweden.
`bo.wahlberg@ee.kth.se`

Jan C. Willems
Department of Electrical Engineering (ESAT-SISTA),Katholieke Universiteit Leuven,
Kasteelpark Arenberg 10, B-3001 Leuven, Belgium.
`Jan.Willems@esat.kuleuven.be`

Henry P. Wynn
Department of Statistics, London School of Economics and Political Science,
Houghton Street, London WC2A 2AE, UK.
`h.wynn@lse.ac.uk`

Yutaka Yamamoto
Department of Applied Analysis and Complex Dynamical Systems, Graduate School
of Informatics, Kyoto University, Kyoto 606-8501, Japan.
`yy@i.kyoto-u.ac.jp`

Coefficients of Variations in Analysis of Macro-policy Effects: An Example of Two-Parameter Poisson-Dirichlet Distributions

Masanao Aoki[*]

Department of Economics, Univ. California, Los Angeles
aoki@econ.ucla.edu

Summary. A class of two-parameter Poisson-Dirichlet distributions have non-vanishing coefficient of variation. This phenomenon is also known as non-self averaging stochastic multi-sector endogenous growth model. This model is used to raise questions on the use of means in assessing effectiveness of macroeconomic or other macro-policy effects. The coefficients of variations of the number of total sectors, and of sectors of a given size all remain positive as the model size grows unboundedly.

Keywords: random combinatorial structure, self-averaging, thermodynamic limit, coefficients of variation.

1   Introduction

This paper discusses a new class of simple stochastic multi-sector growth models in which macroeconomic variables such as the number of sectors or gross outputs have non-vanishing coefficients of variations. This is called non-self averaging in physics literature. For example, as the sizes of models grow as time passes[1], the coefficients of variation of the number of sectors in the model does not converge to zero, but remain positive. This indicates that the model is influenced by history, and is non self-averaging in the language of statistical physics. We show that the class of one-parameter Poisson-Dirichlet models, also known as Ewens models in population genetics, is self-averaging, that is, its coefficient of variations tends to zero as time passes, but its extension to two-parameter Poisson-Dirichlet models by Pitman [6] is not self-averaging.

This fact has an important implication on the effectiveness of macroeconomic policies based on the expected values of model performances, because the actual values of some performance index do not cluster around the expected values when the macroeconomic variable is not self-averaging.

---

[*] Fax number 1-310-825-9528, Tel. no. 1-310-825-2360, aoki@econ.ucla.edu. The author thanks M. Sibuya for useful discussions.

[1] Thus, this model is different from those models which are inhabited by an infinite number of agents from the beginning.

A. Chiuso et al. (Eds.): Modeling, Estimation and Control, LNCIS 364, pp. 1–4, 2007.

2 The Model

Consider an economy composed of several sectors. Different sectors are made up of different types of agents or productive units. The model sectors are thus heterogeneous. Counting the sizes of sectors in some basic units, when the economy is of size n, there are K_n sectors, that is K_n types of agents or productive units in the economy. The number K_n as well as the sizes of individual sectors, n_i, $i = 1, \ldots, K_n$, are random variables.

Time runs continuously. Over time, one of the existing sectors grows by one unit at a rate which is proportional to $(n_i - \alpha)/(n + \theta)$, $i = 1, \ldots K_n$, where α is a parameter between 0 and 1, and θ is another parameter, $\theta + \alpha > 0$. The rate at which a new sector emerges in the economy is equal to $1 - \sum_{i=1}^{k}(n_i - \alpha)(n + \theta) = (\theta + k\alpha)(n + \theta)$.[2] The probability that a new sector emerges is expressible then as

$$q_{\alpha,\theta}(n + 1, k) = \frac{n - k\alpha}{n + \theta} q_{\alpha,\theta}(n, k) + \frac{\theta + (k - 1)\alpha}{n + \theta} q_{\alpha,\theta}(n, k - 1). \tag{1}$$

where $q_{\alpha,\theta}(n, k) := \Pr(K_n = k)$.

Eq. (1) states that the economy composed of k sectors increases in size by one unit either by one of the existing sectors growing by one unit, or by a new sector of size one emerging. We assume that a new sector always begins its life with a single unit. We can restate it as

$$\Pr(K_{n+1} = k + 1 | K_n = k) = \frac{k\alpha + \theta}{n + \theta}, \text{ and } \Pr(K_{n+1} = k | K_n = k) = \frac{n - k\alpha}{n + \theta}. \tag{2}$$

Note that more new sectors are likely to emerge in the economy as the numbers of sectors grow.

3 Asymptotic Properties of the Number of Sectors

We next examine how the number of sectors behaves as the size of the model grows unboundedly. We know how it behaves when α is zero. It involves Stirling number of first kind, see [1], for example. With positive α, the generalized Stirling number of the first kind, $c(n, k; \alpha)$, is involved in its expression. Dropping the subscripts α, θ from Pr, we write

$$\Pr(K_n = k) = \frac{\theta^{[k,\alpha]}}{\alpha^k \theta^{[n]}} c(n, k; \alpha), \tag{3}$$

where $\theta^{[k,\alpha]} := \theta(\theta + \alpha) \cdots (\theta + (k - 1)\alpha)$, and $\theta^{[n]} := \theta^{[n,1]} = \theta(\theta + 1) \cdots (\theta + n - 1)$. See [3] or [8] for their properties.

Define $S_\alpha(n, k) = c(n, k; \alpha)/\alpha^k$. It satisfies a recursion equation

$$S_\alpha(n, k) = (n - k\alpha)S_\alpha(n, k) + S_\alpha(n, k - 1). \tag{4}$$

This function generalizes the power-series relation for $\theta^{[n]} = \sum_1^n c(n, k)\theta^k$, to $\theta^{[n]} = \sum S_\alpha(n, k)\theta^{[k,\alpha]}$. See [1] for example.

---

[2] Our θ is $\beta - \alpha$ in [5].

4   The Coefficients of Variation

4.1   The Number of Sectors

Yamato and Sibuya [8] have calculated the moments of K_n^r, $r = 1, 2, \ldots$, recursively. For example they derive a recursion relation

$$E(K_{n+1}) = \frac{\theta}{n+\theta} + (1 + \frac{\alpha}{n+\theta})E(K_n)$$

from which they obtain

$$E[\frac{K_n}{n^\alpha}] \sim \frac{\Gamma(\theta+1)}{\alpha\Gamma(\theta+\alpha)} \tag{5}$$

by applying the asymptotic expression for the Gamma function

$$\frac{\Gamma(n+a)}{\Gamma(n)} \sim n^a. \tag{6}$$

They also obtain the expression for the variance of K_n/n^α as

$$var(K_n/n^\alpha) \sim \frac{\Gamma(\theta+1)}{\alpha^2}\gamma(\alpha,\theta), \tag{7}$$

where $\gamma(\alpha,\theta) := (\theta+\alpha)/(\Gamma(\theta+\alpha)) - \Gamma(\theta+1)/[\Gamma(\theta+\alpha)]^2$.

The expression for the coefficient of variation of K_n normalized by n^α then is given by

$$limC.V.(K_n/n^\alpha) = \frac{\Gamma(\theta+\alpha)}{\Gamma(\theta+1)}\sqrt{\gamma(\alpha,\theta)}. \tag{8}$$

Note that the expression $\gamma(\alpha,\theta)$ is zero when α is zero, and positive otherwise. We state this result as

Proposition. The limit of the coefficient of variation is positive with positive α, and it is zero only with $\alpha = 0$.

In other words, models with $o < \alpha < 1$ are non self-averaging. Past events influence the path of the growth of this model, i.e., the model experiences non ergodic growth path.

4.2   The Number of Sectors of Specified Size

Let $a_j(n)$ be the number of sectors of size j when the size of the economy is n. From the definitions, note that $K_n = \sum_j a_j(n)$, and $\sum_j ja_j(n)_n$, where j ranges from 1 to n.

The results in [8] can be used to show that the limit of the coefficient of variation of $a_j(n)/n^\alpha$ as n goes to infinity has the same limiting behavior as K_n/n^α, i.e., zero for $\alpha = 0$, and positive for $0 < \alpha < 1$.

5   Discussion

This short note shows that the one-parameter Poisson-Dirichlet model, known as Ewens model in the population genetics literature, is self-averaging, but its extension, two-parameter Poisson-Dirichlet models are not. The behavior of the latter models is history or sample-path dependent. Given such macroeconomic models, the usual practice of minimizing the means of some performance index is not satisfacotry. Performance indices of such models may be fat-tailed, and minimizing the means may not be satisfactory.

We discuss how this type of models is important in macroeconomics and finance modeling.

The two-parameter models are significant because their moments are related to those of the Mittag-Leffler distribution in a simple way, and as Darling-Kac theorem implies, [4], any analysis involving first passages, occupation times, waiting time distributions and the like are bound to involve the Mittag-Leffler functions. In other words, Mittag-Leffler functions are generic in examining model behaviors as the model sizes grow unboundedly.

One straightforward way to link the moments of K_n/n^α to the generalized Mittag-Leffler function $g_{\alpha,\theta}(x) := \frac{\Gamma(\theta+1)}{\Gamma(\mu+1)} x^\mu g_\alpha(x)$, where $\mu := \theta/\alpha$, and where g_α is a probability density with moments

$$\int_0^\infty x^p g_\alpha(x)dx = \frac{\Gamma(p+1)}{\Gamma(p\alpha+1)}, \tag{9}$$

for $p = 0, 1, \ldots$, is to apply the method of moments, [2].

Using the Laplace transform of the Mittag-Leffler function, Mainardi and his associate and colleagues have discussed fractional calculus, and fractional master equations, with applications to financial problems in mind, Mainardi et al. For example see [7].

The class of models in this note may thus turn out to be important not only in finance but also in macroeconomics.

References

1. Aoki, M., (2002) Modeling Aggregate Behavior and Fluctuations in Economics : Stochastic Views of Interacting Agents, (Cambridge Univ. Press, New York).
2. Breiman, L., (1992) Probability, (Siam, Philadelphia).
3. Charalambides, Ch., (2002) Enumerative Combinatorics, (Chapman & Hall/CRC, London).
4. Darling, D. A., and M. Kac (1957) On occupation-times for Markov processes, Transactions of American Mathematical Society, 84, 444-458.
5. Feng, S., and F. M. Hoppe, (1998) Large deviation principles for some random combinatorial structures, The Annals of Applied Probability, 8, 975–994.
6. Pitman, J., (1999) Characterizations of Brownian motion, bridge, meander and excursion by sampling at independent uniform times, Electronic J. Probability 4, Paper 11, 1–33.
7. Scalas, E., (2006) The application of continuous-time random walks in finance and economics, Physica A 362, 225-239.
8. Yamato, H., and M. Sibuya, (2000) Moments of some statistics of Pitman Sampling formula, Bulletin of Informatics and Cybernetics, 32, 1–10, 2000.

How Many Experiments Are Needed to Adapt?[*]

Sergio Bittanti[1], Marco C. Campi[2], and Maria Prandini[1]

[1] Dipartimento di Elettronica e Informazione - Politecnico di Milano,
   piazza Leonardo da Vinci 32, 20133 Milano, Italia
   `{bittanti,prandini}@elet.polimi.it`
[2] Dipartimento di Elettronica per l'Automazione - Università di Brescia,
   via Branze 38, 25123 Brescia, Italia
   `marco.campi@ing.unibs.it`

Summary. System design in presence of uncertainty calls for experimentation, and a question that arises naturally is: how many experiments are needed to come up with a system meeting certain performance requirements?

This contribution represents an attempt to answer this fundamental question. Results are confined to a specific set-up where adaptation is performed according to a worst-case perspective, but many considerations and reflections are central to adaptation in general.

1 Introduction

Given a system $\mathcal{S}$, consider the problem of designing a device $\mathcal{D}$ that achieves some desired behavior when interacting with $\mathcal{S}$. The specification of the 'desired behavior' depends on the intended use of the device, and is usually expressed in terms of some signal $s_{\mathcal{D}}(\omega)$, with reference to certain operating conditions $\omega \in \Omega$ of interest (Figure 1).

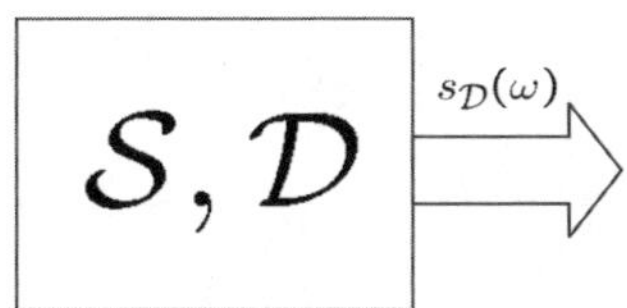

Fig. 1. Characterization through signal $s_{\mathcal{D}}(\omega)$ of device $\mathcal{D}$ while interacting with system $\mathcal{S}$ in the operating condition ω

Example 1 (simulator). Suppose that the device should act as a simulator of the system when the system input u takes on value in a given class of signals U. In this case, $\omega = u$ and the desired behavior for the device can be expressed in terms of the multidimensional signal $s_{\mathcal{D}}(u) = (y(u), y_{\mathcal{D}}(u))$, where $y_{\mathcal{D}}(u)$ and $y(u)$ represent the outputs of

[*] This work is supported by MIUR (Ministero dell'Istruzione, dell'Università e della Ricerca) under the project *Identification and adaptive control of industrial systems* and by CNR - IEIIT.

A. Chiuso et al. (Eds.): Modeling, Estimation and Control, LNCIS 364, pp. 5–14, 2007.
springerlink.com   © Springer-Verlag Berlin Heidelberg 2007

the device and of the system fed by the same input signal $u \in U$ (see Figure 2). Signal $s_{\mathcal{D}}(u)$ should be such that $y_{\mathcal{D}}(u) \simeq y(u)$, for every operating condition of interest, that is for every $u \in U$. ∎

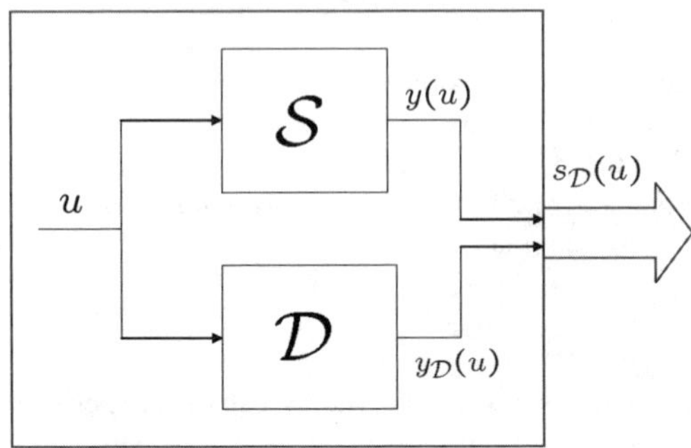

Fig. 2. Device $\mathcal{D}$ acting as a simulator of system $\mathcal{S}$

Example 2 (disturbance compensator). Suppose that the output of system $\mathcal{S}$ is affected by some additive disturbance and the device $\mathcal{D}$ is introduced for compensating the disturbance according to the feedforward scheme in Figure 3. In this case the operating condition is defined by the disturbance realization d. If we denote by $y_{\mathcal{D}}(d)$ the controlled output of the system when the disturbance realization is d, then the desired behavior can be expressed in terms of the signal $s_{\mathcal{D}}(d) = y_{\mathcal{D}}(d)$ and $s_{\mathcal{D}}(d)$ should be small for every d in some set D. ∎

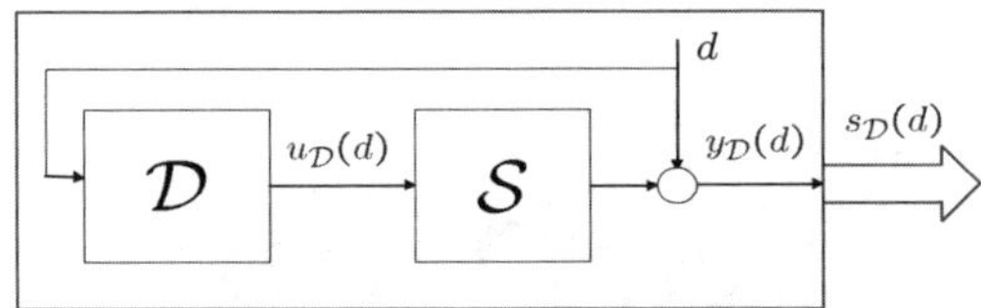

Fig. 3. Device $\mathcal{D}$ acting as a disturbance compensator for system $\mathcal{S}$

Devising a suitable $\mathcal{D}$ for a system $\mathcal{S}$ requires knowledge of some sort on $\mathcal{S}$. Most literature in science and engineering relies on a model-based approach, namely it is assumed that a mathematical model for $\mathcal{S}$ is a-priori available. Alternatively, the knowledge on $\mathcal{S}$ can be accrued through experimentation. This latter approach, considered herein, is referred to as 'adaptive design', [1, 2, 3, 4, 5], since the problem is to adapt $\mathcal{D}$ on the basis of experiments in the face of the lack of a-priori knowledge on system $\mathcal{S}$.

In adaptive design, one fundamental question to ask is:

> *How extensively do we need to experiment in order to come up with a device meeting certain performance requirements?*

This fundamental –and yet largely unanswered– question is the theme this contribution is centered around.

In this paper, a *worst-case perspective* with respect to the possible operating conditions is adopted, and we provide an answer to the above question in this specific set-up. For one answer, many more are the answers that this contribution is incapable to provide, which will also be enlightened along our way.

2 Worst-Case Approach to Adaptation

2.1 Worst-Case Performance

Suppose that the performance of device $\mathcal{D}$ operating in condition ω is quantified by a cost $c(s_\mathcal{D}(\omega))$. Then, the worst-case performance achieved by $\mathcal{D}$ over the set Ω of operating conditions is

$$\max_{\omega \in \Omega} c(s_\mathcal{D}(\omega)),$$

and, correspondingly, one wants to design

$$\mathcal{D}^\star = \arg \min_\mathcal{D} \max_{\omega \in \Omega} c(s_\mathcal{D}(\omega)). \tag{1}$$

$c^\star$ denotes the worst-case performance of device $\mathcal{D}^\star$, that is $c^\star = \max_{\omega \in \Omega} c(s_{\mathcal{D}^\star}(\omega))$.

In e.g. the simulator Example 1, $\omega = u$ and one can take $c(s_\mathcal{D}(u)) = \|y(u) - y_\mathcal{D}(u)\|_2$, the 2-norm of the error signal $y(u) - y_\mathcal{D}(u)$. $c^\star$ can then be interpreted as an upper bound to the largest 2-norm discrepancy between the system behavior and the behavior of the simulator $\mathcal{D}^\star$ in the same operating condition:

$$\|y(u) - y_{\mathcal{D}^\star}(u)\|_2 \le c^\star, \ \forall u \in U.$$

In the disturbance compensator Example 2, a sensible cost is the 2-norm $c(s_\mathcal{D}(d)) = \|y_\mathcal{D}(d)\|_2$. Then, the best disturbance compensator $\mathcal{D}^\star$ satisfies:

$$\|y_{\mathcal{D}^\star}(d)\|_2 \le c^\star, \ \forall d \in D.$$

In many cases, the device $\mathcal{D}$ is parameterized by a vector $\gamma \in \Re^k$, in which case we write $\mathcal{D}_\gamma$ to indicate device $\mathcal{D}$ with parameter γ, and hence designing a device corresponds to selecting a value for γ. Then, with the shorthand

$$J_\gamma(\omega) := c(s_{\mathcal{D}_\gamma}(\omega)),$$

the min-max optimization problem (1) can be rewritten as the following robust optimization program with $k + 1$ optimization variables:

$$\text{RP}: \min_{\gamma, c \in \Re^{k+1}} c \quad \text{subject to:} \tag{2}$$

$$J_\gamma(\omega) \le c, \ \forall \omega \in \Omega.$$

Note that, given a γ, the slack variable c represents an upper bound on the cost $J_\gamma(\cdot)$ achieved over Ω by the device with parameter γ. By solving (2) we seek that $\gamma^\star$ that corresponds to the smallest upper bound $c^\star$.

2.2   Adaptive Design

In model-based design, the cost $J_\gamma(\omega)$ can be evaluated based on the model, and then $\gamma^\star$ is found by solving the robust optimization program (2). Instead, when system $\mathcal{S}$ is unknown or only partially known, the cost $J_\gamma(\omega)$ cannot be explicitly computed so that the constraints in (2) are not known. However, one can conceive of evaluating the constraints experimentally. What exactly this means is discussed in the sequel.

Each constraint is associated with an operating condition $\omega \in \Omega$. To evaluate experimentally a constraint in a specific condition, say $\hat{\omega} \in \Omega$, that is to determine experimentally the domain of feasibility in the (γ, c)'s space where the constraint $J_\gamma(\hat{\omega}) \leq c$ holds, one should run a set of experiments, all in the $\hat{\omega}$ condition, each of which performed with a different device $\mathcal{D}_\gamma$, $\gamma \in \Re^k$. In this way $s_{\mathcal{D}_\gamma}(\hat{\omega})$ is measured for every γ and $J_\gamma(\hat{\omega})$ can be computed. An objection to this way of proceeding is that it would require in principle to test the performance achieved with every and each device $\mathcal{D}_\gamma$ in place. It is an interesting fact that in many situations the overwhelming experimental effort involved in testing many times with different $\mathcal{D}_\gamma$'s can be avoided, and just one single experiment is enough for the purpose of computing $J_\gamma(\hat{\omega})$.

Take e.g. the simulator Example 1. In this example, if $\hat{u}$ is injected into $\mathcal{S}$, signal $\hat{y} = \mathcal{S}[\hat{u}]$ can be collected, along with signal $\hat{u}$ itself. Based on this single experiment, one can then compute $y(\hat{u}) - y_{\mathcal{D}_\gamma}(\hat{u}) = \hat{y} - \mathcal{D}_\gamma[\hat{u}]$ for all γ's, where $\mathcal{D}_\gamma[\hat{u}]$ is obtained by filtering $\hat{u}$ with $\mathcal{D}_\gamma$, an operation that can be executed as an off-line post-process of signal $\hat{u}$. After $y(\hat{u}) - y_{\mathcal{D}_\gamma}(\hat{u})$ has been computed, the constraint $\|y(\hat{u}) - y_{\mathcal{D}_\gamma}(\hat{u})\|_2 = J_\gamma(\hat{u}) \leq c$ is evaluated.

The same conclusion that one experiment is enough can also be drawn for Example 2 whenever both the system and the device are linear. Indeed, swapping the order of $\mathcal{S}$ and $\mathcal{D}_\gamma$, we have:

$$y_{\mathcal{D}_\gamma}(d) = \mathcal{S}[\mathcal{D}_\gamma[d]] + d = \mathcal{D}_\gamma[\mathcal{S}[d]] + d. \tag{3}$$

If we run an experiment in which disturbance $\hat{d}$ is measured and this disturbance is also injected as input to the system (i.e. $\mathcal{D}$ is set to 1 during experimentation in the scheme of Figure 3), from the measured system output $\hat{y} = \mathcal{S}[\hat{d}] + \hat{d}$ and from $\hat{d}$ itself we can then determine

$$\begin{aligned}
y_{\mathcal{D}_\gamma}(\hat{d}) &= \mathcal{D}_\gamma[\mathcal{S}[\hat{d}]] + \hat{d} \quad \text{(using (3))} \\
&= \mathcal{D}_\gamma[\hat{y} - \hat{d}] + \hat{d},
\end{aligned}$$

where computation of $\mathcal{D}_\gamma[\hat{y} - \hat{d}]$ is executed off-line similarly to the simulator example. By computing $\|y_{\mathcal{D}_\gamma}(\hat{d})\|_2 = J_\gamma(\hat{d})$ constraint $J_\gamma(\hat{d}) \leq c$ is then evaluated.

In the sequel we shall assume that one single experiment in condition $\hat{\omega}$ suffices to determine constraint $J_\gamma(\hat{\omega}) \leq c$. This assumption is not fulfilled in all applications of the adaptive scheme, and further discussion on this point is provided in Section 5.

Remark 1. The reader may have noticed that lack of knowledge, for which adaptation is required, can enter the problem in different ways. In Example 1, it was system $\mathcal{S}$ to be unknown. In the disturbance compensator Example 2, again uncertainty stayed with the system $\mathcal{S}$, but even the set D for d could be unknown.

The seemingly different nature of the uncertainty in S and in D can be leveled off by adopting a more abstract behavioral perspective, [6], where the system is just seen as a set of behaviors, i.e. of possible realizations of system signals. In such framework, uncertainty simply corresponds to say that the set of behaviors defining the system is not a-priori known. ∎

We are now facing the central issue this contribution is centered around, that is: an exact solution of the robust optimization program (2) requires to consider as many experiments as the number of elements in Ω, normally an *infinite* number. The impossibility to carry out this task suggests introducing approximate schemes where only a *finite* number of ω's, that is a *finite* number of experiments, is considered. Thus, we can at this point more precisely spell out the question we posed at the end of Section 1, and ask:

> *How many experiments do we need to perform to come up with a design that approximates the solution $\mathcal{D}^\star$ of (2) to a desired level of accuracy?*

3 The Experimental Effort Needed for Adaptation

The fact that one concentrates on a finite number of operating conditions only may appear naive. The interesting fact is that this way of proceeding can be cast within a solid mathematical theory providing us with guarantees on the level of accuracy obtained.

Fix an integer N, and let $\omega^{(1)}, \omega^{(2)}, \ldots, \omega^{(N)} \in \Omega$ be the operating conditions of N experiments run on the system to evaluate the N corresponding constraints for the robust program (2). The robust optimization problem restricted to the N experienced scenarios $\omega^{(i)}$, $i = 1, 2, \ldots, N$, reduces to the following finite optimization problem referred in the sequel to as 'scenario program':

$$\mathrm{SP}_N : \min_{\gamma, c \in \Re^{k+1}} c \quad \text{subject to:} \tag{4}$$

$$J_\gamma(\omega^{(i)}) \leq c, \; i = 1, 2, \ldots, N.$$

As for the selection of the scenarios $\omega^{(i)}$, $i = 1, 2, \ldots, N$, we suppose that they are extracted from set Ω according to some probability distribution P that reflects the likelihood of the different ω situations. This is naturally the case in the disturbance compensator Example 2, assuming the environment randomly selects the disturbance realizations according to an invariant scheme. If the scenarios are selected by the designer of the experiment, like u in Example 1, probability P is artificially introduced to describe the likelihood of the different operating conditions.

Let $(\gamma_N^\star, c_N^\star)$ be the solution of SP_N. $c_N^\star$ quantifies the performance of the device with parameter $\gamma_N^\star$ over the extracted operating conditions $\omega^{(1)}, \omega^{(2)}, \ldots, \omega^{(N)}$. Moreover, we clearly have $c_N^\star \leq c^\star$, the optimal cost with all the constraints in place, that is, for the extracted scenarios, we have designed a very efficient device, in actual effects one that even outperforms device $\mathcal{D}^\star$. We cannot be satisfied with this sole result, however, since, due to the limited number of scenarios, there is no guarantee whatsoever with respect to the much larger multitude of possible operating conditions, all those that have not been seen when performing the design of $\gamma_N^\star$. Hence, the following question arises naturally: what can we claim regarding the performance of the designed device for all other operating conditions $\omega \in \Omega$, those that were not experienced while doing the design according to SP_N in (4)? Answering this question is necessary to provide accuracy guarantees and to pose the method on solid grounds.

The posed question is of the 'generalization type' in a learning-theoretic sense: we want to know how the solution $(\gamma_N^\star, c_N^\star)$ generalizes from experienced operating conditions to unexperienced ones. For ease of explanation, we shall henceforth concentrate on robust optimization problems of *convex-type*, since this case can be handled in the light of a powerful theory that has recently appeared in the literature of robust optimization, [7, 8]. The non-convex case can be dealt with along a more complicated approach and is not discussed herein.

RESULT: *Select a 'violation parameter' $\epsilon \in (0, 1)$ and a 'confidence parameter' $\beta \in (0, 1)$.*

If N satisfies

$$\sum_{i=0}^{k} \binom{N}{i} \epsilon^i (1 - \epsilon)^{N-i} \leq \beta, \tag{5}$$

then, with probability no smaller than $1 - \beta$, the solution $(\gamma_N^\star, c_N^\star)$ to (4) satisfies all constraints of problem (2) with the exception of those corresponding to a set of operating conditions whose probability is at most ϵ.

Bound (5) can be found in [9], a contribution still in the general vein of the theoretical approach opened up in [7, 8].

Let us try to understand in detail the meaning of this result. If we neglect for a moment the part associated with the confidence parameter β, then, the result simply says that, by extracting a number N of operating conditions as given by (5) and running the corresponding N experiments to evaluate the constraints appearing in (4), the solution $(\gamma_N^\star, c_N^\star)$ to (4) violates the constraints corresponding to other, unexperienced, operating conditions with a probability that does not exceed a *user-chosen* level ϵ. This means that the so-determined $c_N^\star$ provides an upper bound for the cost $J_{\gamma_N^\star}(\omega)$ valid for every operating condition $\omega \in \Omega$ with the exclusion of at most an ϵ-probability set.

As for the probability $1 - \beta$, one should note that $(\gamma_N^\star, c_N^\star)$ is a random quantity because it depends on the randomly extracted operating conditions $\omega^{(1)}, \omega^{(2)}, \ldots, \omega^{(N)}$. It may happen indeed that these conditions are not representative enough (one could even extract N times the same operating condition!). In this case no generalization is expected, and the fraction of operating conditions violated by $(\gamma_N^\star, c_N^\star)$ will be larger than ϵ. Parameter β controls the probability of extracting unrepresentative operating conditions, and the final result that $(\gamma_N^\star, c_N^\star)$ violates at most an ϵ-fraction of operating conditions holds with probability $1 - \beta$. One important practical fact is that, due to the structure of the equation in (5), β can be set to be so small (say $\beta = 10^{-6}$) that it is virtually zero for any practical purpose, and this does not lead to a significant increase in the value of N (see also the numerical example in Section 4).

For the reader's convenience, the discussion in this section is summarized in a recipe for a practical implementation of the overall adaptive design scheme.

PRACTICAL RESULT: *Select a violation parameter $\epsilon \in (0,1)$, let $\beta = 10^{-6}$, and compute the least integer N satisfying (5). Run N random experiments and compute the corresponding N constraints for problem (4).*
Then, the solution $\gamma_N^\star$ of (4) achieves performance $c_N^\star$ on all operating conditions but an ϵ fraction of them, and, moreover, $c_N^\star$ is 'better than the best', in the sense that $c_N^\star \leq c^\star$.

Before closing the section, the following final remark is worth making in the light of equation (5):

> *The number of experiments N that are needed to adapt the device does not depend on the system complexity; it instead only depends on the complexity of the device $\mathcal{D}_\gamma$ through the size k of its parametrization γ.*

Thus reality can be any complex and still we can evaluate the experimental effort by only looking at the device being designed.

4   A Numerical Example

We consider the problem of inverting the nonlinear characteristic between input u and output $y(u,d)$ of a system affected by an additive output disturbance d (Figure 4), over the range of values $U = [0,1]$ for u (input-output equalization).

The device is fed by $y(u,d)$ and produces output $y_{\mathcal{D}_\gamma}(u,d) = \gamma_1 y(u,d)^2 + \gamma_2 y(u,d) + \gamma_3$. The performance of the device with parameter $\gamma = (\gamma_1,\,\gamma_2,\,\gamma_3) \in \Re^3$ is given by $\max_{u,d \in U \times D} J_\gamma(u,d)$, where $J_\gamma(u,d) = |y_{\mathcal{D}_\gamma}(u,d) - u|$ and D is the (unknown) range of values for d. In words, this performance expresses the largest deviation off the perfect equalization line $y_{\mathcal{D}} = u$.

We chose $\epsilon = 0.1$, $\beta = 10^{-6}$, and according to (5) N was 205.
The scenario program (4) is in this case

$$\min_{\gamma, c \in \Re^4} c \quad \text{subject to:} \tag{6}$$

$$|\gamma_1 y(u^{(i)}, d^{(i)})^2 + \gamma_2 y(u^{(i)}, d^{(i)}) + \gamma_3 - u^{(i)}| \leq c, \; i = 1, 2, \ldots, 205,$$

where $u^{(1)}$, $u^{(2)}$, $\ldots$, $u^{(205)}$ are random values for u independently extracted from U according to the uniform distribution P_u over $[0,1]$, and $d^{(1)}$, $d^{(2)}$, $\ldots$, $d^{(205)}$ are random values for d independently created by the environment during experimentation according to some (unknown) stationary distribution P_d.

The 205 constraints in (6) can be evaluated by running 205 experiments on the system where the output samples $y^{(i)} = y(u^{(i)}, d^{(i)})$, $i = 1, 2, \ldots, 205$, are collected together with $u^{(i)}$, $i = 1, 2, \ldots, 205$. Figure 5 shows the outcomes of the experiments.

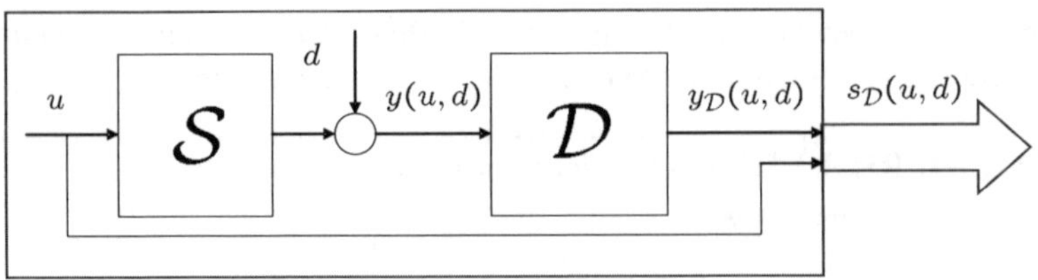

Fig. 4. Inverting a nonlinear characteristic through a device

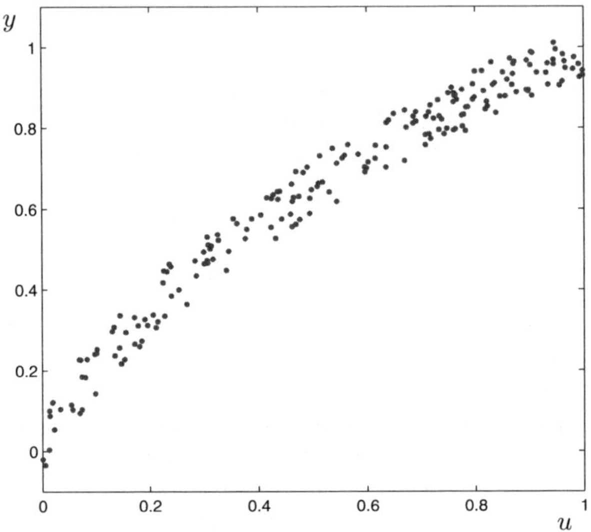

Fig. 5. Outcome of the experiments: samples of input u and output $y(u,d)$

Note that the collected output data present some dispersion due to the presence of the additive disturbance d.

By solving (6) we obtained $\gamma_{205}^\star = (0.424,\ 0.650,\ -0.081)$ and $c_{205}^\star = 0.108$.

$c_{205}^\star$ is the maximum equalization error for the extracted scenarios. In Figure 6, we plot the input and equalized output pairs $(u^{(i)}, y_{\mathcal{D}_{\gamma_{205}^\star}}(u^{(i)}, d^{(i)}))$, $i = 1, 2, \ldots, 205$, and the region $u \pm c_{205}^\star := \{(u, y) : u - c_{205}^\star \leq y \leq u + c_{205}^\star,\ u \in U\}$. $u \pm c_{205}^\star$ is the strip of minimum width centered around the perfect equalization line $y_{\mathcal{D}} = u$ that contains all the 205 input and equalized output pairs.

In the light of the practical result at the end of the previous section, device $\gamma_{205}^\star$ carries a guarantee that the equalized output $y_{\mathcal{D}_{\gamma_{205}^\star}}(u, d)$ differs from u of at most $c_{205}^\star = 0.108$ for all u's and d's except for a subset of probability $P = P_u \times P_d$ smaller than or equal to 0.1; moreover, the region of equalization $u \pm c_{205}^\star$ is contained within $u \pm c^\star$. This result holds irrespectively of D and P_d, which are unknown to the designer of the device.

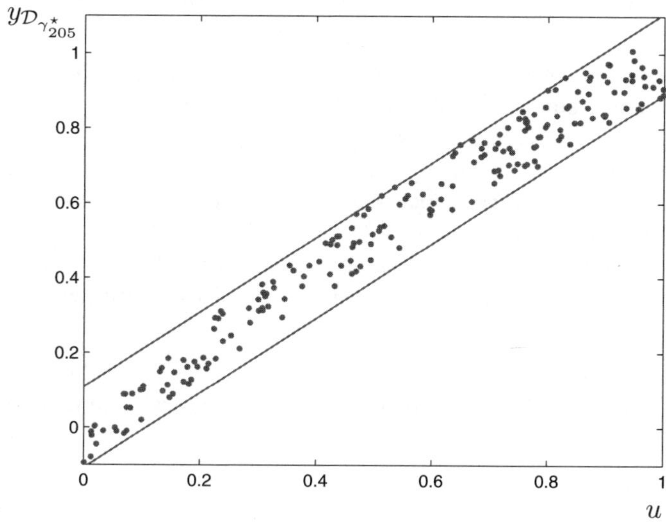

Fig. 6. Input u and equalized output $y_{\mathcal{D}_{\gamma^{\star}_{205}}}(u, d)$ for the extracted scenarios, and the region of equalization $u \pm c^{\star}_{205}$

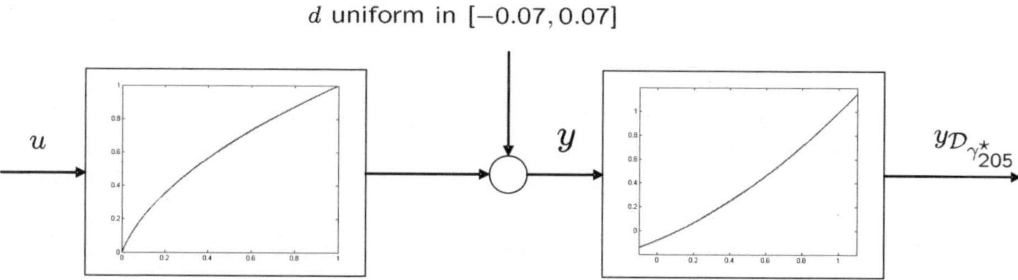

Fig. 7. Actual nonlinear characteristic and disturbance characteristics, along with the designed device

The actual nonlinear characteristic and disturbance d used to generate the data in Figure 5 are shown in Figure 7 together with the designed device with parameter $\gamma^{\star}_{205}$. In this example, the parameter of the device could have been designed so as to exactly invert the nonlinear characteristic. However, the obtained $\gamma^{\star}_{205}$ is different from such a choice, because the device aims at inverting the nonlinear characteristic between u and y while also reducing the effect of d on the reconstructed value for the input u.

5   Conclusions

The main goal of this contribution is that of attracting the reader's attention to the fundamental issue of evaluating the experimental effort needed to perform adaptive design, and some answers have been provided in a specific worst-case context.

Many are the aspects that our discussion has left unsolved, and open to further investigation:

- it is not always the case that one experiment provides all the information needed to evaluate a constraint. In the disturbance compensator example, for instance, if either the system or the device are not linear it is not possible to swap their order, and constraint evaluation calls for many experiments with virtually all possible devices in place. More generally, more experiments are needed when the input to the system depends on the device being designed.
- a perspective different from the worst-case approach can be used for adaptive design. For example, device quality could be assessed by its average performance, [10, 11, 12], rather than its worst-case performance over the set of operating conditions of interest.

Addressing these problems is a difficult task that requires much additional effort.

References

1. Sastry S., Bodson M. (1994) Adaptive Control: Stability, Convergence, and Robustness. Prentice-Hall.
2. Astrom K.J., Wittenmark B. (1994) Adaptive Control. Addison-Wesley.
3. Bittanti S., Picci G. eds. (1996) Identification, Adaptation, Learning. The science of learning models from data. Springer-Verlag, Berlin, Computer and Systems Science Series, Vol. 153.
4. Landau I.D., Lozano R., M'Saad M. (1998) Adaptive Control. Springer-Verlag.
5. Haykin S. (2002) Adaptive Filter Theory. Prentice Hall.
6. Polderman J.W., Willems J.C. (1998) Introduction to Mathematical Systems Theory: A Behavioral Approach. Springer Verlag, New York.
7. Calafiore G., Campi M.C. (2005) Uncertain convex programs: randomized solutions and confidence levels. Math. Program., Ser. A 102: 25–46.
8. Calafiore G., Campi M.C. (2006) The scenario approach to robust control design. IEEE Trans. on Automatic Control 51(5):742–753.
9. Campi M.C., Garatti S. (2007) The exact feasibility of randomized solutions of convex programs. Internal report, University of Brescia, Italy.
10. Vapnik V.N. (1998) Statistical Learning Theory. John Wiley & Sons.
11. Vidyasagar M. (2001) Randomized algorithms for robust controller synthesis using statistical learning theory. Automatica 37(10):1515-1528.
12. Campi M.C., Prandini M. (2003) Randomized algorithms for the synthesis of cautious adaptive controllers. Systems & Control Letters 49(1):21-36.

A Mutual Information Based Distance for Multivariate Gaussian Processes*

Jeroen Boets, Katrien De Cock, and Bart De Moor**

K.U.Leuven, Dept. of Electrical Engineering (ESAT-SCD)
Kasteelpark Arenberg 10, B-3001 Leuven, Belgium
{jeroen.boets,katrien.decock,bart.demoor}@esat.kuleuven.be

Dedicated to Giorgio Picci on the occasion of his 65th birthday.

Summary. In this paper a new distance on the set of multivariate Gaussian linear stochastic processes is proposed based on the notion of mutual information. The definition of the distance is inspired by various properties of the mutual information of past and future of a stochastic process. For two special classes of stochastic processes this mutual information distance is shown to be equal to a cepstral distance. For general multivariate processes, the behavior of the mutual information distance is similar to the behavior of an *ad hoc* defined multivariate cepstral distance.

1 Introduction

This paper is concerned with realization and identification of linear stochastic processes, topics that are central in Giorgio Picci's research interests. With his work in the last decennia he is one of the great inspirators for the development of subspace identification for stochastic processes, to which he also contributed several papers [24, 27]. Within our research group quite some work was done in subspace identification in the nineties [33, 34]. Through this way, Giorgio, we would like to thank you for the countless interesting insights you shared with us and other researchers, but especially for your great friendship. Ad multos annos!

* Research supported by Research Council KUL: GOA AMBioRICS, CoE EF/05/006 Optimization in Engineering (OPTEC), several PhD/postdoc & fellow grants; Flemish Government: FWO: PhD/postdoc grants, projects, G.0407.02 (support vector machines), G.0197.02 (power islands), G.0141.03 (Identification and cryptography), G.0491.03 (control for intensive care glycemia), G.0120.03 (QIT), G.0452.04 (new quantum algorithms), G.0499.04 (Statistics), G.0211.05 (Nonlinear), G.0226.06 (cooperative systems and optimization), G.0321.06 (Tensors), G.0302.07 (SVM/Kernel), research communities (ICCoS, ANMMM, MLDM); IWT: PhD Grants, McKnow-E, Eureka-Flite2; Belgian Federal Science Policy Office: IUAP P6/04 (DYSCO, Dynamical systems, control and optimization, 2007-2011) ; EU: ERNSI.

** Jeroen Boets is a research assistant with the Institute for the Promotion of Innovation through Science and Technology in Flanders (IWT-Vlaanderen) at the K.U.Leuven, Belgium. Dr. Katrien De Cock is a postdoctoral researcher at the K.U.Leuven, Belgium. Prof. Dr. Bart De Moor is a full professor at the K.U.Leuven, Belgium.

A. Chiuso et al. (Eds.): Modeling, Estimation and Control, LNCIS 364, pp. 15–33, 2007.
springerlink.com

In some of our recent work [8, 9] we have established a nice framework with interesting relations between notions from three different disciplines: system theory, information theory and signal processing. These relations are illustrated in a schematic way in Figure 1. The processes considered in the framework are scalar Gaussian linear time-invariant (LTI) stochastic processes. Centrally located in Figure 1 are the principal angles and their statistical counterparts, the canonical correlations. These notions will be explained in Section 3. Through a first link in the figure, expressions are obtained for the mutual information of past and future of a process as a function of its model parameters, by computing the canonical correlations between past and future of the process. Secondly, the notion of subspace angles between two stochastic processes allows to find new expressions for an existing cepstral distance as a function of the model description of the processes. And finally, the definition of a distance between scalar stochastic processes based on mutual information was proven to result in exactly this same cepstral distance.

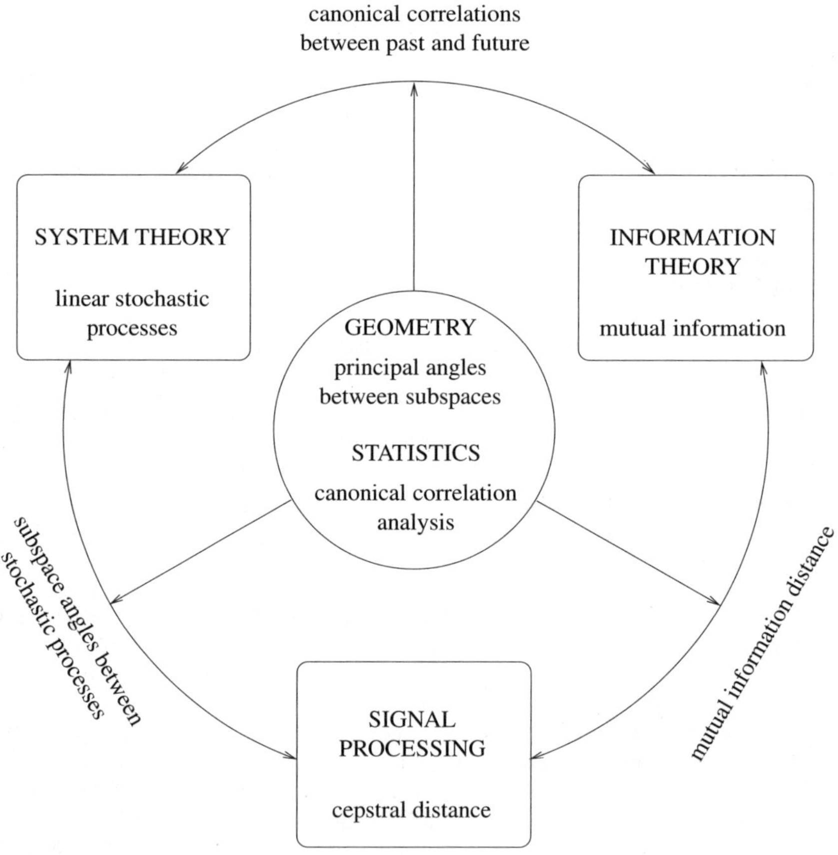

Fig. 1. A schematic representation of the relations between system theory, information theory and signal processing for scalar stochastic processes

In this paper we wish to give a start to the extension of the framework in Figure 1 to multivariate processes. We mainly focus on one aspect of the figure, namely the mutual information distance. More specifically, we define in this paper a new mutual information based distance on the set of *multivariate* Gaussian LTI stochastic processes.

The idea of defining a distance for this kind of processes is not new. Many distances have been considered in the past, both for scalar and multivariate processes. Specifically for scalar processes a lot of distances are defined directly on the basis of the power spectrum, the log-power spectrum or the power cepstrum of the processes [3,13,14,18]. A difficulty with these distances is that some of them can not be generalized in a trivial manner to multivariate processes. Cepstral distances for instance in their definition involve some definition of the logarithm of the power spectrum of the processes.

Several of the distances defined for both scalar and multivariate stochastic processes are based on information-theoretic measures. By considering a stochastic process as an infinite-dimensional random variable, one can define e.g. the (asymptotic) Kullback-Leibler (K-L) divergence, Chernoff divergence and Bhattacharyya divergence of two processes [22, 25, 29, 30, 31, 32]. Often, the processes are assumed to be Gaussian, in which case computationally tractable formulas can be derived.

Mutual information is an information-theoretic measure too. However, it is not applicable in the same sense as the above measures. The difference is that the mutual information of two random variables does not measure the similarity (or dissimilarity) of their probability densities. Instead it is a measure for the *dependence* of two random variables. Since the goal in this paper is to achieve a distance on the set of stochastic processes (without assuming information on their mutual dependencies), several intermediate steps must be taken. These steps are explained in the paper and are inspired by previous work in [6, 8, 9] (see Figure 1).

Distances between stochastic processes or time series have been used in many different areas. Among the most common are speech recognition [3, 13, 14], biomedical applications [2, 12, 23] and video processing [4, 11]. The distances are typically applied in a clustering or classification context.

The paper is organized as follows. In Section 2 we describe the model class we work with: Gaussian LTI stochastic dynamical models. Section 3 recalls the notions of principal angles between two subspaces, canonical correlations and mutual information of two random variables, and applies these notions in the context of stochastic processes. In Section 4 a new distance between multivariate Gaussian processes is proposed based on the notion of mutual information, and its properties are investigated. Section 5 shows several additional relations that hold in the case of scalar processes. In Section 6 we investigate whether the newly defined distance admits a *cepstral nature* by defining an ad hoc power cepstrum and cepstral distance for *multivariate* stochastic processes. Section 7 states the conclusions of the paper and some remaining open problems.

2   Model Class

In this paper we consider stochastic processes $y = \{y(k)\}_{k \in \mathbb{Z}}$ whose first and second order statistics can be described by the following state space equations:

$$\begin{cases} x(k+1) = Ax(k) + Bu(k) \,, \\ \quad\; y(k) = Cx(k) + Du(k) \,, \end{cases}$$

(1)

$$E\left\{u(k)\right\} = 0 \,, \quad E\left\{u(k)u^\top(l)\right\} = I_p \delta_{kl} \,.$$

(2)

with I_p the identity matrix of dimension p and δ_{kl} the Kronecker delta, being 1 for $k = l$ and 0 otherwise. The variable $y(k) \in \mathbb{R}^p$ is the value of the process at time k and is called the output of the model (1)-(2). The state process $\{x(k)\}_{k\in\mathbb{Z}} \in \mathbb{R}^n$ is assumed to be stationary, which implies that A is a stable matrix (all of its eigenvalues lie strictly inside the unit circle). The unobserved input process $\{u(k)\}_{k\in\mathbb{Z}} \in \mathbb{R}^p$ is a stationary and ergodic (normalized) white noise process. Both x and u are auxiliary processes used to describe the process y in this representation. The matrix $D \in \mathbb{R}^{p\times p}$ is assumed to be of full rank. We assume throughout this paper that u and consequently also y is a *Gaussian* process. This means that the process y is fully described by (1)-(2).

The infinite controllability and observability matrix of the model (1) are defined as:

$$\begin{aligned} \mathcal{C} &= \left(B\; AB\; A^2 B\; \cdots \right), \\ \Gamma &= \left(C^\top\; (CA)^\top\; (CA^2)^\top\; \cdots \right)^\top, \end{aligned}$$

respectively. The model (1) is assumed to be minimal, meaning that $\mathcal{C}$ and Γ are of full rank n. The Gramians corresponding to $\mathcal{C}$ and Γ are the unique and positive definite solution of the controllability and observability Lyapunov equation, respectively:

$$\begin{aligned} \mathcal{C}\mathcal{C}^\top &= P = APA^\top + BB^\top \,, \\ \Gamma^\top\Gamma &= Q = A^\top QA + C^\top C \,. \end{aligned}$$

(3)

The controllability Gramian P is also equal to the state covariance matrix, i.e. $P = E\left\{x(k)x^\top(k)\right\}$.

The model (1) is further assumed to be minimum-phase, meaning that its zeros (eigenvalues of $A - BD^{-1}C$) lie strictly inside the unit circle. The inverse model can then be derived from (1) by rewriting it as

$$\begin{cases} x(k+1) = (A - BD^{-1}C)x(k) + BD^{-1}y(k) \,, \\ \quad\; u(k) = \qquad\quad -D^{-1}Cx(k) + \quad D^{-1}y(k) \,, \end{cases}$$

(4)

and is denoted with a subscript $(\cdot)_z$:

$$(A_z, B_z, C_z, D_z) = (A - BD^{-1}C, BD^{-1}, -D^{-1}C, D^{-1}).$$

Analogously, the controllability and observability matrices and Gramians of the inverse model (4) are denoted by $\mathcal{C}_z$, Γ_z, P_z and Q_z. The matrix Q_z, for instance, is the solution of

$$Q_z = (A - BD^{-1}C)^\top Q_z (A - BD^{-1}C) + C^\top D^{-\top} D^{-1}C \,.$$

(5)

Along with the descriptions (1) and (4), a transfer function can be defined from u to y and from y to u, respectively:

$$h(z) = C(zI - A)^{-1}B + D \,,$$

(6)

$$h^{-1}(z) = -D^{-1}C(zI - (A - BD^{-1}C))^{-1}BD^{-1} + D^{-1} \,.$$

Modulo a similarity transformation of the state space model (A, B, C, D) into $(T^{-1}AT, T^{-1}B, CT, D)$ with nonsingular T, there is a one-to-one correspondence between the descriptions (1) and (6). From each of both descriptions, augmented with (2), the second order statistics of the process y can be derived, i.e. its autocovariance sequence

$$\Lambda(s) = E\left\{y(k)y^\top(k-s)\right\} = \begin{cases} CPC^\top + DD^\top & s = 0, \\ CA^{s-1}G & s > 0, \\ G^\top(A^\top)^{|s|-1}C^\top & s < 0, \end{cases} \tag{7}$$

with $G = E\left\{x(k+1)y^\top(k)\right\} = APC^\top + BD^\top$, or equivalently its spectral density function

$$\Phi(z) = \sum_{s=-\infty}^{+\infty} \Lambda(s)z^{-s} = h(z)h^\top(z^{-1}). \tag{8}$$

As stated before, Gaussian processes (which we assume) are fully described by their first and second order statistical properties. Therefore a zero-mean process $\{y(k)\}_{k\in\mathbb{Z}}$ is also fully described by (7) or (8). From equation (8) it can thus be seen that $h(z)$ is not uniquely defined for the process y since the transfer functions $h(z)$ and $h(z)V$ with V a unitary $p \times p$ matrix correspond to the same spectral density function $\Phi(z)$. This is the only non-uniqueness in $h(z)$ under the given assumptions and must be kept in mind while we denote a process in this paper by *one of its* foursomes (A, B, C, D) or *one of its* transfer functions $h(z)$.

We also define doubly infinite block Hankel matrices of data:

$$Y = \left(\begin{array}{ccc} \vdots & \vdots & \vdots & \ddots \\ y(-2) & y(-1) & y(0) & \cdots \\ y(-1) & y(0) & y(1) & \cdots \\ \hline y(0) & y(1) & y(2) & \cdots \\ y(1) & y(2) & y(3) & \cdots \\ \vdots & \vdots & \vdots & \end{array}\right) = \left(\frac{Y_p}{Y_f}\right), \tag{9}$$

corresponding to the processes $y = \{y(k)\}_{k\in\mathbb{Z}}$, $y_p = \{y(-k)\}_{k\in\mathbb{N}_0}$ and $y_f = \{y(k)\}_{k\in\mathbb{N}}$, where the subscript p stands for 'past' and f for 'future'. The block Hankel matrices U, U_p and U_f are analogously defined for the processes u, u_p and u_f.

3 Principal Angles, Canonical Correlations and Mutual Information

In this section the definitions of principal angles between two subspaces, canonical correlations of two random variables and their mutual information are recalled in Sections 3.1, 3.2 and 3.3 respectively. In Section 3.4 these notions are applied in the context of the stochastic processes defined in the previous section. Attention is drawn in particular to the mutual information of past and future of the output process y.

3.1 Principal Angles and Directions

The principal angles between two subspaces [21] are a generalization of the angle between two vectors. Suppose we are given two linear subspaces S_1 and S_2 of the ambient vector space $\mathbb{R}^n$ of dimension $d_1 < n$ and $d_2 < n$, respectively. A natural extension of the one-dimensional case is to choose a unit vector u_1 from S_1 and a unit vector v_1 from S_2 such that the angle between u_1 and v_1 is minimized. The vectors u_1 and v_1 so obtained, are called the first principal directions and the angle between them is the first principal angle θ_1. Next, choose a unit vector $u_2 \in S_1$ orthogonal to u_1 and $v_2 \in S_2$ orthogonal to v_1 such that the angle θ_2 between them is minimized. This is the second principal angle and u_2 and v_2 are the corresponding principal directions. Continue in this way until $\min(d_1, d_2)$ angles and corresponding principal vectors have been found. This informal description is now formalized.

Definition 1. Principal angles and directions
The principal angles $0 \leq \theta_1 \leq \theta_2 \leq \ldots \theta_{\min(d_1,d_2)} \leq \pi/2$ between the subspaces S_1 and S_2 of the ambient space $\mathbb{R}^n$ of dimension d_1 and d_2, respectively, and the corresponding principal directions $u_i \in S_1$ and $v_i \in S_2$ are defined recursively as

$$\cos\theta_1 = \max_{\substack{u \in S_1 \\ v \in S_2}} u^\top v = u_1^\top v_1 \, ,$$

$$\cos\theta_k = \max_{\substack{u \in S_1 \\ v \in S_2}} u^\top v = u_k^\top v_k \, , \; for \; k = 2, \ldots, \min(d_1, d_2) \, ,$$

subject to $\|u\| = \|v\| = 1$ and for $k > 1$: $u^\top u_i = 0$ and $v^\top v_i = 0$, where $i = 1, \ldots, k - 1$.

Let $A \in \mathbb{R}^{p \times n}$ be of rank d_1 and $B \in \mathbb{R}^{q \times n}$ of rank d_2. Then, the ordered set of $\min(d_1, d_2)$ principal angles between the row spaces of A and B is denoted by

$$\left(\theta_1, \theta_2, \ldots, \theta_{\min(d_1,d_2)}\right) = [A \lhd B] \, .$$

In case A and B are of full row rank with $p \leq q$, the squared cosines of the principal angles between $\mathrm{row}(A)$ and $\mathrm{row}(B)$ are equal to the eigenvalues of $(AA^\top)^{-1}AB^\top (BB^\top)^{-1}BA^\top$:

$$\cos^2 [A \lhd B] = \lambda\left((AA^\top)^{-1}AB^\top(BB^\top)^{-1}BA^\top\right) \, . \tag{10}$$

3.2 Canonical Correlations

In canonical correlation analysis [16] the interrelation of two sets of random variables is studied. It is the statistical interpretation of the geometric tool of principal angles between and principal directions in linear subspaces. The aim is to find two bases of random variables, one in each set, that are internally uncorrelated but that have maximal correlations between the two sets. The resulting basis variables are called the canonical variates and the correlation coefficients between the canonical variates are the canonical correlations.

Let V be a zero-mean p-component and W a zero-mean q-component real random variable with joint covariance matrix $Q = E\left\{\begin{pmatrix} V \\ W \end{pmatrix}(V^\top\ W^\top)\right\} = \begin{pmatrix} Q_v & Q_{vw} \\ Q_{wv} & Q_w \end{pmatrix}$. In case Q_v and Q_w are full rank matrices, and $p \leq q$, the p squared canonical correlations of V and W, which we denote by $\mathrm{cc}^2(V, W)$, can be obtained as the eigenvalues of $Q_v^{-1} Q_{vw} Q_w^{-1} Q_{wv}$:

$$\mathrm{cc}^2(V, W) = \lambda(Q_v^{-1} Q_{vw} Q_w^{-1} Q_{wv}) \,. \tag{11}$$

3.3 Mutual Information

Let V be a zero-mean p-component and W a zero-mean q-component random variable. If V and W are mutually dependent, then observing W reduces the uncertainty (or entropy) in V. Otherwise formulated, we gain information about V by observing W. Thus, the variable W must contain information about V. For the same reason V must also contain information about W. Both amounts of information are equal and are quantified as the *mutual information of V and W*, denoted by $I(V; W)$.

Definition 2. The mutual information of two continuous random variables [7]
Let V and W be random variables with joint probability density function $f(v, w)$ and marginal densities $f_V(v)$ and $f_W(w)$, respectively. Then, the mutual information of V and W is defined as

$$I(V; W) = \iint f(v, w) \log \frac{f(v, w)}{f_V(v) f_W(w)} \, dv \, dw \,,$$

if the integral exists.

In case of two zero-mean jointly Gaussian random variables and denoting the covariance matrix of $\begin{pmatrix} V \\ W \end{pmatrix}$ by $Q = \begin{pmatrix} Q_v & Q_{vw} \\ Q_{wv} & Q_w \end{pmatrix}$, this expression can be rewritten as

$$I(V; W) = -\frac{1}{2} \log \frac{\det Q}{\det Q_v \det Q_w} \,,$$

under the assumption that Q_v and Q_w are of full rank. In this case $I(V; W)$ is also related to the canonical correlations of V and W, here denoted by σ_k ($k = 1, \dots, \min(p, q)$), as can be derived using equation (11):

$$I(V; W) = -\frac{1}{2} \log \prod_{k=1}^{\min(p,q)} (1 - \sigma_k^2) \,. \tag{12}$$

3.4 Application to Stochastic Processes

In this section we apply the notions defined in the previous sections to the stochastic processes y_p, y_f, u_p and u_f. A stochastic process, e.g. $\{y(k)\}_{k \in \mathbb{Z}}$, can be seen as an infinite-dimensional random variable consisting of the (ordered) concatenation of the random variables $\ldots$, $y(-2)$, $y(-1)$, $y(0)$, $y(1)$, $\ldots$ We can thus associate with the process y the random variable

$$
\mathcal{Y} = \begin{pmatrix} \vdots \\ y(-2) \\ y(-1) \\ \hline y(0) \\ y(1) \\ \vdots \end{pmatrix} = \begin{pmatrix} \mathcal{Y}_p \\ \mathcal{Y}_f \end{pmatrix} ,
$$

$\mathcal{Y}_p$ and $\mathcal{Y}_f$ being associated with the processes y_p and y_f, and analogously $\mathcal{U}$, $\mathcal{U}_p$ and $\mathcal{U}_f$ for the processes u, u_p and u_f. This way we can compute the canonical correlations and the mutual information for any pair of these processes.

Canonical Correlations

Since we are dealing with stationary and ergodic zero-mean processes, it is readily seen from equations (10) and (11) that the canonical correlations between any two of the processes u, u_p, u_f, y, y_p and y_f are equal to the cosines of the principal angles between the row spaces of the corresponding block Hankel matrices defined in (9), e.g.:

$$
\mathrm{cc}(\mathcal{U}_f, \mathcal{Y}_f) = \cos\left([U_f \lhd Y_f]\right) . \tag{13}
$$

In [8, Chap. 3] the canonical correlations of each pair of these processes were computed. Formulas were derived for the canonical correlations between the past and future output process:

$$
\mathrm{cc}^2(\mathcal{Y}_p, \mathcal{Y}_f) = \lambda\left(P(Q_z^{-1} + P)^{-1}\right), 0, 0, \ldots ,
$$

as well as for the canonical correlations between u_f and y_f:

$$
\mathrm{cc}^2(\mathcal{U}_f, \mathcal{Y}_f) = \lambda\left((I_n + Q_z P)^{-1}\right), 1, 1, \ldots ,
$$

where P and Q_z each follow from a Lyapunov equation (see (3)-(5)). We denote the non-trivial correlations of y_p and y_f by ρ_k, and those of u_f and y_f by τ_k, as follows:

$$
\begin{aligned}
\rho_k^2 &= \lambda\left(P(Q_z^{-1} + P)^{-1}\right) & (k = 1, \ldots, n), \\
\tau_k^2 &= \lambda\left((I_n + Q_z P)^{-1}\right) & (k = 1, \ldots, n).
\end{aligned} \tag{14}
$$

It can be shown that $\rho_k^2 + \tau_k^2 = 1$, for $k = 1, \ldots, n$. These results together with the canonical correlations of the other pairs of processes are summarized in Table 1.

Mutual Information of Past and Future of a Process

Using the relation (12) for Gaussian processes, we can compute from Table 1 the mutual information of each pair of processes. A pair of processes that has at least one canonical correlation equal to 1 does not have a finite amount of mutual information.

Table 1. Overview of the canonical correlations of each pair of processes, where k goes from 1 to n

	$\mathcal{U}_p$	$\mathcal{Y}_p$	$\mathcal{U}_f$	$\mathcal{Y}_f$
$\mathcal{U}_p$	$1,1,\ldots$	$1,1,\ldots$	$0,0,\ldots$	$\rho_k,0,0,\ldots$
$\mathcal{Y}_p$	$1,1,\ldots$	$1,1,\ldots$	$0,0,\ldots$	$\rho_k,0,0,\ldots$
$\mathcal{U}_f$	$0,0,\ldots$	$0,0,\ldots$	$1,1,\ldots$	$\sqrt{1-\rho_k^2},1,1,\ldots$
$\mathcal{Y}_f$	$\rho_k,0,0,\ldots$	$\rho_k,0,0,\ldots$	$\sqrt{1-\rho_k^2},1,1,\ldots$	$1,1,\ldots$

Looking at relation (13) between canonical correlations and principal angles we can say that these processes intersect, since they have a principal angle equal to zero. Conversely, processes that are orthogonal to each other (all canonical correlations equal to 0 or all principal angles equal to $\pi/2$) have mutual information equal to zero. This is for instance the case for u_p and u_f, past and future of the white noise process u. However, processing this white noise u through the filter $h(z)$ (in general) introduces a time correlation in the resulting process y, which appears as a certain amount of mutual information between its past y_p and future y_f, denoted interchangeably by I_{pf}, $I_{\mathrm{pf}}\{y\}$ or $I_{\mathrm{pf}}\{h(z)\}$:

$$I_{\mathrm{pf}} = I(y_p; y_f) = -\frac{1}{2}\log \prod_{k=1}^{n}(1-\rho_k^2) = -\frac{1}{2}\log \prod_{k=1}^{n}\tau_k^2 = \frac{1}{2}\log\det\left(I_n + Q_z P\right) .$$

(15)

Note that ρ_k, τ_k ($k = 1,\ldots,n$) and consequently also I_{pf} are unique for a given stochastic process, since P and Q_z do not change when $h(z)$ is right-multiplied by a unitary matrix, and a similarity transformation of the state space model does not alter the eigenvalues of the product $Q_z P$. So if we write $I_{\mathrm{pf}}\{h(z)\}$ or $\rho_k\{h(z)\}$, this must not be understood as a characteristic of the transfer function $h(z)$ but rather as a characteristic of the process y with spectral density $\Phi(z) = h(z)h^{\top}(z^{-1})$.

Properties of I_{pf}

The mutual information I_{pf} of past and future of a stochastic process y is the amount of information that the past provides about the future and vice versa. Through (15) it is closely connected to the canonical correlations of y_p and y_f. The problem of characterizing this dependence of past and future of a stationary process has received a great deal of attention because of its implications for the prediction theory of Gaussian processes (see [17, 19, 20]). Inspired by the use of canonical correlation analysis in stochastic realization theory [1], a stochastic model reduction technique based on the mutual information of the past and the future has been proposed by Desai and Pal [10], which is also used in stochastic subspace identification [27, 34]. Li and Xie used the past-future mutual information for model selection and order determination problems in [26]. We now state some of the properties of I_{pf}.

(a) $I_{\mathrm{pf}} = 0 \Leftrightarrow h(z) = D$ (see (1))

Since y is Gaussian, $I_{\mathrm{pf}} = 0$ is equivalent with y_p and y_f being uncorrelated, thus $\Lambda(s) = 0_p$ for $s \neq 0$. From stochastic realization theory then follows that $h(z)$ has order zero.

(b) $I_{\mathrm{pf}} \in [0, +\infty)$

This follows from relation (15) and the fact that $\rho_k \in [0, 1)$. Indeed, in [15] it is shown that the number of unit canonical correlations of y_p and y_f is equal to the number of zeros of $h(z)$ on the unit circle. Since $h(z)$ is assumed to be minimum-phase (see Section 2), this number is zero.

(c) I_{pf} (strictly) increases with each increase of a canonical correlation ρ_k ($k = 1, \ldots, n$).

This follows immediately from relation (15) and property (b).

(d) $I_{\mathrm{pf}}\{h(z)\} = I_{\mathrm{pf}}\{Th(z)\}$ for a nonsingular constant matrix $T \in \mathbb{R}^{p \times p}$.

This follows from the definition of canonical correlations or principal angles, since left-multiplying the output variables $y(k)$ ($k \in \mathbb{Z}$) with T does not change the row spaces of Y_p and Y_f. Consequently, the canonical correlations ρ_k and the mutual information I_{pf} do not change.

(e) $I_{\mathrm{pf}}\{h(z)\} = I_{\mathrm{pf}}\{h^{-\top}(z)\}$

Equation (14) shows that the past-future canonical correlations ρ_k ($k = 1, \ldots, n$) only depend on the eigenvalues of the product matrix $Q_z P$. Noting that the state space description of the transpose of the inverse model is given by $h^{-\top}(z) = (A_z^\top, C_z^\top, B_z^\top, D_z^\top)$, it can be seen from (3) that the controllability Gramian of $h^{-\top}(z)$ is given by Q_z, while the observability Gramian of its inverse model $h^\top(z) = (A^\top, C^\top, B^\top, D^\top)$ is equal to P. Consequently, the canonical correlations ρ_k and the mutual information I_{pf} are equal for the transfer functions $h(z)$ and $h^{-\top}(z)$. This invariance property does not, in general, hold for $h(z)$ and $h^{-1}(z)$ since the eigenvalues of $Q_z P$ are usually not equal to those of $Q P_z$.

(f) For $\Phi(z) = \begin{pmatrix} \Phi_1(z) & 0_{p_1 \times p_2} \\ 0_{p_2 \times p_1} & \Phi_2(z) \end{pmatrix}$, it holds that $I_{\mathrm{pf}}\{y\} = I_{\mathrm{pf}}\{y_1\} + I_{\mathrm{pf}}\{y_2\}$.

In this case the p_1-variate process y_1 and the p_2-variate process y_2, constituting the process y, are completely uncorrelated. Therefore, the canonical correlations of y_p and y_f are on the one hand the canonical correlations between y_{1_p} and y_{1_f}, and on the other hand the canonical correlations between y_{2_p} and y_{2_f}: $\rho_k\{y\}$ ($k = 1, \ldots, n_1 + n_2$) is the union of $\rho_k\{y_1\}$ ($k = 1, \ldots, n_1$) and $\rho_k\{y_2\}$ ($k = 1, \ldots, n_2$), with n_1 and n_2 the orders of the processes y_1 and y_2. The result then follows from relation (15).

Properties (a)-(c) indicate that I_{pf} measures the amount of correlation that exists between y_p and y_f, being zero for a white noise process and increasing with each increase of a correlation ρ_k between y_p and y_f. This suggests that I_{pf} can be used as a measure for the amount of *dynamics* in the process y where dynamics are defined in terms of the correlation or the dependence that exists between all future values and all past values of the process at any time instant.

4   A Distance Between Multivariate Gaussian Processes

In this section we define a new distance between multivariate Gaussian processes based on the notion of mutual information. In Section 4.1 the distance is defined and its metric properties are investigated, while in Section 4.2 we show a way to compute the distance.

4.1   Definition and Metric Properties

We propose as a new distance on the set of multivariate Gaussian processes: the *mutual information distance*, denoted by $d_{\mathrm{mi}}(y_1, y_2)$.

Definition 3. The mutual information distance between two Gaussian processes
The mutual information distance between two Gaussian linear stochastic processes y_1 and y_2 with transfer function descriptions $h_1(z)$ and $h_2(z)$ is denoted by $d_{\mathrm{mi}}(y_1, y_2)$ and is defined as

$$d_{\mathrm{mi}}^2(y_1, y_2) = I_{\mathrm{pf}}\left\{h_{12}(z)\right\}, \quad \text{with } h_{12}(z) = \begin{pmatrix} h_1^{-1}(z)h_2(z) & 0_p \\ 0_p & h_2^{-1}(z)h_1(z) \end{pmatrix}.$$

The first thing to note is that the mutual information distance $d_{\mathrm{mi}}(y_1, y_2)$ is a property of the processes y_1 and y_2, and not of the particular transfer functions $h_1(z)$ and $h_2(z)$. Indeed, substituting $\{h_1(z), h_2(z)\}$ by the equivalent $\{h_1(z)V_1, h_2(z)V_2\}$ with V_1, V_2 constant unitary matrices (see (8)), corresponds to left- and right-multiplying $h_{12}(z)$ by a constant unitary matrix. This has no influence on $I_{\mathrm{pf}}\{h_{12}\}$ (see property (d) in Section 3.4).

Following the discussion at the end of Section 3.4, $d_{\mathrm{mi}}(y_1, y_2)$ can be interpreted as a measure for the amount of dynamics in the process y_{12} associated with the transfer function $h_{12}(z)$. It is clear that $d_{\mathrm{mi}}\{y_1, y_1\} = 0$ since $h_{12}(z)$ is in that case a constant matrix and y_{12} is consequently white noise. This also clarifies why the 'ratio' of $h_1(z)$ and $h_2(z)$ is found in $h_{12}(z)$, instead of for instance the difference. From Definition 3 it is also immediately seen that $d_{\mathrm{mi}}(y_1, y_2) = d_{\mathrm{mi}}(g(z)y_1, g(z)y_2)$ for arbitrary transfer functions $g(z)$ satisfying the conditions stated in Section 2 (e.g. being square, stable and minimum-phase). Filtering the processes y_1 and y_2 by a common filter $g(z)$ does not change their mutual information distance.

The following properties hold for the mutual information distance:

1. $d_{\mathrm{mi}}(y_1, y_2) \geq 0$
2. $d_{\mathrm{mi}}(y_1, y_2) = 0 \Leftrightarrow h_2(z) = h_1(z)T$ with T a constant square nonsingular matrix. This follows from property (a) in Section 3.4.
3. $d_{\mathrm{mi}}(y_1, y_2) = d_{\mathrm{mi}}(y_2, y_1)$ is symmetric. This follows immediately from Definition 3.

Examples have shown that $\mathrm{d}_{\mathrm{mi}}(y_1, y_2)$ does not in general satisfy the triangle inequality[1]. The distance thus satisfies only two of the four properties of a true metric (non-negativity and symmetry). However, if we define a set of equivalence classes of stochastic processes, where two processes with transfer functions $h_1(z)$ and $h_2(z)$ are equivalent if and only if there exists a constant square nonsingular matrix T such that $h_2(z) = h_1(z)T$, then the mutual information distance $\mathrm{d}_{\mathrm{mi}}(y_1, y_2)$ defined on this set of equivalence classes, satisfies all metric properties but the triangle inequality. It is then called a *semimetric*.

4.2   Computation

From property (f) in Section 3.4 it follows that

$$\mathrm{d}^2_{\mathrm{mi}}(y_1, y_2) = I_{\mathrm{pf}} \left\{ h_1^{-1}(z)h_2(z) \right\} + I_{\mathrm{pf}} \left\{ h_2^{-1}(z)h_1(z) \right\}. \tag{16}$$

Using this property we now show a way to compute $\mathrm{d}_{\mathrm{mi}}(y_1, y_2)$ making use of the state space descriptions of $h_1(z)$ and $h_2(z)$ of orders n_1 and n_2 respectively. Equations (15) and (16) show that we need to compute the controllability and observability Gramians of both $h_1^{-1}(z)h_2(z)$ and $h_2^{-1}(z)h_1(z)$. This can be easily done by solving the Lyapunov equations (3) from the state space descriptions of both transfer functions. As an example we give a possible state space description of $h_1^{-1}(z)h_2(z)$ denoted by $(A_{12}, B_{12}, C_{12}, D_{12})$:

$$A_{12} = \begin{pmatrix} A_2 & 0_{n_2 \times n_1} \\ B_{z_1}C_2 & A_{z_1} \end{pmatrix}, \ B_{12} = \begin{pmatrix} B_2 \\ B_{z_1}D_2 \end{pmatrix}, \ C_{12} = \begin{pmatrix} D_{z_1}C_2 & C_{z_1} \end{pmatrix}, \ D_{12} = D_{z_1}D_2 \,,$$

with $(A_{z_1}, B_{z_1}, C_{z_1}, D_{z_1}) = (A_1 - B_1 D_1^{-1}C_1, B_1 D_1^{-1}, -D_1^{-1}C_1, D_1^{-1})$. The procedure concerning $h_2^{-1}(z)h_1(z)$ is analogous. Afterwards it remains to compute (16) using (15) and (3).

5   Special Case of Scalar Processes

The only relation in Figure 1 that holds for both scalar and multivariate Gaussian processes is the one between the mutual information distance and the past-future canonical correlations, which can be seen in (15). In the case of scalar processes y_1 and y_2 it follows from property (e) in Section 3.4 that (16) can be rewritten as $\mathrm{d}^2_{\mathrm{mi}}(y_1, y_2) = 2I_{\mathrm{pf}} \left\{ \frac{h_1(z)}{h_2(z)} \right\} = 2I_{\mathrm{pf}} \left\{ \frac{h_2(z)}{h_1(z)} \right\}$. In this case the mutual information distance is also related to so-called *subspace angles between stochastic processes* and to a cepstral distance, as was mentioned in the introduction (see Figure 1). We will shortly recall these two results in Sections 5.1 and 5.2. Based on these relations, several additional expressions for $\mathrm{d}_{\mathrm{mi}}(y_1, y_2)$ can be derived for the scalar case. For more details on this we refer to [8, Chap. 6].

---

[1] In the case of scalar processes or processes with diagonal spectral density function $\Phi(z)$, however, it can be shown that the triangle inequality *is* satisfied (see Sections 5.2 and 6.1 respectively).

5.1   Relation with Subspace Angles Between Scalar Stochastic Processes

Consider the situation in Figure 2 where the single-input single-output models $h_1(z)$ of order n_1 and $h_2(z)$ of order n_2 are driven by a common white noise source $\{u(k)\}_{k\in\mathbb{Z}} \in \mathbb{R}$. It can be shown that in this case only $n_1 + n_2$ canonical correlations between the future y_{1_f} and y_{2_f} of the processes y_1 and y_2 can be different from 1. If we denote these correlations by ν_k $(k = 1,\ldots,n_1 + n_2)$, then the following relation was proven in [8]:

$$\mathrm{d}^2_{\mathrm{mi}}(y_1, y_2) = -\log \prod_{k=1}^{n_1+n_2} \nu_k^2 = -\log \prod_{k=1}^{n_1+n_2} \cos^2 \psi_k\,, \tag{17}$$

where the angles ψ_k $(k = 1,\ldots,n_1 + n_2)$ are the $n_1 + n_2$ largest principal angles between the row spaces of the block Hankel matrices Y_{1_f} and Y_{2_f}. They are called the *subspace angles between $h_1(z)$ and $h_2(z)$*, denoted by $[h_1(z) \lhd h_2(z)]$. They can be expressed as the principal angles between subspaces immediately derived from the models:

$$[h_1(z) \lhd h_2(z)] = \left[\begin{pmatrix} \mathcal{C}^{(1)} \\ \mathcal{O}_z^{(2)^\top} \end{pmatrix} \lhd \begin{pmatrix} \mathcal{O}_z^{(1)^\top} \\ \mathcal{C}^{(2)} \end{pmatrix}\right]. \tag{18}$$

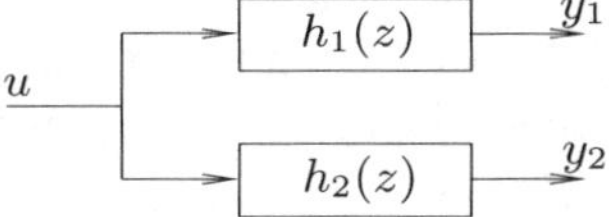

Fig. 2. Setup for the definition of subspace angles between two scalar processes

5.2   Relation with a Cepstral Distance

The power cepstrum of a scalar process y is defined as the inverse Fourier transform of the logarithm of the power spectrum of y:

$$\log \Phi(e^{j\theta}) = \sum_{k=-\infty}^{+\infty} c(k)e^{-jk\theta}\,, \tag{19}$$

where $c(k)$ is the kth cepstral coefficient of y. The sequence $\{c(k)\}_{k\in\mathbb{Z}}$ contains the same information as $\Phi(z)$ and thus also fully characterizes the zero-mean Gaussian process y. The sequence is real and even, i.e. $c(k) = c(-k)$, and can be expressed in terms of the model parameters:

$$c(k) = \begin{cases} \log D^2 & k = 0\,, \\ \sum_{i=1}^{n} \dfrac{\alpha_i^{|k|}}{|k|} - \sum_{i=1}^{n} \dfrac{\beta_i^{|k|}}{|k|} & k \neq 0\,, \end{cases} \tag{20}$$

where the poles of $h(z)$ are denoted by $\alpha_1, \ldots, \alpha_n$ and the zeros by $\beta_1, \ldots, \beta_n$. Based on the cepstral coefficients, a *weighted cepstral distance* was defined in [28]:

$$\mathrm{d}^2_{\mathrm{cep}}(y_1, y_2) = \sum_{k=0}^{+\infty} k(c_1(k) - c_2(k))^2 \,, \tag{21}$$

with c_1 and c_2 the cepstra of the processes y_1 and y_2 and 'cep' referring to 'cepstral'. Based on (18), this distance $\mathrm{d}_{\mathrm{cep}}$ was proven in [8, Chap. 6] (and differently also in [20]) to be equal to the mutual information distance d_{mi}, i.e.:

$$\mathrm{d}_{\mathrm{mi}}(y_1, y_2) = \mathrm{d}_{\mathrm{cep}}(y_1, y_2) \,. \tag{22}$$

This obviously proves that d_{mi} for scalar processes satisfies the triangle inequality. Referring to the discussion in Section 4.1 we can thus say that d_{mi} is a true metric on the set of equivalence classes of scalar stochastic processes, where two processes y_1 and y_2 are equivalent if and only if $h_2(z) = ah_1(z)$ for a non-zero real number a.

6   The Cepstral Nature of the Mutual Information Distance

The equality (22) of d_{mi} and $\mathrm{d}_{\mathrm{cep}}$ was formulated for *scalar* stochastic processes. In the case of *multivariate* processes, one would first need a definition of the power cepstrum of a multivariate process. No such definition is known to the authors of this paper. Therefore, we introduce in Section 6.1 a multivariate power cepstrum and a corresponding weighted cepstral distance, denoted by $\mathrm{d}_{\mathrm{cep}}$.

Even with this new definition, the relation (22) does not hold for general multivariate processes. However, it turnes out experimentally that d_{mi} has a *cepstral character*. This is explained in Section 6.2.

6.1   Multivariate Power Cepstrum and Cepstral Distance

No definition of the power cepstrum of a *multivariate* process y is known to the authors of this paper. Therefore, in analogy with (19), we propose to define the power cepstrum of a multivariate process y as the inverse Fourier transform of the *matrix* logarithm of the power spectrum of y:

$$\log \Phi(e^{j\theta}) = \sum_{k=-\infty}^{+\infty} c(k)e^{-jk\theta} \,, \tag{23}$$

where $c(k) \in \mathbb{R}^{p \times p}$ is the kth cepstral coefficient *matrix* of y. The sequence $\{c(k)\}_{k \in \mathbb{Z}}$ is real and even, and again contains the same information as $\Phi(z)$ and thus also fully characterizes the zero-mean Gaussian process y. However, no analytical expressions as in (20) are known to us for these multivariate cepstral coefficients, although in principle they could be calculated from the state space description (8) of $\Phi(z)$ by expanding the Laurent series of $\log \Phi(z)$ around the origin.

We now define in analogy with (21) a multivariate weighted cepstral distance as

$$\mathrm{d}^2_{\mathrm{cep}}(y_1, y_2) = \sum_{k=0}^{+\infty} k\|c_1(k) - c_2(k)\|^2_{\mathrm{F}} \,, \tag{24}$$

with c_1 and c_2 the cepstra of the multivariate processes y_1 and y_2, and $\|\cdot\|_{\mathrm{F}}$ the Frobenius norm of a matrix. For scalar processes this distance coincides with the previously defined distance (21). No relation with the mutual information distance as in (22) for scalar processes holds for multivariate processes, except for diagonal $\Phi_1(z), \Phi_2(z)$ where it is easily shown that

$$\mathrm{d}^2_{\mathrm{mi}}(y_1, y_2) = \sum_{i=1}^{p} \mathrm{d}^2_{\mathrm{mi}}(y_{1,i}, y_{2,i}) = \sum_{i=1}^{p} \mathrm{d}^2_{\mathrm{cep}}(y_{1,i}, y_{2,i}) = \mathrm{d}^2_{\mathrm{cep}}(y_1, y_2) \,,$$

with $y_{1,i}$ ($i = 1, \ldots, p$) the uncorrelated scalar processes constituting y_1, and analogously for $y_{2,i}$ ($i = 1, \ldots, p$). The first equality follows from Definition 3 and property (f) in Section 3.4. The second equality follows from relation (22) for scalar processes.

The distance (24) can be computed based on the model descriptions of the processes y_1 and y_2. These allow to compute exact values of $\log \Phi(e^{j\theta})$ where θ varies over a discretization of the interval $[0, 2\pi]$. After applying the inverse fast Fourier transform (IFFT) to obtain estimates of the cepstral coefficients, one can further approximate (24) by replacing $+\infty$ in the formula by a finite L.

6.2   The Cepstral Nature of the Mutual Information Distance

For scalar processes, several simulation experiments were performed in [5] in order to compare the behavior of the cepstral distance $\mathrm{d}_{\mathrm{cep}}$, which is equal to d_{mi} because of (22), with the behavior of the $\mathbf{H}_2$ distance, denoted by $\mathrm{d}_{\mathrm{h}_2}$:

$$\mathrm{d}^2_{\mathrm{h}_2}(h_1(z), h_2(z)) = \|h_1(z) - h_2(z)\|^2_{\mathrm{h}_2} = \frac{1}{2\pi} \int_0^{2\pi} \|h_1(e^{j\theta}) - h_2(e^{j\theta})\|^2_{\mathrm{F}} d\theta \,. \tag{25}$$

In order to make $\mathrm{d}_{\mathrm{h}_2}$ a distance between processes instead of between transfer functions, we agree to fix the transfer function description of a stochastic process. We always choose the D-matrix of a model (1) or (6) to be D_{chol}, the unique Cholesky factor of $DD^\top$, which is invariant for a given stochastic process.

In this section we focus on two aspects that showed in the scalar case a difference in behavior between the cepstral distance and the $\mathbf{H}_2$ distance:

1. The influence of poles of $h_1(z)$ and $h_2(z)$ approaching the unit circle.
2. The influence of poles of $h_2(z)$ approaching the unit circle (with fixed zeros), compared to the influence of zeros of $h_2(z)$ approaching the unit circle (with fixed poles). Poles and zeros of $h_1(z)$ are kept fixed.

In order to understand why we choose these two experimental settings, one should notice an important difference between $\mathrm{d}_{\mathrm{h}_2}$ in (25) and $\mathrm{d}_{\mathrm{cep}}$ in (21) and (24), namely the presence of the *logarithm* of the power spectrum in the definition of the cepstrum (19) and (23). For the scalar case this has the following consequences:

1. High peaks in the spectrum of $h_i(z)$ (corresponding to poles close to the unit circle) have a greater influence on $d_{h_2}(h_1, h_2)$ than on $d_{cep}(h_1, h_2)$.
2. Deep valleys in the spectrum of $h_i(z)$ (corresponding to zeros close to the unit circle) have a greater influence on $d_{cep}(h_1, h_2)$ than on $d_{h_2}(h_1, h_2)$.

It can be shown that cepstral distances in the scalar case are equally dependent on the poles and zeros of $h_i(z)$: the distance between two models is equal to the distance between the inverses of the two models. The distance d_{h_2}, on the other hand, is much less sensitive to the depth of a valley than to the height of a peak in the spectrum of $h_i(z)$.

It turns out that, in the *multivariate* case, the mutual information distance d_{mi} and the cepstral distance d_{cep} have several characteristics in common, whereas the $\mathbf{H}_2$ distance d_{h_2} behaves very differently:

1. The distance $d_{h_2}(h_1, h_2)$ grows much faster than $d_{cep}(h_1, h_2)$ and $d_{mi}(h_1, h_2)$ as the poles of $h_1(z)$ and $h_2(z)$ approach the unit circle. This means that d_{h_2} is more sensitive to high peaks in the spectrum of $h_i(z)$ than d_{cep} and d_{mi}. The distances d_{cep} and d_{mi} evolve quite similarly to each other.
2. The distance $d_{h_2}(h_1, h_2)$ grows much faster in case $h_2(z)$ has fixed zeros but poles approaching the unit circle, than in case $h_2(z)$ has fixed poles but zeros approaching the unit circle. For both the distances $d_{cep}(h_1, h_2)$ and $d_{mi}(h_1, h_2)$, on the other hand, the evolution of the distance in case of poles approaching the unit circle is very similar to the evolution in case of zeros approaching the unit circle. This means that d_{h_2} is much more sensitive to high peaks than to deep valleys in the spectrum of $h_i(z)$, whereas d_{cep} and d_{mi} are more or less equally sensitive. The distances d_{cep} and d_{mi} also evolved quite similarly to each other.

With these conclusions we do not claim that one of the distances is *better* than the others. We only wish to point out some differences between them. On the basis of these differences one can choose which distance to use in a specific application.

7   Conclusions and Open Problems

7.1   Conclusions

In this paper we defined the mutual information distance on the set of multivariate Gaussian linear stochastic processes, based on the notion of mutual information of past and future of a stochastic process and inspired by the various properties of this notion. We demonstrated how it can be computed from the state space description of the processes and showed that it is a semimetric on a set of equivalence classes of stochastic processes. For two special classes of stochastic processes, namely scalar processes and processes with diagonal spectral density function, a link exists between the mutual information distance and a previously defined scalar cepstral distance.

The mutual information distance shows a behavior similar to an *ad hoc* defined multivariate cepstral distance and dissimilar from the $\mathbf{H}_2$ distance: it does not inflate when poles of the models are approaching the unit circle and it is more sensitive to differences in zeros than the $\mathbf{H}_2$ distance.

7.2   Open Problems

In this paper a possible extension for multivariate processes was considered of the theory for scalar processes described in Section 5 and Figure 1. The proposed Definition 3 of a multivariate distance however only involves the notion of mutual information and not the notions of subspace angles or cepstral distances between stochastic processes. Thus there remain quite some challenges and issues to be investigated concerning a comparable theory for multivariate stochastic processes.

Furthermore, it would be nice to have more rigorous evidence for the conclusions drawn in Section 6.2.

Multivariate Power Cepstrum and Cepstral Distance
No definition of the power cepstrum of a multivariate process is known to the authors of this paper. Therefore, we introduced an ad hoc definition (23) in Section 6.1. For these cepstral coefficients, however, no analytical expressions are known comparable to e.g. (20) for the scalar coefficients. This topic needs further investigation.

Based on the definition of a multivariate power cepstrum one can define distances in the cepstral domain. In this paper one possible approach was considered in (24) in analogy with (21). But this is clearly not the only possibility.

Subspace Angles Between Multivariate Stochastic Processes
The definition of subspace angles between scalar stochastic processes based on Figure 2 is not readily extendable to multivariate processes. The non-uniqueness of the transfer function description of a multivariate process (see the discussion below (8)) also causes non-uniqueness in the definition of the subspace angles between two multivariate processes. Further investigation is necessary to find a good way to circumvent this problem.

Relations Between System Theory, Information Theory and Signal Processing
Looking at Figure 1 for scalar processes, it is very tempting to look for similar relations in the case of multivariate processes. The two previous topics described the lack of a definition of subspace angles and cepstral distances between multivariate processes. A possible guideline in the search for these definitions could be the attempt to establish a relation with the distance d_{mi} similar to (17) and (22) for scalar stochastic processes. Alternatively, the search for definitions of subspace angles and cepstral distances between multivariate processes could also be guided by the search for a direct link between both, not necessarily through d_{mi}.

References

1. H. Akaike. Markovian representation of stochastic processes by canonical variables. *SIAM Journal on Control*, 13(1):162–173, 1975.
2. C. W. Anderson, E. A. Stolz, and S. Shamsunder. Multivariate autoregressive models for classification of spontaneous electroencephalografic signals during mental tasks. *IEEE Transactions on Biomedical Engineering*, 45(3):277–286, March 1998.

3. M. Basseville. Distance measures for signal processing and pattern recognition. *Signal Processing*, 18(4):349–369, December 1989.

4. A. Bissacco, A. Chiuso, Y. Ma, and S. Soatto. Recognition of human gaits. In *Proceedings of the IEEE International Conference on Computer Vision and Pattern Recognition (CVPR 01)*, volume II, pages 52–58, Kauai, Hawaii, December 2001.

5. J. Boets, K. De Cock, and B. De Moor. Distances between dynamical models for clustering time series. In *Proceedings of the 14th IFAC Symposium on System Identification (SYSID 2006)*, pages 392–397, Newcastle, Australia, March 2006.

6. J. Boets, K. De Cock, M. Espinoza, and B. De Moor. Clustering time series, subspace identification and cepstral distances. *Communications in Information and Systems*, 5(1):69–96, 2005.

7. T. M. Cover and J. A. Thomas. *Elements of Information Theory*. Wiley Series in Telecommunications. Wiley, New York, 1991.

8. K. De Cock. *Principal Angles in System Theory, Information theory and Signal Processing*. PhD thesis, K.U.Leuven, Leuven, Belgium, May 2002. Available as "ftp://ftp.esat.kuleuven.be/pub/SISTA/decock/reports/phd.ps.gz".

9. K. De Cock and B. De Moor. Subspace angles between ARMA models. *Systems & Control Letters*, 46(4):265–270, July 2002.

10. U. B. Desai, D. Pal, and R. D. Kirkpatrick. A realization approach to stochastic model reduction. *International Journal of Control*, 42(4):821–838, 1985.

11. G. Doretto, A. Chiuso, Y. N. Wu, and S. Soatto. Dynamic textures. *International Journal of Computer Vision*, 51(2):91–109, 2003.

12. W. Gersch. Nearest neighbor rule in classification of stationary and nonstationary time series. In D. F. Findley, editor, *Applied Time Series Analysis II*, pages 221–270. Academic Press, New York, 1981.

13. R. M. Gray, A. Buzo, A. H. Gray, Jr., and Y. Matsuyama. Distortion measures for speech processing. *IEEE Transactions on Acoustics, Speech, and Signal Processing*, ASSP–28(4):367–376, August 1980.

14. A. H. Gray, Jr. and J. D. Markel. Distance measures for speech processing. *IEEE Transactions on Acoustics, Speech, and Signal Processing*, ASSP–24(5):380–391, October 1976.

15. E. J. Hannan and D. S. Poskitt. Unit canonical correlations between future and past. *The Annals of Statistics*, 16(2):784–790, June 1988.

16. H. Hotelling. Relations between two sets of variates. *Biometrika*, 28:321–372, 1936.

17. I. A. Ibragimov and Y. A. Rozanov. *Gaussian Random Processes*. Springer, New York, 1978.

18. F. Itakura and T. Umezaki. Distance measure for speech recognition based on the smoothed group delay spectrum. In *Proceedings of the IEEE International Conference on Acoustics, Speech and Signal Processing (ICASSP87)*, volume 3, pages 1257–1260, 1987.

19. N. P. Jewell and P. Bloomfield. Canonical correlations of past and future for time series: definitions and theory. *The Annals of Statistics*, 11(3):837–847, 1983.

20. N. P. Jewell, P. Bloomfield, and F. C. Bartmann. Canonical correlations of past and future for time series: bounds and computation. *The Annals of Statistics*, 11(3):848–855, 1983.

21. C. Jordan. Essai sur la géométrie à n dimensions. *Bulletin de la Société Mathématique*, 3:103–174, 1875.

22. Y. Kakizawa, R. H. Shumway, and M. Taniguchi. Discrimination and clustering for multivariate time series. *Journal of the American Statistical Association*, 93:328–340, 1998.

23. K. Kalpakis, D. Gada, and V. Puttagunta. Distance measures for effective clustering of ARIMA time-series. In *Proceedings of the 2001 IEEE International Conference on Data Mining (ICDM'01)*, pages 273–280, San Jose, CA, November-December 2001.

24. T. Katayama and G. Picci. Realization of stochastic systems with exogenous inputs and subspace identification methods. *Automatica*, 35(10):1635–1652, 1999.

25. D. Kazakos and P. Papantoni-Kazakos. Spectral distance measures between Gaussian processes. *IEEE Transactions on Automatic Control*, 25(5):950–959, 1980.

26. L. Li and Z. Xie. Model selection and order determination for time series by information between the past and the future. *Journal of time series analysis*, 17(1):65–84, 1996.

27. A. Lindquist and G. Picci. Canonical correlation analysis, approximate covariance extension, and identification of stationary time series. *Automatica*, 32(5):709–733, 1996.

28. R. J. Martin. A metric for ARMA processes. *IEEE Transactions on Signal Processing*, 48(4):1164–1170, April 2000.

29. M. S. Pinsker. *Information and Information Stability of Random Variables and Processes*. Holden–Day, San Francisco, 1964. Originally published in Russian in 1960.

30. F. C. Schweppe. On the Bhattacharyya distance and the divergence between Gaussian processes. *Information and Control*, 11(4):373–395, 1967.

31. F. C. Schweppe. State space evaluation of the Bhattacharyya distance between two Gaussian processes. *Information and Control*, 11(3):352–372, 1967.

32. R. H. Shumway and A. N. Unger. Linear discriminant functions for stationary time series. *Journal of the American Statistical Association*, 69:948–956, December 1974.

33. P. Van Overschee and B. De Moor. Subspace algorithms for the stochastic identification problem. *Automatica*, 29:649–660, 1993.

34. P. Van Overschee and B. De Moor. *Subspace Identification for Linear Systems: Theory – Implementation – Applications*. Kluwer Academic Publishers, Boston, 1996.

Differential Forms and Dynamical Systems

Christopher I. Byrnes

Electrical and Systems Engineering, Washington University in St. Louis
`chrisbyrnes@wustl.edu`

Summary. One of the modern geometric views of dynamical systems is as vector fields on a manifold, with or without boundary. The starting point of this paper is the observation that, since one-forms are the natural expression of linear functionals on the space of vector fields, the interaction between the two makes some aspects of the study of equilibria and periodic orbits more tractable, at least in certain cases.

1 Introduction

"For Bourbaki, Poincaré was the devil incarnate. For students of chaos and fractals, Poincaré is of course God on Earth." – M.H. Stone

Stone knew the Bourbaki well; during WWII he was entrusted by A. Weil with all of the volumes written by Bourbaki at that time. He also shared, to some extent, their view that all of mathematics should be deducible in a unified way form a small number of basic principles. On the other hand, he was a student of G. D. Birkhoff who researched dynamical systems in a fashion much closer to Poincaré than to Bourbaki.

Nonlinear dynamics and nonlinear control are both full of great concepts, great constructions and an amazing array of more special methods. However, in both fields it is typically true that the more general a result is, the less often one can use it directly - despite the fact that in some cases a general result at least shifts the burden of analysis to something a bit more tractable. Personally, however, I still wonder what dynamical systems would look like had Poincaré known about Lyapunov theory.

Of course, one of the great general feats is the Poincaré-Bendixson Theory for planar dynamical systems, classifying limit sets, i.e. either an $\omega-$limit or an $\alpha-$ limit, as containing either an equilibrium or being a periodic orbit. In Section 2, we review criteria for the existence of equilibria or periodic orbits for planar dynamical systems, in the context of differential forms and their calculus.

In Section 3, we briefly discuss the Principle of the Torus, which can be thought of as a higher dimensional analogue of the existence of periodic orbits for dynamical systems evolving on a Poincaré annulus containing no equilibrium. The hypotheses, however, are a little onerous - especially the assumption that there should exist a cross-section for the flow that is homeomorphic to the disk.

A. Chiuso et al. (Eds.): Modeling, Estimation and Control, LNCIS 364, pp. 35–44, 2007.
springerlink.com

G.D. Birkhoff, in his 1927 book entitled "Dynamical Systems" and reprinted in 1996 as [1], gave necesssary and sufficient conditions for the existence of a cross-section. In modern terminology, it amounts to the existence of closed one-form having a property that can be checked infinitesimally, just as in Lyapunov theory. We review and reformulate this in Section 4, paying special attention to the kinds of domains which support the hypotheses.

In Section 5, we summarize these observations, yielding the recent result by Brockett and the author on necessary and sufficient conditions for the existnece of periodic orbits in autonomous systems. In Section 6, we present results concerning the stability and robustness of periodic orbits. Both sections rely heavily on the use of differential forms.

I want to conclude by congratulating mio fratello, Giorgio Picci, on his 65th birthday. I hope that this paper is a modest tribute to his work, especially how natural the ideas underlying his work are and the consequent elegance.

2   Planar Dynamical Systems

Suppose $\dot{x} = f(x)$ is a differential equation, defining a dynamical system evolving in the plane $\mathbb{R}^2$. Following the Poincarè-Bendixson Theorem, we are interested in results concerning either existence of equilibria or, in case of the lack thereof, existence of periodic orbits. The following well-known result can, in fact, be used for both provided, of course, the invariance hypothesis is satisfied.

Theorem 1 (*Brouwer's Fixed Point Theorem*). *If a continuous map f leaves a disk $\mathbb{D}^n$ invariant, then f has a fixed point, $f(x_0) = x_0$, for some $x_0 \in \mathbb{D}^n$.*

In the figure below we depict this situation for $n = 2$.

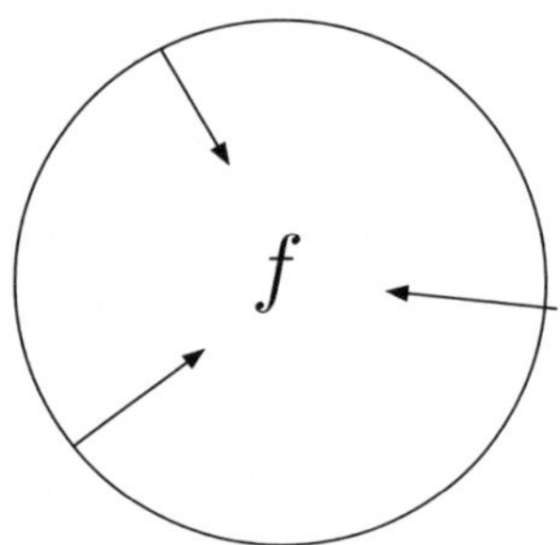

There are many well known proofs of Theorem 1, with our favorite being based on a theorem of M. Hirsch, see [2]. For general n, if f is C^1 then the flow Φ_t of the differential equation of the dynamical system also leaves $\mathbb{D}^n$ invariant and the fixed points of the time-t map, for $t << \infty$ can be shown to be equilibrium.

Corollary 1. *If a C^1 map f leaves a disk $\mathbb{D}^n$ invariant, then the vector field f has an equilbrium, $f(x_0) = 0$, for some $x_0 \in \mathbb{D}^n$.*

Since we will state and prove a more general theorem in our present case of interest, $n = 2$, we need not give proofs of these statements

Indeed, consider a bounded (open) domain M_0 in $\mathbb{R}^2$ with a smooth boundary ∂M_0. We then define $M = M_0 \bigcup \partial M_0$ consisting of an outer boundary ∂M_o and the inner boundaries ∂M_i, $0 \le i \le g$, of g holes as depicted below.

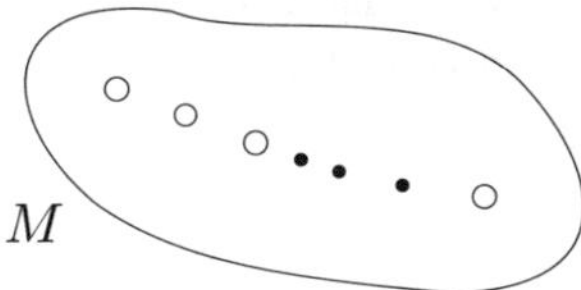

Motivated by classical reasons, we shall call g the genus of M.

We shall state this as an equilibrium theorem, and prove it in the case that the vector field has hyperbolic equilibiria.

Theorem 2 (Birkhoff's Fixed Point Theorem). *If M has genus g and if the C^1 vector field f points inward on the boundary ∂M of M, then f has at least $|1 - g|$ equilibria.*

Proof. Let $C \subset M$ be an oriented closed curve. Write $f = \begin{pmatrix} f_1 \\ f_2 \end{pmatrix}$ and define the Poincaré index relative to C and f as

$$\mathrm{ind}_C(f) = \frac{1}{2\pi} \int_C \underbrace{\frac{f_2 df_1 - f_1 df_2}{\|f\|^2}}_{d(\ )=0}$$

As a matter of fact , if we set

$$\omega = \frac{f_2 df_1 - f_1 df_2}{\|f\|^2}$$

and consider the Gauss map $F = \frac{f}{\|f\|}$ on C defined by f, then $F : C \to S^1$ and

$$\omega = \frac{F^* d\theta}{2\pi}$$

so that, in particular, the the Poincaré index is integer valued and $d\omega = 0$ (see [3] esp. pp. $50 - 51$).

Now, Gauss made the observation that if the vector fields f_1, f_2 always point outwards (or inwards), then $\lambda f_1 + (1 - \lambda) f_2$ is a jointly continuous deformation of one field into the other, which is never zero on C, and therefore the integer value of the index will be constant. In particular, since f points in the direction of the negative of the outward normal n on ∂M_0, we have

$$\mathrm{ind}_{\partial M_o}(f) = -1.$$

Similarly, f points in the outward normal direction on each of the interior boundaries ∂M_i so that

$$\mathrm{ind}_{\cup \partial M_i}(f) = g.$$

Now suppose f has isolated equilibria $x_1, \cdots, x_N$. Consider the closed curves $C_i = \{x : \|x - x_i\| = \epsilon\}$, where $\epsilon << \infty$ so that each C_i and its interior lie in M_o. Then f is defined on $M - \cup_{i=1}^{N} C_i$ and therefore so is ω and $d\omega$.

Recall that Green's Theorem states that if α is a smooth-one form on an oriented two manifold N with smooth boundary ∂N, then

$$\int_N d\alpha = \int_{\partial N} \alpha$$

Taking $N = M$ and $\alpha = \omega$, we have

$$\text{ind}_{\partial M}(f) - \sum_{i=1}^{n} \text{ind}_{C_i}(f) = \int_{M - \cup_{i=1}^{N} C_i} d\omega = 0.$$

Therefore

$$|g - 1| = \left| \sum_{i=1}^{N} \text{ind}_{C_i}(f) \right| \leq \sum_{i=1}^{N} |\text{ind}_{C_i}(f)| = N.$$

The case of $g = 1$ hole is special, since there does not have to be an equilibrium. The corresponding closed domain with one hole is usually called a Poincaré annulus, A^2, if there in fact no equilibrium. In this case, the Poincaré-Bendixson Theorem implies the following result.

Theorem 3. *Suppose f points inward on ∂A^2. If f has no equilibria, then f has a periodic orbit.*

While we will take this as a given, we are also very interested in the construction, and topology, of cross-sections for such a flow.Recall that a cross-section is, in this case, a curve C, homeomorphic to a closed interval for which the flow is transverse and returns in finite time. The following figure depicts a flow on a Poincaré annulus and several cross-sections.

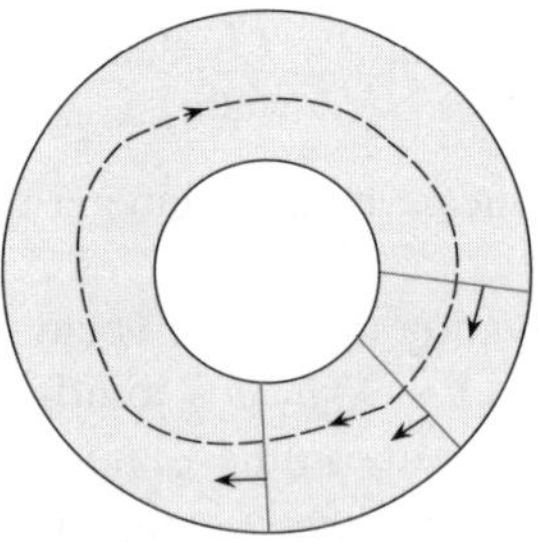

Choosing a cross-section C, let $t_1 = \inf\{t > 0 : x(t; x_0) \in C\}$ and define the Poincaré map $\mathcal{P} : C \to C$ via

$$\mathcal{P}(x_0) = x(t_1, x_0).$$

By Theorem 1, $\mathcal{P}$ has a fixed point, which of course implies f has a periodic orbit. Actually, as the following figure suggests

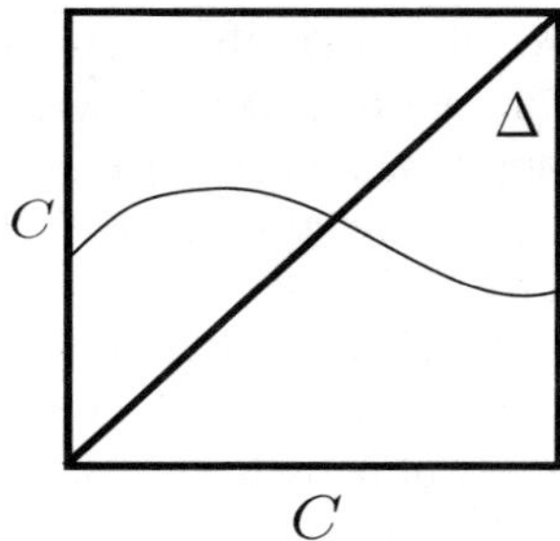

For $n = 1$, there is are many elementary proofs. For example, if graph(f) does not intersect the diagonal then, because it is connected, it must lie in one of the two connected sets we obtain by removing the diagonal, Δ, from $C \times C$. If it lies in the upper "triangle,' then f is undefined at one end-point of C. If it lies in the lower "triangle," then f is undefined at the other end-point.

Actually, there is a nice exercise in ([4], p.247) which considers the annulus bounded by the circles of radius 1 and 2, positively invariant under a vector field f which is everywhere transverse to the cross-sections $\theta = $ constant, and asks for a proof of the existence of a periodic orbit. This of course is a very convenient family of (foliation by) cross-sections for the flow and one would like to know whether, in general, crosss-sections are the level sets of some Lyapunov-like function. The answer is no, as we can deduce in the plane using Poincaré's Method of Tangential Curves [5].

Poincaré's Method assumes a bounded domain D, on which is defined a family of curves, $P(x, y) = $ constant, and a vector field which is tranverse to these cross-sections at most points and tangent at all others (we will actually make this precise in higher dimensions). The set of points at which f is tangent is called the tangent locus.

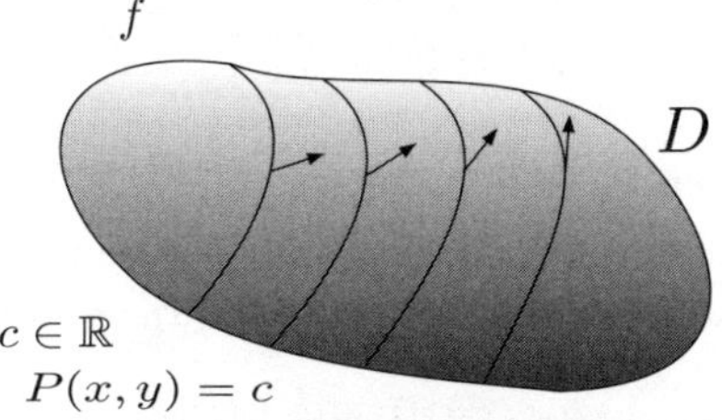

The Method of Tangential Curves asserts that if the tangent locus contains no equilibrium or periodic orbits then D contains no periodic orbits. There are two aspects of Poincaré's Method that are particularly noteworthy.

The first is that, by writing V instead of P, and thinking about $\dot{V} = 0$ instead of the tangent locus, Poincaré's Method can be seen as a precursor to LaSalle's Theorem, which is its higher dimensional generalization. More explicitly, consider $\dot{x} = f(x)$, $x \in \mathbb{R}^n$, and a compact subset $D \subset \mathbb{R}^n$. Suppose there exists V such that $L_f V(x) \geq 0$. Then, for any trajectory remaining in D any ω-limit x^* remains in D and satisfies $L_f V(x^*) = 0$. For $P = V$ and $n = 2$, this would imply that the tangent locus contains $\omega-$limits and hence either equilibria or periodic orbits, contrary to hypothesis, /therefore every trajectory must leave D.

The second is that this shows that a family of cross-sections in a Poincaré annulus, such as the family $\theta = $ constant discussed above, can never be the level sets of a Lyapunov-like function, or any function, because periodic orbits do exist. Indeed, for example, θ is not a single-valued function. Rather, we should be using the closed, not exact, one form $d\theta$ which gives rise to cross-sections described by $d\theta = 0$. In fact, thinking about both equilibria and periodic orbits, one observes that exact one-forms are very useful for analyzing equilibria, while closed, non-exact one-forms can be shown to be extremely useful for the study of periodic orbits.

3   The Principle of the Torus for Autonomous Systems

The quest for finding a higher dimensional generalization of the Poincaré -Bendixson Theorem, or the the existence of the right analogue of the Poincaré annulus, has had to deal with more complicated limiting behavior and more complicated topology. One such generalization, the Principle of the Torus [6], is worth mentioning as a gateway to a more general approach and because it has found a fair amount of application in engineering and science [7], and also captures some of the features encountered in forced oscillations of hyperbolically stable autonomous systems.

Given a dynamical system $\dot{x} = f(x)$ in $\mathbb{R}^n$, the Principle assumes the existence of an invariant toroidal region R and a cross-section C for the vector field f, with C is homeomorphic to a disk. this situation is depicted below.

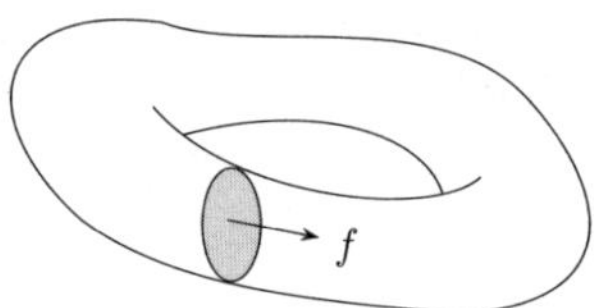

In particular, $\forall\, x_0 \in C \,\exists\, t > 0$ so that $x(t; x_0) \in C$. Then, the Poincaré map $\mathcal{P} : C \to C$ is defined and is continuous because f is transverse to C.

By Theorem 1, $\mathcal{P}$ has a fixed point x_0, i.e., $x(t_1, x_0) = x_0$ and therefore $x(\cdot, x_0)$ is periodic.

While it does sometime occur in practice, the hypothesis that there is a cross-section, which is also homeomorphic to a disk, is quite strong and hard to check. Supressing, for the moment, the topological properties of C, one should ask whether there are criteria for the existence of cross-sections. As we saw in the previous section, the cross-sections will not in general be level sets of smooth functions. This can be seen again in this context for a standard solid n -torus, with coordinates $(\rho.\theta)$, with differential equation

$$\dot{\rho} = 0, \dot{\theta} = 1$$

for which a family of cross-sections is defined by $d\theta = 0$. In particular, this family cannot be defind as the level sets of a smooth real-valued function. That this situation is indeed the general case will be addressed in the next section.

4 Lyapunov-Like Differential Forms for the Existence of Cross Sections

Again, we consider $\dot{x} = f(x)$ evolving in $\mathbb{R}^n$. In his 1927 classic [1], G. D. Birkhoff shows that a necessary condition for a local cross-section C to exist is that C should be part of a family of cross-section defined by an angular variable $\phi = $ constant, which he constructs from the flow and which should be increasing along the flow (see esp. pp. $143 - 145$). He expresses this in terms of the pair of constraints

$$\dot{\phi} = \sum_{i=1}^{i=n} a_i f_i > 0 \text{ and } \frac{\partial a_i}{\partial x_j} = \frac{\partial a_j}{\partial x_i}, \; i, j = 1, 2, \cdots, n$$

and claims that their existence will define a cross-section by setting ϕ to be equal to the constant 0.

In modern terminology, the functions a_i are the components of a 1-form $\omega = \sum_{i=1}^{n} a_i dx_i$ such that

$\Leftrightarrow \; d\omega = 0$
$\Leftrightarrow \; \langle \omega, f \rangle > 0.$
$\omega = 0$ defines a cross-section.

Some care, however has to be given as to the domain of these objects, especially ω. We first consider the planar case, following the notation set in Section 2.

If the genus of M is not 1 then, by Theorem 2, f has at least $|g - 1|$ zeros, so that $\langle \omega, f \rangle > 0$ is untenable and there cannot be an increasing angular variable. From the classification of compact orientable two dimensional manifolds by their genus, we deduce the same conclusion for compact 3-dimensional manifolds-with-boundary in $\mathbb{R}^3$. Therefore, angular variables in the sense of Birkhoff will only exist on the solid 2-torus.

This low-dimensional analysis gives some credence to standard assumptions concerning a positively invariant manifold which is a "toroidal" region. We shall take a more general approach.

Definition 1. Consider a compact orientable manifold M, perhaps with smooth boundary. We call M a circular manifold provided M is contractible to a simple closed curve $\gamma \subset M$.

Examples: circles, compact cylinders, solid tori.

Remark: A circular manifold is connected, since it is contractible to a connected space. More generally, A circular manifold is a $K(\mathbb{Z}; 1)$; i.e., an Eilenberg-Maclane space with $\pi_1(M) = (\gamma) = \mathbb{Z}$ and $\pi_i(M) = \{0\}$ for $i \geq 2$.

Remark: If $n = 1$, then either $M \simeq [a, b]$ or $M \simeq S^1$. Only the latter is contractible to a circle. In particular, in this case, M has nonempty boundary. Suppose conversely that M is a circular manifold without boundary. If M is orientable then the volume form on M represents a nonzero class in $H_{dR}^n(M)$, by Stokes' Theorem, and therefore M is only a circular manifold when $n = 1$. If M is not orientable , applying the same reasoning to its orientable double cover $\widetilde{M}$ we find that $M \simeq \mathbb{R}P^1$ and is therefore also a circle. Summarizing, a circular manifold has $\partial(M) \neq \emptyset$ if, and only if, $n \geq 2$.

5 Necessary and Sufficient Conditions for Existence of Periodic Orbits

Theorem 4 (Brockett-Byrnes). *A necessary and sufficient condition for the existence of a periodic orbit of a smooth vector field f is:*

1. *existence of a positively inverse circular manifold M^n,*
2. *a closed one form ω for which $\langle \omega, f \rangle > 0$ on M^n and for which ω never vanishes on ∂M^n.*

Proof.
Necessity: If γ exists then there exists a closed one-form $d\theta$ on γ such that

$$\int_\gamma \alpha = 1$$

Sufficiency: Choose a generator γ_1 for $\pi_1(M) \simeq \mathbb{Z}$. We can assume $\int_{\gamma_1} w = 1$. We now construct the "period map" $J : M \to S^1$. Fix $P \in M$. For any $Q \in M$ choose a path γ_2 from P to Q and consider

$$\widetilde{J}_{\gamma_2}(Q) = \int_{\gamma_2} w.$$

If γ_3 is another such path

$$\gamma_3 - \gamma_2 \sim \ell\,\gamma_1 \text{ so } \widetilde{J}_{\gamma_3}(Q) = \widetilde{J}_{\gamma_2}(Q) + \ell.$$

That is, in $\mathbb{R}$, $\widetilde{J}_{\gamma_3}(Q) \equiv \widetilde{J}_{\gamma_2}(Q)$ mod $\mathbb{Z}$. Therefore we may define

$$J : M \to S^1 = \mathbb{R}/\mathbb{Z} \text{ via } J(Q) = \int_P^Q w.$$

It will turn out that $J^{-1}(\theta)$ is connected (because $\int_{\gamma_1} w = 1$) and is therefore a leaf of $w = 0$. Choose $\theta_0 \in S^1$ $L_0 = J^{-1}(\theta_0)$. Choose $x_0 \in L_0$. Since

$$\langle w, f \rangle > 0 \text{ we have } \int_{J(\Phi_t(x_0))} d\theta > 0$$

and so there exists $T > 0$ such that

$$\mathcal{P}(X_0) = \Phi_T(x_0) \in L_0.$$

Therefore L_0 is a cross-section for the flow.

We conclude by proving that $\mathcal{P}$ has a fixed point, using the homotopy of $L_0 \subset M \to S^1$ and the Lefschitz fixed point theorem.

6   Stability and Robustness of Periodic Orbits

Our proof of the existence of periodic orbits computed the Lefschitz number of $\mathcal{P}$ using the Lefschitz Fixed Point Theorem in homology. This theorem can also be proved using differential forms.

In fact, the Lefschetz number computes the oriented intersection number of the graph of $\mathcal{P}(x)$ with the diagonal (the graph of the identity map) in $L_0 \times L_0$.

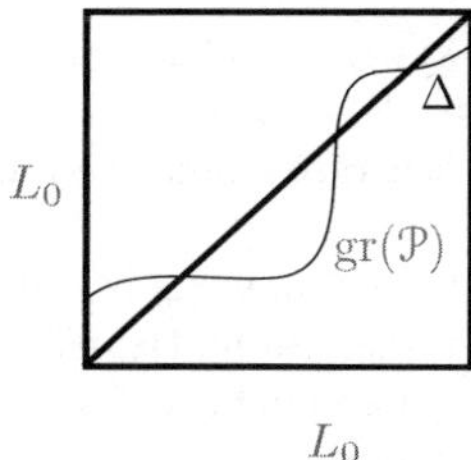

For C^∞ objects, a transverse (hyperbolic) intersection number is computed as

$$\det(I - D\mathcal{P}(x_0))$$

at an intersection point $(x_0, \mathcal{P}(x_0))$. Therefore, if we have only hyperbolic intersections, then we must have

Corollary 2 (An index formula). *Under the hypotheses above, if $\mathcal{P}$ has only hyperbolic fixed points then they are finite in number and*

$$\sum_\gamma sign\ det(I - D\mathcal{P}_\gamma) = 1$$

where the sum is over distinct periodic orbits.

We note that, if x_0 is asymptotically stable (and hyperbolic)

$$\mathrm{sign}\ \det(I - D\mathcal{P}(x_0)) = 1.$$

Corollary 3. *The local exponential stability of every orbit implies the global uniqueness of a single stable orbit.*

Proof. If each periodic orbit is exponentially orbitally stable, the index formula reduces to

$$\#\gamma \cdot 1 = 1 \ \text{ or } \ \#\gamma = 1.$$

Because our conditions are open in the C^∞ topology, they are also robust with respect to sufficiently small perturbations.

Corollary 4. *Suppose $\dot{x} = f_0(x)$, $x \in \mathbb{R}^n$, has an asymptotically stable periodic orbit γ_0. Then, for any jointly continuous perturbation*

$$\dot{x} = f(x, \mu)$$

where for some μ_0, $f(x, \mu_0) = f_0(x)$, there must exist a periodic solution, for each $\mu \sim \mu_0$.

Corollary 5 (Averaging). *If 0 is a locally exponentially stable equilibrium for the system $\dot{x} = f(x) + \epsilon g(x, t, \epsilon)$ when $\epsilon = 0$, and $g(x, t + T, \epsilon) = g(x, t, \epsilon)$ then for $\epsilon \ll \infty$ there exists a locally exponentially stable periodic orbit $\gamma_\epsilon(t)$ of period T whose amplitude is $\mathcal{O}(\epsilon)$.*

References

1. Birkhoff, (1966) G.D. Dynamical systems, 2nd edition. American Mathematical Society Colloquium Publications, Vol. IX, Providence, R.I.
2. Milnor J.(1997) Topology from the differentiable viewpoint. Princeton University Press, Princeton
3. Guckenheimer J. and Holmes P. (1983) Nonlinear oscillations, dynamical systems, and bifurcations of vector fields, Springer-Verlag, Berlin Heidelberg New York
4. Hirsch, M. and Smale, S. (1974) Differential equations, dynamical systems, and linear algebra. Academic Press, New York London.
5. Yanqian, Ye et al. (1986) Theory of limit cycles, Trans. Math. Monographs 66, AMS Providence.
6. Pliss V.A. (1966) Nonlocal problems of the theory of oscillations. Academic Press, New York London.
7. B. Li (1981) Periodic Orbits of Autonomous Ordinary Differential Equations: Theory and Applications, Nonlinear Analysis, Theory, Methods and Applications, Vol. 5, 931-958.

An Algebraic Framework for Bayes Nets of Time Series

Peter E. Caines[1] and Henry P. Wynn[2]

[1] Department of Electrical and Computer Engineering, McGill University, 3480 University Street, Montreal, QC H3A 2A7, Canada
peterc@cim.mcgill.ca
[2] Department of Statistics, London School of Economics and Political Science, London, UK
h.wynn@lse.ac.uk

Summary. Graphical models in which every node holds a time-series were analysed in Caines et al. [13], [14] using the notion of lattice conditional independence (LCI) due to Anderson et al. [1], [2]. Under certain feedback free (or causality) conditions, LCI imposes a special zero structure on the stochastic realizations of those processes generated by state space systems; this structure comes directly from the transitive directed acyclic graph (TDAG) which is in one-to-one correspondence with the Boolean Hilbert lattice of the LCI formulation. In this paper, these properties of sets of stochastic processes are generalized to the setting of infinite Bayes nets of time series; this formulation contains as a special case conditionally independent processes embedded in chain recurrent spatial structures.

Keywords: Bayes nets, stochastic processes, time series, conditional independence, feedback free conditions, causality, stochastic realization, chained systems, recurrence.

1 Introduction

The subject of this paper is the construction Bayes nets, or graphical models, in which every node is a time series. Among the first work in this growing filed is that of Dahlhouse [15] and Eichler [12].

The key condition for the whole theory of graphical models is conditional independence, so this is necessarily the starting point for any extension to infinite dimensional processes. The theory presented in (Caines et al. [13], [14]) and in this paper provides a dynamic version of static conditions for conditional independence since the basic definition in the process case requires the conditional independence property to hold recursively in time *up to time t*. As will be seen in the next section, the viewpoint adopted in this paper is that the theory of stochastic realization developed by Akaike [8], [9], Lindquist and Picci [4], [5], [3], and others (see e.g. Caines [7], Chapter 4), forms an appropriate framework to capture the properties of the conditional independence of groups of process. But this is the case only if the so-called feedback free, or causality, relations hold pairwise between each conditioning process and each of the members of the set of processes it renders conditionally independent.

However conditional independence is only half the story. The next step is to find a structure which preserves the utility of the state space formulations but allows one

A. Chiuso et al. (Eds.): Modeling, Estimation and Control, LNCIS 364, pp. 45–57, 2007.
springerlink.com

to first build reasonably complex graphical models, and, second, to capture both time dependence and the cross-process dependence. Broadly speaking, the Lattice Conditionally Independence (LCI) in the Gaussian case (or, the Lattice Conditionally Orthogonality (LCO) property in the more general second order case) is most appropriate for expressing such properties since it respects a generalized "arrow of time" both in the space orderings of a given family of processes and with respect to time.

In this paper, Section 2 introduces the notions of conditional independence, the feedback free property and gives the first stochastic realization theorem; Section 3 presents the relevant lattice projectors and introduces the notion of a transitive directed acyclic graph (TDAG) of a set of processes; then, in Section 4, a class of countably infinite TDAG structures is presented which has sufficient regularity to permit the construction of what may be termed spatio-temporal stochastic realizations.

2   Conditional Independence and Stochastic Realization

For a stochastic process X, the *process up to time t* is defined as the finite or infinite segment of the process given by $X^{(t)} = \{\ldots, X_{t-1}, X_t\}$.

Definition 2.1 (Caines et al. [13], [14]).   A triple of multivariate stochastic processes $\{X_t, Y_t, Z_t; t = \ldots - 1, 0, 1, \ldots\}$ is said to be *conditionally independent up to time t* if

$$X^{(t)} \perp\!\!\!\perp Y^{(t)} \mid Z^{(t)},$$

and is said to be *conditionally independent* if this relation holds for all t.   □

We motivate this notion with the following simple example.

Example 2.1.   Consider the unique stationary trivariate Gaussian process $\{X_t, Y_t, Z_t\}$ generated by the $AR(1)$ linear system:

$$
\begin{aligned}
X_t - a_{11}X_{t-1} - a_{12}Y_{t-1} - a_{13}Z_{t-1} &= \varepsilon_{1,t} \\
Y_t - a_{21}X_{t-1} - a_{22}Y_{t-1} - a_{23}Z_{t-1} &= \varepsilon_{2,t} \\
Z_t - a_{31}X_{t-1} - a_{32}Y_{t-1} - a_{33}Z_{t-1} &= \varepsilon_{3,t}
\end{aligned}
$$

where (i) $\{\varepsilon_t\} = \{\varepsilon_{1,t}, \varepsilon_{2,t}, \varepsilon_{3,t}\}$ is a white noise process, that is to say, is a Gaussian, zero mean, unit variance process independent over time and index, and (ii) the system is asymptotically stable in the sense that (in an obvious notation) the eigenvalues of A are asymptotically stable (i.e. have absolute value strictly less than 1).

It is straightforward to check that if $a_{21} = a_{31} = a_{32} = a_{12} = 0$, the processes up to time t for all t are conditionally independent. There are several ways to show this: (i) via conditional expectations, (ii) via the covariance generating functions (i.e. the spectral density matrix for the joint process $\{X_t, Y_t, Z_t\}$), and (iii) via the explicit construction of the semi-infinite projections, which are the analogues of the projections P_J described in Section 3.2 below.

A fourth method is to explicitly solve for X_t and Y_t using the shift operator z defined on any process V via $zV_t = V_{t-1}$:

$$X_t = \frac{\varepsilon_{1,t} + a_{13} z Z_t}{1 - a_{11} z}$$

$$Y_t = \frac{\varepsilon_{2,t} + a_{23} z Z_t}{1 - a_{22} z}$$

Now if Z_s, $s \leq t$, is fixed we see that X_t and Y_t only depend on $\varepsilon_{1,s}$, $s \leq t$, and $\varepsilon_{2,s}$, $s \leq t$, respectively, which are mutually independent and are independent of Z_s, $s \leq t$, which is a function only of $\varepsilon_{3,s}$, $s \leq t$. We note that in this example, Z at the instant t does not appear in the expressions above for X_t and Y_t. $\square$

To generalise these facts we assume first that $\{X_1, \ldots, X_d\} = \{\{X_{1,t}, \ldots, X_{d,t}\}, t \in \mathbf{Z}\}$ is a collection of multivariate zero mean second order stochastic processes.

Definition 2.2. A pair of second order processes (X_i, X_j), $i \neq j$, is said to be *feedback free* if and only if $X_{t,i}|X_j = X_{t,i}|X_j^{(t)}, t \in \mathbf{Z}$, where "$|$" represents orthogonal projection. In case the equality holds with $|X_j^{(t)}$ replaced by $|X_j^{(t-1)}$, (X_i, X_j), $i \neq j$, is said to be *(strongly) feedback free*. $\square$

(Note that orthogonal projection is identical to conditional expectation in the zero mean Gaussian case).

It follows (see Theorem 2.1, Chapter 10, Caines [7] for this and other characterizations) that a pair of wide sense stationary processes (X_i, X_j), $i \neq j$, (assumed to possess a spectral density) is feedback free if and only if the spectral density matrix has a factorization where the first matrix factor is asymptotically stable and inverse asymptotically stable:

$$G_{X_i, X_j}(z) = \begin{bmatrix} A & B \\ 0 & C \end{bmatrix} \begin{bmatrix} A^* & 0 \\ B^* & C^* \end{bmatrix}(z), \tag{1}$$

where A^* is the conjugate of A, etc. The strongly feedback free case corresponds to the additional condition $B(0) = 0$.

Let the feedback free property hold between the conditioning process and each of the other processes in a conditional independence triple; then a temporal dependence is imposed on the conditional independence structure as is shown in the next result where X_i denotes the whole process which, in the doubly infinite case would be $\{X_{i,t}\}_{-\infty}^{\infty}$.

Theorem 2.1 (Caines et al. [13]). Subject to the assumption that the pairs $\{X_i, X_k\}$ and $\{X_j, X_k\}$ are feedback free, the following two types of conditional independence are equivalent,

1. $X_i \perp\!\!\!\perp X_j | X_k$,
2. $X_i^{(t)} \perp\!\!\!\perp X_j^{(t)} | X_k^{(t+s)}$, $s \in \{0, 1, \ldots\}, t = \ldots -1, 0, 1, \ldots$.

Subject to the strong feedback free condition, $(t + s)$ above is replaced by $(t + s - 1)$. $\square$

The next result shows that under the condition that the pairs of processes under consideration are feedback free, it is the case that processes generated by state-space systems are conditionally independent if and only if they have a stochastic realization with a specific structure.

Theorem 2.2 (Caines et al. [14]). Let the three jointly full rank Gaussian stochastic processes $\{X_{i,t}, X_{j,t}, X_{k,t}; t \geq 0\}$ be such that the joint process possesses a finite dimensional stochastic realization on $t = 0, 1, \ldots$. Then (i) each pair $\{X_i, X_k\}$ and $\{X_j, X_k\}$ is feedback free, and (ii) the conditional independence relations $X_i^{(t)} \perp\!\!\!\perp X_j^{(t)} \mid X_k^{t+s}$, $s = 0, 1, \ldots$; $t = 0, 1, \ldots$, hold if and only if $\{X_{i,t}, X_{j,t}, X_{k,t}\}$ has a (not necessarily time invariant) stochastic state-space realization on $t = 0, 1, \ldots$ of the form:

$$\begin{bmatrix} s_{i,t+1} \\ s_{j,t+1} \\ s_{k,t+1} \end{bmatrix} = \begin{bmatrix} F_{ii} & 0 & F_{ik} \\ 0 & F_{jj} & F_{ik} \\ 0 & 0 & \tilde{F}_{kk} \end{bmatrix} \begin{bmatrix} s_{i,t} \\ s_{j,t} \\ s_{k,t} \end{bmatrix} + \begin{bmatrix} M_{ii} & 0 & M_{ik} \\ 0 & M_{jj} & M_{jk} \\ 0 & 0 & M_{kk} \end{bmatrix} \begin{bmatrix} u_{i,t} \\ u_{j,t} \\ u_{k,t} \end{bmatrix}$$

$$\begin{bmatrix} X_{i,t} \\ X_{j,t} \\ X_{k,t} \end{bmatrix} = \begin{bmatrix} H_{ii} & 0 & H_{ik} \\ 0 & H_{jj} & H_{jk} \\ 0 & 0 & H_{kk} \end{bmatrix} \begin{bmatrix} s_{i,t} \\ s_{j,t} \\ s_{k,t} \end{bmatrix} + \begin{bmatrix} N_{ii} & 0 & N_{ik} \\ 0 & N_{jj} & N_{jk} \\ 0 & 0 & N_{kk} \end{bmatrix} \begin{bmatrix} u_{i,t} \\ u_{j,t} \\ u_{k,t} \end{bmatrix},$$

where (i) the joint system input and observation noise process $u_t; t = 0, 1, \ldots$ is a full rank zero mean orthogonal Gaussian process with block diagonal (instantaneous) covariance (ii) u is independent of the Gaussian zero mean initial condition s_0, and (iii) $s(0)$ has a covariance matrix Σ which possesses a factorization $\Sigma = \Sigma^{\frac{1}{2}} \Sigma^{\frac{T}{2}}$, where $\Sigma^{\frac{1}{2}}$ has the same block structure as the system matrices. The replacement of the feedback free condition by the strong feedback free condition is obtained by imposing the additional constraints that the matrices $H_{ik}, H_{jk}, N_{ik}, N_{jk}$ are zero and Σ is block diagonal. $\qquad\square$

3 Lattice Conditionally Independence and Stochastic Realization

3.1 Lattices of Subspaces

One of the purpose of this paper, as mentioned in the introduction, is to describe an algebraic framework which is suitable for the extension of the notion of finite Bayes nets of time series to the case of infinite but countable sets of nodes. We will see that there are a number of essentially isomorphic structures by which this framework can be described. It turns out, first, that the state-space formulation of conditional independence given in the last section is a special case of this structure and, second, that we can build stochastic realizations of a more general class of graphical models of time series using an extension deriving from one of a set of equivalent algebraic structures. This is the transitive directed acyclic graph (TDAG) structure and its implementation will be given in the next section.

We now give a brief summary of some relevant, closely related, algebraic structures.

Boolean Lattices of Subspaces. let H be a separable Hilbert space with a countable basis $\{\varepsilon_i\}$, $i \in \mathbb{N}$. We shall call the index set for the basis the *basis index set*.

Let U and V be a two subspaces of a separable Hilbert space H and let P_U and P_V be the associated projectors. Then we have that the projector for $U \cap V$ is the product $P_U P_V$ if and only if P_U and P_V commute: $P_U P_V = P_V P_U$. Taking $U \cup V$ to mean $\mathrm{span}(U, V) = U + V$ the corresponding projection is then

$$P_{U \cup V} = P_U + P_V - P_{U \cap V} = P_U + P_V - P_U P_V \qquad (2)$$

(see Kadison and Ringrose [10]).

We are interested in the special Boolean lattice of subspaces of H is a countable collection of subspaces $\{U_J | J \in \mathcal{L}\}$ closed under $\cap$ and $\cup$. In our case each subspace is uniquely indexed by its representation in terms of a countable basis $\{e_i\}$, and we write $U_J = \mathrm{span}\{e_i,\ i \in J\}$, where J is a possibly infinite, but countable set of indices.

Countable Subset Lattice. By the Stone representation theorem, any complete, atomic Boolean lattice is isomorphic to a lattice of subsets of some set (Johnstone [16]). In our case, these sets are simply subsets of the index set J under the union and intersection operations. We call this lattice $B(\mathcal{L})$. Thus for any countable set J of indices

$$V_J = \cup_{i \in J} U_i \to J,$$

and for index sets J and K

$$U_J \cup U_K \to J \cup K, \quad U_J \cap U_K \to J \cap K$$

We can simplify the notation somewhat by writing P_J for the projector P_{U_J}, associated with the subspace U_J.

Countable Transitive Acyclic Directed Graphs. To each Boolean lattice, and the equivalent subset lattice, we associate a transitive directed acyclic graph (TDAG), $\mathcal{G}(E, V)$. The vertex set E is the basic index set, defined above. There is a directed chain from vertex i to vertex j if every index set J in $B(\mathcal{L})$ which contains j contains i. Then we have the edge $i \to j$ if there is a chain from i to j with no intermediate vertices. Transitivity follows from the fact that $i \to j$ and $j \to k$ implies that any K containing k contains j and therefore contains i. The inverse construction is also straightforward.

Conditional Orthogonality. If I, J and K are disjoint sets of integers such that $P_{I \cup K} P_{J \cup K} = P_{J \cup K} P_{I \cup K}$ we write $I \perp J | K$. Thus the Boolean lattice gives a collection of conditional orthogonality statements. These can be written in terms of projections also as

$$(P_{I \cup K} - P_K)(P_{J \cup K} - P_K) = 0,$$

The projections $P_{I \cup K} - P_K$ and $P_{J \cup K} - P_K$ may be termed the "innovation" subspaces associated to the pairs of spaces (I, K) and (J, K).

3.2 Lattice Conditionally Orthogonal Stochastic Hilbert Spaces

Consider a full rank collection of multivariate zero mean second order stochastic processes $\{X_{1,t}, \ldots, X_{d,t}\}$ and consider its associated Hilbert space. In case $\{X_1, \ldots, X_d\}$

is Gaussian the projections are conditional expectations. We initially consider the case where each $X_{i,t}$ is univariate. An orthonormal basis corresponds to a basis of uncorrelated unit variance random variables $\{e_{i,t}; \quad t = \ldots, -1, 0, 1, \ldots; \quad i = 1, \ldots, d\}$ and these are independent in the Gaussian case. Associated with such a basis is a family of rank 1 projectors $\{P_{i,t}; i = 1, \ldots, d\}$; for instance, if $X_{i,t} = a_i^T e_t$, then $P_{i,t} = a_i[a_i^T a_i]^{-1} a_i^T$. Of course, the orthonormal basis is not unique and a special basis may play a special role so that the projectors of interest may depend on the basis of interest; this is the situation with the innovations basis which appeared implicitly in Theorem 2.1.

The finite dimensional Lattice Conditional Orthogonality (LCO) condition is straightforward to define. We can consider it here first in the case of a single time point. There are many equivalent statements and here we simply list some of them. We take first take the case $d = 3$ and consider conditional independence. The condition $X_1 \perp X_2 | X_3$, which we read as X_1 *conditionally orthogonal to* X_2 *given* X_3, then has the following equivalent statements:

1. $\{\Gamma^{-1}\}_{12} = 0$
2. If P_{13} and P_{23} are the orthogonal projectors onto $\mathrm{span}(X_1, X_3)$ and $\mathrm{span}(X_2, X_3)$, respectively, then $P_{13} P_{23} = P_{23} P_{13}$.
3. If I is the identity projection in $\mathrm{span}(X_1, X_2, X_3)$ and P_3 is the projector on $\mathrm{span}(X_3)$ then
$$P_{13} + P_{23} - P_3 = I$$
4. $(P_{13} - P_3)(P_{23} - P_3) = 0$
5. There is an orthornormal basis for $\mathrm{span}(X_1, X_2, X_3)$ such that in this basis P_{13} and P_{23} take the form
$$P_{13} = \begin{bmatrix} 1 & 0 & 0 \\ 0 & 0 & 0 \\ 0 & 0 & 1 \end{bmatrix}, \quad P_{23} = \begin{bmatrix} 0 & 0 & 0 \\ 0 & 1 & 0 \\ 0 & 0 & 1 \end{bmatrix}$$

It is interesting, even in this simple case, to see the subset lattice at work on the diagonal elements of the projectors. In binary arithmetic

$$101 + 011 - 001 = 111,$$

or in terms of the indicator, that is Boolean variables:

$$I_{13 \cup 23} = I_{13} + I_{23} - I_3$$

The LCO condition is attributed to an index sets $J \subset \{1, \ldots, d\}$ such that all the corresponding projectors P_J commute. Then by an identical argument as used above the subspace lattice can be read off from the subset lattice of indices. The associated TDAG is then exactly as describe above and the random variables form a special acyclic Bayesian graphical model with the global Markov property. For the conditional independence of X_1, X_2 given X_3, the lattice has elements $123, 13, 23, 3$ and we adjoin $\emptyset$. The TDAG is

$$3 \to 1, \quad 3 \to 2.$$

We shall say that the collection of jointly wide sense stationary zero mean Gaussian processes $X_1, \ldots, X_d$ is *LCI-$\mathcal{G}$-compatible* if the conditions associated with the procedure below are satisfied:

1. For the given TDAG $\mathcal{G}(E, V)$, or the equivalent distributive Boolean lattice $\mathcal{L}$, associate to every vertex V_i a stochastic process X_i.
2. For every conditional independence statement that can written down from $\mathcal{G}$, or the corresponding Boolean lattice $\mathcal{L}$, require that the corresponding feedback free condition holds. Note that we need to group processes together according special index sets. Thus, for an index set J we say $X_J = \{X_{j,t},\ j \in J\}$. If our conditional independence is to be

$$X_J \amalg X_K | X_{J \cap K}, \quad J, K \in \mathcal{L},$$

then we require that pairs $(X_J, X_{J \cap K})$ and $(X_K, X_{J \cap K})$ are each feedback free.
3. Finally, verify that each of the up-to-t version of conditional independence statements given by the graph $\mathcal{G}$ holds (or the equivalent condition in Theorem 2.2). With an obvious notation: for every $J, K \in \mathcal{L}$

$$X_J^{(t)} \amalg X_K^{(t)} | X_{J \cap K}^{(t)}$$

To summarize: for $X_1, \ldots, X_d$ to be LCI-$\mathcal{G}$-compatible all those pairs of feedback free conditions and conditional independence conditions must hold which can be listed from the original TDAG, $\mathcal{G}$, or equivalently the Boolean lattice $\mathcal{L}$. To complete the analysis we need to generalize Theorem 3.2 to the LCO/LCI case. We do this by extending the zero structures of the representations in Theorem 2.2.

Definition 3.1. A square matrix A is said to be of *block form* if it can be partitioned into blocks B_{ij}, $(i, j = 1, ..., d)$, such that B_{ij} is $n_i \times n_j$, $(i, j = 1, \ldots, d)$ and the blocks are stacked in standard index order: larger i means lower position, larger j means further to the right. We say that a square matrix A has a zero structure *compatible with a transitive directed graph* $\mathcal{G}(E, V)$ with d vertices if A is of block form such that $B_{ij} = 0$ whenever $j \to i$ is *not* in the edge set E of $\mathcal{G}$. $\square$

Looking at the structure of the stochastic realization in Theorem 2.2, we see that the upper triangular form with zero blocks in the requisite positions is compatible with the graph $1 \leftarrow 3 \to 2$, because the zero blocks are the blocks $21, 31, 32, 12$, corresponding to the missing directed edges. It is evident how this structure generalizes to a collection of processes $X_1, \ldots, X_d$ and we shall term as $\mathcal{G}$-*compatible* the corresponding Gaussian stochastic realization generalizing that which appears in Theorem 2.2. This terminology also clearly extends to spectral factors and to their inverses, when they exist.

We state the main result in a brief form to avoid cumbersome notation and omit the proof, which is an extension of that for Theorem 2.2.

Theorem 3.1. Let $\{X_1, \ldots, X_d\}$ be a set of zero mean jointly full rank Gaussian stochastic processes generated by a finite dimensional state space system and let $\mathcal{G}$ be a TDAG with d vertices. Then $X_1, \ldots, X_d$ is LCI-$\mathcal{G}$-compatible if and only if $\{X_1, \ldots, X_d\}$ has a Gaussian stochastic realization which is $\mathcal{G}$-compatible. $\square$

In the light of Theorem 3.1, the nature of the lattice of projectors associated with a set of processes $\{X_1, \ldots, X_d\}$ satisfying the hypotheses of the theorem above becomes evident. Precisely as stated earlier, the projector associated with any subset S of the processes $\{X_1, \ldots, X_d\}$ up to a given instant t will be an orthogonal projection into the span of the (linear functions of the) innovations processes up to the instant t which are mapped into S by the dynamical transitions of the stochastic realization.

In the particular case of three processes as presented in Theorem 2.2, the projectors can be directly read off from the stochastic realization. For example, in the simple case of the projector $P_{\{X_{i,0}, X_{k,0}; X_{i,1}, X_{k,1}\}}$ associated with the indicated argument processes, it can be seen to be a 6×6 block with identity matrices in the first, third, fourth and sixth block positions on the diagonal and zeroes elsewhere. This corresponds, for two time instants, to the static scalar three dimensional case in item 5 of Section 3.2.

In case $\{X_1, \ldots, X_d\}$ is a set of jointly wide-sense stationary Gaussian stochastic processes we may state the following version of Theorem 3.1.

Theorem 3.2. Let $\mathcal{G}$ be a TDAG with d vertices and let $\{X_1, \ldots, X_d\}$ be a set of jointly stationary zero mean Gaussian stochastic processes which possesses a rational spectral density matrix $\{\Phi(z); z \in C\}$, which is non-singular together with its inverse on the unit circle.

Then $X_1, \ldots, X_d$ is $LCI\text{-}\mathcal{G}$-compatible if and only if $\{X_1, \ldots, X_d\}$ possesses an asymptotically stable and inverse asymptotically stable Gaussian stochastic realization which is $\mathcal{G}$-compatible and for which the state space process is stationary.

Furthermore, this is the case if and only if $\{\Phi(z); z \in C\}$ possesses spectral factors which are (i) asymptotically stable and inverse asymptotically stable, (ii) $\mathcal{G}$-compatible. $\qquad\qquad\square$

In the strong feedback free case, the stochastic realization is further restricted in analogy with the specialization in Theorem 3.1 and the spectral factor property (1) has the additional condition: (iii) $\{\Psi(z); z \in C\}$ has off diagonal terms vanishing at $z = 0$.

The last part of this theorem may be paraphrased as stating that for a set $X_1, \ldots,$ X_d of wide sense stationary Gaussian processes satisfying the standing hypotheses of the theorem, if $X_1, \ldots, X_d$ is $LCI\text{-}\mathcal{G}$-compatible then the zero structure arising from $\mathcal{G}$ is displayed by the zero structure of specific spectral factors Ψ

$$G(z) = \Psi\Psi^*(z) \tag{3}$$

of the spectral density of the joint process.

In both of the examples below each node represents a univariate $AR(1)$ process:

$$X^{(t)} = \Phi X^{(t-1)} + \varepsilon^{(t)}$$

and we use Theorem 3.2 to claim a particular form for Φ.

Example 3.1. Consider the graph $\{4 \to 1,\ 4 \to 2,\ 4 \to 3,\ 2 \to 1, 3 \to 1\}$, then

$$\Phi = \begin{bmatrix} a_{11} & a_{12} & a_{13} & a_{14} \\ 0 & a_{22} & 0 & a_{24} \\ 0 & 0 & a_{33} & a_{34} \\ 0 & 0 & 0 & a_{44} \end{bmatrix}$$

Example 3.2. In this case we take the graph:

$$5 \rightarrow 3, 5 \rightarrow 1, 3 \rightarrow 1, 5 \rightarrow 4, 5 \rightarrow 2, 4 \rightarrow 2$$

The form of Φ is then

$$\Phi = \begin{bmatrix} a_{11} & 0 & a_{13} & 0 & a_{15} \\ 0 & a_{22} & 0 & a_{24} & a_{25} \\ 0 & 0 & a_{33} & 0 & a_{35} \\ 0 & 0 & 0 & a_{44} & a_{45} \\ 0 & 0 & 0 & 0 & a_{55} \end{bmatrix}$$

We can confirm the conditional independence of the up-to-t series by elimination. For example to confirm

$$X_1 \perp\!\!\!\perp X_2 \mid (X_3, X_4, X_5)$$

we eliminate $\varepsilon_3, \varepsilon_4, \varepsilon_5$ from the equation

$$X = (I - z\Phi)^{-1}\varepsilon \tag{4}$$

to obtain equations for $X_{1,t}$ and $X_{2,t}$ in terms of $X_{3,t}, X_{4,t}, X_{5,t}, \varepsilon_{1,t}, \varepsilon_{2,t}$.

$$X_{1,t} = \frac{a_{13}zX_{3,t} + a_{15}zX_{5,t} + \varepsilon_{1,t}}{1 - a_{11}z}$$

$$X_{2,t} = \frac{a_{24}zX_{4,t} + a_{25}zX_{5,t} + \varepsilon_{2,t}}{1 - a_{22}z}$$

When the matrix appearing in the $AR(1)$ representation is asymptotically stable the relevant spectral factorization can be easily written down since $\Psi(z) = (I - z\Phi)^{-1}$ is itself the required strong spectral factor. We note that the zero structure of Ψ is the same as that of Φ. This is a generic fact: the $\mathcal{G}$ compatible matrices (with non-zero diagonals and prescribed dimensions) form a group in that the product of any two such matrices is of the same form and so are the inverses. In the simple examples given here we see that the infinite moving average, which derives from an $AR(1)$ in the LCI class, has a structure in which $X^{(i)}$ is a function only of present and past innovations which can be traced backwards from its node in the TDAG.

As an example we compute the Moving Average representation for $X_{3,t}$ in Example 3.2 as

$$X_{3,t} = \frac{e_{3,t} + z(-a_{55}e_{3,t} + a_{35}e_{5,t})}{(1 - a_{33}z)(1 - a_{55}z)},$$

confirming the dependence only on $e_{3,t}$ and $e_{5,t}$: the innovations reachable by reverse arrows.

4   Spatially Patterned Infinite Bayes Nets

Evidently, combining the "spatial" dependence of the TDAG with the linearly ordered nature of time (a linear TDAG) for each separate process gives a "super" TDAG, which might be regarded as a generalized time structure. This leads to the notion that the TDAG/LCI is an appropriate setting to carry out a generalisation of the theory presented so far; this in turn leads to an algebraic Bayes net theory for certain spatially countably infinite configurations of processes.

We initiate such investigations in this paper by introducing one particularly simple structured system which we shall be termed a *chain recurrent TDAG*. The constituent elements of such a system are triples $\{(X_{k,t}, Y_{k,t}, Z_{k,t}; t \geq 0); k \in \mathbf{Z}\}$ of jointly distributed zero mean Gaussian multivariate processes attached to the nodes of a countably infinite TDAG identified with the integers, $\mathbf{Z}$. The "root" process Z_k is placed at the node $k \in \mathbf{Z}$; a directed edge connects it to a node bearing the "leaf" process Y_k and finally a distinct edge from k leads to the node bearing the "leaf" process X_k. Henceforth we shall not distinguish between a node and the process attached to it.

The sequential structure of the spacial configuration is obtained in the following way: for each integer $k \in \mathbf{Z}$, the leaf process X_k is identified with the root process Z_{k+1} of the subsequent triple; on the other hand there is no directed edge leading from the node Y_k. Clearly, the generalization with respect to the theory in Section 3 is that the finite TDAGs of that section are now replaced by a countably infinite tree structure.

In order to formulate the appropriate notions of the feedback free and conditional independence properties on a chained recurrent TDAG we make the following definitions.

For a process $Q_j \in \{(X_k, Y_k, Z_k); k \in \mathbf{Z}\}$ the *spatial antecedents* Q_j^a consist of Q_j itself together with all nodes (i.e. processes) of the same type (i.e. X, Y or Z) which may connected to Q_j by finite connected paths of directed edges; correspondingly, the *spatial future* Q_j^f consists of all nodes (i.e. processes) of the same type to which Q_j may be linked by finite connected paths of directed edges.

For example, the set of spatial antecedents of Z_k, $k \in \mathbf{Z}$, is the set $\{Z_j; j \leq k, j \in \mathbf{Z}\}$, while that of Y_k is $\{Y_j; j \leq k, j \in \mathbf{Z}\}$.

Then the triple of processes $P, Q, R \in \{(X_k, Y_k, Z_k); k \in \mathbf{Z}\}$ in a chained recurrent TDAG is said to be *spatially conditionally independent* at node k if

$$P_k^a \perp\!\!\!\perp Q_k^a \mid R_k^a,$$

and is said to be *spatially conditionally independent* if this relation holds for all $k \in \mathbf{Z}$.

Furthermore, a pair of zero mean Gaussian stochastic processes

$$P, Q \in \{(X_k, Y_k, Z_k); k \in \mathbf{Z}\},$$

in a chained recurrent TDAG is said to be *spatially feedback free* if and only if, for all $k \in \mathbf{Z}$,

$$P_k | Q_k^a \cup Q_k^f = P_k | Q_k^a,$$

where "|" represents conditional expectation in the Gaussian case. $\square$

Next we define the space

$$\bigoplus \mathbf{H} \equiv \bigoplus \{\mathbf{H}\{(X, Y, Z)_k; k \in \mathbf{Z}\}\}_{j=-\infty}^{\infty}.$$

which is the space of doubly infinite sequences of copies of the Hilbert space spanned by the second order processes X, Y, Z.

Taking the definitions above, and employing analogous proof methods to those used for Theorems 3.1, 3.2, we may link the notions developed so far in the following way.

Consider the family $\mathbf{F}$ of zero mean Gaussian processes $\{(X_{k,t}, Y_{k,t}, Z_{k,t}; t \geq 0); k \in \mathbf{Z}\}$ indexed by $\{k \in \mathbf{Z}\}$ associated to the chained recurrent TDAG $\mathbf{T}(\mathbf{F})$. Assume the temporal process $\{(X_{k,t}, Y_{k,t}, Z_{k,t}); t \geq 0\}$ is full rank and wide sense stationary for each $k \in \mathbf{Z}$ and that the spatial process $\{(X_{k,t}, Y_{k,t}, Z_{k,t}); k \in \mathbf{Z}\}$ is full rank wide sense stationary for each $t \geq 0$.

Then the family $\mathbf{F}$ satisfies the spatial feedback free and conditional independence conditions if and only if it possesses a spatially indexed infinite impulse response (equivalently infinite moving average) representation in the chained recurrent form shown in (5) below; in (5) the vector entries and the operator entries in the spatio-temporal operator $\mathbf{W}$ are the formal z transforms of (i) processes, and (ii) non-anticipative impulse responses respectively, and the innovations processes $(\varepsilon^x, \varepsilon^y, \varepsilon^z)$ are mutually independent for all values of the space and time variable. Furthermore, the block-zero pattern of the spatial impulse response operator $\mathbf{W}$ in (5) is in one-to-one relation with that of the chain recurrent TDAG $\mathbf{T}(\mathbf{F})$ and $\mathbf{W}$ is Toelpitz with respect to diagonal shifts respecting the zero structure.

$$
\begin{bmatrix}
* \\
Y_{l+1}(z) \\
Z_{l+1}(z) \\
Y_l(z) \\
Z_l(z) \\
Y_{l-1}(z) \\
Z_{l-1}(z) \\
*
\end{bmatrix}
\equiv
\begin{bmatrix}
* \\
Y_{l+1}(z) \\
X_l(z) \\
Y_l(z) \\
X_{l-1}(z) \\
Y_{l-1}(z) \\
X_{l-2}(z) \\
*
\end{bmatrix}
= \mathbf{W}(z)\varepsilon(z)
\tag{5}
$$

$$
=
\begin{bmatrix}
* & * & 0 & * & 0 & * & 0 & * & 0 & * \\
* & 0 & W^y_{l+1}(z) & W^{yx}_l(z) & 0 & W^{yx}_{l-1}(z) & 0 & W^{yx}_{l-2}(z) & 0 & * \\
* & 0 & 0 & W^x_l(z) & 0 & W^x_{l-1}(z) & 0 & W^x_{l-2}(z) & 0 & * \\
* & 0 & 0 & 0 & W^y_l(z) & W^{yx}_{l-1}(z) & 0 & W^{yx}_{l-2}(z) & 0 & * \\
* & 0 & 0 & 0 & 0 & W^x_{l-1} & 0 & W^x_{l-2}(z) & 0 & * \\
* & 0 & 0 & 0 & 0 & 0 & W^y_{l-1} & W^{yx}_{l-2}(z) & 0 & * \\
* & 0 & 0 & 0 & 0 & 0 & 0 & W^x_{l-2}(z) & 0 & * \\
* & 0 & 0 & 0 & 0 & 0 & 0 & 0 & * & *
\end{bmatrix}
\begin{bmatrix}
* \\
* \\
\varepsilon^y_{l+1}(z) \\
\varepsilon^x_l(z) \\
\varepsilon^y_l(z) \\
\varepsilon^x_{l-1}(z) \\
\varepsilon^y_{l-1}(z) \\
\varepsilon^x_{l-2}(z) \\
* \\
*
\end{bmatrix}
$$

In case the formal transfer functions in the operator above are all rational, then there exists a stochastic realization of the temporal process $\{(X_t, Y_t, Z_t); t \geq 0\}$ (taking values in R^{2n+m} for each k component) whose state space is a shift invariant subspace of $\bigoplus \mathbf{H}$ and whose infinite input matrix, and infinite (diagonal block wise asymptotically stable) state transition matrix and infinite output matrix all possess the block zero pattern of (5) and hence of the TDAG $\mathbf{T}(\mathbf{F})$.

As is to be expected from the role of Teoplitz matrices in theory of wide sense stationary time series, the spatio-temporal operator $\mathbf{W}$ has a Teoplitz form. In (5) the diagonal shift invariance of the operator $\mathbf{W}$ corresponds to the stationarity of the spatial process on the chain recurrent TDAG $\mathbf{T}(\mathbf{F})$. We may take the formal ω transform of the representation (5) of the family $\mathbf{F}$ so that powers of ω index spatial shifts of the triple of temporal processes $\{(X_{k,t}, Y_{k,t}, Z_{k,t}); t \geq 0\}$ and to spatial shifts in the action of the spatial impulse response matrix of $\mathbf{W}$. This permits us to write (5) in the compact form

$$\begin{bmatrix} Y(z,\omega) \\ Z(z,\omega) \end{bmatrix} \equiv \begin{bmatrix} Y(z,\omega) \\ X(z,\omega) \end{bmatrix} = \begin{bmatrix} W^y(z,\omega) & W^{yx}(z,\omega) \\ 0 & W^x(z,\omega) \end{bmatrix} \begin{bmatrix} \varepsilon^y(z,\omega) \\ \varepsilon^x(z,\omega) \end{bmatrix} \qquad (6)$$

The representation (6) clearly reveals the feedback free structure of the spatio-temporal processes $(Y, Z) \equiv (Y, X)$ associated to the chain recurrent TDAG $\mathbf{T}(\mathbf{F})$. It may be seen that, subject to conditions ensuring that the operator $\mathbf{W}$ is invertible, the transformation between the infinite moving average and infinite autoregressive space-time representations of $(Y, Z) \equiv (Y, X)$ becomes evident and is susceptible to analysis. (We note that in the finite TDAG case the invertibility condition was made explicit in terms of the full rank condition on the (rational) spectral density matrix on the unit circle of the vector of all processes on the TDAG.) Furthermore, the sequential operation of linear operations on the processes attached to $\mathbf{T}(\mathbf{F})$ may be formalized using representations of the form (6) in the manner of the Concatenation Theorem of Caines ([7], Chapter 2).

Projectors and the Subset Lattice
It is possible using the Teoplitz structures above to describe the algebraic machinery of the Boolean lattice $\mathcal{L}$ and the subset lattice $B(\mathcal{L})$. The association is that the structure of the observation space, and its conditional independence structure subject to the feedback free condition, is given by the associated TDAG $\mathbf{T}$. Hence the associated set of projectors has the subset lattice $B(\mathcal{L})$ derived from $\mathbf{T}$ as described in Section 3. Evidently, for arbitrary countably infinite TDAGs there will not necessarily be any particular structure to exploit. However, as has been illustrated in the chain recurrent TDAG case with its associated Toeplitz structure, there exist infinite TDAGs with regularity properties for which the analysis and computation of the lattice of projectors could well be tractable.

References

1. Andersson S. A. and Perlman M. D., *Lattice models for conditional independence in a multivariate normal distribution*, Ann. Statist. **21** (1993), 1318–1358.
2. Andersson S. A. and Madsen J., *Symmetry and lattice conditional independence in a multivariate normal distribution*, Ann. Statist. **26** (1998), 525–572.
3. Lindquist A. and Picci G., *A hardy space approach to the stochastic realization problem*, Proceedings of the 1978 Conference Decision and Control (1978), 933–939.
4. ______ , *On the stochastic realization problem*, SIAM J. Control Optim. **17** (1979), no. 3, 365–389.
5. ______ , State space models for Gaussian stochastic processes in *Stochastic Systems: The Mathematics of Filtering and Identification and Applications*, pp. 169–204, Pub: Reidel, Dordrecht, 1981, Ed: M. Hazewinkel and J. C. Willems.

6. Perlman M. D. Andersson S. A., Madigan D. and Trigg C. M., *On the relationship between conditional independence models determined by finite distributive lattices and directed acyclic graph*, J. Stat. Plan. Inf **48** (1995), 25–46.
7. Caines P. E., Linear Stochastic Systems, New York; Chichester: Wiley, 1988.
8. Akaike H., *Stochastic theory of minimal realization*, IEEE Trans. Autom. Control **19** (1974), 667–674.
9. ______ , *Markovian representation of stochastic processes by canonical variables*, SIAM J. Control Optim. **13(1)** (1975), 162–173.
10. R.V. Kadison and J.R Ringrose, *Fundamentals of the theory of operator algebras, volume 1*, Graduate studies in mathematics, vol. 15, American Mathematical Society, San Diego, 1983.
11. L. Lauritzen, Graphical Models, Clarendon Press, Oxford, 1996.
12. Eichler M, *A graphical approach for evaluating effective connectivity in neural systems*, Phil. Trans. Roc. Soc. B **360** (2005), 953–967.
13. Caines P. E. Deardon R. and Wynn H., *Conditional orthogonality and conditional stochastic realization*, Directions in Mathematical Systems Theory and Optimization, Lecture Notes in Control and Inform. Sc, A.Rantzer and C.I. Byrnes (Eds.) **286** (2003), 71–84.
14. ______ , *Bayes nets of time series: stochastic realisations and projections*, Optimal Design and Related Areas in Optimization and Statistics, Springer Lecture Notes, L. Pronzato and A. Zhigljavsky (Eds.) (2007), 153–164.
15. Dahlhaus R., *Graphical interaction models for time series*, Metrika **51** (2000), 151–72.
16. Johnstone P. T., Stone Spaces, Cambridge studies in advanced mathematics, vol. 3, American Mathematical Society, Cambridge, 1982.

A Birds Eye View on System Identification

Manfred Deistler

Institut für Wirtschaftsmathematik
Forschungsgruppe Ökonometrie und Systemtheorie
Technische Universität Wien
Argentinierstraße 8 / 105-2 A-1040 Wien
Manfred.Deistler@tuwien.ac.at

Summary. System identification is concerned with obtaining good models from data, i.e. with data driven modeling. In this contribution the aim is to explain and discuss ideas, general approaches and theories underlying identification of linear systems. Identification of linear systems is a nonlinear problem and is "prototypical" also for many parts of identification of nonlinear systems.

1 Introduction

The art of identification is to find a good model from, in general, noisy data. This is an important problem in many areas of application. Often the task of identification is so complex, that it cannot be performed with the naked eye and systematic approaches have to be used. This is done, partly under quite different perspectives, in statistics, econometrics, system theory and the field of inverse problems.

The main steps in identifications are:

- Specification of the model class, i.e. of the class of all a priori feasible candidate systems. In this step the a priori information concerning the phenomenon to be modeled is incorporated. This typically includes, for instance, the selection of (candidates for) the input-variables or assumptions on the relation between the variables.
- Specification of the class of observations, data preprocessing.
- Identification in the narrow sense: An identification procedure is a rule, in the automatic case a function, attaching a system from the model class to the data. In this step the emphasis is on the development of procedures and algorithms on one side and on their evaluation on the other side.

Here only identification from equally spaced, discrete time, time series data $y_t = (y_t^{(i)})^{i=1\ldots s} \in \mathbb{R}$, $t = 1 \ldots T$ is considered. For explanation of time series data, dynamic systems are often natural candidates.

In this contribution the focus is on what we call the main stream theory for identification of linear systems (see [6], [7]). We add a few remarks on alternative model classes and approaches for identification of linear systems and on identification of nonlinear systems.

A. Chiuso et al. (Eds.): Modeling, Estimation and Control, LNCIS 364, pp. 59–71, 2007.
springerlink.com

The mainstream theory of identification deals with the following setting:

- The model class consists of linear, time-invariant, finite dimensional, causal and stable systems only. The classification of the variables into inputs and outputs is given a priori.
- Uncertainty is modeled by the use of stochastic models for noise. In particular here the noise is assumed to be stationary with a rational spectral density. These assumptions on the noise are in a sense standard, but are nevertheless not innocent. The have been criticized on grounds of not being justified in a number of applications (see e.g. [25]). In our opinion, stochastic noise models are at least an important "test bed" for evaluating identification procedures.
- The observed inputs are free of noise and uncorrelated with the noise process.
- The approach to estimation is semi-nonparametric in the following sense: In general the parameter space for describing system- and noise parameters will be not finite-dimensional, since e.g. systems of arbitrarely high orders are considered. In this approach the model class is broken down into subclasses such that each subclass has a finite- dimensional parameter- space. Estimation then consists of two steps: The model selection step, where the subclass is estimated by a vector of integers, characterizing this subclass. Once the subclass is obtained, its parameter-space is a subset of a suitable Euclidian space and estimation is concerned with estimating a parameter, which is a vector of real-valued entries, in this space.
- For the statistical analysis, emphasis is laid on asymptotic properties (consistency, asymptotic normality and asymptotic efficiency), mainly because finite sample properties are hard to obtain analytically.

We consider the following three "modules" in the theory of system identification:

- Structure theory: Here an idealized problem is considered, as we commence from the stochastic processes generating the data or their population second moments rather than from the data. In the ergodic case one could also say that we commence from an infinite, rather than from a finite data string. The relation between "external behavior" (as described e.g. by the population second moments of the observations) and "internal" (system and noise-) parameters is analysed. Identifiabilty, realization- and parametrization theory are important parts of structure theory.
- Estimation of real-valued parameters for a given subclass: Here we commence from a given subclass whose parameter space is a subset of an Euclidean space and in addition contains a non-void open subset of this space. Estimators are often found from general principles, here in particular from optimizing a likelihood-type criterion function over the parameter-space.
- Model selection: In general the orders or the relevant inputs are not known a priori and have to be determined from the data. One way of doing this is e.g. estimation of integers characterizing the orders by information criteria like AIC or BIC, or, more generally by using a criterion defining a trade-off between the quality of fit to the data achievable in a certain model-subclass and the complexity of this sub-class.

2   Structure Theory

As has been stated already, structure theory is concerned with an analysis of the relation between external behavior and internal parameters. Such an analysis turns out to be important for a deeper understanding of many identification procedures. For the linear mainstream case, the relation between the population second moments of the observations or equivalently the transfer-functions (and noise covariance matrices) and the system (and noise) parameters is considered.

Main model classes for linear systems are:

- $AR(X)$ models
- $ARMA(X)$ models
- State space models

In many applications $AR(X)$ systems still dominate for a number of reasons. Main advantages of (unrestricted), $AR(X)$ models are:

- There are no problems of non-identifiability; in more general terms structure theory is so simple, that for standard situations it does not have to be considered separately.
- Least squares estimators are of maximum likelihood type; they are explicitly given, fast to calculate and asymptotically efficient.

Things are different in case of "structural" a priori restrictions (i.e. if restrictions on the parameter space are imposed by a priori knowledge); but nevertheless, also then $AR(X)$ system identification is "easier" compared to the $ARMA(X)$ or state space case.

On the other hand $AR(X)$ models are less flexible than $ARMA(X)$ and state space models in the sense that, in general, more parameters are needed to achieve the same quality of approximation.

In this contribution, we will mainly consider the case where we have no observed inputs.

Here the focus is on state space systems, but we also consider the $ARMA(X)$ case. A state space system in innovations representation is of the form

$$x_{t+1} = Ax_t + B\epsilon_t(+Lz_t) \tag{1}$$

$$y_t = Cx_t + \epsilon_t(+Dz_t) \tag{2}$$

where y_t are the s-dimensional outputs, x_t is the n-dimensional state, (ϵ_t) is, in general unobserved, s-dimensional white noise (i.e. $E\epsilon_t = 0$, $E\epsilon_s\epsilon_t' = \delta_{st}\Sigma$, where E denotes expectation and δ_{st} is the Kronecker symbol) and z_t are the m-dimensional observed inputs. The random variables y_t, x_t, ϵ_t and z_t are defined over an underlying probability space $(\Omega, \mathcal{A}, \mathcal{P})$. $A \in \mathbb{R}^{n \times n}$, $B \in \mathbb{R}^{n \times s}$, $L \in \mathbb{R}^{n \times m}$, $C \in \mathbb{R}^{s \times n}$ and $D \in \mathbb{R}^{s \times m}$ are parameter matrices.

Throughout we assume that the *stability condition*

$$|\lambda_{max}(A)| < 1 \tag{3}$$

where λ_{max} denotes an eigenvalue of maximum modulus, and the *miniphase condition*

$$|\lambda_{max}(A - BC)| \leq 1 \tag{4}$$

hold. The steady state solution of (1) (2) is given by

$$y_t = C(Iz^{-1} - A)^{-1}(B\epsilon_t(+Lz_t)) + \epsilon_t(+Dz_t) \tag{5}$$

Here z is used for a complex variable as well as for the backward shift on the integers $\mathbb{Z}$, i.e. $z(y_t|t \in \mathbb{Z}) = (y_{t-1}|t \in \mathbb{Z})$.

In addition, throughout we assume that $Ez_s\epsilon_t' = 0$ holds and that Σ is nonsingular.

$ARMA(X)$ systems are (vector-) difference equations of the form

$$a(z)y_t = b(z)\epsilon_t(+d(z)z_t) \tag{6}$$

where

$$a(z) = \Sigma_{j=0}^p a_j z^j \quad ; \quad b(z) = \Sigma_{j=0}^q b_j z^j \quad ; \quad d(z) = \Sigma_{j=0}^r d_j z^j \quad ;$$

$$a_j; b_j \in \mathbb{R}^{s \times s} \quad ; \quad d_j \in \mathbb{R}^{s \times m}$$

We assume that the *stability condition*

$$\det a(z) \neq 0 \qquad |z| \leq 1 \tag{7}$$

and the *miniphase condition*

$$\det b(z) \neq 0 \qquad |z| < 1 \tag{8}$$

hold, and again we assume

$$Ez_s\epsilon_t' = 0$$

and that Σ is nonsingular. The steady state solution then is given by

$$y_t = a^{-1}(z)[b(z)\epsilon_t(+d(z)z_t)] \tag{9}$$

Note that by (3) or (7) (and by assuming stationarity for (z_t)) the infinite sums in (5) and (9) respectively, i.e.

$$y_t = \Sigma_{j=0}^\infty k_j \epsilon_{t-j}(+\Sigma_{j=0}^\infty l_j z_{t-j}) \tag{10}$$

where, e.g,

$$k_j = CA^{j-1}B \quad , \quad j > 0 \quad , \quad k_0 = I \tag{11}$$

and

$$k(z) = \Sigma_{j=0}^\infty k_j z^j = a^{-1}(z)b(z) \tag{12}$$

converge e.g. in the mean squares sense. In addition (y_t) and (x_t) are stationary processes.

From now onwards, we will, for the sake of brevity of notation, unless the contrary is stated explicitly, restrict ourselves to the case, where there are no observed inputs. Then the external behavior of (1), (2) or (6) is described by the covariance function

$\gamma : \mathbb{Z} \to \mathbb{R}^{s \times s}$, $\gamma(t) = E y_t y_0'$ of the process (y_t) or equivalently by its spectral density $f : [-\pi, \pi] \to \mathbb{C}^{s \times s}$ defined by

$$f(\lambda) = (2\pi)^{-1} \Sigma_{t=-\infty}^{\infty} e^{-i\lambda t} \gamma(t) \tag{13}$$

From (10), f is given by

$$f(\lambda) = (2\pi)^{-1} k(e^{-i\lambda}) \Sigma k^*(e^{-i\lambda}) \tag{14}$$

where $*$ denotes the conjugate transpose. Throughout we assume $k(0) = I$. This implies no restriction for f and establishes a one-to-one relation between f and (k, Σ).

Under our assumptions,

- Every state space system (1) (2) and every $ARMA$ system (6) has a rational transfer function $k(z)$ which is analytic in a disk containing the closed unit disk (and thus is causal and stable) and which satisfies $\det k(z) \neq 0$, $|z| < 1$.
- Conversely, for every rational transfer function $k(z)$ which is analytic in a disk containing the closed unit disk and which satisfies $\det k(z) \neq 0$, $|z| < 1$ and $k(0) = I$ there is a stable and miniphase state space -, and a stable and miniphase $ARMA$ representation.

Thus, in particular, state space- and $ARMA$ representations are two alternative ways to describe the same class of external (input/output) behaviors $k(z)$. Note that the assumption $k(0) = I$ is a normalizing condition defining Σ. We have ([14], chapter 1):

Any rational and a.e. nonsingular spectral density matrix f may be uniquely factorized as in (14) where $k(z)$ is rational, analytic whithin a circle containing the closed unit disk, $\det k(z) \neq 0$, $|z| < 1$ and $k(0) = I$ and where $\Sigma > 0$.

Consider the following set of $s \times s$ transfer functions: $U_A = \{k | k$ is rational, $k(0) = I$, $k(z)$ has no poles for $|z| \leq 1$ and no zeros for $|z| < 1\}$. By $M(n) \subset U_A$ we denote the set of all transfer functions of order n (to be more precise, the set of all transfer functions corresponding to minimal state space systems with state dimension n). By T_A we denote the set of all triples (A, B, C), where s is fixed but n is arbitrary, satisfying (3) and (4), by $S(n) \subset T_A$ the subset of all (A, B, C) for fixed n and by $S_m(n) \subset S(n)$ the subset of all minimal (A, B, C). We define the mapping $\pi : T_A \to U_A$ such that $\pi(A, B, C) = C(Iz^{-1} - A)^{-1} B + I$ (also defined by (11)).

Now, T_A is not a "good" parameter space because:

- T_A is not finite dimensional.
- π is (surjective but) not injective, i.e. we do not have identifiability.
- There exists no continuous selection, in the sense that there is no continuous mapping attaching to every $k \in U_A$ a unique element from the equivalence class $\pi^{-1}(k)$.

Here U_A is endowed with the so called pointwise topology T_{pt} [14] which corresponds to the relative topology in the product space $(\mathbb{R}^{s \times s})^{\mathbb{N}}$ for the coefficients $(k_j | j \in \mathbb{N})$.

In order to obtain "good" parameter spaces, U_A and T_A are broken into bits, U_α and T_α say, $\alpha \in I$ such that

- π restricted to T_α, $\pi/T_\alpha : T_\alpha \to U_\alpha$ is bijective. Injectivity of π/T_α implies identifiability.

- U_α is finite dimensional in the sense that $U_\alpha \subset \cup_{i=1}^n M(i)$ for some n. Usually, taking into account the restrictions in T_α, T_α is reparametrized by expressing the $(A, B, C) \in T_\alpha$ by their "free" parameters, τ say. We use T_α also for this set of free parameters τ and we assume that this T_α contains an open set in an embedding Euclidian space $\mathbb{R}^{d_\alpha}$. The mapping $\Psi_\alpha : U_\alpha \to T_\alpha$: $\Psi_\alpha(\pi(\tau)) = \tau \; \forall \tau \in T_\alpha$ is called a *parametrization*.
- The parametrization $\Psi_\alpha : U_\alpha \to T_\alpha$ is a homeomorphism; this is an assumption of well-posedness.
- U_α is T_{pt}-open in its closure $\overline{U}_\alpha$
- $\cup_{\alpha \in I} U_\alpha$ is a cover of U_A

Usually, I is a set of vectors of integers (multiindices) characterizing the bits U_α and T_α. Note that not all approaches used have the desirable properties listed above.

Completely analogous statements hold for the $ARMA$ case, where (using the same symbols) the mapping π is defined by $\pi(a, b) = a^{-1}b$.

The most common approaches are:

- *Canonical forms* defining decompositions of $M(n)$, such as *echelon forms* [14] or *balanced realizations*. Here $M(n)$ is decomposed into sets U_α of different dimension. Echelon forms for state space and $ARMA$ systems have "nice" free parameters in terms of elements of (A, B, C), and of (a, b) and define a very simple bijection between state space and $ARMA$ parameters. Balanced realizations (which only exist for state space systems) have "nice" parameter spaces, but the free parameters are rather complicated transformations of the elements of (A, B, C).
- The overlapping description of the manifold $M(n)$ by local coordinates ([14]).
- The "full parametrization" for state space systems. Here $S(n) \subset \mathbb{R}^{n^2+2ns}$ or $S_m(n)$ are used as parameter spaces for $\overline{M}(n)$ (the closure of $M(n)$ in U_A) or $M(n)$ respectively. Clearly in this case we do not have identifiability. For $k \in M(n)$, the classes of observationally equivalent (A, B, C), $\pi^{-1}(k) \cap S(n)$ are n^2-dimensional manifolds.
- Data driven local coordinates, $DDLC$, for state space systems. Here $S_m(n)$ is reparametrized in terms of coordinates that separately describe the tangent space to the manifold of observationally equivalent (minimal) systems corresponding to an initial estimator at a suitably chosen point and its $2ns$-dimensional orthocomplement [18], [20]. The orthocomplement then is taken as the new parameter space.
- $ARMA$ systems with prescribed column-degrees ([5])
- $ARMA$ parametrizations commencing from writing k as $c^{-1}p$ where c is a least common denominator polynomial for k and where the degrees of c and p serve as integer valued parameters.

3   Estimation for a Given Subclass

Here we commence from the data $y_t, t = 1 \ldots T$ and we assume that U_α is given. We in addition assume that we have identifiability and that the parametrization $\Psi_\alpha : U_\alpha \to T_\alpha$ has the desirable properties listed above.

Let $\tau \in T_\alpha \subset \mathbb{R}^{d_\alpha}$ denote the vector of free parameters for U_α and let $\sigma \in \underline{\Sigma} \subset \mathbb{R}^{\frac{n(n+1)}{2}}$ denote the vector formed by the on and above diagonal elements of Σ. $\underline{\Sigma}$ corresponds to the set of symmetric positive definite matrices. We assume that the overall parameter space is of the form $\Theta = T_\alpha \times \underline{\Sigma}$.

Many identification procedures, at least asymptotically, commence from the sample second moments of the observations:

$$\hat{\gamma}(s) = T^{-1} \Sigma_{t=1}^{T-s} y_{t+s} y_t{}', \qquad s \geq 0$$

Now, $\hat{\gamma}$ can be directly realized as an MA system, "typically" of order $T.s$. By $\check{k}_T$ we denote the corresponding transfer function. Clearly, in many cases, its order is too high. "Typical" identification procedures therefore consist of two steps:

- A "projection" or model reduction step, where e.g. $\check{k}_T$ is approximated by an element $\hat{k}_T$ say, in U_α $(\overline{U}_\alpha)$. From a statistical point of view, this is the essential information concentration step and the statistical properties depend on the way the approximation is defined.
- A realization step, where $\hat{k}_T \in U_\alpha$ is realized by $\tau \in T_\alpha$. This step is important for a number of reasons, for instance from a numerical point of view, however certain statistical properties do not depend on this step.

One may distinguish between two types of estimation procedures, namely:

- Optimization based procedures (M-estimators), which are obtained from optimizing a criterion function over the parameterspace and where the estimators are not given explicitly

and

- Direct procedures, such as instrumental variable methods or subspace methods, where the estimators are explicit functions of the data

The most common criterion function is the Gaussian (log) likelihood function, which (when multiplied by $-2T^{-1}$) is (up to a constant) of the form

$$\hat{L}_T(\theta) = T^{-1} logdet \Gamma_T(\theta) + T^{-1} y'(T) \Gamma_T(\theta)^{-1} y(T) \tag{15}$$

where $y(T) = (y_1', \ldots, y_T')'$ is the stacked vector of observations, $\theta = (\tau', \sigma')'$ is the vector of system and noise parameters, $y(T; \theta)$ is a stacked vector of random variables formed from the outputs of systems with system and noise parameters θ, (in an analogous way as $y(T)$) and finally $\Gamma_T(\theta) = Ey(T; \theta)y(T; \theta)'$. The Gaussian maximum likelihood estimator (MLE) then is defined by

$$\hat{\theta}_T = argmin_{\theta \in \Theta} \hat{L}_T(\theta)$$

It is well known that, although the likelihood is written down as if the observations were Gaussian the asymptotic properties of the MLE do not depend on the Gaussianity of the observations. In addition, for the likelihood (15), the Gaussian distribution is assumed to come from a stationary process; transients in the observations do not influence the asymptotic properties of the MLE.

There exist a number of alternative criterion functions such as the Whittle Likelihood of Ljung's prediction error criterion [17] which (in most cases) give asymptotically equivalent estimators. The Whittle likelihood is of the form:

$$\hat{L}_{w,T}(k,\sigma) = logdet\Sigma + (2\pi)^{-1}\int_{-\pi}^{\pi} tr[(k(e^{-i\lambda})\Sigma k^*(e^{-i\lambda}))^{-1}I(\lambda)]d\lambda \qquad (16)$$

where tr denotes the trace and I is the periodogramm, i.e. the Fourier transform of $\hat{\gamma}$. Formula (16) shows the approximation of I by $k \in U_\alpha$ in a clear way.

In maximum likelihood estimation a number of observations are important:

- For "natural" parameter spaces, the likelihood function is not necessarily semi-continuous and thus the existence of its optimum is not guaranteed (see [10]).
- In general, the MLE is not given by an explicit function of the data; thus the estimators are obtained by a numerical optimization procedure.
- $\hat{L}_T$ depends on τ only via the corresponding transfer function k, thus (with a slight sloppyness in notation) we may define a "coordinate-free" likelihood function $\hat{L}_T(k,\sigma)$.
- Neither T_α nor U_α are closed sets and boundary points may occur in optimizing the likelihood function (see [14]).

As far as consistency of the MLE's is concerned, the first correct proofs have been given in [12] (for the univariate case) and [11], [8], for a general result see also [14]. Coordinate free consistency says that for $k_0 \in \overline{U}_\alpha$ (where k_0 denotes the true system) and if $limT^{-1}\Sigma_{t=1}^{T-s}\varepsilon_{t+s}\varepsilon_t' = \delta_{0,s}\Sigma_0$ a.e. we have for the MLE's $\hat{k}_T \to k_0$ a.e. and $\hat{\Sigma}_T \to \Sigma_0$ a.e. The proof uses the basic idea of [24] developed for the i.i.d. case. The specific additional difficulties are not only due to the fact that the observations are dependent, but also due to the fact that the "natural" parameter spaces are not compact. As can be shown

$$lim_{T\to\infty}\hat{L}_T(k,\sigma) = L(k,\sigma) = \qquad (17)$$

$$logdet\Sigma + (2\pi)^{-1}\int_{-\pi}^{\pi} tr[(k(e^{-i\lambda})\Sigma k^*(e^{-i\lambda}))^{-1}$$

$$(k_0(e^{-i\lambda})\Sigma_0 k_0^*(e^{-i\lambda}))]d\lambda \quad a.e.$$

holds, where the subscript 0 again denotes the true quantities and the asymptotic likelihood L has a unique minimum at k_0, Σ_0. (Note the similarity of (16) and L). Clearly pointwise convergence in (17) alone does not ensure convergence of the optima, but in our case, the latter can be shown, in particular since $(\hat{k}_T, \hat{\Sigma}_T)$ can be shown to enter a compact set. For a detailed proof of the consistency result, see [14], chapter 4.

Even for $k_0 \notin \overline{U}_\alpha$, the MLE's have a "generalized consistency" property, as they converge a.e. to the set D of minimizes of L over $\overline{U}_\alpha \times \underline{\Sigma}$.

Now consistency "in coordinates", i.e. for the parameter estimators $\hat{\tau}_T = \psi_\alpha(\hat{k}_T)$ follows directly from the consistency of $\hat{k}_T$ and the continuity of ψ_α (and from the openness of U_α in $\overline{U}_\alpha$), if $\tau_0 \in T_\alpha$ holds.

Under additional assumptions (see [14]) asymptotic normality can be shown by using the idea explained in [4], extended to the stationary case.

For actual calculation of estimators, the usual procedure consists of a consistent explicit estimator, e.g. a subspace procedure, to obtain an initial estimator in the first step and one Gauß-Newton step is order to obtain an asymptotically efficient estimator.

If T_β denotes a parameterspace obtained by a diffeomorphic mapping from T_α, then the transformation of the asymptotic distributions of the MLE's is straightforward; nevertheless the choice of parameterspaces turns of to be important from a numerical point of view, where it is taken into account that optimization has to be performed over a grid.

Explicit estimation procedures are usually numerically fast and reliable, but are in many cases not asymptotically efficient. Recently so called subspace identification procedures [1], [15], [23] have attracted a lot of attention. Subspace estimators are for state space systems and they are based on a realization algorithm combined with a model reduction step. Usually the model reduction step is performed by omitting the smaller eigenvalues in a singular value decomposition. For the case of observed inputs, subspace procedures turn out be more intricate. The statistical properties of classes subspace procedures have been investigated e.g. in [9], [3] and [2].

4   Model Selection

Here we confine our discussion to the problem of estimating the model order n. In many cases information criteria defining a trade-off between the quality of fit achievable in a certain model-subclass and the complexity of this subclasses are used for this purpose. Note that we here have a situation which is "closure nested", i.e. $n_1 < n_2$ implies $\overline{M}(n_1) \subset \overline{M}(n_2)$ and the dimension of $M(n_1)$ is smaller than the dimension of $M(n_2)$. In particular criteria of the form

$$A(n) = logdet\,\hat{\Sigma}_T(n) + 2ns.c(T)T^{-1} \tag{18}$$

where $\hat{\Sigma}_T(n)$ is the MLE of Σ_0 over $\overline{M}(n) \times \underline{\Sigma}$ and $c(T)$ is a prescribed positive function of T, are frequently used. For $c(T) = 2$ we obtain the AIC, for $c(T) = logT$, the BIC criterium. The corresponding order estimate $\hat{n}_T$ is obtained by minimizing $A(n)$ (in a certain range of integers).

The statistical properties of such estimators have been analysed by Hannan [13], see also [14], chapter 5. In particular (under suitable additional assumptions) $\hat{n}_T$ is (strongly) consistent if

$$lim_{T \to \infty} \frac{c(T)}{T} = 0$$

and

$$liminf_{T \to \infty} \frac{c(T)}{loglogT} > 0$$

hold (and thus BIC gives consistent estimators) and AIC does not give consistent estimators, and asymptotically leads to overestimation of the true order n.

Taking a closer look, things turn out to be more subtle. One may argue, as has been done, e.g. in [22], that in most cases order estimation is only an intermediate goal. Shibata showed that under certain assumptions, in particular if the true system is of infinite order, an autoregressive spectral estimate based on AIC and MLE is asymptotically optimal.

Here we concentrate on two issues. The first one may be entitled "decomposition into subclass is in the eye of the beholder". Consider e.g. AR models for $s = 1$ of the form

$$y_t + a_1 y_{t-1} + a_2 y_{t-2} = \epsilon_t$$

Then, considering only the system parameters, "usually" we have the following parameterspaces:

$T = \{(a_1, a_2) \in \mathbb{R}^2 | 1 + a_1 z + a_2 z^2 \neq 0 | z | \leq 1\}$
$T_0 = \{(0,0)\}$
$T_1 = \{(a_1, 0) | |a_1| < 1, a_1, a_1 \neq 0\}$
$T_2 = T - (T_0 \cup T_1)$

and we want to make a data driven choice between T_0, T_1 and T_2.

Now, from the Bayesian derivation of BIC [21] we see that, in order to obtain BIC, T_0, T_1 and T_2 must have strictly positive prior probabilities; thus BIC has to be justified by some kind of a priori knowledge. For instance a flat prior on T would suggest just to use the MLE over T, rather than to do model selection in a first step. In addition other prior distributions may give positive prior probabilities to other low dimensional subsets of T and thus result in an other decomposition of T.

The second issue is concerned with properties of post model selection estimators, i.e. with properties of estimators for real-valued parameters taking into account the uncertainty coming from model selection. One may argue, that, if a (strongly) consistent model selection criterion is used, then the true model order is known from a certain sample size onwards and thus the asymptotic variance of the estimators for τ after model selection is the same as in the case where the true order is a priori known. As has been shown in [16] this argument is grossly misleading, because it is pointwise in the parameterspace and does not hold uniformly there.

5   Linear Non-mainstream Cases

In a number of important cases, the systems are linear, but the models or their identification is not in the mainstream setting. We do not intend to give a survey on such cases here, but we only make a few remarks. Important special cases are:

- Linear systems with time-varying parameters. There several different approaches to this problem, such as systems with slowly varying parameters, where the variation of coefficients is described by an autoregression, or systems with structural changes, which may be triggered by a random mechanism, such as Markov switching models, or smooth transition models.
- Identification in closed loop.
- The wide area of symmetric modeling, where no a priori distinction between inputs and outputs is made, errors-in-variables and linear dynamic factor models; the latter are used in particular for high dimensional time series.

- Unstable systems, in particular integration and cointegration, which is of great importance for econometrics.
- Long memory

6   Nonlinear Systems

Of course there is a wide range of nonlinear systems and identification of nonlinear systems is a word like "non-elephant zoology" (also in the sense that linear systems are "huge animals"). Again, as in section 5, we do not intend to give a survey on this topic, we only make a few remarks on this field. Identification of nonlinear systems consists of a number of only weakly connected subareas. The most important of these subareas are:

- The asymptotic theory for M-estimation for parametric classes of nonlinear dynamic systems, see e.g. [19]. Identification of linear systems is a nonlinear problem, since the mapping from data to parameters is nonlinear. Identification of nonlinear systems uses partly the same ideas as identification of linear systems. The main problem in the setting of nonlinear systems is, that there is no general structure theory available, and thus, for instance identifiability is often assumed rather than shown.
- Neural nets are a frequently used model class also for time series. Recurrent neural nets are a particular class of dynamic nonlinear models. Identification of neural nets is a semi-nonparametric problem.
- Nonparametric estimation for nonlinear time series models, e.g. estimation of nonlinear autoregressions

$$y_t = g(y_{t-1}, \ldots, y_{t-p}) + \epsilon_t \tag{19}$$

by kernel methods is, an area which has received substantial attention in the last two decades. The systems (19) can be generalized by replacing ϵ_t by a model for the conditional variance of y_t. Another important class in this area are so called additive models, which are $MISO$ models, where the effects of the single inputs are nonlinear but additive, i.e. there is no interaction effect of different single inputs. Of central interest for nonparametric estimation is the choice of design parameters, such as the bandwidth of a kernel, and, in asymptotic theory e.g. the rates of convergence.
- Special classes of nonlinear systems, justified by "physical" a priori knowledge or "stylized facts" in data, such as $GARCH$-type models or stochastic volatility models for explaining or forecasting conditional variances, in particular for finance data, have attracted a substantial number of researchers recently.
- Chaotic systems and their identification have been considered in the last 25 years, but the number of convincing success stories in applications seems to be limited.

7   Present State and Future Developments

Theory and methods in system identification have reached a certain state of maturity. There is a large body of methods and theory available serving the needs for many applications, but nevertheless, in many cases, identification is still not a standard task and needs a special design by an expert.

On the other hand, the areas and the number of applications, are increasing rapidly. Application fields like medicine, biology or finance are "boom areas" and pose a number of new and interesting questions.

The development of system identification now is more driven by demand from applications than by developments in theory, i.e. there is "demand pull" rather than "theory push".

One can also observe an increasing fragmentation corresponding to different data structures, model classes and prior knowledge in different fields of application. The development of theory and methods is also done by different, not very much interacting communities, like econometrics, system- and control theory or signal processing.

System identification is in a certain sense an enabling technology and in many cases not visible for non-experts.

There are still major open problems in system identification, such as

- large parts of identification of nonlinear systems
- Identification of spatio-temporal systems
- Identification for large data sets data and for high dimensional time series
- Improved model selection and regularization procedures
- Further automatization
- Hybrid procedures
- The use of symbolic computation

Summing up, system identification is still an interesting area, in particular new applications pose new challenges. The field has shifting boundaries and the question arises, whether in the future there will be still a substantial common body of theory and methods. Besides the danger of fragmentation, for certain parts of the field, there is also the danger of becoming selfreferential and not relevant for applications.

References

1. H. Akaike, Canonical Correlations Analysis of Time Series and the Use of an Information Criterion, in: R.H. Mehra and D.G. Lainiotis, eds., *System Identification: Advances and Case Studies,* Academic Press, New York, 27 − 96, 1976.
2. D. Bauer, Comparing the CCA Subspace Method to Pseudo Maximum Likelihood Methods in the Case of No Exogenous Inputs, *Journal of Time Series Analysis*, 26, 631-668, 2005.
3. A. Chiuso and G. Picci, The Asymptotic Variance of Subspace Estimates, *Journal of Econometrics,* 118, 292-312, 2003.
4. H. Cramer, *Mathematical Methods of Statistics,* Princeton University Press, Princeton, 1946.
5. M. Deistler, *The Properties of the Parametrization of ARMAX Systems and Their Relevance for Structural Estimation and Dynamic Specification,* Econometrica 51, 1983, 1187 − 1208.
6. M. Deistler, *System Identification - General Aspects and Structure,* in G. Goodwin (ed.) *System Identification, and, Adaptive Control* (Festschrift for B.D.O. Anderson), Springer, London, 2001, 3 − 26.
7. M. Deistler, *Linear Models for Multivariate Time Series Analysis,,* in: "Handbook of Time Series Analysis", Matthias Wintherhalder, Bjoern Schelter, Jens Timmer, eds., Wiley-VCH, Berlin, 2006, 283 − 306.

8. M. Deistler, W. Dunsmuir and E.J. Hannan, Vector Linear Time Series Models: Corrections and Extensions, *Adv. Appl. Probab.,* 10,1978, 360 − 372.

9. M. Deistler, K. Peternell and W. Scherrer, Consistency and Relative Efficiency of Subspace Methods, *Automatica,* 31, 1865-1875, 1995.

10. M. Deistler and B.M. Poetscher, The Behaviour of the Likelihood Function for ARMA Models, *Adv. Appl. Probab.,* 16, 1984, 843 − 865.

11. W. Dunsmuir and E.J. Hannan, Vector linear time series models, *Adv. Appl. Probab.,* 8, 1976, 339 − 364.

12. E.J.Hannan, The Asymptotic Theory of Linear Time Series Models, *J. Appl. Probab.* 10, 1973, 130 − 145.

13. E.J.Hannan, Estimating the dimension of a linear system, *J. Multivariate Anal.* 11, 459-473, 1981.

14. E.J. Hannan and M. Deistler, *The Statistical Theory of Linear Systems,* Wiley, New York, 1988.

15. W.E. Larimore, System Identification, Reduced Order Filters and Modeling via Canonical Variate Analysis, in: H.S. Rao and P. Dorato, eds., *Proceedings of the* 1983 *American Control Conference,* 445-451, 1983.

16. H. Leeb and B.M. Poetscher, Model Selection and Inference: Facts and Fiction, *Econometric Theory*, 21, 21 − 59.

17. L. Ljung, *System Identification. Theory for the User,* Prentice Hall; 2nd ed., 1998.

18. T. McKelvey and A. Helmersson, System Identification using an Over-Parametrized Model Class - Improving the Optimization Algorithm, in: *Proceedings 36th IEEE Conference on Decision and Control*, San Diego, USA, 2984-2989, 1997.

19. B.M. Poetscher and I. Prucha, *Dynamic Nonlinear Econometric Models: Asymptotic Theory*, Springer, New York, 1993

20. T. Ribarits, M. Deistler and McKelvey, An Analysis of the Parametrization by Data Driven Local Coordinates for Multivariable Linear Systems, *Automatica*, 40, 789-803, 2004.

21. G. Schwarz, Estimating the dimension of a model, *Ann. Statist.*, 6, 461-464, 1978.

22. R. Shibata, An optimal autoregressive spectral estimate, *Ann. Statist.*, 9, 300-306, 1981.

23. P. Van Overschee and B. De Moor, *Subspace Identification for Linear Systems: Theory, Implementation, Applications,* Kluwer Academic Press, Boston, 1996.

24. A. Wald, Note on the consistency of the maximum likelihood estimate, *Ann. Math. Stast.,* 20, 595 − 601, 1949.

25. J.C. Willems, Thoughts on System Identification, in: B.A. Francis and J.C. Willems, eds. *Control of Uncertain Systems: Modelling, Approximation and Design (Festschrift for K. Glover)*, Springer Lecture Notes in Control and Information Sciences, 389 − 417, 2006.

Further Results on the Byrnes-Georgiou-Lindquist Generalized Moment Problem[*]

Augusto Ferrante[1], Michele Pavon[2], and Federico Ramponi[1]

[1] Dipartimento di Ingegneria dell'Informazione, Università di Padova,
via Gradenigo 6/B, 35131 Padova, Italy
`augusto@dei.unipd.it, rampo@dei.unipd.it`
[2] Dipartimento di Matematica Pura ed Applicata, Università di Padova,
via Trieste 63, 35131 Padova Italy
`pavon@math.unipd.it`

Summary. In this paper, we consider the problem of finding, among solutions of a moment problem, the best Kullback-Leibler approximation of a given *a priori* spectral density. We present a new complete existence proof for the dual optimization problem in the Byrnes-Lindquist spirit. We also prove a descent property for a matricial iterative method for the numerical solution of the dual problem. The latter has proven to perform extremely well in simulation testbeds.

1 Introduction

The concept of *positive real function*, originating in Networks Theory with Brune in 1930, has proven to be one of the deepest and most unifying ones of electrical engineering. Names such as Foster, Cauer, Bode, Darlington, Youla, Popov, Kalman, Yakubovich, Faurre, B.D.O. Anderson, J.C. Willems spring to mind. The quest first posed by Kalman in 1964 for a realization theory for stochastic systems could rely on these precious foundations. The strict sense version of the *stochastic realization problem* (also called *Markovian representation problem*) has roots also in the theory of Markov processes and in mathematical statistics (Bahadur transitive sufficient statistics, etc.). Giorgio Picci was one of the pioneers (together with McKean [39], Akaike [1, 2] and Ruckebusch [46]) in the early-mid seventies in this field. Perhaps, among the forerunners, he was the only one that drew equal inspiration from both of these areas, due to the fact that he had equal interest in the concepts of Systems Theory and of Mathematical Statistics [43, 44, 45]. Perhaps, this can be recognized as one of the main characteristics of all of Giorgio's rather diversified scientific production, which lays at the interface between stochastic systems theory and mathematical statistics. Another salient aspect of his research work is the taste for profound, foundational questions that continues today, for instance, in his work on subspace methods identification.

One of us (M.P.) had the privilege to witness the early days of the great Lindquist-Picci collaboration [33]- [38], and to receive continuous, generous help from Giorgio

[*] Work partially supported by the MIUR-PRIN Italian grant "New Techniques and Applications of Identification and Adaptive Control".

A. Chiuso et al. (Eds.): Modeling, Estimation and Control, LNCIS 364, pp. 73–83, 2007.
springerlink.com   © Springer-Verlag Berlin Heidelberg 2007

both as a student and as a young researcher. The best way I can pay a personal tribute to Giorgio as a scientist is to observe that, although I have interacted and continue to interact with a large number and spectrum of colleagues, whenever there is a deep question on the table, I still go or address people to Giorgio, precisely as I did more than thirty years ago.

It is a joy to celebrate his devotion to science and research *per se* that continues unabated to this day, and his manifold, benchmark contributions to stochastic systems theory.

In this paper, we study a generalized moment problem for spectral densities in the spirit of Byrnes-Georgiou-Lindquist [4, 6, 7, 8, 9, 10, 11, 17, 21, 22, 23, 24, 25, 26, 27, 40]. This problem includes as special case the *covariance extension problem* and Nevanlinna-Pick interpolation for positive real functions. It also includes as special case a maximum entropy problem which has been shown to be related to a special one-step ahead Wiener-Kolmogorov prediction problem [27]. It features a Kullback-Leibler type index. It lays, therefore, very much in the center of the field in which Giorgio has been active for some forty years.

In the Byrnes-Georgiou-Lindquist approach, the smooth parametrization of the solutions to the generalized moment problem occurs in a convex optimization setting. The Kullback-Leibler criterion, where optimization is performed with respect to the second argument, is employed as cost index. Other distances between spectra have been recently investigated in [18, 19, 28, 29]. The contribution of this paper is twofold: On the one hand we provide a detailed and complete existence proof for the dual optimization problem (Section 5) in the Lindquist-Byrnes spirit [11]. On the other hand, we show that the matricial algorithm proposed in [41] for the numerical solution of the dual problem may be viewed as a modified steepest descent method (Section 6).

2   A Generalized Moment Problem

Let $\mathcal{S}_+(\mathbb{T})$ be the family of bounded, coercive, spectral density functions on the unit circle. Thus, a measurable, bounded function Φ belongs to $\mathcal{S}_+(\mathbb{T})$ if the values of Φ are real and bounded away from zero. Notice that $\Phi \in \mathcal{S}_+(\mathbb{T})$ if and only if $\Phi^{-1} \in \mathcal{S}_+(\mathbb{T})$. Let $\Psi \in \mathcal{S}_+(\mathbb{T})$ represent an *a priori* estimate of the spectrum of an underlying zero-mean, wide-sense stationary stochastic process $\{y(k), k \in \mathbb{Z}\}$. Suppose we can estimate the asymptotic covariance Σ of the n-dimensional stationary process $\{x_k; k \in \mathbb{Z}\}$ satisfying

$$x_{k+1} = Ax_k + By_k, \quad k \in \mathbb{Z}, \tag{1}$$

where A is a *stability matrix* and the pair (A, B) is reachable. The rational transfer function

$$G(z) = (zI - A)^{-1}B, \qquad A \in \mathbb{C}^{n \times n}, \ B \in \mathbb{C}^{n \times 1}, \tag{2}$$

models a bank of filters. When Ψ is not consistent with Σ, we need to find $\Phi \in \mathcal{S}_+(\mathbb{T})$ that is closest to Ψ in a suitable sense among spectra satisfying

$$\int G(e^{i\vartheta})\Phi(e^{i\vartheta})G^*(e^{i\vartheta}) = \Sigma, \tag{3}$$

where star denotes transposition plus conjugation. The Hermitian matrix Σ is assumed to be positive definite and integration takes place on $[-\pi, \pi]$ with respect to the normalized Lebesgue measure $d\vartheta/2\pi$.

The question of existence of $\Phi \in \mathcal{S}_+(\mathbb{T})$ satisfying (3) and, when existence is granted, the parametrization of all solutions to (3), may be viewed as a generalized moment problem. We refer to [30] and references therein for a full discussion on the importance and on the manifold applications of such problem. Here, we only mention that, by suitably choosing the matrices A and B, this problem reduces the celebrated *covariance extension problem*, see [30] for details. Existence of $\Phi \in \mathcal{S}_+(\mathbb{T})$ satisfying constraint (3) is a nontrivial issue. It was shown in [23,24] that such family is nonempty if and only if there exists $H \in \mathbb{C}^{1 \times n}$ such that

$$\Sigma - A\Sigma A^* = BH + H^*B^*. \tag{4}$$

For simplicity of notation, we reformulate the constraint by normalizing Σ to the identity. Indeed, the set of solutions to (3) does not change if we replace (Σ, A, B) with $(I, A' = \Sigma^{-1/2}A\Sigma^{1/2}, B' = \Sigma^{-1/2}B)$. Notice that in this way G is replaced by $G' := \Sigma^{-1/2}G$. Thus constraint (3) from now on reads

$$\int G\Phi G^* = I. \tag{5}$$

3   Kullback-Leibler Criterion

In [30], the following problem is considered:

Problem 1. Given $\Psi \in \mathcal{S}_+(\mathbb{T})$, find $\hat{\Phi}$ that solves

$$\text{minimize} \quad \mathbb{S}(\Psi\|\Phi) \tag{6}$$

$$\text{over} \quad \left\{ \Phi \in \mathcal{S}(\mathbb{T}) \mid \int G\Phi G^* = I \right\}, \tag{7}$$

where $\mathbb{S}(\Psi\|\Phi)$ is the Kullback-Leibler index:

$$\mathbb{S}(\Psi\|\Phi) = \int \Psi \log\left(\frac{\Psi}{\Phi}\right).$$

As is well known, this pseudo-distance originates in hypothesis testing, where it represents the mean information for observation for discrimination of an underlying probability density from another [32, p.6]. It also plays a central role in information theory, identification, stochastic processes, etc., see e.g. [3, 12, 13, 15, 20, 31, 42, 48] and references therein. It is also known in these fields as *divergence*, *relative entropy*, *information distance*. etc. If

$$\int \Phi = \int \Psi,$$

we have $\mathbb{S}(\Psi\|\Phi) \geq 0$. The choice of $\mathbb{S}(\Psi\|\Phi)$ as a distance measure, even for spectra that have different zeroth moment, is discussed in [30, Section III]. This choice is essentially based on the possibility of rescaling Ψ. Clearly, in this way the optimization problem amounts to approximating the "shape" of the a priori spectrum. In this spirit, Georgiou has recently investigated other distances between *rays* of power spectra, [28,29]. This is of course sensible to pursue in several engineering applications such as speech processing or prediction problems.

4   Optimality Conditions and the Dual Problem

The variational analysis in [30] is outlined as follows (see also [41]). Let

$$\mathcal{L}'_+ := \{\Lambda = \Lambda^* \in \mathbb{C}^{n \times n} : G^*\Lambda G > 0, \forall e^{i\vartheta} \in \mathbb{T}\}. \tag{8}$$

For $\Lambda \in \mathcal{L}'_+$ consider the *Lagrangian function*

$$L(\Phi, \Lambda) = \mathbb{S}(\Psi\|\Phi) + \mathrm{tr}\left(\Lambda\left(\int G\Phi G^* - I\right)\right)$$

$$= \mathbb{S}(\Psi\|\Phi) + \int G^*\Lambda G\Phi - \mathrm{tr}\,(\Lambda), \tag{9}$$

where "tr" denotes the trace operator. Consider the *unconstrained* minimization of the strictly convex functional $L(\Phi, \Lambda)$:

$$\mathrm{minimize}\{L(\Phi, \Lambda)|\Phi \in \mathcal{S}(\mathbb{T})\} \tag{10}$$

This is a convex optimization problem.

Theorem 1. *Suppose* $\hat{\Lambda} \in \mathcal{L}'_+$ *is such that*

$$\int G\frac{\Psi}{G^*\hat{\Lambda}G}G^* = I. \tag{11}$$

Then $\hat{\Phi}$ *given by*

$$\hat{\Phi} = \frac{\Psi}{G^*\hat{\Lambda}G} \tag{12}$$

is the unique solution of Problem 1.

Thus, the original Problem 1 is now reduced to finding $\hat{\Lambda} \in \mathcal{L}'_+$ satisfying (11). We define the linear operator $\Xi : \mathcal{H} \to \mathcal{C}(\mathbb{T})$ by $\Xi(\Lambda) = G^*\Lambda G$, where $\mathcal{H}$ is the set of Hermitian matrices of dimension $n \times n$ and $\mathcal{C}(\mathbb{T})$ is the set of continuous functions on $\mathbb{T}$. Observe that if $\Lambda \in \mathcal{L}'_+$ satisfies (11), then, for all $\Lambda_K \in \ker \Xi$, the sum $\Lambda + \Lambda_K$ also belongs to $\mathcal{L}'_+$ and satisfies (11). Hence, we may restrict the search for $\hat{\Lambda}$ to the orthogonal complement of $\ker \Xi$ i.e. to the range of the adjoint operator $\Gamma := \Xi^*$. It is easy to see that the operator Γ is defined by

$$\Gamma(\Phi) = \int G\Phi G^*, \qquad \Phi \in \mathcal{C}(\mathbb{T}). \tag{13}$$

Thus, by setting

$$\mathcal{L}_+ := \mathcal{L}'_+ \cap [\ker(\Xi)]^\perp = \mathcal{L}'_+ \cap \operatorname{Range}(\Gamma) \qquad (14)$$

we are reduced to find $\hat{\Lambda} \in \mathcal{L}_+$ satisfying (11). This is accomplished via duality theory. Consider the dual functional

$$\Lambda \to \inf\{L(\Phi,\Lambda)|\Phi \in \mathcal{S}(\mathbb{T})\}.$$

For $\Lambda \in \mathcal{L}_+$, the dual functional takes the form

$$\Lambda \to L\left(\frac{\Psi}{G^*\Lambda G}, \Lambda\right) = \int \Psi \log G^*\Lambda G - \operatorname{tr}(\Lambda) + \int \Psi. \qquad (15)$$

Consider now the maximization of the dual functional (15) over the set $\mathcal{L}_+$. Let, as in [30],

$$J_\Psi(\Lambda) := -\int \Psi \log G^*\Lambda G + \operatorname{tr}(\Lambda). \qquad (16)$$

The dual problem is then equivalent to

$$\text{minimize} \quad \{J_\Psi(\Lambda)|\Lambda \in \mathcal{L}_+\}. \qquad (17)$$

The dual problem is a strictly convex optimization problem [30]. Byrnes and Lindquist have shown in [11] that J_Ψ has a unique minimum point in $\mathcal{L}_+$. This result implies that, under assumption (4), there exists a (unique) $\hat{\Lambda} \in \mathcal{L}_+$ satisfying (11). Such a $\hat{\Lambda}$ then provides the optimal solution of the primal problem (6)-(7) via (12). In the next section, we provide a new detailed proof of this result inspired by that of [11].

5 An Existence Theorem

Let us consider the closure of $\mathcal{L}_+$, given by

$$\overline{\mathcal{L}}_+ = \{\Lambda = \Lambda^* \in \mathbb{C}^{n \times n} : \Lambda \in \operatorname{Range}(\Gamma),\ G^*\Lambda G \geq 0, \forall e^{i\vartheta} \in \mathbb{T}\}. \qquad (18)$$

On the convex set $\overline{\mathcal{L}}_+$, we define the sequence of functions

$$J_\Psi^n(\Lambda) := \operatorname{tr}(\Lambda) - \int \Psi \log\left(G^*\Lambda G + \frac{1}{n}\right). \qquad (19)$$

Lemma 1. *The pointwise limit $J_\Psi^\infty(\Lambda) = \lim_n J_\Psi^n(\Lambda)$ exists and defines a lower semi-continuous, convex function on $\overline{\mathcal{L}}_+$ with values in the extended reals.*

Proof. For each n, J_Ψ^n is a continuous, convex function on the closed convex set $\overline{\mathcal{L}}_+$. Hence epi (J_Ψ^n), the epigraph of J_Ψ^n, is a closed, convex subset of $\mathbb{C}^{n \times n} \times \mathbb{R}$. Moreover, for $\Lambda \in \overline{\mathcal{L}}_+$, $J_\Psi^n(\Lambda) < J_\Psi^{n+1}(\Lambda)$. Hence, J_Ψ^∞ is well defined and in fact $J_\Psi^\infty(\Lambda) = \sup_n J_\Psi^n(\Lambda)$. It follows that epi $(J_\Psi^\infty) = \cap_n \text{epi } (J_\Psi^n)$ is also closed and convex. We conclude that J_Ψ^∞ is lower semicontinuous and convex on $\overline{\mathcal{L}}_+$. $\blacksquare$

Lemma 2. *Assume (4). Then,*

1. *J_Ψ^∞ is bounded below on $\overline{\mathcal{L}}_+$;*
2. *$J_\Psi^\infty(\Lambda) = J_\Psi(\Lambda)$ on $\mathcal{L}_+$;*
3. *$J_\Psi^\infty(\Lambda)$ is finite on all of $\overline{\mathcal{L}}_+ \setminus \{0\}$.*

Proof. By (4), there exists $\Phi_1 \in \mathcal{S}_+(\mathbb{T})$ satisfying (5), namely $\int G\Phi_1 G^* = I$. Hence, $\mathrm{tr}\,(\Lambda)$ can be written as $\mathrm{tr}\,(\Lambda \int G\Phi_1 G^*) = \int G^* \Lambda G\Phi_1$, and we get

$$
\begin{aligned}
J_\Psi^n(\Lambda) &= \int \left[G^* \Lambda G\Phi_1 - \Psi \log\left(G^* \Lambda G + \frac{1}{n} \right) \right] \\
&= \int \Phi_1 \left[G^* \Lambda G - \frac{\Psi}{\Phi_1} \log\left(G^* \Lambda G + \frac{1}{n} \right) \right].
\end{aligned}
$$

Since the function $x - \beta \log(x + \frac{1}{n})$ with $\beta > 0$ attains its minimum at $x = \beta - \frac{1}{n}$, we get

$$
J_\Psi^n(\Lambda) = \int \Phi_1 \left[G^* \Lambda G - \frac{\Psi}{\Phi_1} \log\left(G^* \Lambda G + \frac{1}{n} \right) \right] \geq \int \psi - \frac{1}{n} \int \Phi_1 - \mathbb{S}(\Psi \| \Phi_1).
$$

We conclude that $J_\Psi^\infty \geq \int \psi - \mathbb{S}(\Psi \| \Phi_1)$ on all of $\overline{\mathcal{L}}_+$. To establish 2, notice that, by Beppo Levi's theorem,

$$
J_\Psi^\infty(\Lambda) := \mathrm{tr}\,(\Lambda) - \int \lim_{n\to\infty} \Psi \log\left(G^* \Lambda G + \frac{1}{n} \right), \quad \Lambda \in \overline{\mathcal{L}}_+. \tag{20}
$$

To prove 3, observe that for $0 \neq \Lambda \in \partial\mathcal{L}_+$, the boundary of $\mathcal{L}_+$, $G^*\overline{\Lambda}G$ is a nonzero rational spectral density so that $\log G^*\overline{\Lambda}G$ is integrable over $\mathbb{T}$ [47, pag. 64]. Since Ψ is bounded, also $\Psi \log G^*\overline{\Lambda}G$ is integrable. ∎

In view of these lemmata, we extend $J_\Psi(\Lambda)$ to all of $\overline{\mathcal{L}}_+$ by setting $J_\Psi(\Lambda) := J_\Psi^\infty(\Lambda)$ on $\partial\mathcal{L}_+$. Notice that, by (20), J_Ψ is finite and given by (16) on $\overline{\mathcal{L}}_+ \setminus \{0\}$, and it is $+\infty$ in $\Lambda = 0$.

Lemma 3. *Assume (4). Then*

$$
\lim_{\|\Lambda\|\to+\infty} J_\Psi(\Lambda) = +\infty. \tag{21}
$$

Proof. Recall that by (4), there exists $\Phi_1 \in \mathcal{S}_+(\mathbb{T})$ satisfying (5), and, consequently, $\mathrm{tr}\,(\Lambda) = \int G^* \Lambda G\Phi_1 > 0, \forall \Lambda \in \mathcal{L}_+$. Suppose Λ_k is a sequence of matrices in $\mathcal{L}_+$ such that $\lim_{k\to\infty} \|\Lambda_k\| = +\infty$. Define the normalized sequence $\Lambda_k^0 := \frac{\Lambda_k}{\|\Lambda_k\|}$ (of course, we can assume $\Lambda_k \neq 0, \forall k$). Since $\mathrm{tr}\, \Lambda_k^0 > 0$,

$$
\eta := \liminf_{k\to+\infty} \mathrm{tr}\, \Lambda_k^0 \geq 0.
$$

Consider a sub-sequence such that the limit of its trace is η. This subsequence contains a convergent sub-subsequence $\{\Lambda_{k_m}^0\}$ since Λ_k^0 belongs to the surface of the unit ball, which is compact. Let $\Lambda_\infty := \lim_{m\to\infty} \Lambda_{k_m}^0$. Since $G^* \Lambda_n^0 G > 0$ on $\mathbb{T}$, $G^* \Lambda_\infty G \geq 0$

on $\mathbb{T}$. Moreover, $\Lambda_\infty \in \mathrm{Range}\,\Gamma$, since $\mathrm{Range}\,\Gamma$ is finite-dimensional, and hence closed. This implies that $G^*\Lambda_\infty G$ cannot be identically zero. In fact, if so, $\Lambda_\infty \in \mathcal{L}_+ = \ker(\Xi) = \mathrm{Range}\,\Gamma^\perp$. Then $\Lambda_\infty \in \mathrm{Range}\,\Gamma \cap \mathrm{Range}\,\Gamma^\perp = \{0\}$, which is a contradiction since $\|\Lambda_\infty\| = 1$. Thus

$$\eta = \lim_{n \to \infty} \mathrm{tr}\,\Lambda_n^0 = \mathrm{tr}\,\Lambda_\infty = \int G^*\Lambda_\infty G \Phi_1 > 0 \qquad (22)$$

Hence, there exists a K such that $\mathrm{tr}\,\Lambda_k^0 > \eta/2$ for all $k \geq K$. Finally, since $G^*\Lambda_k^0 G \leq G^*G$, we obtain:

$$\begin{aligned}
\liminf_{k \to \infty} J_\Psi(\Lambda_k) &= \liminf_{k \to \infty} \|\Lambda_k\| \mathrm{tr}\,\Lambda_k^0 - \int \Psi \log \|\Lambda_k\| G^*\Lambda_k^0 G \\
&= \liminf_{k \to \infty} \|\Lambda_k\| \mathrm{tr}\,\Lambda_k^0 - (\textstyle\int \Psi) \log \|\Lambda_k\| - \int \Psi \log G^*\Lambda_k^0 G \\
&\geq \liminf_{k \to \infty} \|\Lambda_k\| \mathrm{tr}\,\Lambda_k^0 - (\textstyle\int \Psi) \log \|\Lambda_k\| - \int \Psi \log G^*G \\
&\geq \liminf_{k \to \infty} \|\Lambda_k\| \frac{\eta}{2} - (\textstyle\int \Psi) \log \|\Lambda_k\| - \int \Psi \log G^*G \\
&= \liminf_{k \to \infty} \frac{\eta}{2} \left(\|\Lambda_k\| - \frac{\int \Psi}{\eta/2} \log \|\Lambda_k\| \right) - \int \Psi \log G^*G \\
&= +\infty.
\end{aligned}$$

$\blacksquare$

Theorem 2. *Assume that the feasibility condition (4) is satisfied. Then the problem of minimizing the functional $J_\Psi(\Lambda) = \mathrm{tr}\,\Lambda - \int \Psi \log G^*\Lambda G$ over $\mathcal{L}_+$ admits a unique solution $\hat{\Lambda} \in \mathcal{L}_+$.*

Proof. In view of Lemma 1, Lemma 2 and Lemma 3, the functional J_Ψ is inf-compact on the closed set $\overline{\mathcal{L}}_+$, and therefore it admits a minimum point $\hat{\Lambda}$ there. We show next that $\hat{\Lambda} \in \mathcal{L}_+$. Of course, $\hat{\Lambda}$ is not the zero matrix since $J_\Psi(0) = +\infty$. Let $0 \neq \overline{\Lambda} \in \partial\mathcal{L}_+$. By Lemma 2, $J_\Psi(\overline{\Lambda})$ is finite. Observe that, by (4), $I \in \mathcal{L}_+$. By convexity of $\overline{\mathcal{L}}_+$, it then follows that $\overline{\Lambda} + \epsilon(I - \overline{\Lambda}) \in \overline{\mathcal{L}}_+, \forall \epsilon \in [0, 1]$. We compute the one-sided directional derivative or hemidifferential

$$J'_{\Psi_+}(\overline{\Lambda}; I - \overline{\Lambda}) := \lim_{\epsilon \searrow 0} \left[\frac{J_\Psi(\overline{\Lambda} + \epsilon(I - \overline{\Lambda})) - J_\Psi(\overline{\Lambda})}{\epsilon} \right]$$
$$= \mathrm{tr}\,(I - \overline{\Lambda}) + \int \Psi - \int \frac{G^*G\Psi}{G^*\overline{\Lambda}G} = -\infty. \qquad (23)$$

Hence, $\overline{\Lambda}$ cannot be a minimum point. We conclude that $\hat{\Lambda} \in \mathcal{L}_+$. $\blacksquare$

6 A Descent Method for the Dual Problem

In general, the optimal solution of the dual problem needs to be computed numerically. This is a delicate problem because of the unboundedness of the gradient of J_Ψ at the boundary of $\mathcal{L}_+$, see (23). The approaches proposed in [30] and references therein involve some preliminary reparametrization of $\mathcal{L}_+$, which may imply loss of global convexity.

In [41], a different matricial iterative method was proposed that appears to be very fast and numerically robust. This method does not restrict the search of $\hat{\Lambda}$ to $\mathcal{L}_+$ and indeed it normally converges to a $\hat{\Lambda} \notin \mathrm{Range}(\Gamma)$. We show below that this method may be viewed as a modified *gradient descent method* with fixed step size. This method is described as follows.

Let

$$\mathcal{M} := \{M \in \mathcal{L}'_+ : \ 0 \le M \le I, \ \mathrm{tr}\,[M] = 1\}, \tag{24}$$

$$\mathcal{M}_+ := \{M \in \mathcal{M} : \ M > 0\}. \tag{25}$$

For $M \in \mathcal{M}$, define the map Θ by

$$\Theta(M) := \int M^{1/2} G \left[\frac{\Psi}{G^* M G} \right] G^* M^{1/2}. \tag{26}$$

Theorem 3. [41]. *The map Θ maps $\mathcal{M}$ into $\mathcal{M}$ and $\mathcal{M}_+$ into $\mathcal{M}_+$.*

Consider the following iterative algorithm.

Algorithm. Let $M_0 = \frac{1}{n} I$. Note that $M_0 \in \mathcal{M}_+$. Define the sequence $\{M_k\}_{k=0}^\infty$ by

$$M_{k+1} := \Theta(M_k). \tag{27}$$

Notice that, by Theorem 3, $M_k \in \mathcal{M}_+$ for all k. Moreover, since $M_k \in \mathcal{M}, \forall k$, the sequence is bounded. Hence it has at least one accumulation (limit) point in the closure $\overline{\mathcal{M}}$ of $\mathcal{M}$.

Theorem 4. *Suppose that the sequence $\{M_k\}_{k=0}^\infty$ has a limit $\hat{M} \in \mathcal{M}_+$. Then $\hat{M} \in \mathcal{L}'_+$ and satisfies (11), and therefore provides the optimal solution of the approximation problem via (12).*

Notice that even when the sequence generated by (27) converges to a singular matrix $\hat{M} \in \mathcal{M}$, it is still possible, though not guaranteed, that such a matrix solves the original problem. We next show that the algorithm may be viewed as a modified gradient descent method. To this aim, rewrite (27) as

$$M_{k+1} = M_k + M_k^{1/2} \left[\int \frac{G \Psi G^*}{G^* M_k G} - I \right] M_k^{1/2}. \tag{28}$$

Proposition 1. *Define*

$$\Delta M_k := M_k^{1/2} \left[\int \frac{G \Psi G^*}{G^* M_k G} - I \right] M_k^{1/2}, \tag{29}$$

so that (28) reads $M_{k+1} = M_k + \Delta M_k$. Then, ΔM_k is a descent direction at M_k for J_Ψ.

Proof. Let

$$\nabla J_\Psi(M_k) = I - \int \frac{G\Psi G^*}{G^* M_k G}$$

denote the "gradient" of J_Ψ at M_k. Then,

$$< \nabla J_\Psi(M_k), \Delta M_k > = \mathrm{tr}\ (\nabla J_\Psi(M_k)\Delta M_k) = -\mathrm{tr}\ \left(M_k^{1/4}\nabla J_\Psi(M_k)M_k^{1/4}\right)^2.$$

By Theorem3, $M_k > 0$, for all k. It follows that $\mathrm{tr}\ (\nabla J_\Psi(M_k)\Delta M_k) < 0$, unless $\nabla J_\Psi(M_k) = 0$ in which case M_k is a fixed point of the iteration which solves the dual problem by Theorem(4). ∎

One could implement the matricial iteration as

$$M_{k+1} = M_k + \alpha_k \Delta M_k, \tag{30}$$

where $0 < \alpha_k \leq 1$ is determined through backstepping, see e.g. [5]. Our extensive simulation (see e.g. [41]), however, shows that convergence in fact occurs with $\alpha_k \equiv 1$! Indeed, the algorithm appears to perform numerically very well. In fact, at each step the integral (26) may be computed very precisely and efficiently via a spectral factorization technique that only requires to solve an algebraic Riccati equation and a Lyapunov equation, both of dimension n. We have performed an extensive number of simulations where the sequence generated by (27) never failed to converge. In a very small number of cases, we have observed convergence toward a singular matrix which, however, satisfied (11), and therefore provided the optimal solution of the approximation problem.

References

1. H. Akaike, *Markovian representation of stochastic processes by canonical variables*, SIAM J. Contr. vol. **13**, pp. 162-173, 1975
2. H. Akaike, *Stochastic theory of minimal realization*, IEEE Trans. Aut. Contr. vol. **AC-19**, pp. 667-674, 1974
3. A. Barron, *Entropy and the central limit theorem*, Ann. Probab. vol. **14**, pp. 336-342, 1986.
4. A.Blomqvist, A.Lindquist and R.Nagamune, *Matrix-valued Nevanlinna-Pick interpolation with complexity constraint: An optimizaiton approach*, IEEE Trans. Aut. Control vol. **48**, pp. 2172-2190, 2003.
5. S. Boyd, L. Vandenberghe, *Convex Optimization*, Cambridge University Press, Cambridge, UK, 2004.
6. C. I. Byrnes, T. Georgiou, and A. Lindquist, *A new approach to spectral estimation: A tunable high-resolution spectral estimator*, IEEE Trans. Sig. Proc. vol. **49**, pp. 3189-3205, 2000.
7. C. I. Byrnes, T. Georgiou, and A. Lindquist, *A generalized entropy criterion for Nevanlinna-Pick interpolation with degree constraint*, IEEE Trans. Aut. Control vol. **46**, pp. 822-839, 2001.
8. C. I. Byrnes, T. Georgiou, A. Lindquist and A. Megretski, *Generalized interpolation in H-infinity with a complexity constraint*, Trans. American Math. Society vol. **358(3)**, pp. 965-987, 2006 (electronically published on December 9, 2004).

9. C. I. Byrnes, S. Gusev, and A. Lindquist, *A convex optimization approach to the rational covariance extension problem*, SIAM J. Control and Opimization vol. **37**, pp. 211-229, 1999.
10. C. I. Byrnes, S. Gusev, and A. Lindquist, *From finite covariance windows to modeling filters: A convex optimization approach*, SIAM Review vol. **43**, pp. 645-675, 2001.
11. C. I. Byrnes and A. Lindquist, *The generalized moment problem with complexity constraint*, Integral Equations and Operator Theory vol. **56(2)**, pp. 163-180, 2006 (published online March 29, 2006).
12. E. Carlen and A. Soffer, *Entropy production by convolution and central limit theorems with strong rate information*, Comm. Math. Phys. vol. **140**, pp. 339-371, 1991.
13. T. M. Cover and J. A. Thomas, *Information Theory*, Wiley, New York, 1991.
14. H. Cramér, *Mathematical methods of statistics*, Princeton Univ. Press, Princeton, 1946.
15. I. Csiszár, *Maxent, mathematics and information theory*, in Proc. 15th Inter. Workshop on Maximum Entropy and Bayesian Methods, K.M. Hanson and R.N. Silver eds., Kluver Academic, pp. 35-50, 1996.
16. J. C. Doyle, B. A. Francis and A. R. Tannenbaum, *Feedback Control Theory*, Macmillan, New York, 1992.
17. P. Enquist, *A homotopy approach to rational covariance extension with degree constraint*, Int. J. Appl. Math. and Comp. Sci. vol. **11**, pp. 1173-1201, 2001.
18. A. Ferrante, M. Pavon and F. Ramponi, *Constrained spectrum approximation in the Hellinger distance*, preprint Oct. 2006. To appear in Proc. of ECC07 Conf. 2007.
19. A. Ferrante, M. Pavon and F. Ramponi, *Hellinger vs. Kullback-Leibler multivariable spectrum approximation*, submitted. 2007.
20. H. Föllmer, *Random fields and diffusion processes*, in École d'Été de Probabilités de Saint-Flour XV-XVII, edited by P. L. Hennequin, Lecture Notes in Mathematics, Springer-Verlag, New York, vol. **1362**, pp. 102-203, 1988.
21. T. Georgiou, *Realization of power spectra from partial covariance sequences*, IEEE Trans. on Acoustics, Speech, and Signal Processing vol. **35**, pp. 438-449, 1987.
22. T. Georgiou, *The interpolation problem with a degree constraint*, IEEE Trans. on Aut. Control vol. **44**, pp. 631-635, 1999.
23. T. Georgiou, *Spectral estimation by selective harmonic amplification*, IEEE Trans. Aut. Control vol. **46**, pp. 29-42, 2001.
24. T. Georgiou, *The structure of state covariances and its relation to the power spectrum of the input*, IEEE Trans. Aut. Control vol. **47**, pp. 1056-1066, 2002.
25. T. Georgiou, *Spectral analysis based on the state covariance: the maximum entropy spectrum and linear fractional parameterization*, IEEE Trans. Aut. Control vol. **47**, pp. 1811-1823, 2002.
26. T. Georgiou, *Solution of the general moment problem via a one-parameter imbedding*, IEEE Trans. Aut. Control vol. **50**, pp. 811-826, 2005.
27. T. Georgiou, *Relative entropy and the multivariable multidimensional moment problem*, IEEE Trans. Inform. Theory vol. **52**, pp. 1052-1066, 2006.
28. T. Georgiou, *Distances between power spectral densities*, arXiv e-print `math.OC/0607026`.
29. T. Georgiou, *An intrinsic metric for power spectral density functions*, arXiv e-print `math.OC/0608486`.
30. T. Georgiou and A. Lindquist, *Kullback-Leibler approximation of spectral density functions*, IEEE Trans. Inform. Theory vol. **49**, pp. 2910-2917, 2003.
31. E. T. Jaynes, *Papers on Probability, Statistics and Statistical Physics*, R.D. Rosenkranz ed., Dordrecht, 1983.
32. S. Kullback, *Information Theory and Statistics 2nd ed.*, Dover, Mineola NY, 1968.
33. A. Lindquist and G.Picci, On the stochastic realization problem, *SIAM J. Control and Optimization* **17** (1979), 365-389.

34. A. Lindquist and G. Picci, Forward and backward semimartingale models for Gaussian processes with stationary increments, *Stochastics* **15** (1985), 1-50.

35. A. Lindquist and G. Picci, Realization theory for multivariate stationary Gaussian processes, *SIAM J. Control and Optimization* **23** (1985), 809-857.

36. A. Lindquist and G. Picci, A geometric approach to modeling and estimation of linear stochastic systems, *J. Mathematical Systems, Estimation, and Control* **1** (1991), 241-333.

37. A. Lindquist and G. Picci, Geometric methods for state space identification, in *Identification, Adaptation, Learning: The Science of Learning Models from Data*, S. Bittanti and G. Picci (editors), Nato ASI Series (Series F, Vol 153), Springer, 1996, 1–69.

38. A. Lindquist and G. Picci, Canonical correlation analysis, approximate covariance extension, and identification of stationary time series, *Automatica* **32** (1996), 709–733.

39. H. P. McKean Jr., *Brownian motion with a several-dimensional time*, Th. Probab. Applic. vol. **8**, pp. 335-354, 1963

40. R. Nagamune, *A robust solver using a continuation method for Nevanlinna-Pick interpolation with degree constraint*, IEEE Trans. Aut. Control vol. **48**, pp. 113-117, 2003.

41. M. Pavon and A. Ferrante, *On the Georgiou-Lindquist approach to constrained Kullback-Leibler approximation of spectral densities*, IEEE Trans. Aut. Control vol. **51**, pp. 639-644, 2006.

42. M. Pavon and F. Ticozzi, *On entropy production for controlled Markovian evolution*, J. Math. Phys., vol. **47**, 063301, 2006.

43. G. Picci, *Stochastic realization of Gaussian processes*, Proc. IEEE vol. **64**, pp. 112-122, 1976.

44. G. Picci, *Some connections between the theory of sufficient statistics and the identifiability problem*, SIAM J. Appl. Math. vol. **33**, pp. 383-398, 1977.

45. G. Picci, *On the internal structure of finite-state stochastic processes*, in Recent developements in Variable Structure Systems, R. Mohler and A. Ruberti Eds. eds., Springer Lecture Notes in Economics and Mathematical Systems, vol. 162, pp. 288–304, 1978.

46. G. Ruckebush, *Representations markoviennes de processus gaussiens stationnaires*, Thèse 3ème cycle, Paris VI, 1975.

47. Yu. A. Rozanov, *Stationary Random Processes*, Holden-Day, San Francisco, 1967.

48. V. Vedral, *The role of relative entropy in quantum information theory*, Rev. Mod. Phys vol. **74**, pp. 197-, 2002.

Factor Analysis and Alternating Minimization

Lorenzo Finesso[1] and Peter Spreij[2]

[1] Institute of Biomedical Engineering, CNR-ISIB, Padova, Italy
   `lorenzo.finesso@isib.cnr.it`
[2] Korteweg-de Vries Institute for Mathematics, Universiteit van Amsterdam,
   Amsterdam, The Netherlands
   `spreij@science.uva.nl`

Dedicated to Giorgio Picci on the occasion of his 65th birthday.
Happy Birthday Giorgio!

1 Introduction

Factor analysis, in its original formulation, deals with the linear statistical model

$$Y = HX + \varepsilon \tag{1}$$

where H is a deterministic matrix, X and ε independent random vectors, the first with
dimension smaller than Y, the second with independent components. What makes
this model attractive in applied research is the *data reduction* mechanism built in it.
A large number of observed variables Y are explained in terms of a small number of
unobserved (latent) variables X perturbed by the independent noise ε. Under normality
assumptions, which are the rule in the standard theory, all the laws of the model are
specified by covariance matrices. More precisely, assume that X and ε are zero mean
independent normal vectors with $\mathbb{C}\mathrm{ov}(X) = P$ and $\mathbb{C}\mathrm{ov}(\varepsilon) = D$, where D is diagonal.
It follows from (1) that $\mathbb{C}\mathrm{ov}(Y) = HPH^\top + D$.

Building a factor analysis model of the observed data requires the solution of a difficult algebraic problem. Given Σ_0, the covariance matrix of Y, find the triples (H, P, D)
such that $\Sigma_0 = HPH^\top + D$. Due to the structural constraint on D, which is assumed
to be diagonal, the existence and unicity of a factor analysis model are not guaranteed.
As it turns out, the right tools to deal with this situation come from the theory of stochastic realization, see [5] (trying to spot the master's hand) for an early contribution
on the subject.

In the present paper we make a first attempt at understanding how to build an optimal
approximate factor analysis model. The criterion we have chosen to evaluate the distance between covariances is the I-divergence between the corresponding normal laws.
The algorithm that we propose for the construction of the best approximation is inspired
by the alternating minimization procedure of [4] and [6].

A. Chiuso et al. (Eds.): Modeling, Estimation and Control, LNCIS 364, pp. 85–96, 2007.
springerlink.com   © Springer-Verlag Berlin Heidelberg 2007

2   The Model

Consider two independent, zero mean, normal vectors X and ε of respective dimensions k and n. We will assume that $\mathrm{Cov}(X) = I$, the identity matrix, and $\mathrm{Cov}(\varepsilon) = D > 0$, a diagonal matrix. Let H be an $n \times k$ matrix (in this paper $k < n$) and let the random vector Y be defined by

$$Y = HX + \varepsilon. \tag{2}$$

Under these assumptions (2) is called a factor analysis (FA) model of size k for the vector Y. Notice that allowing $\mathrm{Cov}(X) = P > 0$ does not produce a more general model, as a square root of P can always be absorbed in H. We will say that a normal vector Y admits a FA model of size k if it is equal in distribution to $HX + \varepsilon$ for some X and ε as above, i.e. if its covariance Σ_0 can be written as $\Sigma_0 = HH^\top + D$. Not every normal vector Y admits a FA model, the hard constraint being imposed by the diagonal structure of D. A probabilistic interpretation stems from $\mathrm{Cov}(Y|X) = D$ (see equation (28) of the Appendix) i.e. the n components of Y are conditionally independent given the $k < n$ components of some vector X. In Remark 1 of the next section the condition for the existence of a FA model is slightly reformulated.

Although the construction of an exact FA model is not always possible, one can search for a best approximate model, according to some criterion. In this paper we opt for minimizing the I-divergence (Kullback-Leibler distance) between normal laws. Recall that given two probability measures $\mathbb{P}_1$ and $\mathbb{P}_2$, defined on the same measurable space, such that $\mathbb{P}_1 \ll \mathbb{P}_2$, the I-divergence of $\mathbb{P}_1$ with respect to $\mathbb{P}_2$ is defined as

$$D(\mathbb{P}_1||\mathbb{P}_2) = \mathbb{E}_{\mathbb{P}_1} \log \frac{d\mathbb{P}_1}{d\mathbb{P}_2}.$$

If $\mathbb{P}_1$ and $\mathbb{P}_2$ are normal measures on the same space $\mathbb{R}^n$, with zero means and strictly positive covariance matrices Σ_1 and Σ_2 respectively, the I-divergence $D(\mathbb{P}_1||\mathbb{P}_2)$ takes the explicit form

$$D(\mathbb{P}_1||\mathbb{P}_2) = \frac{1}{2} \log \frac{|\Sigma_2|}{|\Sigma_1|} + \frac{1}{2} \mathrm{tr}(\Sigma_2^{-1}\Sigma_1) - \frac{n}{2}. \tag{3}$$

Since the I-divergence only depends on the covariance matrices, we usually write $D(\Sigma_1||\Sigma_2)$ instead of $D(\mathbb{P}_1||\mathbb{P}_2)$.

The approximate factor analysis problem can be posed as follows:

Problem 1. Given the positive covariance matrix $\Sigma_0 \in \mathbb{R}^{n \times n}$ and the integer $k < n$ minimize

$$D(\Sigma_0||HH^\top + D) = \frac{1}{2} \log \frac{|HH^\top + D|}{|\Sigma_0|} + \frac{1}{2} \mathrm{tr}((HH^\top + D)^{-1}\Sigma_0) - \frac{n}{2}.$$

over all pairs (H, D) where $H \in \mathbb{R}^{n \times k}$ and $D > 0$ is of size n and diagonal.

Notice that $D(\Sigma_1||\Sigma_2)$, computed as in (3), can be considered as a divergence between two positive definite matrices, without referring to normal distributions. Hence Problem 1 also has a meaning, when one refrains from assumptions like normality.

Existence of the minimum is guaranteed by the following

Proposition 1. *There exist matrices $H^* \in \mathbb{R}^{n \times k}$ and $D^* > 0$ of size n and diagonal minimizing the I-divergence in Problem 1.*

The proof is deferred to section 4.2, since it uses later results.

In order to construct an algorithm for the solution of Problem 1 we will imitate the approach of [6]. The algorithm will therefore be derived by a relaxation technique, lifting the original minimization problem to a higher dimensional space. In the larger space a double minimization problem equivalent to Problem 1 can be formulated, leading in a natural way to an alternating minimization algorithm.

3 Lifting of the Original Problem

In this section we will embed Problem 1 into a higher dimensional space. First we introduce the relevant sets of covariances. Given $k < n$ we denote by

$$\boldsymbol{\Sigma} = \{\Sigma \in \mathbb{R}^{(n+k)\times(n+k)} : \Sigma = \begin{pmatrix} \Sigma_{11} & \Sigma_{12} \\ \Sigma_{21} & \Sigma_{22} \end{pmatrix} > 0\}. \tag{4}$$

where Σ_{11} is $n \times n$. Two subsets of $\boldsymbol{\Sigma}$ will play a special role.

$$\boldsymbol{\Sigma}_0 = \{\Sigma \in \boldsymbol{\Sigma} : \Sigma_{11} = \Sigma_0\}. \tag{5}$$

where Σ_0 is a given covariance. We also consider the subset

$$\boldsymbol{\Sigma}_1 = \{\Sigma \in \boldsymbol{\Sigma} : \Sigma = \begin{pmatrix} HH^\top + D & HQ \\ (HQ)^\top & Q^\top Q \end{pmatrix}\}, \tag{6}$$

where $H \in \mathbb{R}^{n \times k}, Q \in \mathbb{R}^{k \times k}$ invertible, $D > 0$ diagonal. Elements of $\boldsymbol{\Sigma}_1$ will often be denoted by $\Sigma(H, D, Q)$.

Remark 1. Notice that a normal vector Y, with $\mathbb{C}\text{ov}(Y) = \Sigma_0$, admits a FA model of size k iff $\boldsymbol{\Sigma}_0 \cap \boldsymbol{\Sigma}_1 \neq \emptyset$. Supposing that this is the case, take $\Sigma \in \boldsymbol{\Sigma}_0 \cap \boldsymbol{\Sigma}_1$ then, for some (H, D, Q), one has

$$\Sigma = \begin{pmatrix} \Sigma_0 & HQ \\ (HQ)^\top & Q^\top Q \end{pmatrix} = \begin{pmatrix} HH^\top + D & HQ \\ (HQ)^\top & Q^\top Q \end{pmatrix} > 0.$$

is a *bonafide* covariance of a normal vector V of dimension $n + k$. Partition $V^\top = (Y^\top, Z^\top)^\top$. It is easy to verify that $\mathbb{C}\text{ov}(Y) = \Sigma_0 = HH^\top + D$ is the same as $\mathbb{C}\text{ov}(HX + \varepsilon)$ for some X standard normal and ε normal, independent from X, and with diagonal covariance D.

The lifted minimization problem can be posed as follows

Problem 2

$$\min_{\Sigma' \in \boldsymbol{\Sigma}_0, \Sigma_1 \in \boldsymbol{\Sigma}_1} D(\Sigma' \| \Sigma_1)$$

which can be viewed as an iterated minimization problem over each of the variables. The two resulting partial minimization problems will be investigated in the following sections. In section 3.3 we will show the connection between Problems 1 and 2. More precisely, we will prove

Proposition 2. *Let Σ_0 be given. It holds that*

$$\min_{H,D} D(\Sigma_0 \| H H^\top + D) = \min_{\Sigma' \in \boldsymbol{\Sigma}_0, \Sigma_1 \in \boldsymbol{\Sigma}_1} D(\Sigma' | \Sigma_1).$$

3.1   The First Partial Minimization Problem

In this section we consider the first of the two partial minimization problems. Here we minimize, for a given positive definite matrix $\Sigma \in \boldsymbol{\Sigma}$, the divergence $D(\Sigma' \| \Sigma)$ over $\Sigma' \in \boldsymbol{\Sigma}_0$. The unique solution to this problem can be computed analytically and follows from

Lemma 1. *Let (Y, X) be a random vector distributed according to some $Q = Q^{Y,X}$ and let $\boldsymbol{\mathcal{P}}$ the set of all distributions $P = P^{Y,X}$ whose marginal $P^Y = P_0$, for some fixed $P_0 \ll Q^Y$. Then $\min_{P \in \boldsymbol{\mathcal{P}}} D(P\|Q) = D(P^*\|Q)$ where P^* is given by the Radon-Nikodym derivative*

$$\frac{dP^*}{dQ} = \frac{dP_0}{dQ^Y}.$$

Moreover,

$$D(P^*\|Q) = D(P_0\|Q^Y). \tag{7}$$

and, for any other $P \in \boldsymbol{\mathcal{P}}$, one has the Pythagorean law

$$D(P\|Q) = D(P\|P^*) + D(P^*\|Q). \tag{8}$$

Proof. First we show that (7) holds. Recall that Y has law P_0 under P^*, then

$$D(P^*\|Q) = \mathbb{E}_{P^*} \log \frac{dP^*}{dQ} = \mathbb{E}_{P^*} \log \frac{dP_0}{dQ^Y} = \mathbb{E}_{P_0} \log \frac{dP_0}{dQ^Y} = D(P_0\|Q^Y).$$

To show that P^* is a minimizer it is clearly sufficient to prove that (8) holds.

$$\begin{aligned}
D(P\|Q) &= \mathbb{E}_P \log \frac{dP}{dP^*} + \mathbb{E}_P \log \frac{dP^*}{dQ} \\
&= D(P\|P^*) + \mathbb{E}_P \log \frac{dP_0}{dQ^Y} \\
&= D(P\|P^*) + \mathbb{E}_{P_0} \log \frac{dP_0}{dQ^Y} = D(P\|P^*) + D(P^*\|Q),
\end{aligned}$$

where we used the fact that all $P \in \boldsymbol{\mathcal{P}}$ have marginal $P^Y = P_0$. $\square$

Remark 2. The law P^* is easily characterized in terms of the problem data P_0 and Q noticing that the marginal $P^{*Y} = P_0$ and the conditional $P^{*X|Y} = Q^{X|Y}$.

We now apply Lemma 1 to the case of normal laws and solve the first partial minimization. See also [2] for a different proof.

Proposition 3. *Let Q and P_0 be zero mean normal laws with strictly positive covariances $\Sigma \in \boldsymbol{\Sigma}$ and $\Sigma_0 \in \mathbb{R}^{n \times n}$ respectively. Then, $\min_{\Sigma' \in \boldsymbol{\Sigma}_0} D(\Sigma' \| \Sigma)$ is attained by the zero mean normal law P^* with covariance*

$$\Sigma^* = \begin{pmatrix} \Sigma_0 & \Sigma_0 \Sigma_{11}^{-1} \Sigma_{12} \\ \Sigma_{21} \Sigma_{11}^{-1} \Sigma_0 & \Sigma_{22} - \Sigma_{21} \Sigma_{11}^{-1} (\Sigma_{11} - \Sigma_0) \Sigma_{11}^{-1} \Sigma_{12} \end{pmatrix} > 0.$$

Moreover,

$$D(\Sigma^* \| \Sigma) = D(\Sigma_0 \| \Sigma_{11}).$$

Proof. This follows from Remark 2. A direct computation gives

$$\begin{aligned} \Sigma_{12}^* = \mathbb{E}_{P^*} XY^\top &= \ \mathbb{E}_{P^*}(\mathbb{E}_{P^*}[X|Y]Y^\top) \\ &= \mathbb{E}_{P^*}(\mathbb{E}_Q[X|Y]Y^\top) = \ \mathbb{E}_{P^*}(\Sigma_{21} \Sigma_{11}^{-1} YY^\top) \\ &= \Sigma_{21} \Sigma_{11}^{-1} \mathbb{E}_{P_0} YY^\top = \ \Sigma_{21} \Sigma_{11}^{-1} \Sigma_0. \end{aligned}$$

Likewise, we have

$$\begin{aligned} \Sigma_{22}^* = \mathbb{E}_{P^*} XX^\top &= \mathbb{C}\mathrm{ov}_{P^*}(X) \\ &= \mathbb{C}\mathrm{ov}_{P^*}(X|Y) + \mathbb{E}_{P^*}(\mathbb{E}_{P^*}[X|Y]\mathbb{E}_{P^*}[X|Y]^\top) \\ &= \mathbb{C}\mathrm{ov}_Q(X|Y) + \mathbb{E}_{P^*}(\mathbb{E}_Q[X|Y]\mathbb{E}_Q[X|Y]^\top) \\ &= \Sigma_{22} - \Sigma_{21} \Sigma_{11}^{-1} \Sigma_{12} + \mathbb{E}_{P^*}(\Sigma_{21} \Sigma_{11}^{-1} Y (\Sigma_{21} \Sigma_{11}^{-1} Y)^\top) \\ &= \Sigma_{22} - \Sigma_{21} \Sigma_{11}^{-1} \Sigma_{12} + \mathbb{E}_{P_0}(\Sigma_{21} \Sigma_{11}^{-1} YY^\top \Sigma_{11}^{-1} \Sigma_{12}) \\ &= \Sigma_{22} - \Sigma_{21} \Sigma_{11}^{-1} \Sigma_{12} + \Sigma_{21} \Sigma_{11}^{-1} \Sigma_0 \Sigma_{11}^{-1} \Sigma_{12}. \end{aligned}$$

Notice that, since $\Sigma > 0$ by assumption,

$$\Sigma_{22}^* - \Sigma_{21}^* (\Sigma_{11}^*)^{-1} \Sigma_{12}^* = \Sigma_{22} - \Sigma_{21} \Sigma_{11}^{-1} \Sigma_{12} > 0$$

which, together with the assumption $\Sigma_0 > 0$, shows that $\Sigma^* > 0$.

The last relation, $D(\Sigma^* \| \Sigma) = D(\Sigma_0 \| \Sigma_{11})$, reflects equation (7). $\qquad\square$

3.2   The Second Partial Minimization Problem

In this section we turn to the second partial minimization problem. Here we minimize, for a given positive definite matrix $\Sigma \in \boldsymbol{\Sigma}$, the divergence $D(\Sigma \| \Sigma_1)$ over $\Sigma_1 \in \boldsymbol{\Sigma}_1$.

Clearly this problem cannot have a unique solution in terms of the matrices H and Q. Indeed, if U is a unitary $k \times k$ matrix and $H' = HU$, $Q' = U^\top Q$, then $H'H'^\top = HH^\top$, $Q'^\top Q' = Q^\top Q$ and $H'Q' = HQ$. Nevertheless, the optimal matrices $HH^\top$, HQ and $Q^\top Q$ are unique, as we will see in Proposition 4. First we need to introduce some notation and conventions. If P is a positive definite matrix, we denote by $P^{1/2}$ any matrix satisfying $(P^{1/2})^\top (P^{1/2}) = P$, and by $P^{-1/2}$ its inverse. If M is any square matrix, we denote by $\Delta(M)$ the diagonal matrix

$$\Delta(M)_{ii} = M_{ii}.$$

Recall that we denote by $\Sigma(H, D, Q)$ a typical element of $\boldsymbol{\Sigma}_1$.

Proposition 4. *Given $\Sigma \in \boldsymbol{\Sigma}$ the $\min_{\Sigma_1 \in \boldsymbol{\Sigma}_1} D(\Sigma\|\Sigma_1)$ is attained at a Σ_1^* such that $\Sigma_1 \in \boldsymbol{\Sigma}_1$ is solved by*

$$Q^* = \Sigma_{22}^{1/2},$$
$$H^* = \Sigma_{12}\Sigma_{22}^{-1/2},$$
$$D^* = \Delta(\Sigma_{11} - \Sigma_{12}\Sigma_{22}^{-1}\Sigma_{21}).$$

Thus the minimizing matrix $\Sigma_1^* = \Sigma(H^*, D^*, Q^*)$ becomes

$$\Sigma_1^* = \begin{pmatrix} \Sigma_{12}\Sigma_{22}^{-1}\Sigma_{21} + \Delta(\Sigma_{11} - \Sigma_{12}\Sigma_{22}^{-1}\Sigma_{21}) & \Sigma_{12} \\ \Sigma_{21} & \Sigma_{22} \end{pmatrix}. \tag{9}$$

Moreover, the Pythagorean law

$$D(\Sigma\|\Sigma(H, D, Q)) = D(\Sigma\|\Sigma_1^*) + D(\Sigma_1^*\|\Sigma(H, D, Q)) \tag{10}$$

holds for any $\Sigma(H, D, Q) \in \boldsymbol{\Sigma}_1$, and therefore Σ_1^* is unique.

Proof. It is sufficient to show the validity of (10). We first compute

$$2D(\Sigma\|\Sigma(H, D, Q)) - 2D(\Sigma\|\Sigma_1^*).$$

It follows from Lemma A.2 that $|\Sigma(H, D, Q)| = |D| \times |Q^\top Q|$. In view of equation (3) the above difference becomes

$$\log|D| + \log|Q^\top Q| - \log|D^*| - \log|Q^{*\top}Q^*| + \mathrm{tr}\big(\Sigma(H, D, Q)^{-1}\Sigma\big) - \mathrm{tr}\big(\Sigma_1^{*-1}\Sigma\big). \tag{11}$$

Using Corollary A.1, we compute

$$\Sigma(H, D, Q)^{-1} = \begin{pmatrix} D^{-1} & -D^{-1}HQ^{-\top} \\ -Q^{-1}H^\top D^{-1} & Q^{-1}(H^\top D^{-1}H + I)Q^{-\top} \end{pmatrix}, \tag{12}$$

and hence we get that

$$\begin{aligned}
\mathrm{tr}\big(\Sigma(H, D, Q)^{-1}\Sigma\big) &= \mathrm{tr}\big(D^{-1}(\Sigma_{11} - HQ^{-\top}\Sigma_{21})\big) \\
&\quad + \mathrm{tr}\big(-Q^{-1}H^\top D^{-1}\Sigma_{12} + Q^{-1}(H^\top D^{-1}H + I)Q^{-\top}\Sigma_{22}\big) \\
&= \mathrm{tr}\big(D^{-1}(\Sigma_{11} - 2HQ^{-\top}\Sigma_{21}) + Q^{-1}(H^\top D^{-1}H + I)Q^{-\top}\Sigma_{22}\big). \tag{13}
\end{aligned}$$

Apply now Lemma A.2 to (9) and write $\Delta = \Delta(\Sigma_{11} - \Sigma_{12}\Sigma_{22}^{-1}\Sigma_{21})$, to get

$$\Sigma_1^{*-1} = \Sigma(H^*, D^*, Q^*)^{-1} = \begin{pmatrix} \Delta^{-1} & -\Delta^{-1}\Sigma_{12}\Sigma_{22}^{-1} \\ -\Sigma_{22}^{-1}\Sigma_{21}\Delta^{-1} & \Sigma_{22}^{-1}\Sigma_{21}\Delta^{-1}\Sigma_{12}\Sigma_{22}^{-1} + \Sigma_{22}^{-1} \end{pmatrix}.$$

Therefore

$$\mathrm{tr}\big(\Sigma_1^{*-1}\Sigma\big) = \mathrm{tr}\big(\Delta^{-1} \times (\Sigma_{11} - \Sigma_{12}\Sigma_{22}^{-1}\Sigma_{21})\big) + \mathrm{tr}\, I_k = \mathrm{tr}\big(\Delta^{-1}\Delta\big) + k = n + k. \tag{14}$$

Combining equations (11), (13), and (14), we find that

$$
\begin{aligned}
D(\Sigma\|\Sigma(H,D,Q)) - D(\Sigma\|\Sigma_1^*) = \\
\log|D| + \log|Q^\top Q| - \log|D^*| - \log|Q^{*^\top}Q^*| \\
+ \operatorname{tr}\big(D^{-1}(\Sigma_{11} - HQ^{-\top}\Sigma_{21})\big) \\
+ \operatorname{tr}\big(-Q^{-1}H^\top D^{-1}\Sigma_{12} + Q^{-1}(H^\top D^{-1}H + I)Q^{-\top}\Sigma_{22}\big) \\
- (n+k).
\end{aligned} \tag{15}
$$

We proceed with the computation of $2D(\Sigma_1^*\|\Sigma(H,D,Q))$.

$$
\begin{aligned}
2D(\Sigma(H^*,D^*,Q^*)\|\Sigma(H,D,Q)) = \\
\log|D| + \log|Q^\top Q| - \log|D^*| - \log|Q^{*^\top}Q^*| - (n+k) \\
+ \operatorname{tr}\big(\Sigma(H,D,Q)^{-1}\Sigma(H^*,D^*,Q^*)\big).
\end{aligned} \tag{16}
$$

Combining equations (9), (12), and $\operatorname{tr}\big(D^{-1}(\Sigma_{12}\Sigma_{22}^{-1}\Sigma_{21} + \Delta)\big) = \operatorname{tr}\big(D^{-1}\Sigma_{11}\big)$, we obtain

$$
\begin{aligned}
\operatorname{tr}\big(\Sigma(H,D,Q)^{-1}\Sigma(H^*,D^*,Q^*)\big) = \operatorname{tr}\big(D^{-1}\Sigma_{11}\big) \\
- 2\operatorname{tr}\big(D^{-1}HQ^{-\top}\Sigma_{21}\big) + \operatorname{tr}\big(Q^{-1}(H^\top D^{-1}H + I)Q^{-\top}\Sigma_{22}\big).
\end{aligned} \tag{17}
$$

Insertion of (17) into (16) and a comparison with (15) yields the result. $\qquad\square$

Remark 3. Notice that the matrix $H^*H^{*^\top}$ is strictly dominated by Σ_{11} (in the sense of positive matrices). This easily follows from $\Sigma_{11} - H^*H^{*^\top} = \Sigma_{11} - \Sigma_{12}\Sigma_{22}^{-1}\Sigma_{21} > 0$, and the assumption $\Sigma > 0$. By the same token $D^* > 0$.

3.3   The Link to the Original Problem

We now establish the connection between the lifted problem and the original Problem 1.

Proof of Proposition 2. Let $\Sigma_1 = \Sigma(H,D,Q)$ and denote by $\Sigma^* = \Sigma^*(\Sigma_1)$, the solution of the first partial minimization over $\boldsymbol{\Sigma}_0$. We have, for all $\Sigma' \in \boldsymbol{\Sigma}_0$,

$$
\begin{aligned}
D(\Sigma'\|\Sigma_1) &\geq D(\Sigma^*\|\Sigma_1) \\
&= D(\Sigma_0\|HH^\top + D) \\
&\geq \min_{H,D} D(\Sigma_0\|HH^\top + D),
\end{aligned}
$$

where we used Proposition 1 to write min on the RHS. It follows that

$$
\inf_{\Sigma'\in\boldsymbol{\Sigma}_0,\Sigma_1\in\boldsymbol{\Sigma}_1} D(\Sigma'\|\Sigma_1) \geq \min_{H,D} D(\Sigma_0\|HH^\top + D).
$$

Conversely, let (H^*, D^*) be the minimizer of $(H,D) \mapsto D(\Sigma_0\|HH^\top + D)$, pick an arbitrary invertible Q^*, and let $\Sigma^* = \Sigma(H^*, D^*, Q^*)$ be the corresponding element

92 L. Finesso and P. Spreij

in Σ_1. Furthermore, let $\Sigma^{**} \in \Sigma_0$ be the minimizer of $\Sigma \mapsto D(\Sigma\|\Sigma^*)$ over Σ_0. Then

$$
\min_{H,D} D(\Sigma_0\|HH^\top + D) = D(\Sigma_0\|H^*H^{*\top} + D^*)
$$
$$
\geq D(\Sigma^{**}\|\Sigma^*)
$$
$$
\geq \inf_{\Sigma' \in \Sigma_0, \Sigma_1 \in \Sigma_1} D(\Sigma'\|\Sigma_1),
$$

which shows the opposite inequality. Finally, to show that we can replace the infima with minima also in the lifted problem, notice that (see Proposition 3) $D(\Sigma^{**}\|\Sigma^*) = D(\Sigma_0\|H^*H^{*\top} + D^*)$. $\qquad\square$

4 Alternating Minimization Algorithm

In this section we combine the two partial minimization problems above to derive an iterative algorithm for Problem 1. It turns out that this algorithm is also instrumental in proving the existence of a solution to Problem 1.

4.1 The Algorithm

We suppose that the given matrix Σ_0 is strictly positive definite. Pick the initial values H_0, D_0, Q_0 such that H_0 is of full rank, $D_0 > 0$ is diagonal, Q_0 and $H_0 H_0^\top + D_0$ are invertible.

At the t-th iteration the matrices H_t, D_t and Q_t are available. Start solving the first partial minimization problem with $\Sigma = \Sigma(H_t, D_t, Q_t)$. Use the resulting matrix as data for the second partial minimization, the solution of which gives the update rules

$$
Q_{t+1} = \Big(Q_t^\top Q_t - Q_t^\top H_t^\top (H_t H_t^\top + D_t)^{-1} H_t Q_t
$$
$$
+ Q_t^\top H_t^\top (H_t H_t^\top + D_t)^{-1} \Sigma_0 (H_t H_t^\top + D_t)^{-1} H_t Q_t \Big)^{1/2}, \qquad (18)
$$
$$
H_{t+1} = \Sigma_0 (H_t H_t^\top + D_t)^{-1} H_t Q_t Q_{t+1}^{-1}, \qquad (19)
$$
$$
D_{t+1} = \Delta(\Sigma_0 - H_{t+1} H_{t+1}^\top). \qquad (20)
$$

In (18) there is some freedom in computing the square root that determines Q_{t+1}. Properly choosing the square root will result in the disappearance of Q_t from the algorithm. This is an attractive feature, since Q_t only serves as an auxiliary variable. One can write the RHS of equation (18), before taking the square root, as

$$
Q_t^\top (I - H_t^\top (H_t H_t^\top + D_t)^{-1}(H_t H_t^\top + D_t - \Sigma_0)(H_t H_t^\top + D_t)^{-1} H_t) Q_t
$$

and denoting

$$
R_t = I - H_t^\top (H_t H_t^\top + D_t)^{-1}(H_t H_t^\top + D_t - \Sigma_0)(H_t H_t^\top + D_t)^{-1} H_t \qquad (21)
$$

a possible square root is given by

$$R_t^{1/2} Q_t.$$

Notice that R_t only involves the iterates H_t and D_t. The update equation (18) can therefore be rewritten as

$$H_{t+1} = \Sigma_0 (H_t H_t^\top + D_t)^{-1} H_t R_t^{-1/2}. \tag{22}$$

The final version of the algorithm is given by equations (20),(21), and (22) which, for clarity, we present as

Algorithm 1

$$R_t = I - H_t^\top (H_t H_t^\top + D_t)^{-1} (H_t H_t^\top + D_t - \Sigma_0)(H_t H_t^\top + D_t)^{-1} H_t, \tag{23}$$

$$H_{t+1} = \Sigma_0 (H_t H_t^\top + D_t)^{-1} H_t R_t^{-1/2}, \tag{24}$$

$$D_{t+1} = \Delta(\Sigma_0 - H_{t+1} H_{t+1}^\top). \tag{25}$$

In order to avoid taking a square root at each step one can introduce the matrices $K_t = H_t Q_t$ and $P_t = Q_t^T Q_t$ and write the updates for K_t and P_t. Equations (18), (19), and (20) easily give

Algorithm 2

$$K_{t+1} = \Sigma_0 (K_t P_t^{-1} K_t^\top + D_t)^{-1} K_t, \tag{26}$$

$$P_{t+1} = P_t - K_t^\top (K_t P_t^{-1} K_t^\top + D_t)^{-1} (K_t P_t^{-1} K_t^\top + D_t - \Sigma_0)(K_t P_t^{-1} K_t^\top + D_t)^{-1} K_t,$$

$$D_{t+1} = \Delta(\Sigma_0 - K_{t+1} P_{t+1}^{-1} K_{t+1}^\top).$$

After the final iteration, the T-th say, one can take $H_T = K_T Q_T^{-1}$, where Q_T is a square root of P_T.

Notice that in both Algorithm 1 and 2 it is required to invert $n \times n$ matrices (like e.g. $(H_t H_t^\top + D_t)^{-1}$). Applying corollary A.1 one gets $(H_t H_t^\top + D_t)^{-1} H_t = D_t^{-1} H_t (I + H_t^\top D_t^{-1} H_t)$. Hence, we can replace e.g. (22) with

$$H_{t+1} = \Sigma_0 D_t^{-1} H_t (I + H_t^\top D_t^{-1} H_t)^{-1} R_t^{-1/2}. \tag{27}$$

By the same token one can write

$$K_{t+1} = \Sigma_0 D_t^{-1} K_t (P_t + K_t^\top D_t^{-1} K_t)^{-1} P_t$$

to replace (26).

Some properties of the algorithm are summarized in the next proposition.

Proposition 5. *For Algorithm 1 the following hold for all t.*

(a) $D_t > 0$ and $(D_t)_{ii} \leq (\Sigma_0)_{ii}$.
(b) R_t is invertible.
(c) If H_0 is of full column rank, so is H_t.
(d) $H_t H_t^\top \leq \Sigma_0$.

(e) If $\Sigma_0 = H_0 H_0^\top + D_0$ then the algorithm stops.

(f) The objective function decreases at each iteration. More precisely, let $\Sigma_{0,t}$ be the solution of the first partial minimization with data $\Sigma_t = \Sigma(H_t, D_t, Q_t)$. Then

$$D(\Sigma_0||H_{t+1}H_{t+1}^\top) - D(\Sigma_0||H_t H_t^\top) = -\Big(D(\Sigma_{t+1}||\Sigma_t) + D(\Sigma_{0,t}||\Sigma_{0,t+1})\Big).$$

(g) The limit points (H, D) of the algorithm satisfy the relations

$$H = (\Sigma_0 - HH^\top)D^{-1}H,$$
$$D = \Delta(\Sigma_0 - HH^\top).$$

Proof. (a) This follows from Remark 3.

(b) Use the identity $I - H_t^\top(H_t H_t^\top + D_t)^{-1}H_t = (I + H_t^\top D_t^{-1}H_t)^{-1}$ and the assumption $\Sigma_0 > 0$.

(c) Use the assumption $\Sigma_0 > 0$, (a), and (b).

(d) Again from Remark 3 and the construction of the algorithm as a combination of the two partial minimization problems.

(e) This is a triviality upon noticing that one can take $R_t = I$ in this case.

(f) It follows from a concatenation of Lemma 1 and Proposition 4. Notice that we can express the decrease as the sum of two I-divergences, since the Pythagorean law holds for both partial minimizations.

(g) We consider Algorithm 2 first. Assume that all variables converge. Then, from (26), the limit points K, P, D satisfy the relation $K = \Sigma_0 D^{-1}K(P + K^\top D^{-1}K)^{-1}P$. Postmultiplication by $P^{-1}(P + K^\top D^{-1}K)$ yields, after rearranging terms, $K = (\Sigma_0 - KP^{-1}K^\top)D^{-1}K$. Let now Q be a square root of P and $H = KQ^{-1}$ to get the first relation. The rest is trivial. $\qquad\square$

4.2 Proof of Proposition 1

Let D_0 and H_0 be arbitrary and perform one step of the algorithm to get matrices D_1 and H_1. It follows from Proposition 5 that $D(\Sigma_0||H_1 H_1^\top + D_1) \leq D(\Sigma_0||H_0 H_0^\top + D_0)$. Moreover, $H_1 H_1^\top \leq \Sigma_0$ and $D_1 \leq \Delta(\Sigma_0)$. Hence the search for a minimum can be confined to the set of matrices (H, D) satisfying $HH^\top \leq \Sigma_0$ and $D \leq \Delta(\Sigma_0)$. Next, we claim that it is also sufficient to restrict the search for a minimum to all matrices (H, D) such that $HH^\top + D \geq \varepsilon I$ for some sufficiently small $\varepsilon > 0$. Indeed, if the last inequality is violated, then $HH^\top + D$ has an eigenvalue less than ε. Write the Jordan decompositions $HH^\top + D = U\Lambda U^\top$, and let $\Sigma_U = U^\top \Sigma_0 U$. Then $D(\Sigma_0||HH^\top + D) = D(\Sigma_U||\Lambda)$, as one easily verifies. Denoting by λ_i the eigenvalues of $HH^\top + D$ and letting σ_{ii} be the diagonal elements of Σ_U, we can write $D(\Sigma_U|\Lambda) = -\frac{1}{2}\log|\Sigma_U| + \frac{1}{2}\sum_i \log \lambda_i - \frac{n}{2} + \frac{1}{2}\sum_i \frac{\sigma_{ii}}{\lambda_i}$. Let λ_{i_0} be a minimum eigenvalue and take ε smaller than the minimum of all σ_{ii}, which is positive, since Σ_0 is strictly positive definite. Then the contribution for $i = i_0$ in the summation to the divergence $D(\Sigma_U||\Lambda)$ is at least $\log \varepsilon + 1$, which tends to infinity for $\varepsilon \to 0$. This proves the claim. So, we have shown that a minimizing pair (H, D) has to satisfy $HH^\top \leq \Sigma_0$, $D \leq \Delta(\Sigma_0)$, and $HH^\top + D \geq \varepsilon I$, for some $\varepsilon > 0$. In other words we have to minimize the I-divergence over a compact set on which it is clearly continuous. This proves Proposition 1. $\qquad\square$

References

1. T.W. Anderson (1984), *An introduction to multivariate statistical analysis*, Second ed., Wiley.
2. E. Cramer (1998), Conditional iterative proportional fitting for Gaussian distributions, *J. Multivariate Analysis*, **65(2)**, 261–276.
3. E. Cramer (2000), Probability measures with given marginals and conditionals: I-projections and conditional iterative proportional fitting, *Statistics & Decisions*, **18(3)**, 311–329.
4. I. Csiszár and G. Tusnády (1984), Information geometry and alternating minimization procedures, *Statistics & Decisons, supplement issue* **1**, 205-237.
5. L. Finesso and G. Picci (1984), Linear statistical models and stochastic realization theory. *Analysis and optimization of systems, Part 1 (Nice, 1984)*, 445–470, Lecture Notes in Control and Inform. Sci., 62, Springer, Berlin.
6. L. Finesso and P. Spreij (2006), Nonnegative matrix factorization and I-divergence alternating minimization, *Linear Algebra and its Applications*, **416**, 270–287.

A   Appendix

For ease of reference we collect here some standard formulas for the normal distribution and some matrix algebra.

A.1   Multivariate Normal Distribution

Let $(X^\top, Y^\top)^\top$ be a zero mean normal vector with covariance matrix

$$\Sigma = \begin{pmatrix} \Sigma_{XX} & \Sigma_{XY} \\ \Sigma_{YX} & \Sigma_{YY} \end{pmatrix}.$$

Assume that Σ_{YY} is invertible. The conditional law of X given Y is normal with $\mathbb{E}[X|Y] = \Sigma_{XY}\Sigma_{YY}^{-1}Y$ and

$$\mathbb{C}\mathrm{ov}[X|Y] = \Sigma_{XX} - \Sigma_{XY}\Sigma_{YY}^{-1}\Sigma_{YX}. \tag{28}$$

A.2   Partitioned Matrices

Lemma 2. *Let A, D be square matrices. Assume invertibility where required.*

$$\begin{pmatrix} A & C \\ B & D \end{pmatrix} = \begin{pmatrix} I & CD^{-1} \\ 0 & I \end{pmatrix} \begin{pmatrix} A - CD^{-1}B & 0 \\ 0 & D \end{pmatrix} \begin{pmatrix} I & 0 \\ D^{-1}B & I \end{pmatrix},$$

$$\begin{pmatrix} A & C \\ B & D \end{pmatrix} = \begin{pmatrix} I & 0 \\ BA^{-1} & I \end{pmatrix} \begin{pmatrix} A & 0 \\ 0 & D - BA^{-1}C \end{pmatrix} \begin{pmatrix} I & A^{-1}C \\ 0 & I \end{pmatrix},$$

$$\begin{pmatrix} A & C \\ B & D \end{pmatrix}^{-1} =$$

$$\begin{pmatrix} (A - CD^{-1}B)^{-1} & -(A - CD^{-1}B)^{-1}CD^{-1} \\ -D^{-1}B(A - CD^{-1}B)^{-1} & D^{-1}B(A - CD^{-1}B)^{-1}CD^{-1} + D^{-1} \end{pmatrix}.$$

Corollary 1

$$(D - BAC)^{-1} = D^{-1} + D^{-1}B(A^{-1} - CD^{-1}B)^{-1}CD^{-1}.$$

Proof. For Lemma 2 a check will suffice. The Corollary follows using the two decompositions of the Lemma with A replaced by A^{-1} and comparing the two expressions of the lower right block of the inverse matrix. $\qquad\square$

Tensored Polynomial Models

Paul A. Fuhrmann* and Uwe Helmke

[1] Department of Mathematics
     Ben-Gurion University of the Negev
     Beer Sheva, Israel
[2] Universität Würzburg
     Institut für Mathematik
     Würzburg, Germany

Dedicated to Giorgio Picci on his 65th birthday

Summary. The theme of the present paper is the study of two different versions of tensor products of functional models and present some applications to various problems related to system theory. Among those are stability of higher order systems, tangent spaces, spaces of intertwining maps, invariant factors of tensor products of linear transformations and the solvability of Sylvester equations.

Keywords: Tensor products, polynomial models, Sylvester equation, invariant factors.

1 Introduction

The theory of polynomial and rational models, initiated by the author in Fuhrmann [1976], proved to be a very powerful tool for a variety of system problems as well as for unifying various approaches to linear system theory. Recently, there has been growing interest in a variety of multilinear problems, including potential extensions to bilinear systems, Yang-Mills instantons from physics, and structured linear matrix equations. The latter include the analysis of classes of linear equations of which the Sylvester, Lyapunov and Stein equations are special cases. To meet these challenges, our intention in the present paper is to extend the theory to tensored polynomial and rational models. The original motivation for this stemmed from an algebraic approach to the derivation of classical stability criteria, using polynomials in two variables, see Kalman [1969] which was extended in Willems and Fuhrmann [1992] to the multivariable case. Another study in which tensored models were studied is the characterization of the tangent space of the space of rational functions given in Helmke and Fuhrmann [1998]. Further motivation comes from recent work on quadratic differential forms, see Willems and Trentelman [1998].

Because of space limitations, we only describe some of the principal results concerning two types of tensored models. In the last section we give some applications, in particular to Sylvester and Lyapunov equations. These results are then applied tothe

---

* Partially supported by the ISF under Grant No. 1282/05.

A. Chiuso et al. (Eds.): Modeling, Estimation and Control, LNCIS 364, pp. 97–112, 2007.
springerlink.com   © Springer-Verlag Berlin Heidelberg 2007

analysis of the classic case. The results described are not new, see de Souza and S.P. Bhattacharyya [1981] and Heinig and Rost [1984] and the further references therein. However, the method of proof is new and has the advantage of greater clarity. The full details will appear in a subsequent publication.

2 Tensored Models

2.1 Preliminaries

In the following, the focal point of our interest will be the study of tensor products of polynomial and rational models. Now both models carry two structures, one being that of a vector space over a field $\mathbb{F}$, the other of a module over the polynomial ring $\mathbb{F}[z]$. Since the tensor product depends very much on the ring used, the result is that we have to study both tensor products, each leading to different applications. If M_1, M_2 are modules over a commutative ring R, we will denote by $M_1 \otimes_R M_2$ the R-module defined by the relevant tensor product.

We specialize our discussion to polynomial modules. Given the field $\mathbb{F}$ the polynomial ring $\mathbb{F}[z]$ is a rank 1 module over itself but an infinite dimensional vector space (module) over $\mathbb{F}$. The module of polynomial vectors $\mathbb{F}[z]^p$ itself is isomorphic to the tensor product $\mathbb{F}[z] \otimes \mathbb{F}^p$. In this situation we have $\mathbb{F}[z]^p \otimes_{\mathbb{F}} \mathbb{F}[z]^m \simeq \mathbb{F}[z]^{p \times m}$, as well as $\mathbb{F}[z]^p \otimes_{\mathbb{F}[z]} \mathbb{F}[z]^m \simeq \mathbb{F}[z]^{p \times m}$. In the following we will actually use these isomorphisms as identifying the above tensor products. In the case of the tensor product taken over the field $\mathbb{F}$, the product is that of two vector spaces and there is no collapsing. It will prove fruitful to consider this tensor product as that of two polynomial modules with different variables, and have $\mathbb{F}[z]^p \otimes_{\mathbb{F}} \mathbb{F}[w]^m = \mathbb{F}[z, w]^{p \times m}$.

In the following, we will need several more spaces. By $\mathbb{F}((z^{-1}, w^{-1}))$ we denote the field of truncated Laurent series in the variables z, w and by $\mathbb{F}((z^{-1}, w^{-1}))^{p \times m}$ the module of all $p \times m$ matrices with entries in $\mathbb{F}((z^{-1}, w^{-1}))$, i.e. the set of all series $\sum_{i=-\infty}^{s} \sum_{j=-\infty}^{t} Q_{ij} z^i w^j$. We shall routinely use the isomorphism $\mathbb{F}((z^{-1}, w^{-1}))^{p \times m} \simeq \mathbb{F}^{p \times m}((z^{-1}, w^{-1}))$. By $\mathbb{F}[z, w]$ we denote the ring of polynomials in the commuting variables z, w and by $\mathbb{F}[[z^{-1}, w^{-1}]]$ the ring of formal power series in z^{-1}, w^{-1}. We denote by $\mathbb{F}[z, w]^{p \times m}$ the space of $p \times m$ polynomial matrices. We will find it useful to use $\mathbb{F}[[z^{-1}, w]^{p \times m}$ to denote the subspace of $\mathbb{F}((z^{-1}, w^{-1}))^{p \times m}$ of matrices whose entries are formal power series in z^{-1} and polynomial in w. $\mathbb{F}[z, w^{-1}]]^{p \times m}$ is similarly defined. In the same vein, $\mathbb{F}((z^{-1}, w^{-1}]]^{p \times m}$ denotes the space of $p \times m$ matrix functions whose entries are truncated Laurent series in the variable z and formal power series in the variable w, i.e. have the representation, with s an integer, $\sum_{i=-\infty}^{s} \sum_{j=-\infty}^{0} Q_{ij} z^i w^j$.

The direct sum $\mathbb{F}((z^{-1}))^m = \mathbb{F}[z]^m \oplus z^{-1}\mathbb{F}[[z^{-1}]]^m$ is replaced in this setting by the following direct sum representation.

$$\begin{aligned}
\mathbb{F}((z^{-1}, w^{-1}))^{p \times m} &= \mathbb{F}[z, w^{-1}))^{p \times m} \oplus z^{-1}\mathbb{F}[[z^{-1}, w^{-1}))^{p \times m} \\
&= \mathbb{F}((z^{-1}, w]^{p \times m} \oplus \mathbb{F}((z^{-1}, w^{-1}]]^{p \times m} w^{-1} \\
&= \mathbb{F}[z, w]^{p \times m} \oplus z^{-1}\mathbb{F}[[z^{-1}, w]^{p \times m} \\
&\quad \oplus \mathbb{F}[z, w^{-1}]]^{p \times m} w^{-1} \oplus z^{-1}\mathbb{F}[[z^{-1}, w^{-1}]]^{p \times m} w^{-1}
\end{aligned} \tag{1}$$

To these direct sum representations correspond, respectively, the following projection identities.

$$I = \pi_+^z \otimes I + \pi_-^z \otimes I = I \otimes \pi_+^w + I \otimes \pi_-^w$$
$$= \pi_+^z \otimes \pi_+^w + \pi_-^z \otimes \pi_+^w + \pi_+^z \otimes \pi_-^w + \pi_-^z \otimes \pi_-^w \tag{2}$$

Throughout, we will use $\tilde{A}$ to denote the transpose of a matrix A.

2.2   Tensored Polynomial and Rational Models

A one variable polynomial model X_D with $D(z) \in \mathbb{F}[z]^{p \times p}$ nonsingular is isomorphic to the quotient module $\mathbb{F}[z]^p / D\mathbb{F}[z]^p$, see Fuhrmann [1976] and subsequent papers. Thus, to begin with, our interest will focus on the study of the tensor products of quotient modules. In general, if M_1, M_2 are R modules, with R a commutative ring and $N_i \subset M_i$ are submodules, then the quotient spaces M_i/N_i have a natural R module structure. Let N be the submodule generated in $M_1 \otimes_R M_2$ by $N_1 \otimes_R M_2$ and $M_1 \otimes_R N_2$. Then we have $M_1/N_1 \otimes_R M_2/N_2 \simeq (M_1 \otimes_R M_2)/N$. Because the tensor product is taken over a ring, considerable collapsing can occur. As an example, consider the ring of polynomials $\mathbb{F}[z]$ and the quotient modules $\mathbb{F}[z]/d_i\mathbb{F}[z]$. Since $d\mathbb{F}[z] = d_1\mathbb{F}[z] + d_2\mathbb{F}[z]$, with d is the greatest common divisor of d_1 and d_2, we have

$$\mathbb{F}[z]/d_1\mathbb{F}[z] \otimes_{\mathbb{F}[z]} \mathbb{F}[z]/d_2\mathbb{F}[z] = \mathbb{F}[z]/(d_1\mathbb{F}[z] + d_2\mathbb{F}[z]) = \mathbb{F}[z]/d\mathbb{F}[z].$$

Let now $D_1 \in \mathbb{F}[z]^{p \times p}$ and $D_2 \in \mathbb{F}[z]^{m \times m}$ be nonsingular polynomial matrices. Noting that $D_1(z)\mathbb{F}[z]^{p \times m} + \mathbb{F}[z]^{p \times m}D_2(z)$ is the submodule of $\mathbb{F}[z]^{p \times m}$ generated by $D_1(z)\mathbb{F}[z]^{p \times m}$ and $\mathbb{F}[z]^{p \times m}D_2(z)$, this generalizes to the vectorial case as

$$(\mathbb{F}[z]^p / D_1(z)\mathbb{F}[z]^p) \otimes_{\mathbb{F}[z]} (\mathbb{F}[z]^m / \mathbb{F}[z]^m D_2(z)) \simeq$$
$$\simeq \mathbb{F}[z]^{p \times m} / (D_1(z)\mathbb{F}[z]^{p \times m} + \mathbb{F}[z]^{p \times m}D_2(z)) \tag{3}$$

However, if we take the tensor product of the two polynomial models over the field F, then we consider only the vector space structure, there is no collapsing and we have

$$\dim \mathbb{F}[z]^p / D_1(z)\mathbb{F}[z] \otimes_{\mathbb{F}} \mathbb{F}[z]^m / D_2(z)\mathbb{F}[z]^m = \deg \det D_1 \cdot \deg \det D_2. \tag{4}$$

Equivalently, using the isomorphism $X_D \simeq \mathbb{F}[z]^p / D(z)\mathbb{F}[z]^p$, we have

$$\dim(X_{D_1(z)} \otimes_{\mathbb{F}} X_{D_2(z)}) = \deg \det D_1 \cdot \deg \det D_2.$$

Next, we analyze the tensor products of quotient modules of polynomial modules in two distinct variables, i.e. we consider $\mathbb{F}[z]^p / D_1(z)\mathbb{F}[z]$ and $\mathbb{F}[w]^m / D_2(w)\mathbb{F}[w]^m$. Since there is no common ring to these modules, we take the tensor product over the underlying field $\mathbb{F}$. The submodule, or rather subspace, generated by $D_1(z)\mathbb{F}[z]^p$ and $\mathbb{F}[w]^m D_2(w)$ in $\mathbb{F}[z, w]^{p \times m} = \mathbb{F}[z]^p \otimes_{\mathbb{F}} \mathbb{F}[w]^m$, is clearly

$$D_1(z)\mathbb{F}[z]^p \otimes_{\mathbb{F}} \mathbb{F}[w]^m + \mathbb{F}[z]^p \otimes_{\mathbb{F}} \mathbb{F}[w]^m D_2(w) = D_1(z)\mathbb{F}[z, w]^{p \times m} + \mathbb{F}[z, w]^{p \times m}D_2(w).$$

which implies

$$(\mathbb{F}[z]^p / D_1(z)\mathbb{F}[z]^p) \otimes_{\mathbb{F}} (\mathbb{F}[w]^m / \mathbb{F}[w]^m D_2(w)) \simeq$$
$$\simeq \mathbb{F}[z, w]^{p \times m} / (D_1(z)\mathbb{F}[z, w]^{p \times m} + \mathbb{F}[z, w]^{p \times m}D_2(w)).$$

Now the quotient module $\mathbb{F}[z,w]^{p\times m}/(D_1(z)\mathbb{F}[z,w]^{p\times m}+\mathbb{F}[z,w]^{p\times m}D_2(w))$ can be represented in a concrete way as a two variable polynomial model. To this end, we introduce tensored module structures on the module of truncated matrix Laurent series in two variables z,w, i.e. on $\mathbb{F}((z^{-1},w^{-1}))$. For $A(z)\in\mathbb{F}((z^{-1}))^{p\times p}$ and $\overline{A}(w)\in\mathbb{F}((w^{-1}))^{m\times m}$, we define the **Kronecker product** of A and $\overline{A}$ by $(A(z)\otimes\overline{A}(w))F(z,w)=A(z)F(z,w)\overline{A}(w)$. Clearly $A(z)\otimes\overline{A}(w)$ is an $\mathbb{F}$-linear map. There are many derivatives to this definition. In particular, we will look at the restriction to polynomial spaces i.e. to maps $D(z)\otimes\overline{D}(w):\mathbb{F}[z,w]^{p\times m}\longrightarrow\mathbb{F}[z,w]^{p\times m}$, where $D(z),\overline{D}(z)$ are nonsingular polynomial matrices.

We define now two maps $\pi_{D(z)\otimes_\mathbb{F}\overline{D}(w)}:\mathbb{F}[z,w]^{p\times m}\longrightarrow\mathbb{F}[z,w]^{p\times m}$ and $\pi_{D(z)\otimes\overline{D}(z)}:\mathbb{F}[z]^{p\times m}\longrightarrow\mathbb{F}[z]^{p\times m}$ by $\pi_{D(z)\otimes\overline{D}(w)}F(z,w)=(D(z)\otimes\overline{D}(w))(\pi_-^z\otimes\pi_-^w)(D(z)\otimes\overline{D}(w))^{-1}F(z,w)=(\pi_{D(z)}\otimes\pi_{\overline{D}(w)})F(z,w)$, and $\pi_{D(z)\otimes\overline{D}(z)}F(z)=(D(z)\otimes\overline{D}(z))\pi_-(D(z)\otimes\overline{D}(z))^{-1}F(z)=(\pi_{D(z)}\otimes\pi_{\overline{D}(z)})F(z)=D(z)[\pi_-(D(z)^{-1}F(z)\overline{D}(z)^{-1})]\overline{D}(z)$, respectively. Clearly, $\pi_-^z\otimes\pi_-^w$ is a projection map in $\mathbb{F}((z^{-1},w^{-1}))^{p\times m}$ and π_- a projection map in $\mathbb{F}((z^{-1}))^{p\times m}$. Hence $\pi_{D(z)\otimes\overline{D}(w)}$ is a projection map in $\mathbb{F}[z,w]^{p\times m}$.

Proposition 1

1. *Let $D(z)\in\mathbb{F}[z]^{p\times p}$ and $\overline{D}(w)\in\mathbb{F}[w]^{m\times m}$ be nonsingular polynomial matrices. Then*

   a) *The maps $\pi_{D(z)\otimes I}$, $\pi_{I\otimes\overline{D}(w)}$ and $\pi_{D(z)}\otimes\pi_{\overline{D}(w)}$ are all projections and we have*

$$\pi_{D(z)\otimes I}\pi_{I\otimes\overline{D}(w)}=\pi_{I\otimes\overline{D}(w)}\pi_{D(z)\otimes I}=\pi_{D(z)}\otimes\pi_{\overline{D}(w)}=\pi_{D(z)\otimes\overline{D}(w)}. \tag{5}$$

   b) *We have*

$$\mathrm{Ker}\,\pi_{D(z)\otimes I}=D(z)\mathbb{F}[z,w]^{p\times m}$$
$$\mathrm{Ker}\,\pi_{I\otimes\overline{D}(w)}=\mathbb{F}[z,w]^{p\times m}\overline{D}(w) \tag{6}$$
$$\mathrm{Ker}\,\pi_{D(z)\otimes\overline{D}(w)}=D(z)\mathbb{F}[z,w]^{p\times m}+\mathbb{F}[z,w]^{p\times m}\overline{D}(w)$$

   c) *A polynomial matrix $Q(z,w)$ is in $X_{D(z)\otimes\overline{D}(w)}$ if and only if*

$$D(z)^{-1}Q(z,w)\overline{D}(w)^{-1}$$

   is strictly proper in both variables.

   d) *We have the isomorphism*

$$X_{D(z)\otimes\overline{D}(w)}\simeq X_{D(z)}\otimes_\mathbb{F}X_{\overline{D}(w)} \tag{7}$$

   e) *We have*

$$\dim X_{D(z)\otimes\overline{D}(w)}=\deg\det D\cdot\deg\det\overline{D}. \tag{8}$$

2. *Let $D(z) \in \mathbb{F}[z]^{p \times p}$ and $\overline{D}(z) \in \mathbb{F}[z]^{m \times m}$ be nonsingular. Then*
   a) *We have $Q(z) \in X_{D(z) \otimes \overline{D}(z)}$ if and only if $D(z)^{-1}Q(z)\overline{D}(z)^{-1}$ is strictly proper.*
   b) *We have*

$$\operatorname{Ker} \pi_{D(z) \otimes \overline{D}(z)} = D(z)\mathbb{F}[z]^{p \times m}\overline{D}(z). \tag{9}$$

   c) *Given two nonsingular polynomial matrices $D(z) \in \mathbb{F}[z]^{p \times p}$ and $\overline{D}(w) \in \mathbb{F}[w]^{m \times m}$. Let $d_1, \ldots, d_p$ be the invariant factors of D ordered so that $d_i | d_{i-1}$ and $\overline{d}_1, \ldots, \overline{d}_m$, the invariant factors of $\overline{D}$ similarly ordered. Let $\delta_i = \deg d_i$, $= 1, \ldots, p$ and $\overline{\delta}_i = \deg \overline{d}_i$. If $n = \deg \det D$ and $\overline{n} = \deg \det \overline{D}$ then it is clear that $n = \sum_{i=1}^{p} \delta_i$ and $\overline{n} = \sum_{i=1}^{m} \overline{\delta}_i$.*
   Then we have

$$\dim X_{D(z) \otimes \overline{D}(z)} = m \deg \det D + p \deg \det \overline{D}. \tag{10}$$

Definition 1. *Let $D(z) \in \mathbb{F}[z]^{p \times p}$ and $\overline{D}(w) \in \mathbb{F}[w]^{m \times m}$ be nonsingular polynomial matrices.*

1. *We define the* **tensored polynomial model** $X_{D(z) \otimes \overline{D}(w)}$ *by*

$$X_{D(z) \otimes \overline{D}(w)} = \operatorname{Im} \pi_{D(z) \otimes \overline{D}(w)} \tag{11}$$

2. *We define the* **Kronecker product polynomial model** $X_{D(z) \otimes \overline{D}(z)}$ *by*

$$X_{D(z) \otimes \overline{D}(z)} = \operatorname{Im} \pi_{D(z) \otimes \overline{D}(z)} \tag{12}$$

Proposition 1 says that a tensored polynomial model, in the sense of (11), is the tensor product of polynomial models. As a result the dimension formula (8) is multiplicative. This is no longer true if we use Kronecker tensored polynomial models in the sense of (12). We note that $\det(D(z) \otimes \overline{D}(z)) = (\det D)^m (\det \overline{D})^p$ implies $\deg \det(D(z) \otimes \overline{D}(z)) = p \deg \det \overline{D} + m \deg \det D$. Hence, the dimension formula is additive.

In analogy with the introduction of tensored polynomial models, we introduce next the tensored rational models. Given two nonsingular polynomial matrices $D(z) \in \mathbb{F}[z]^{p \times p}$ and $\overline{D}(w) \in \mathbb{F}[w]^{m \times m}$, we define a map $\pi^{D(z) \otimes \overline{D}(w)} : z^{-1}\mathbb{F}[[z^{-1}, w^{-1}]]^{p \times m}w^{-1} \longrightarrow z^{-1}\mathbb{F}[[z^{-1}, w^{-1}]]^{p \times m}w^{-1}$ by

$$\pi^{D(z) \otimes \overline{D}(w)} H(z, w) = (\pi_-^z \otimes \pi_-^w)(D(z) \otimes \overline{D}(w))^{-1}(\pi_+^z \otimes_{\mathbb{F}} \pi_+^w)(D(z) \otimes \overline{D}(w))H(z, w) \tag{13}$$

We define the **tensored rational model** $X^{D(z) \otimes \overline{D}(w)}$ by

$$X^{D(z) \otimes \overline{D}(w)} = \operatorname{Im} \pi^{D(z) \otimes \overline{D}(w)}.$$

Clearly, we have $H(z, w) \in X^{D(z) \otimes \overline{D}(w)}$ if and only if $D(z)H(z, w)\overline{D}(w) \in \mathbb{F}[z, w]^{p \times m}$, i.e. it is a polynomial in both variables.

2.3   Module Structures on Tensored Models

So far, on $\mathbb{F}[z]^p \otimes_{\mathbb{F}} \mathbb{F}[w]^m = \mathbb{F}[z, w]^{p \times m}$ we have only the vector space structure. Of course, given any linear transformation X in $\mathbb{F}[z, w]^{p \times m}$, there exists an induced $\mathbb{F}[\zeta]$-module structure defined by

$$\zeta \cdot F(z, w) = X F(z, w) \tag{14}$$

and hence

$$p(\zeta) \cdot F(z, w) = p(X)F(z, w), \qquad p \in \mathbb{F}[\zeta]. \tag{15}$$

two interesting special cases are given by

$$XF(z, w) = (z \otimes I - I \otimes w)F(z, w) = zF(z, w) - F(z, w)w, \tag{16}$$

and

$$XF(z, w) = (z \otimes w)F(z, w) = zF(z, w)w. \tag{17}$$

These special module structures are significant as they lead eventually to the analysis of the Sylvester and Stein equations.

The $\mathbb{F}[z, w]$-module structure on $X_{D(z) \otimes \overline{D}(w)}$ is defined, for

$$p(z, w) = \sum_{i=1}^{k} \sum_{j=1}^{l} p_{ij} z^{i-1} w^{j-1} \in \mathbb{F}[z, w]$$

by

$$p(z, w) \cdot Q(z, w) = \sum_{i=1}^{k} \sum_{j=1}^{l} p_{ij} \pi_{D(z) \otimes \overline{D}(w)} z^{i-1} Q(z, w) w^{j-1} \tag{18}$$

This implies that for $p(z, w) = \sum_{i=1}^{s} p_i(z) q_i(w) \in \mathbb{F}[z, w]$, we have

$$p(z, w) \cdot Q(z, w) = \sum_{i=1}^{s} \pi_{D(z) \otimes \overline{D}(w)} [(p_i(z)Q(z, w)q_i(w)] \tag{19}$$

It is easily shown that this is independent of the representation

$$p(z, w) = \sum_{i=1}^{s} p_i(z) q_i(w).$$

Similarly, we define an $\mathbb{F}[z, w]$-module structure on the tensored rational model $X^{D(z) \otimes \overline{D}(w)}$ by letting, for $p(z, w) = \sum_{i=1}^{k} \sum_{j=1}^{l} p_{ij} z^{i-1} w^{j-1} \in \mathbb{F}[z, w]$ and $H(z, w) \in X^{D(z) \otimes \overline{D}(w)}$

$$p(z, w) \cdot H(z, w) = \pi^{D(z) \otimes \overline{D}(w)} [\sum_{i=1}^{k} \sum_{j=1}^{l} p_{ij} z^{i-1} H(z, w) w^{j-1}] \tag{20}$$

With the $\mathbb{F}[z, w]$-module structure on $X_{D(z) \otimes \overline{D}(w)}$ and $X^{D(z) \otimes \overline{D}(w)}$, the multiplication map $D(z) \otimes \overline{D}(w) : X^{D(z) \otimes \overline{D}(w)} \longrightarrow X_{D(z) \otimes \overline{D}(w)}$ is an $\mathbb{F}[z, w]$-module isomorphism, i.e. we have $X_{D(z) \otimes \overline{D}(w)} \simeq X^{D(z) \otimes \overline{D}(w)}$.

If we specialize definition (18) to the polynomial $p(z, w) = z - w$, we get, with $Q(z, w) \in X_{D(z) \otimes_{\mathbb{F}} \overline{D}(w)}$,

$$(z - w) \cdot Q(z, w) = \pi_{D(z) \otimes_{\mathbb{F}} \overline{D}(w)} (zQ(z, w) - Q(z, w)w). \tag{21}$$

We refer to this as the **generalized Sylvester operator**. In fact, with $A \in \mathbb{F}^{p \times p}$ and $\overline{A} \in \mathbb{F}^{m \times m}$, if $D(z) = zI - A$ and $\overline{D}(w) = wI - \overline{A}$ then $Q(z, w) \in X_{D(z) \otimes \overline{D}(w)}$ if and only if $Q(z, w) \in \mathbb{F}^{p \times m}$, i.e. $Q(z, w)$ is a constant matrix. In that case we have $X_{D(z) \otimes_{\mathbb{F}} \overline{D}(w)} = \mathbb{F}^{p \times m}$ and

$$(z - w) \cdot Q \pi_{(zI-A) \otimes (wI-\overline{A})}(z - w)Q = AQ - Q\overline{A}, \tag{22}$$

which is the standard **Sylvester operator**. The equation $(z - w) \cdot Q = R$ reduces in this case to the **Sylvester equation** $AQ - Q\overline{A} = R$. An extensive study of the Sylvester equation and its solution by polynomial methods will be undertaken in Section 3.

2.4   Duality

We extend now to the context of tensored models the duality theory for polynomial models as developed in Fuhrmann [1981], which was based on the identification of the dual space to $\mathbb{F}[z]^m$ with $z^{-1}\mathbb{F}[[z^{-1}]]^m$.

In analogy with $(\mathbb{F}[z]^m)^* \simeq z^{-1}\mathbb{F}[[z^{-1}]]^m$, we can identify the dual space of $\mathbb{F}[z]^{p \times m}$ with the space $z^{-1}\mathbb{F}[[z^{-1}]]^{p \times m}$ by letting, for $H \in z^{-1}\mathbb{F}[[z^{-1}]]^{p \times m}$ and $P \in \mathbb{F}[z]^{p \times m}$,

$$[H, P] = (\operatorname{trace} \tilde{H}P)_{-1} = \operatorname{trace}(\tilde{H}P)_{-1}. \tag{23}$$

Here $(X)_{-1}$ denotes the residue of a Laurent series of X, i.e. for $X(z) = \sum_{i=-\infty}^{n} x_i z^i$ we let $(X)_{-1} = x_{-1}$ The availability of this pairing allows us to prove the following.

Proposition 2. *Let $D_1(z) \in \mathbb{F}[z]^{p \times p}$ and $D_2(z) \in \mathbb{F}[z]^{m \times m}$ be nonsingular. Then we have the identification*

$$(X_{\tilde{D}_1(z) \otimes \tilde{D}_2(z)})^* \simeq (\mathbb{F}[z]^{p \times m} / D_1(z)\mathbb{F}[z]^{p \times m} D_2(z))^* = X^{\tilde{D}_1(z) \otimes \tilde{D}_2(z)}, \tag{24}$$

where the dual space is defined with respect to the pairing defined for $H \in X^{D_1(z) \otimes D_2(z)}$ and $P(z) \in \mathbb{F}[z]^{p \times m}$ by (23).

We proceed to extend the duality theory to the context of polynomial spaces in two variables. To this end, we introduce in $\mathbb{F}((z^{-1}, w^{-1}))^{p \times m}$ a bilinear form by defining, for $G, H \in \mathbb{F}((z^{-1}, w^{-1}))^{p \times m}$

$$\begin{aligned}
[H, G] &= \operatorname{trace} \sum_{i=-\infty}^{\infty} \sum_{j=-\infty}^{\infty} \tilde{H}_{-i-1, -j-1} G_{ij} \\
&= \sum_{i=-\infty}^{\infty} \sum_{j=-\infty}^{\infty} \operatorname{trace} \tilde{H}_{-i-1, -j-1} G_{ij}
\end{aligned} \tag{25}$$

Note that the sum defining $[H, G]$ contains only a finite number of nonzero terms. Clearly the form defined in (25) is nondegenerate. It is easy to see that

$$(\mathbb{F}[z, w]^{p \times m})^{\perp} = \mathbb{F}[z, w^{-1}]]^{p \times m} + \mathbb{F}[[z^{-1}, w]^{p \times m}. \tag{26}$$

In particular, we have $\mathbb{F}[z, w]^{p \times m} \subset (\mathbb{F}[z, w]^{p \times m})^{\perp}$. The next result gives a concrete representation of the dual space of $\mathbb{F}[z, w]^{p \times m}$.

Proposition 3. *The dual space of* $\mathbb{F}[z, w]^{p \times m}$ *can be identified with*

$$(z^{-1}\mathbb{F}[[z^{-1}, w^{-1}]]w^{-1})^{p \times m}.$$

We proceed to give a concrete representation of the dual space to a tensored polynomial model. Given a subspace $\mathcal{V}$ of a linear space $\mathcal{X}$, we use the isomorphism $(\mathcal{X}/\mathcal{V})^* \simeq \mathcal{V}^\perp$, as well as the identity $(\mathcal{U} + \mathcal{V})^\perp = \mathcal{U}^\perp \cap \mathcal{V}^\perp$.

Proposition 4. *Let* $D(z) \in \mathbb{F}[z]^{p \times p}$ *and* $\overline{D}(w) \in \mathbb{F}[w]^{m \times m}$ *be nonsingular polynomial matrices. We have*

$$X^*_{D(z) \otimes \overline{D}(w)} = X^{\tilde{D}(z) \otimes \tilde{\overline{D}}(w)}. \tag{27}$$

2.5 Homomorphisms of Tensored Models

The central result in the theory of polynomial and rational models is the characterization of the module homomorphisms. Given $D(z) \in \mathbb{F}[z]^{p \times p}$ and $\overline{D}(z) \in \mathbb{F}[z]^{m \times m}$ nonsingular. With $D(z)$ we associate a polynomial model X_D and similarly for $\overline{D}(z)$. The homomorphisms from $X_{\overline{D}} \longrightarrow X_D$, are the maps intertwining $S_{\overline{D}}$ and S_D, i.e. satisfy $ZS_{\overline{D}} = S_D Z$ are of the form

$$Zf = \pi_D N f, \qquad f \in X_{\overline{D}} \tag{28}$$

where for some polynomial matrices $N, \overline{N} \in \mathbb{F}[z]^{p \times m}$ we have the intertwining relation $N\overline{D} = D\overline{N}$. The polynomial matrices N and $\overline{N}$ are uniquely determined by the homomorphism Z if we require that $D^{-1}N = \overline{N}\overline{D}^{-1}$ be strictly proper. Moreover Z is injective if and only if $\overline{D}, \overline{N}$ are right coprime and Z is surjective if and only if D, N are left coprime. This can be extended to the characterization of homomorphism between tensored models $X_{D_1(z) \otimes_{\mathbb{F}} \overline{D}_2(w)}$ and $X_{\overline{D}_1(z) \otimes_{\mathbb{F}} D_2(w)}$.

Theorem 1. *Given the two tensored polynomial models* $X_{D_1(z) \otimes_{\mathbb{F}} \overline{D}_2(w)}$ *and* $X_{\overline{D}_1(z) \otimes_{\mathbb{F}} D_2(w)}$, *then*

1. A map $Z : X_{D_1(z) \otimes_{\mathbb{F}} \overline{D}_2(w)} \longrightarrow X_{\overline{D}_1(z) \otimes_{\mathbb{F}} D_2(w)}$ *is an* $\mathbb{F}[z, w]$-*homomorphism if and only if there exist appropriately sized polynomial matrices* $N_1, \overline{N}_1, N_2, \overline{N}_2$ *satisfying*

$$\begin{aligned} \overline{N}_1(z)D_1(z) &= \overline{D}_1(z)N_1(z) \\ D_2(w)\overline{N}_2(w) &= N_2(w)\overline{D}_2(w) \end{aligned} \tag{29}$$

in terms of which Z *is given, for* $Q(z, w) \in X_{D_1(z) \otimes_{\mathbb{F}} \overline{D}_2(w)}$, *by*

$$\begin{aligned} ZQ(z, w) &= \pi_{\overline{D}_1(z) \otimes_{\mathbb{F}} D_2(w)}(\overline{N}_1(z) \otimes_{\mathbb{F}} N_2(w))Q(z, w) \\ &= \pi_{\overline{D}_1(z) \otimes_{\mathbb{F}} D_2(w)}\overline{N}_1(z)Q(z, w)\overline{N}_2(w) \end{aligned} \tag{30}$$

2. The map Z *defined by (29) and (30) is*
  a) injective if and only if D_1, N_1 *are right coprime and* D_2, N_2 *are left coprime.*
  b) surjective if and only if $\overline{D}_1, \overline{N}_1$ *are left coprime and* $\overline{D}_2, \overline{N}_2$ *are right coprime.*
  c) bijective if and only if both sets of coprimeness conditions hold.

The availability of isomorphisms for tensored models allows us a reduction in the complexity of their analysis. This is done by reducing both D and $\overline{D}$ to their respective Smith forms Δ and $\overline{\Delta}$, and leads to the isomorphism $X_{D(z)\otimes\overline{D}(w)} \simeq X_{\Delta(z)\otimes\overline{\Delta}(w)}$. The tensored model $X_{\Delta(z)\otimes\overline{\Delta}(w)}$ can be written symbolically as $(X_{d_i(z)\otimes\overline{d}_j(w)})$. Over an algebraically closed field $\mathbb{F}$, the irreducible monic polynomials are of the form $p(z) = (z - \alpha)$. Thus, taking the primary decomposition of the invariant factors, we can further reduce the analysis to tensor products of the form $X_{(z-\alpha)^p\otimes(w-\overline{\alpha})^m}$. Using translations, this can be further reduced to the study of tensor products of nilpotent models. Clearly, the Sylvester map in $X_{z^p\otimes w^m}$ is nilpotent and the cyclic decompositions of the two Sylvester maps are isomorphic as vector spaces. Thus it suffices to study the cyclic decomposition of $X_{z^p\otimes w^m}$. Here we refer also to the closely related work of Dirr, Helmke and Kleinsteuber [2006] that describes such tensor product decompositions via the Clebsch-Gordan decomposition from representation theory.

3 Applications

3.1 The Space of Intertwining Maps

Given linear transformations A and B, acting in linear spaces $\mathcal{X}, \mathcal{Y}$ respectively, we say that a linear transformation $Z : \mathcal{X} \longrightarrow \mathcal{Y}$ **intertwines** A and B if $ZA = BZ$. The set of all linear transformations intertwining A and B is a linear space which we denote by Intw (A, B). Thus an intertwining map Z is an $\mathbb{F}[z]$-module homomorphism with the module structures in $\mathcal{X}, \mathcal{Y}$ being those induced by A and B respectively. A special case of intertwining maps is the **commutant** $\mathcal{C}(T)$ of a linear transformation T, namely the set of all operators Z commuting with T, i.e. satisfying $ZT = TZ$. Thus $\mathcal{C}(T) = $ Intw (T, T). If A, B transform by similarity to PAP^{-1}, RBR^{-1}, then $ZA = BZ$ transforms into $(RZP^{-1})(PAP^{-1}) = (RBR^{-1})(RZP^{-1})$, i.e. we have Intw $(PAP^{-1}, RBR^{-1}) = R\,\text{Intw}\,(A, B)P^{-1}$. Now the map S_D is isomorphic to S_Δ where Δ is the Smith form of D. We can assume without loss of generality that $D = \text{diag}(d_1,\ldots,d_p)$ and $\overline{D} = \text{diag}(\overline{d}_1,\ldots,\overline{d}_m)$, where the d_i are the invariant factors of D ordered so that $d_i|d_{i-1}$ and similarly for $\overline{D}$.

We have the following theorem.

Theorem 2. *Given nonsingular $D(z) \in \mathbb{F}[z]^{p\times p}$ and $\overline{D}(z) \in \mathbb{F}[z]^{m\times m}$. Let $d_1,\ldots,d_p$ be the invariant factors of $D(z)$ ordered so that $d_i|d_{i-1}$ and $\overline{d}_1,\ldots,\overline{d}_m$, the invariant factors of $\overline{D}(z)$ similarly ordered. Let $\delta_i = \deg d_i$ and $\overline{\delta}_j = \deg \overline{d}_j$, $n = \sum_{i=1}^p \delta_i = \deg \det D$, $\overline{n} = \sum_{j=1}^m \overline{\delta}_j = \deg \det \overline{D}$ and $e_{ij} = d_i \wedge \overline{d}_j = g.c.d.(d_i, \overline{d}_j)$. Then*

1. There exists a linear isomorphism

$$X_{D(z)\otimes\overline{D}(z)}/(D(z)X_{I\otimes\overline{D}(z)} + X_{D(z)\otimes I}\overline{D}(z)) \simeq \text{Intw}\,(S_{\overline{D}}, S_D). \tag{31}$$

2. We have

$$\dim X_{D(z)\otimes\overline{D}(z)} = m \deg \det D + p \deg \det \overline{D}. \tag{32}$$

3.

$$\dim[D(z)X_{I\otimes\overline{D}(z)} \cap X_{D(z)\otimes I}\overline{D}(z)] = \sum_{i=1}^p \sum_{j=1}^m \deg e_{ij}. \tag{33}$$

4.

$$\dim[D(z)X_{I\otimes\overline{D}(z)} + X_{D(z)\otimes I}\overline{D}(z)] = nm + p\overline{n} - \sum_{i=1}^{p}\sum_{j=1}^{m}\deg e_{ij}. \qquad (34)$$

5. We have

$$\dim\operatorname{Intw}(S_{\overline{D}}, S_D) = \sum_{i=1}^{p}\sum_{j=1}^{m}\deg e_{ij}. \qquad (35)$$

Corollary 1

1. *Given nonsingular $D(z) \in \mathbb{F}[z]^{p\times p}$ and $\overline{D}(z) \in \mathbb{F}[z]^{m\times m}$ have the same nontrivial invariant factors d_i, ordered so that $d_i|d_{i-1}$. Let $\delta_i = \deg d_i$. Then we have*

$$\dim\operatorname{Intw}(S_{\overline{D}}, S_D) = \sum_{i}(2i - 1)\delta_i. \qquad (36)$$

2. *Let $A \in \mathbb{F}^{n\times n}$ have invariant factors $d_1, \ldots, d_n$ ordered so that $d_i|d_{i-1}$. Then*

$$\dim\mathcal{C}(A) = \sum_{i}(2i - 1)\delta_i. \qquad (37)$$

3. *Given two linear transformations A, B then $\operatorname{Intw}(A, B) = \{0\}$, i.e. there exist no nontrivial intertwining maps, if and only if the minimal polynomials, or equivalently the characteristic polynomials, of A, B are coprime.*

3.2   The Polynomial Sylvester Equation

We proceed now to a more detailed study of the Sylvester equation in the tensored polynomial model framework. We saw, in Section 2, that the classical Sylvester equation $AX - X\overline{A} = C$ corresponds to the equation

$$S_{z-w}Q(z, w) = R(z, w), \qquad (38)$$

with $Q, R \in X_{(zI-A)\otimes_{\mathbb{F}}(wI-\overline{A})}$ necessarily constant matrices.

Theorem 3. *Let $D(z) \in \mathbb{F}[z]^{p\times p}$ and $\overline{D}(w) \in \mathbb{F}[w]^{m\times m}$ be nonsingular. Defining the Sylvester operator $S : X_{D(z)\otimes\overline{D}(w)} \longrightarrow X_{D(z)\otimes\overline{D}(w)}$ by (21), then for $R(z, w) \in X_{D(z)\otimes\overline{D}(w)}$, we have*

1. *The Sylvester equation*

$$S_{z-w}Q(z, w) = R(z, w) \qquad (39)$$

is solvable if and only if there exists polynomial matrices $P(z), \overline{P}(w)$ for which

$$D(z)\overline{P}(z) - P(z)\overline{D}(z) - R(z, z) = 0. \qquad (40)$$

*We will refer to (40) as the **polynomial Sylvester equation**. In that case, the solution is given by*

$$Q(z, w) = \frac{D(z)\overline{P}(w) - P(z)\overline{D}(w) + R(z, w)}{z - w}. \qquad (41)$$

2. *The Sylvester equation (40) has a unique solution for every $R(z, w) \in X_{D(z)\otimes\overline{D}(w)}$ if and only if $d \wedge \overline{d} = 1$, where $d(z) = \det D(z)$ and $\overline{d}(z) = \det \overline{D}(z)$.*

3.3 Solving the Sylvester Equation

Our aim now is to use the analysis of the polynomial Sylvester equation in order to solve the standard Sylvester equation $AX - XB = C$ under the assumption that the characteristic polynomials of A and B are coprime. Most of the results presented in this section are not new, see de Souza and S.P. Bhattacharyya [1981] or Heinig and Rost [1984] and the further references in these papers. However, the method of proof seems to be new and has greater clarity as far as the presentation is concerned. Some of the technique employed, especially the analysis of the relation between Bezoutians and finite section Hankel matrices, have been obtained in Fuhrmann and Helmke [1989], see also Fuhrmann [1996].

Before proceeding, we introduce some notation and recall some known results. We will denote by $E_{ij} \in \mathbb{F}^{p \times m}$ the matrix whose ij entry is one and all other entries are zero. Given a polynomial $q(z) = z^n + q_{n-1}z^{n-1} + \cdots + q_0$, we define the companion matrices, using Kalman's notation, by

$$
C_q^\sharp = \begin{pmatrix} 0 & & & -q_0 \\ 1 & & & \cdot \\ & \cdot & & \cdot \\ & & \cdot & \cdot \\ & & 1 & -q_{n-1} \end{pmatrix}, \qquad
C_q^\flat = \begin{pmatrix} 0 & 1 & & \\ & & \cdot & & \cdot \\ & & & \cdot & \\ & & & & 1 \\ -q_0 & \cdots & & -q_{n-1} \end{pmatrix}. \tag{42}
$$

For the polynomial model X_q we single out two bases, the **standard basis**, namely $\mathcal{B}_{st} = \{1, z, \ldots, z^{n-1}\}$ and the **control basis**, namely $\mathcal{B}_{co} = \{e_1, \ldots, e_n\}$, where the polynomials e_i are defined by

$$
e_i(z) = z^{n-i} + q_{n-1}z^{n-i-1} + \cdots + q_i.
$$

The control basis for $X_{\overline{q}}$ will be denoted by $\mathcal{B}_{\overline{co}}$. It is known, see Fuhrmann [1996], that the companion matrices in (42) are the matrix representations of S_q with respect to the standard basis and the control basis respectively, i.e. we have $C_q^\sharp = [S_q]_{st}^{st}$ and $C_q^\flat = [S_q]_{co}^{co}$. Given the monic polynomial $q(z) = z^n + q_{n-1}z^{n-1} + \cdots + q_0$ and $\overline{q}(z) = z^n + \overline{q}_{n-1}z^{n-1} + \cdots + \overline{q}_0$, we define the upper triangular Hankel matrix H_q by

$$
H_q = B(q, 1) = [I]_{co}^{st} = \begin{pmatrix} q_1 & \cdot\cdot & q_{n-1} & 1 \\ \cdot & \cdot\cdot & \cdot & \\ \cdot & \cdot\cdot & & \\ q_{n-1} & \cdot & & \\ 1 & & & \end{pmatrix}. \tag{43}
$$

The Hankel matrix $H_{\overline{q}}$ is analogously defined.

Given the Sylvester equation $AX - XB = C$ which we want to solve, we begin by solving a special Sylvester equation, namely

$$
C_q^\sharp Y - Y C_{\overline{q}}^\flat = E_{11} \tag{44}
$$

To begin with, we assume that $q, \overline{q}$ are coprime polynomials of the same degree n. Thus $Y \in \mathbb{F}^{n \times n}$. Using the isomorphism

$$
\mathbb{F}^{n \times n} \simeq X_{q(z) \otimes \overline{q}(w)}, \tag{45}
$$

given by

$$(a_{ij}) \mapsto \sum_{i=1}^{n}\sum_{j=1}^{n} a_{ij} z^{i-1} w^{j-1} \tag{46}$$

we can rewrite equation (44) as

$$\pi_{q(z)\otimes\overline{q}(w)}(z-w)f(z,w) = 1. \tag{47}$$

If $p, \overline{p}$ solve the polynomial Sylvester equation (40), i.e. $q(z)\overline{p}(z) - p(z)\overline{q}(z) + 1 = 0$, then this is equivalent to the Bezout equation

$$p(z)\overline{q}(z) - q(z)\overline{p}(z) = 1. \tag{48}$$

Since $y(z,w) \in X_{q(z)\otimes\overline{q}(w)}$, we have an expansion

$$y(z,w) = \frac{q(z)\overline{p}(w) - p(z)\overline{q}(w) + 1}{z-w} = \sum_{i=1}^{n}\sum_{j=1}^{n} \eta_{ij} z^{i-1} w^{j-1}. \tag{49}$$

With $Y = (\eta_{ij})$, we have a solution of (44).

We recall that, given a polynomial $a(z)$ of degree n, the **reverse polynomial** is defined by $a^{\sharp}(z) = a(z^{-1})z^n$.

We state now the main result on the solvability of the Sylvester equation.

Theorem 4. *Let $q, \overline{q}$ be coprime polynomials of degree n. Let $p, \overline{p}$ be the unique solutions of the scalar polynomial Sylvester equation*

$$q(z)\overline{p}(z) - p(z)\overline{q}(z) + 1 = 0, \tag{50}$$

satisfying $\deg p < \deg q$ *and* $\deg \overline{p} < \deg \overline{q}$. *Then*

1. Let $Y = (\eta_{ij})$ be defined by

$$\frac{q(z)\overline{p}(w) - p(z)\overline{q}(w) + 1}{z-w} = \sum_{i=1}^{n}\sum_{j=1}^{n} \eta_{ij} z^{i-1} w^{j-1}, \tag{51}$$

then Y is the unique solution of the Sylvester equation

$$C_q^{\sharp} Y - Y C_{\overline{q}}^{\flat} = E_{11} \tag{52}$$

2. We have the following identification of the matrix Y solving (52), namely

$$Y = [p(S_q)]_{co}^{st}. \tag{53}$$

3. Let $Z = (\zeta_{ij})$ be defined by

$$\frac{q(z)\overline{p}(w) - p(z)\overline{q}(w) + 1}{z-w} = \sum_{i=1}^{n}\sum_{j=1}^{n} \zeta_{ij} e_i(z)\overline{e}_j(w), \tag{54}$$

then Z is the unique solution of the Sylvester equation

$$C_q^{\flat} Z - Z C_{\overline{q}}^{\sharp} = E_{nn}. \tag{55}$$

Moreover, we have

$$Z = [p(S_q)]_{st}^{co}. \tag{56}$$

4. The solutions Y and Z of (54) and (55) respectively are related via

$$Y = H_q Z H_{\overline{q}}. \tag{57}$$

5. Let $p, \overline{p}$ be the solutions of the polynomial Sylvester equation (50). Writing $\frac{p(z)}{q(z)} = \sum_{i=1}^{\infty} \frac{g_i}{z^i}$ and similarly $\frac{\overline{p}(z)}{\overline{q}(z)} = \sum_{i=1}^{\infty} \frac{\overline{g}_i}{z^i}$, we define the finite section Hankel matrices by

$$H_{p/q}^{(n)} = \begin{pmatrix} g_1 & \cdots & g_n \\ \cdot & \cdot\cdot\cdot & \cdot \\ \cdot & \cdot\cdot\cdot & \cdot \\ \cdot & \cdot\cdot\cdot & \cdot \\ g_n & \cdots & g_{2n-1} \end{pmatrix}, \qquad H_{\overline{p}/\overline{q}}^{(n)} = \begin{pmatrix} \overline{g}_1 & \cdots & \overline{g}_n \\ \cdot & \cdot\cdot\cdot & \cdot \\ \cdot & \cdot\cdot\cdot & \cdot \\ \cdot & \cdot\cdot\cdot & \cdot \\ \overline{g}_n & \cdots & \overline{g}_{2n-1} \end{pmatrix} \tag{58}$$

Then we have

$$H_{p/q}^{(n)} = H_{\overline{p}/\overline{q}}^{(n)}. \tag{59}$$

6. The matrix Z defined by (54) is an $n \times n$ Hankel matrix having the representations

$$Z = B(q, \overline{q})^{-1} = H_{p/q}^{(n)} = H_{\overline{p}/\overline{q}}^{(n)}. \tag{60}$$

and

$$Z = H_q^{-1} p(C_q^{\sharp}) = p(C_q^{\flat}) H_q^{-1}. \tag{61}$$

With Z the solution of (55),
7. We define the $n \times n$ shift S and transposition matrix J by

$$S = \begin{pmatrix} 0 & 1 & \cdot & \cdot & \cdot & 0 \\ & \cdot & \cdot & 1 & \cdot & \cdot & \cdot \\ & \cdot & \cdot & \cdot & \cdot & \cdot \\ & \cdot & \cdot & \cdot & 1 & 0 \\ & \cdot & \cdot & \cdot & \cdot & 1 \\ 0 & \cdot & \cdot & \cdot & \cdot & 0 \end{pmatrix}, \qquad J = \begin{pmatrix} 0 & \cdots & 1 \\ 0 & \cdot\cdot & 1 & 0 \\ 0 & \cdots & 0 \\ 0 & 1 & \cdot\cdot & 0 \\ 1 & \cdots & 0 \end{pmatrix} \tag{62}$$

Then, for the solution Y of (52), we have the following representation

$$Y = - \left[q(\tilde{S}) \overline{p}^{\sharp}(S) - p(\tilde{S}) \overline{q}^{\sharp}(S) \right] J \tag{63}$$

*We will refer to this as the **Gohberg-Semencul** formula.*

3.4   Invariant Factors of the Sylvester Map

We saw already that the study of the generalized Sylvester map is reducible to the study of nilpotent scalar tensored polynomial models of the form $X_{z^p \otimes w^m}$. This allows us to study the Sylvester map by polynomial techniques.

Theorem 5. *Given integers p, m with $p \leq m$. Define $S_{z+w} : \mathbb{F}_{p,m}[z, w] \longrightarrow \mathbb{F}_{p,m}[z, w]$ by*

$$S_{z+w} f(z, w) = z f(z, w) + f(z, w) w \mod (z^p, w^m), \tag{64}$$

for $f(z, w) \in \mathbb{F}_{p,m}[z, w]$. Then

1. We have

$$\dim \operatorname{Ker} S_{z+w} = p. \tag{65}$$

and a basis for $\operatorname{Ker} S_{z+w}$ is given by the vectors $h^{(1)}, \ldots, h^{(p)}$ defined by

$$h^{(k)} = \sum_{\nu=0}^{p-k} (-1)^\nu z^{k-1+\nu} w^{m-1-\nu}. \tag{66}$$

2. S_{z+w} has p nontrivial invariant factors.

3. Let $\alpha_1, \ldots, \alpha_k$ be a nonzero solution of the equation $P\alpha = 0$, where P is the full rank $(k-1) \times k$ matrix defined by

$$P_{ij} = \binom{p + m - 2k + 1}{(p - k + 1) - (i - 1) + (j - 1)} \tag{67}$$

Define, for $k = 1, \ldots, p$,

$$g^{(k)} = \sum_{i=1}^{k} \alpha_i z^{i-1} w^{k-i} \tag{68}$$

Then

$$\mathcal{V}_k = \operatorname{span} \{ S_{z+w}^i g^{(k)} | i = 0, \ldots, p + m - 2k + 1 \} \tag{69}$$

is a cyclic invariant subspace of dimension $p + m - 2k + 1$, i.e. we have

$$S_{z+w}^{p+m-2k+1} g^{(k)} = 0 \tag{70}$$

and

$$S_{z+w}^{p+m-2k} g^{(k)} \neq 0. \tag{71}$$

4. A basis for $\mathbb{F}_{p,m}[z, w]$ is given by $\{ S_{z+w}^j g^{(k)} | k = 1, \ldots, p; j = 0, \ldots, p + m - 2k \}$.

5. We have the direct sum decomposition

$$\mathbb{F}_{p,m}[z, w] = \mathcal{V}_1 \oplus \cdots \oplus \mathcal{V}_p. \tag{72}$$

6. The invariant factors of S_{z+w} are $z^{p+m-2j+1}$, $j = 1, \ldots, p$.

The analysis of the Sylvester map based on a square $p \times p$ nilpotent matrix N_p is of course a special case of Theorem 5. However, in this case the space $\mathbb{F}^{p \times p}$ decomposes into the subspace of symmetric matrices $\mathbb{F}_{sym}^{p \times p}$ and the subspace of antisymmetric matrices $\mathbb{F}_{asym}^{p \times p}$, respectively. Each of these subspaces is invariant under the Sylvester map. The following theorem analyses these direct sum decompositions. We use again the isomorphism (45) to go over from matrix to polynomial equations which are easier to handle. The Sylvester map is now represented by S_{z+w} acting in $\mathbb{F}_{p,p}[z, w]$.

Theorem 6

1. *Define the vectors* $h^{(1)}, \ldots, h^{(p)}$ *by*

$$h^{(k)}(z, w) = \sum_{\nu=0}^{p-k} (-1)^{\nu} z^{k-1+\nu} w^{p-1-\nu}, \tag{73}$$

then $h^{(k)}$ *is a symmetric polynomial if and only if* $p-k$ *is even and an antisymmetric polynomial if and only if* $p - k$ *is odd.*

2. *The vectors* $h^{(1)}, \ldots, h^{(p)}$ *form a basis for* Ker S_{z+w}.

3. *Define the vectors* $g^{(k)}, k = 1, \ldots, p$ *by*

$$g^{(k)} = \begin{cases} z^{\frac{k-1}{2}} w^{\frac{k-1}{2}} & k \text{ odd} \\[2ex] z^{\frac{k}{2}} w^{\frac{k}{2}-1} - z^{\frac{k}{2}-1} w^{\frac{k}{2}} & k \text{ even} \end{cases} \tag{74}$$

Then the subspaces $\mathcal{V}_k = \text{span } \{S_{z+w}^j g^{(k)} | j = 0, \ldots, 2p-2k\}$ *are cyclic invariant subspaces.*

4. *The subspaces* $\mathcal{V}_1, \mathcal{V}_3, \ldots$ *contain only even polynomials whereas* $\mathcal{V}_2, \mathcal{V}_4, \ldots$ *contain only odd polynomials*

5. *We have the direct sum decomposition*

$$\mathbb{F}_{p,p}[z, w] = \mathbb{F}_{p,p}^{sym}[z, w] \oplus \mathbb{F}_{p,p}^{asym}[z, w]. \tag{75}$$

Furtermore, we have the finer direct sum decompositions

$$\mathbb{F}_{p,p}^{sym}[z, w] = \begin{cases} \oplus_{i=1}^{\frac{p+1}{2}} \mathcal{V}_{2i-1} & p \text{ odd} \\[2ex] \oplus_{i=1}^{\frac{p}{2}} \mathcal{V}_{2i-1} & p \text{ even} \end{cases} \tag{76}$$

and

$$\mathbb{F}_{p,p}^{asym}[z, w] = \begin{cases} \oplus_{i=1}^{\frac{p-1}{2}} \mathcal{V}_{2i} & p \text{ odd} \\[2ex] \oplus_{i=1}^{\frac{p}{2}} \mathcal{V}_{2i} & p \text{ even} \end{cases} \tag{77}$$

6. *The invariant factors of the sylvester map are* $z^{2j-1}, j = 1, \ldots, p.$

Theorem 5 leads directly to the analysis of a special case, namely that of the Lyapunov map. Let $\mathbb{F}_{sym}^{p \times p}$, $\mathbb{F}_{asym}^{p \times p}$ be the subspaces of $\mathbb{F}^{p \times p}$ of symmetric and antisymmetric matrices respectively. Given $A \in \mathbb{F}^{p \times p}$, we define the **Lyapunov map** $L_A : \mathbb{F}_{sym}^{p \times p} \longrightarrow \mathbb{F}_{sym}^{p \times p}$ by $L_A = S_{A,\tilde{A}} | \mathbb{F}_{sym}^{p \times p}$ or equivalently by

$$L_A(X) = AX + X\tilde{A}, \qquad X \in \mathbb{F}_{sym}^{p \times p} \tag{78}$$

and the **complementary Lyapunov map** $K_A : \mathbb{F}_{asym}^{p \times p} \longrightarrow \mathbb{F}_{asym}^{p \times p}$ by $K_A = S_{A,\tilde{A}} | \mathbb{F}_{asym}^{p \times p}$ or equivalently by

$$K_A(X) = AX + X\tilde{A}, \qquad X \in \mathbb{F}_{asym}^{p \times p}. \tag{79}$$

We have the following.

Theorem 7. *Let $N_p \in \mathbb{F}^{p \times p}$ be the nilpotent matrix defined above. Then*

1. We have

$$\dim \operatorname{Ker} L_N = \begin{cases} \frac{p}{2} & p \ even \\ \frac{p+1}{2} & p \ odd \end{cases} \tag{80}$$

2. We have

$$\dim \operatorname{Ker} K_N = \begin{cases} \frac{p}{2} & p \ even \\ \frac{p-1}{2} & p \ odd \end{cases} \tag{81}$$

3. The Lyapunov map L_N has $p/2$ invariant factors when p is even and $(p+1)/2$ invariant factors when p is odd and they are given by

$$\begin{cases} \{z^{4j-1}\}_1^{p/2} & p \ even \\ \{z^{4j-3}\}_1^{(p+1)/2} & p \ odd \end{cases} \tag{82}$$

4. The complementary Lyapunov map K_N has $p/2$ invariant factors when p is even and $(p-1)/2$ invariant factors when p is odd and they are given by

$$\begin{cases} \{z^{4j-3}\}_1^{p/2} & p \ even \\ \{z^{4j-1}\}_1^{(p-1)/2} & p \ odd \end{cases} \tag{83}$$

References

[2006] G. Dirr, U. Helmke and M. Kleinsteuber, "Lie algebra representations, nilpotent matrices and the C-numerical range", *Lin. Alg. Appl.*, 413, 534-566.

[1976] P.A. Fuhrmann, "Algebraic system theory: An analyst's point of view", *J. Franklin Inst.* 301, 521-540.

[1981] P.A. Fuhrmann, "Duality in polynomial models with some applications to geometric control theory", *IEEE Trans. Aut. Control*, AC-26, (1981), 284-295.

[1996] P.A. Fuhrmann, *A Polynomial Approach to Linear Algebra*, Springer Verlag, New York.

[1984] G. Heinig and K. Rost, *Algebraic Methods for Toeplitz-like Matrices and Operators*, Akademie-Verlag, Berlin.

[1989] U. Helmke and P.A. Fuhrmann, "Bezoutians", *Lin. Alg. Appl.*, vols. 122-124, 1039-1097.

[1998] U. Helmke and P.A. Fuhrmann, "Tangent spaces of rational functions", *Lin. Alg. Appl.*, vol. 271, 1-40.

[1969] R.E. Kalman, "Algebraic characterization of polynomials whose zeros lie in algebraic domains", *Proc. Nat. Acad. Sci.*, 64, 818-823.

[1981] E. de Souza and S.P. Bhattacharyya, "Controllability, observabiliy and the solution of $AX - XB = C$", *Lin. Alg. Appl.*, vol. 39, 167-181.

[1992] J.C. Willems and P.A. Fuhrmann, "Stability theory for high order systems", *Lin. Alg. Appl.*, 167, 131-149.

Distances Between Time-Series and Their Autocorrelation Statistics

Tryphon T. Georgiou

Department of Electrical and Computer Engineering,
University of Minnesota Minneapolis, MN 55455
tryphon@umn.edu

This paper is dedicated to Giorgio Picci on the occasion of his sixty-fifth birthday

Summary. We begin with an interpretation of the L_1-distance between two power spectral densities and then, following an analogous rationale, we develop a natural metric for quantifying distance between respective covariance matrices.

1 Introduction

Consider two discrete-time, stationary, zero-mean, (real-valued for notational convenience) random processes $\mathbf{y}_k$ and $\hat{\mathbf{y}}_k$ ($k \in \mathbb{Z}$) having power spectral densities $f_{\mathbf{y}}(\theta)$ and $f_{\hat{\mathbf{y}}}(\theta)$ ($\theta \in [-\pi, \pi]$), and autocorrelation functions R_ℓ and $\hat{R}_\ell$ ($\ell \in \mathbb{Z}$), respectively, i.e.,

$$R_\ell = E\{\mathbf{y}_k \mathbf{y}_{k+\ell}\} = \frac{1}{2\pi} \int_{-\pi}^{\pi} f(\theta) e^{-j\ell\theta} d\theta,$$

and similarly for the "hatted" quantities. When the power spectrum contains a singular part, then $f(\theta)d\theta$ needs to be replaced by a non-negative finite spectral measure $d\mu(\theta)$.

We are interested in quantifying the distance between respective spectra and statistics for two such random process $\mathbf{y}_k$ and $\hat{\mathbf{y}}_k$. When two vectors

$$\mathrm{R}_n := \begin{bmatrix} R_0\ R_1\ \ldots\ R_{n-1} \end{bmatrix}, \text{ and}$$
$$\hat{\mathrm{R}}_n := \begin{bmatrix} R_0\ R_1\ \ldots\ R_{n-1} \end{bmatrix}$$

of autocorrelation samples are available and need to be compared, one may use any metric in $\mathcal{R}^n$ for that purpose, as for instance $\|\mathrm{R}_n - \hat{\mathrm{R}}_n\|_2 = \sqrt{\sum_k (R_k - \hat{R}_k)^2}$. However, we are not aware of any significance that can be attached to such a distance other than the fact that it is a metric in $\mathcal{R}^n$. Our goal in this paper, is to seek a metric which can be physically motivated.

Similarly, if we are to compare $f_{\mathbf{y}}(\theta)$ and $f_{\hat{\mathbf{y}}}(\theta)$, it appears difficult to motivate the use of an L_2-distance $\|f_{\mathbf{y}}(\theta) - f_{\hat{\mathbf{y}}}(\theta)\|_2$. For one thing, the L_2-distance cannot be generalized to deal with spectral measures when singular parts are present. There are certainly other alternatives. In the speech processing literature in particular there is a plethora of distances that, however, are not metrics [6] but have been motivated by specific needs. Function theoretic alternatives that one can use (e.g., L_p-norms, etc.)

A. Chiuso et al. (Eds.): Modeling, Estimation and Control, LNCIS 364, pp. 113–122, 2007.

including Wasserstein-like transportation measures typically lack a physical interpretation. In a recent study [4,5] a pseudometric was constructed as a geodesic between spectral densities/measures with respect to a rather natural Riemannian metric —this metric quantifies the degradation of predictive-error variance when the predictor is designed based on the wrong choice between two alternatives and the geometry is, in essence, Euclidean but only after we transform spectral densities using the logarithmic map.

In the current paper we focus on the L_1 distance

$$\|f_{\mathbf{y}}(\theta) - f_{\hat{\mathbf{y}}}(\theta)\|_1 := \frac{1}{2\pi} \int_{-\pi}^{\pi} |f_{\mathbf{y}}(\theta) - f_{\hat{\mathbf{y}}}(\theta)| \, d\theta$$

which has also a rather natural interpretation. After a brief discussion of the relevance of the L_1-distance, following a similar rationale, we will develop an analogous metric between finite partial covariance data of the corresponding random processes.

2 Interpretation of the L_1 Distance

Given $\mathbf{y}_k$ and $\hat{\mathbf{y}}_k$ we postulate that there exist two random processes ψ_k and $\hat{\psi}_k$ so that

$$\mathbf{y}_k + \psi_k = \hat{\mathbf{y}}_k + \hat{\psi}_k. \tag{1}$$

Alternatively, we postulate that there exists a random process $\mathbf{z}_k$ and that the two original random processes relate to $\mathbf{z}_k$ via

$$\mathbf{y}_k = \mathbf{z}_k - \psi_k$$
$$\hat{\mathbf{y}}_k = \mathbf{z}_k - \hat{\psi}_k.$$

It is natural to seek such perturbations of minimal total combined variance

$$E\{\psi_k^2\} + E\{\hat{\psi}_k^2\} \tag{2}$$

that is sufficient to "reconcile" the two processes. The combined variance $E\{\psi_k^2\} + E\{\hat{\psi}_k^2\}$ represents the minimal amount of "energy" of perturbations in the two time-series that is needed to render the two indistinguishable. Intuitively, the minimal combined variance which is consistent with the available data quantifies the distance between the two.

Given $f_{\mathbf{y}}, f_{\hat{\mathbf{y}}}$, the optimal choice consists of random processes ψ_k and $\hat{\psi}_k$ such that y_k and $\hat{\psi}_k$ are independent, $\hat{\mathbf{y}}_k$ and $\hat{\psi}_k$ are also independent, and

$$f_{\psi}(\theta) = \begin{cases} f_{\hat{\mathbf{y}}}(\theta) - f_{\mathbf{y}}(\theta) & \text{if } f_{\hat{\mathbf{y}}}(\theta) - f_{\mathbf{y}}(\theta) \geq 0, \\ 0 & \text{otherwise,} \end{cases} \tag{3a}$$

$$f_{\hat{\psi}}(\theta) = \begin{cases} f_{\mathbf{y}}(\theta) - f_{\hat{\mathbf{y}}}(\theta) & \text{if } f_{\hat{\mathbf{y}}}(\theta) - f_{\mathbf{y}}(\theta) \leq 0, \\ 0 & \text{otherwise.} \end{cases} \tag{3b}$$

Then, the power spectrum of the "sum"

$$\mathbf{z}_k := \mathbf{y}_k + \psi_k = \hat{\mathbf{y}}_k + \hat{\psi}_k$$

is simply

$$f_z(\theta) := \max\{f_{\mathbf{y}}(\theta),\, f_{\hat{\mathbf{y}}}(\theta)\},\ \theta \in [-\pi, \pi],$$

and

$$d(f_{\mathbf{y}}, f_{\hat{\mathbf{y}}}) := E\{\psi_k^2\} + E\{\hat{\psi}_k^2\} \tag{4}$$

$$= \frac{1}{2\pi} \int_{-\pi}^{\pi} (f_\psi(\theta) + f_{\hat{\psi}}(\theta)) d\theta$$

$$= \frac{1}{2\pi} \int_{-\pi}^{\pi} |f_{\mathbf{y}}(\theta) - f_{\hat{\mathbf{y}}}(\theta)|\, d\theta$$

$$= \|f_{\mathbf{y}} - f_{\hat{\mathbf{y}}}\|_1. \tag{5}$$

Obviously, this construction extends in the obvious way to the case of not-necessarily absolutely continuous power spectra as well, and the metric includes the measure of any discrepancies between the singular parts of the two spectral measures.[1] Clearly, $d(f_{\mathbf{y}}, f_{\hat{\mathbf{y}}})$ is a metric as seen from (5). Building on a similar rationale, in the next section, we develop a metric for covariance matrices.

3   A Distance Between Covariance Matrices

It is often the case that only a finite segment of the autocorrelation function of time-series $\mathbf{y}_k$ and $\hat{\mathbf{y}}_k$ is available (and even then, possibly uncertain). Thus, it is of interest to consider distances between the partial autocorrelation statistics R and $\hat{\text{R}}$. To this end, we follow the dictum of the previous section and define as a distance measure the minimal combined variance of random processes ψ_k and $\hat{\psi}_k$ for which (1) holds. Naturally, since only partial covariance samples are available, the random processes $\psi_k, \hat{\psi}_k$ and (1) need to be consistent with these data.

First denote by

$$\mathbf{R}_n := \begin{bmatrix} R_0 & R_1 & \dots & R_{n-1} \\ R_{-1} & R_0 & \dots & R_{n-2} \\ \vdots & \vdots & \ddots & \vdots \\ R_{-(n-1)} & R_{-(n-2)} & \dots & R_0 \end{bmatrix}$$

the $n \times n$ covariance matrix corresponding to $\mathbf{y}_k$ and the covariance samples in R_n and, similarly, $\hat{\mathbf{R}}_n$ for the Toeplitz matrix based on $\hat{\text{R}}_n$. If $\mathbf{Q}_n, \hat{\mathbf{Q}}_n$ denote the corresponding finite Toeplitz covariances of the random processes ψ_k and $\hat{\psi}_k$, respectively, for which (1) holds, then

$$\mathbf{R}_n + \mathbf{Q}_n = \hat{\mathbf{R}}_n + \hat{\mathbf{Q}}_n \tag{6}$$

---

[1] It will be interesting to explore the practical significance of other possibilities for quantifying distance such as

$$\frac{\int_{-\pi}^{\pi} (f_\psi(\theta) + f_{\hat{\psi}}(\theta)) d\theta}{\int_{-\pi}^{\pi} f_{\mathbf{z}}(\theta) d\theta} \quad \text{or} \quad \int_{-\pi}^{\pi} \frac{f_\psi(\theta) + f_{\hat{\psi}}(\theta)}{f_{\mathbf{z}}(\theta)}\, \frac{d\theta}{2\pi}.$$

and the minimal sum $Q_0 + \hat{Q}_0$ of the respective variances can serve as a metric quantifying the distance between $\mathbf{R}$ and $\hat{\mathbf{R}}$.

The computation of $\mathbf{Q}_n, \hat{\mathbf{Q}}_n$ minimizing the sum $Q_0 + \hat{Q}_0$, or equivalently minimizing

$$\frac{1}{n}\text{trace}\,(\mathbf{Q}_n + \hat{\mathbf{Q}}_n),$$

is a convex problem –since the positivity constraints are convex. The Toeplitz structure is peripheral, and the idea of defining such metrics extends equally well to non-negative definite Hermitian matrices and to more general positive operators. For notational convenience we develop the framework in the context of real symmetric matrices.

So, we let

$$\mathcal{M} := \{M \in \mathcal{R}^{n \times n} \mid M = M' \geq 0\}$$

be the cone of non-negative symmetric $n \times n$-matrices and

$$\mathcal{T}_n := \{\mathbf{R} \in \mathcal{M} \mid \mathbf{R} \text{ is a Toeplitz matrix}\}$$

be the cone of non-negative Toeplitz matrices in $\mathcal{M}$. We address the case of matrices in $\mathcal{M}$ and define a suitable metric, which is then specialized to $\mathcal{T}_n$.

Proposition 1. *Let $M_1, M_2 \in \mathcal{M}$ and*

$$\tau(M_1, M_2) := \min \left\{ \frac{1}{n}\text{trace}\,(M) \mid M \in \mathcal{M}, \right.$$

$$M \geq M_1 \text{ and } M \geq M_2 \}.$$

Then

$$\delta(M_1, M_2) := 2\tau(M_1, M_2) - \text{trace}\,(M_1) - \text{trace}\,(M_2) \tag{7}$$

defines a metric on $\mathcal{M}$.

Proof. Given $M_1, M_2 \in \mathcal{M}$,

$$\mathcal{C}(M_1, M_2) := \{M \mid M \geq M_1 \text{ and } M \geq M_2\}$$

is a (convex) cone of non-negative definite matrices. It follows that there is an element $M_{12} \in \mathcal{C}(M_1, M_2)$ having minimal trace.

Clearly $\delta(M_1, M_2)$ is symmetric in its arguments and takes positive values unless $M_1 = M_2$, in which case $\delta(M_1, M_2) = 0$. Thus, we only need to prove the triangle inequality. Given $M_i \in \mathcal{M}$ for $i \in \{1, 2, 3\}$, we denote by M_{ik} corresponding minimal elements as above for $i, k \in \{1, 2, 3\}$, and we let

$$\Delta_{ik} := M_{ik} - M_k.$$

These matrices are non-negative by construction, the identities

$$M_i + \Delta_{ki} = M_k + \Delta_{ik}$$

hold, and

$$\delta(M_i, M_k) = \frac{1}{n}\text{trace}\,(\Delta_{ik} + \Delta_{ki})$$

for $i, k \in \{1, 2, 3\}$. But then,

$$M_1 + \Delta_{21} - \Delta_{12} = M_2$$
$$= M_3 + \Delta_{23} - \Delta_{32},$$

and hence,

$$M_1 + \Delta_{21} + \Delta_{32} = M_3 + \Delta_{23} + \Delta_{12}.$$

From the minimal property of Δ_{13} and of Δ_{31} with regard to having the least value for the combined trace so that $M_1 + \Delta_{31} = M_3 + \Delta_{13}$, it follows that

$$\text{trace}\,(\Delta_{13} + \Delta_{31}) \leq \text{trace}\,(\Delta_{21} + \Delta_{32} + \Delta_{23} + \Delta_{12}).$$

Therefore,

$$\delta(M_1, M_2) + \delta(M_2, M_3) = \delta(M_1, M_3),$$

which completes the proof. $\qquad\qquad\square$

We now observe that the steps of the proof of Proposition 1 permit incorporating linear constraints on the structure of elements of $\mathcal{M}$, such as the constraint of all matrices being Toeplitz. Hence, whereas $\delta(\cdot, \cdot)$ may be used directly as a distance measure between elements of $\mathcal{T}_n$, the corresponding minimal-trace perturbations Δ_{ik} may not belong to $\mathcal{T}_n$ in general. But, since the Toeplitz property is a linear constraint, we may define a completely analogous distance measure enforcing such perturbations (if so desired) to be Toeplitz.

Proposition 2. *Let $M_1, M_2 \in \mathcal{T}_n$ and*

$$\tau_{\mathcal{T}}(M_1, M_2) := \min\left\{\frac{1}{n}\text{trace}\,(M) \mid M \in \mathcal{T}_n, \text{ and}\right.$$

$$\left. M \geq M_1, \, M \geq M_2 \right\}.$$

Then

$$\delta_{\mathcal{T}}(M_1, M_2) := 2\tau_{\mathcal{T}}(M_1, M_2) - \text{trace}\,(M_1) - \text{trace}\,(M_2) \qquad (8)$$

defines a metric on $\mathcal{T}_n$.

Proof. The proof follows the steps of the proof of Proposition 1 verbatim, except for the fact that we now constraint all matrices to belong to $\mathcal{T}_n$. $\qquad\square$

Proposition 3. *Let $f_{\mathbf{y}}, f_{\hat{\mathbf{y}}}$ be power spectral densities, i.e., nonnegative and integrable on $[-\pi, \pi]$. Let as before $\mathbf{R}_n, \hat{\mathbf{R}}_n$ denote the corresponding Toeplitz covariance matrices, and let $n \in \{1, 2, \ldots\}$. Then*

$$\lim_{n \to \infty} \delta_{\mathcal{T}}(\mathbf{R}_n, \hat{\mathbf{R}}_n) = \|f_{\mathbf{y}} - f_{\hat{\mathbf{y}}}\|_1.$$

Proof. Clearly

$$\lim_{n \to \infty} \delta_{\mathcal{T}}(\mathbf{R}_n, \hat{\mathbf{R}}_n) \leq \|f_{\mathbf{y}} - f_{\hat{\mathbf{y}}}\|_1$$

since a choice of ψ_k and $\hat{\psi}_k$ with power spectra as in (3a-3b) gives rise to partial co-variance matrices $\mathbf{Q}_n$, $\hat{\mathbf{Q}}_n$, for all n, for which (6) holds. The respective 0th elements Q_0 and $\hat{Q}_0$ remain the same for all n and the left hand side is

$$\|f_{\mathbf{y}} - f_{\hat{\mathbf{y}}}\|_1 = Q_0 + \hat{Q}_0$$

since the power spectra in (3a-3b) have no overlap in their support.

To show the converse inequality, consider the sequence of minimizing $\mathbf{Q}_n$, $\hat{\mathbf{Q}}_n$. These are Toeplitz matrices with bounded entries (since their corresponding 0th element is bounded by $\|f_{\mathbf{y}} - f_{\hat{\mathbf{y}}}\|_1$). Each can be extended to an infinite Toeplitz matrix, and thereby, gives rise to power spectral densities q_n and $\hat{q}_n$ such that the first n Fourier coefficients of $f_{\mathbf{y}} + q_n$ and $f_{\hat{\mathbf{y}}} + \hat{q}_n$ coincide. The spectral densities q_n and $\hat{q}_n$ can be obtained from $\mathbf{Q}_n$, $\hat{\mathbf{Q}}_n$ by any particular positive extension, for instance a "maximum entropy" one. We can take those as pairs, and since they are bounded there exists a subsequence weakly convergent to possibly non-negative measures, $d\mu$ and $d\hat{\mu}$, such that

$$f_{\mathbf{y}} d\theta + d\mu = f_{\hat{\mathbf{y}}} d\theta + d\hat{\mu}$$

since their Fourier coefficients must coincide. If $d\mu$, $d\hat{\mu}$ do have singular parts then these should be identical and the absolutely continuous parts must balance as well, so there exist power spectral densities q and $\hat{q}$ such that

$$f_{\mathbf{y}} + q = f_{\hat{\mathbf{y}}} + \hat{q}. \tag{9}$$

But then,

$$\begin{aligned}
\lim_{n \to \infty} \delta_{\mathcal{T}}(\mathbf{R}_n, \hat{\mathbf{R}}_n) &\geq \lim_{n \to \infty} \frac{1}{2\pi} \int_{-\pi}^{\pi} (q_n(\theta) + \hat{q}_n(\theta)) d\theta \\
&= \frac{1}{2\pi} \int_{-\pi}^{\pi} (d\mu + d\hat{\mu}) \\
&\geq \frac{1}{2\pi} \int_{-\pi}^{\pi} (q + \hat{q}) d\theta \\
&\geq \|f_{\mathbf{y}} - f_{\hat{\mathbf{y}}}\|_1,
\end{aligned}$$

the last inequality from (9). $\square$

3.1   An Example

The metric $\delta_{\mathcal{T}}(\mathbf{R}_n, \hat{\mathbf{R}}_n)$ of the previous section admits no simple expression in terms of the respective eigenvalues. This should be contrasted with its limiting value $d(f_{\mathbf{y}}, f_{\hat{\mathbf{y}}})$ which is the L_1 distance between the corresponding power spectral densities. We highlight this with an example.

Let

$$\mathbf{R}_3 = \begin{bmatrix} 1 & 1 & 1 \\ 1 & 1 & 1 \\ 1 & 1 & 1 \end{bmatrix}$$

and

$$\hat{\mathbf{R}}_3 = \begin{bmatrix} 1 & 1/2 & 1/2 \\ 1/2 & 1 & 1/2 \\ 1/2 & 1/2 & 1 \end{bmatrix}.$$

Then, clearly,

$$\mathbf{Q}_3 = \begin{bmatrix} x & y & y \\ y & x & y \\ y & y & x \end{bmatrix}$$

and

$$\hat{\mathbf{Q}}_3 = \begin{bmatrix} x & v & v \\ v & x & v \\ v & v & x \end{bmatrix}$$

where

$$1 + y = 1/2 + v$$

and x is minimal subject to $\mathbf{Q}_3 \geq 0$ as well as $\hat{\mathbf{Q}}_3 \geq 0$. The last two inequalities imply that

$$0 \leq x \quad \text{as well as}$$
$$-\frac{1}{2}x \leq y, v \leq x.$$

It follows that the optimal choice (minimal x) is

$$\begin{aligned} x &= 1/3 \\ y &= -1/6 \\ v &= 1/3. \end{aligned}$$

Then

$$\delta_{\mathcal{T}}(\mathbf{R}_3, \hat{\mathbf{R}}_3) = 2x = 2/3,$$

while the respective eigenvalues are

$$\mathrm{spec}(\mathbf{R}_3) = \{3, 0, 0\} \text{ and}$$
$$\mathrm{spec}(\hat{\mathbf{R}}_3) = \{1, 1, 1/2\}.$$

There is no simple expression for $\delta_{\mathcal{T}}(\mathbf{R}_3, \hat{\mathbf{R}}_3)$ based solely on knowledge of $\mathrm{spec}(\mathbf{R}_3)$ and $\mathrm{spec}(\hat{\mathbf{R}}_3)$, although $\delta_{\mathcal{T}}(\mathbf{R}_3, \hat{\mathbf{R}}_3)$ can actually be expressed as the absolute sum of the eigenvalues of the difference $\mathbf{R}_3 - \hat{\mathbf{R}}_3$.

The covariance $\mathbf{R}_3$ has a unique extension and corresponds to a measure with unit weight at $\theta = 0$, i.e., a spectral line (Dirac delta) at $\theta = 0$. Assuming that $\hat{\mathbf{R}}_3$ originates from a spectral measure which has a similar weight of amplitude $1/2$ at $\theta = 0$ and a uniform absolutely continuous part of amplitude $1/2$, then

$$\|d\mu - d\hat{\mu}\|_1 = 1/2 + 1/2 = 1$$

adding the L_1-norm of the difference of the absolutely continuous parts with the absolute integral of the discrepancy between the two measures. We leave it as an exercise to the reader to verify that if $\hat{\mathbf{R}}_n$ is as we just assumed, namely $\hat{R}_k = 1/2$ for $k \geq 1$, and similarly, $R_k = 1$ for all k, then $\delta_{\mathcal{T}}(\mathbf{R}_n, \hat{\mathbf{R}}_n) \to 1$ as $n \to \infty$.

4 Approximating Sample Covariances

It is often the case that the autocovariance matrix $\mathbf{R}_n$ of a random process $\mathbf{y}_k$ is estimated in a way that does not guarantee this to be Toeplitz. For instance, it is quite common for $\mathbf{R}_n$ to be estimated by averaging observation samples

$$\hat{\mathbf{R}}_n = \frac{1}{N+1} \sum_{\ell=0}^{N} \begin{bmatrix} y_{1+\ell} \\ \vdots \\ y_{n+\ell} \end{bmatrix} \begin{bmatrix} y_{1+\ell} & \cdots & y_{n+\ell} \end{bmatrix}$$

The estimate $\hat{\mathbf{R}}_n$ is non-negative definite by construction, but may not be Toeplitz. Yet, for purposes of analysis it is often beneficial to approximate $\mathbf{R}_n$ by a Toeplitz one, or possibly, by one with additional structure (e.g., corresponding to a moving average process or, more generally, to the state of a known dynamical system). The problem of seeking such an approximant which is closest to $\mathbf{R}_n$ in $\delta(\cdot, \cdot)$, is readily solvable via convex optimization.

4.1 Comparison with the von Neumann Entropy

In [1], the question was raised as to what are appropriate ways to approximate a given sample covariance with one that abides by a known linear structure. It was proposed that the Kullback-Leibler-von Neuman distance

$$\mathbb{S}(\hat{\mathbf{R}}\|\mathbf{R}) := \operatorname{trace}\left(\hat{\mathbf{R}} \left(\log \hat{\mathbf{R}} - \log \mathbf{R} \right) \right)$$

provides a convenient convex functional for which the optimal approximant is uniquely defined. An academic example was presented in [1] which is recapitulated here as it helps underscore differences with approximation in the sense of minimizing $\delta(\hat{\mathbf{R}}, \mathbf{R})$.

Consider the positive-definite matrix below as the estimated value for a covariance matrix

$$\hat{\mathbf{R}}_3 = \frac{1}{3} \begin{bmatrix} 1.1 & .9 & 1.05 \\ .9 & .8 & .9 \\ 1.05 & .9 & 1.1 \end{bmatrix}.$$

The minimizer of

$$\{\mathbb{S}(\hat{\mathbf{R}}, \mathbf{R}) \mid \mathbf{R} \text{ being Toeplitz}, \mathbf{R} > 0, \operatorname{trace}(\mathbf{R}) = \operatorname{trace}(\hat{\mathbf{R}})\}$$

is unique (see [1]) and given by

$$\mathbf{R}_{3,\mathrm{vN}} = \frac{1}{3} \begin{bmatrix} 1 & .942 & .957 \\ .942 & 1 & .942 \\ .957 & .942 & 1 \end{bmatrix} .$$

It is interesting to point out the the closest Toeplitz matrix to $\hat{\mathbf{R}}$ in the least-squares sense fails to be positive-definite ([1], cf. [2]). On the other hand, the optimal approximant in $\delta(\cdot,\cdot)$-sense can be obtained by observation and is equal to

$$\mathbf{R}_{3,\delta} = \frac{1}{3} \begin{bmatrix} 1.1 & .9 & 1.05 \\ .9 & 1.1 & .9 \\ 1.05 & .9 & 1.1 \end{bmatrix} .$$

In the above, a second subscript indicates the sense in which the matrix approximates $\hat{\mathbf{R}}_3$. Obviously the traces of $\mathbf{R}_{3,\delta}$ and $\hat{\mathbf{R}}_3$ are not the same, in general. However, equality of the traces can be easily imposed as an added linear constraint.

4.2   Structured Covariances

For purposes of illustration, consider a moving average process

$$\mathbf{y}_k = \mathbf{w}_k + \mathbf{w}_{k-1} + \mathbf{w}_{k-2}$$

where $\mathbf{w}_k$ is a zero-mean, unit-variance, Gaussian white noise process. The autocorrelation sequence of $\mathbf{y}_k$ is

$$\begin{bmatrix} R_0 & R_1 & R_2 & R_3 & 0 & \ldots \end{bmatrix} = \begin{bmatrix} 3 & 2 & 1 & 0 & 0 & \ldots \end{bmatrix} .$$

Simulating $\mathbf{y}_k$ over a window $k \in \{0, 1, \ldots, 100\}$, and based on a particular such realization, the corresponding $n \times n$ sample covariance matrix, for $n = 5$, was computed to be

$$\hat{\mathbf{R}}_5 = \begin{bmatrix} 4.0362 & 2.9053 & 1.8043 & 0.4042 & 0.1718 \\ 2.9053 & 4.0547 & 2.9268 & 1.7945 & 0.3800 \\ 1.8043 & 2.9268 & 4.0792 & 2.9143 & 1.7733 \\ 0.4042 & 1.7945 & 2.9143 & 4.0819 & 2.9421 \\ 0.1718 & 0.3800 & 1.7733 & 2.9421 & 4.0237 \end{bmatrix} .$$

Obviously, this matrix is not Toeplitz due to the finiteness of the observation record. The closest Toeplitz approximant to $\hat{\mathbf{R}}_5$, in the sense of the metric $\delta(\cdot,\cdot)$, turns out to be

$$\mathbf{R}_{5,\mathrm{Toeplitz}} = \begin{bmatrix} 4.0677 & 2.9237 & 1.7912 & 0.3979 & 0.1822 \\ 2.9237 & 4.0677 & 2.9237 & 1.7912 & 0.3979 \\ 1.7912 & 2.9237 & 4.0677 & 2.9237 & 1.7912 \\ 0.3979 & 1.7912 & 2.9237 & 4.0677 & 2.9237 \\ 0.1822 & 0.3979 & 1.7912 & 2.9237 & 4.0677 \end{bmatrix}$$

for which

$$\delta(\hat{\mathbf{R}}_5, \mathbf{R}_{5,\mathrm{Toeplitz}}) = 0.0308.$$

Interestingly, $\mathbf{R}_{5,\text{Toeplitz}}$ does not correspond to a moving average process of order 2 (or even, of order $3, 4$) as it can be readily verified by the fact that the trigonometric polynomials, e.g.,

$$\sum_{k=-4}^{4} R_k e^{jk\theta}$$

takes negative values.

The set of covariance matrices which are generated by moving average processes of a given order, is convex and admits a characterization via a set of linear matrix inequalities ([8, 3]). Thus, the closest approximant to $\hat{\mathbf{R}}$ which corresponds to a moving average process of any given order can be readily computed. In particular, if we specify the order to be 2, then the optimal approximant to $\hat{\mathbf{R}}_5$ becomes

$$\mathbf{R}_{5,\text{MA}(2)} = \begin{bmatrix} 4.1184 & 2.7678 & 1.3497 & 0 & 0 \\ 2.7678 & 4.1184 & 2.7678 & 1.3497 & 0 \\ 1.3497 & 2.7678 & 4.1184 & 2.7678 & 1.3497 \\ 0 & 1.3497 & 2.7678 & 4.1184 & 2.7678 \\ 0 & 0 & 1.3497 & 2.7678 & 4.1184 \end{bmatrix}$$

for which

$$\delta(\hat{\mathbf{R}}_5, \mathbf{R}_{5,\text{MA}(2)}) = 0.5017.$$

References

1. T.T. Georgiou (2003) Toeplitz covariance matrices and the von Neumann relative entropy. In: K. Hashimoto, Y. Oishi, and Y. Yamamoto (eds) Control and Modeling of Complex Systems: Cybernetics in the 21st Century. MA: Birkhauser, Boston
2. T.T. Georgiou (2006) Relative Entropy and the multi-variable multi-dimensional Moment Problem, IEEE Trans. on Information Theory, 52(3): 1052-1066.
3. T.T. Georgiou (2006) Decomposition of Toeplitz matrices via convex optimization, IEEE Signal Processing Letters, 13: 537-540.
4. T.T. Georgiou (2007) Distances and Riemannian metrics for spectral density functions, IEEE Trans. on Signal Processing, to appear. Preprint available at: http://arxiv.org/abs/math/0608486/
5. T.T. Georgiou (2007) An intrinsic metric for power spectral density functions, IEEE Signal Processing Letters, to appear. Preprint available at: http://arxiv.org/abs/math/0607026/
6. R. Gray, A. Buzo, A. Gray, and Matsuyama (1980) Distortion measures for speech processing, IEEE Trans. on Acoustics, Speech, and Signal Proc., 28(4): 367-376.
7. P. Stoica and R. Moses (2005) Introduction to Spectral Analysis. Prentice Hall
8. P. Stoica, T. McKelvey, and J. Mari (2000) MA estimation in polynomial time, IEEE Trans. on Signal Processing, 48: 1999-2012.

Global Identifiability of Complex Models, Constructed from Simple Submodels*

Markus Gerdin[1], Torkel Glad[2], and Lennart Ljung[2]

[1] NIRA Dynamics AB, Gothenburg, Sweden
  `markus.gerdin@gmail.com`
[2] Linköping University, S-58183 Linköping Sweden
  `{torkel,ljung}@isy.liu.se`

To Giorgio, with admiration, on the occasion of his 65th birthday

Summary. It is a typical situation in modern modeling that a total model is built up from simpler submodels, or modules, for example residing in a model library. The total model could be quite complex, while the modules are well understood and analysed. A procedure to decide global parameter identifiability for such a collection of model equations of differential-algebraic nature is suggested. It is shown how to make use of the natural modularization of the model structure. Basically, global identifiability is obtained if and only if each module is identifiable, and the connecting signals can be retrieved from the external signals, without knowledge of the values of the parameters.

1 Introduction

Identifiability is a crucial concept in System Identification. It concerns the question of whether the parameters in a model structure can be uniquely retrieved from input-output data. Clearly, being able to assess the identifiability of a structure beforehand, without going through all the estimation labor would be a very helpful technique.

The literature on identifiability and techniques to check identifiability is extensive, see, e.g. [3], Chapter 4, [7, 8], [4]. Identifiability of linear black-box models in terms of canonical forms etc, is basically a solved problem. However even for linear models physically parameterized structures form a highly non-trivial challenge, see e.g. [5], and there are no efficient techniques other than in certain subclasses of problems. The idea of [4] to use differential algebra leads to a well defined algorithm for quite general structures, but it suffers from too high computational complexity in many realistic cases.

In this contribution we shall consider a common situation in defining model structures: Suppose that the model is made of from several interconnected modules. Each of these modules are simple and have well defined identifiability properties, in case they are encountered on their own. The modules are interconnected in well defined ways, but the interconnecting signals are not necessarily measured. The question is then how the

* This work has been supported by the Swedish Foundation for Strategic Research (SSF) through VISIMOD and ECSEL and by the Swedish Research Council (VR) which is gratefully acknowledged.

A. Chiuso et al. (Eds.): Modeling, Estimation and Control, LNCIS 364, pp. 123–133, 2007.

identifiability of the total model can be studied in terms of the module properties and their interconnections. This way of dealing with model structures is typical in modular-based (or object-oriented) modeling environments, such as MODELICA, see e.g. [1]. A more complete discussion on identifiablity of modular models described by algebraic differential equations (DAEs) is given in [2].

2   The Problem

Consider a model structure built up from a collection of sub-models, or *modules*, $M_k, k = 1, \ldots, m$. Each module is a model that describes the relationship between a number of connecting signals $w_{k,i}(t), i = 1, \ldots, n_k$. This relationship involves also a number of internal signal $\ell_{k,i}(t), 1 = 1, \ldots, p_k$ and a number of parameters $\theta_{k,i}, i = 1, \ldots, q_k$. The relations may be expressed as a collection of algebraic or differential equations. With p denoting the differentiation operator, the model for module M_k can be written

$$f_{k,i}(\ell_{k,1}(t), \ldots, \ell_{k,p_k}(t), w_{k,1}(t), \ldots, w_{k,n_k}(t),$$
$$\ldots, \theta_{k,1}, \ldots \theta_{k,q_k}, p) = 0, i = 1, \ldots, r_k$$

With obvious "vectorization" of $\ell_{k,\cdot}, w_{k,\cdot}, \theta_{k,\cdot}$ and $f_{k,\cdot}$ we will write the above equation as

$$f_k(\ell_k(t), w_k(t), \theta_k, p) = 0 \tag{1a}$$

The modules are then connected by describing how the connecting signals w_k interconnect and relate to globally external signals $z(t)$ ("inputs" and "outputs")

$$g(z(t), w_1(t), \ldots, w_m(t)) = 0 \tag{1b}$$

Typically the inputs and outputs will be equal to some of the connecting signals $w_{k,i}$. The distinction between inputs and outputs is immaterial for the present discussion. See Figure 1.

As a simple case, the reader may picture M_k as resistors, capacitors and inductors, with well known relationships (1a) between voltage drops, currents ($\ell_k(t), w_k(t)$) and component parameters θ_k (resistance, capacitance, inductance). An arbitrary RLC-circuit can then be defined by the interconnections (1b) between voltages and currents $w_k(t)$ following Kirchhoff's laws.

Now, the total model is given as (1) with $z(t)$ as external signals, also depicted in Figure 1. The identifiability question for this total model is

- *Given the external signals $z(t), t \in T$, is it possible to uniquely determine all the parameters $\theta_k, k = 1, m$?. That is, can the equations (1) be satisfied for different parameter values for a given $z(t), t \in T$?*

If it indeed is possible to uniquely retrieve the parameter values, we say that the model is *globally identifiable* for the given external signal. (See, e.g. [4] and [2] for more strict definitions).

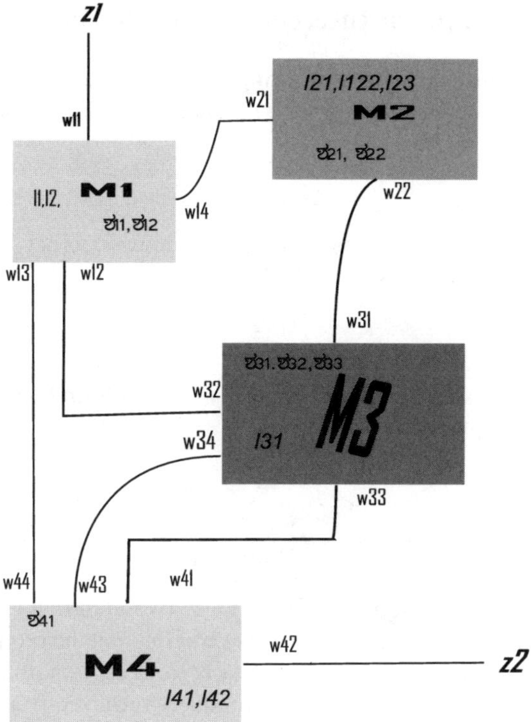

Fig. 1. Interconnected modules of submodels. $w_{k,i}$ is the i:th external connecting signal in module k and z are the global external signals. $\ell_{k,i}$ and $\theta_{k,i}$ (with a strange font) are internal signals and parameters in the modules.

The following result is plausible

Result 1. *The model (1) is globally identifiable if and only if*

a) The external signals $w_k(t)$ can be uniquely retrieved from $z(t)$ and (1), without knowledge of the parameter values θ_k.

b) Each module M_k is globally identifiable for retrieved external signal $w_k(t)$

A formal version of this result will be proved in the course of this contribution, but the result is not difficult to appreciate and understand intuitively. Clearly a) and b) will imply global identifiability, since with retrieved interconnecting signals w_k identifiability of the modules will guarantee that all parameters can be found uniquely. Conversely, if a module is not identifiable with known signals w_k it is clear that its parameters cannot be found when the signals are not known (or reconstructed). It is perhaps somewhat less obvious to realize that w_k can be uniquely retrieved for a globally identifiably model structure. However, the argument is as follows: Suppose the model structure is well defined, so that the model can be simulated for any set of parameter values. Then with the globally identified parameters inserted and the (input part of the) external signals $z(t)$ well defined, all interconnecting signals w_k will be generated in a unique way by such a simulation.

3   A Simple Example of Interconnected Modules

To illustrate the role of the interconnecting signals, let us consider an almost trivial example. Take a module as an integrator with unknown gain. The input to the integrator is w_1 and its output is w_2:

$$\dot{w}_2(t) = \theta_1 w_1(t) \tag{2a}$$

Clearly this module is globally identifiable with known external signals w_1, w_2. Let us now cascade two such modules:

$$\dot{w}_4(t) = \theta_2 w_3(t) \tag{2b}$$

For the cascaded system, we have input $z_1(t) = w_1(t)$ and output $z_2(t) = w_4(t)$ so the interconnecting equations are

$$z_1(t) = w_1(t) \tag{3a}$$
$$w_3(t) = w_2(t) \tag{3b}$$
$$z_2(t) = w_4(t) \tag{3c}$$

The model structure obtained by cascading these two identifiable modules is however not identifiable: From measuring only $z(t)$ we can find out the product $\theta_1 \cdot \theta_2$ but not the individual values. In view of Result 1 in the previous section, this must mean that the interconnecting signal $w_3 = w_2$ cannot be uniquely retrieved from $z(t)$, even knowing that the signals relate via integrators. In fact, we will know that w_2 is proportional to $\dot{z}_2(t)$ but have no way to find the coefficient of proportionality.

Now, if, say θ_1 is known to be $\theta_1 = 1$, the global structure should be identifiable. That means that $w_2(t)$ can be uniquely retrieved, but it is instructive to realize that it has to be done with care. Knowing the double integrator structure and the input $z_1(t)$ we can only conclude that

$$w_2(t) = \int z_1(s)ds + Constant$$

where the constant is due to initial conditions. So w_2 cannot be uniquely retrieved. However, knowing also $z_2(t)$ gives the relation

$$\theta_2 w_2(t) = \dot{z}_2(t) \tag{4}$$

From this we may obtain (differentiate (4), use $\dot{w}_2 = z_1$ and eliminate θ_2 from the obtained two equations):

$$\ddot{z}_2(t) w_2(t) = z_1(t) \dot{z}_2(t) \tag{5}$$

so, as long as $\ddot{z}_2(t)$ is nonzero we can find $w_2(t)$ uniquely. Here are no unknown constants nor initial conditions involved. The simple example also shows that some care must be exercised when examining the retrieval of the interconnecting signals. The condition that $\ddot{z}_2$ is nonzero means that z_1 and $\theta_1 \dot{\theta}_2$ must be nonzero. This is an example of an *excitation condition* on the (input) signals, and particular parameter values that typically is required to assure identifiability.

4 Preliminary Considerations and Tools

Our goal is to use Result 1 to study identifiability of complex model structures (1) in terms of identifiability of the modules M_k. In many applications these modules will be standardized in model libraries so the identifiability of these can be examined once and for all. The crux then is to establish condition a) in the Result, that the interconnecting signals can be uniquely retrieved from the external signals $z(t)$, and the model equations without knowledge of the parameter values. An obvious way is to find a parameter-free relation, such as (5) from which w can be determined from z. How can that be done?

For that we will use the tools of differential algebra, [6] as applied to identifiability in [4]. In short, the idea is that the original set of model equations can be transformed to a new set by differentiating, and performing simple algebraic operations. Under certain conditions the solutions to the new equation set will be identical to the solutions of the original one. So the identifiability analysis can be applied to the new set instead. It is then a matter to let the new set be as well suited as possible for such analysis.

Remark: Strictly speaking the formal result is limited to the case that the f_k are polynomial expressions. However, more general cases can also be handled. Such a case is given in Example 3 below.

More specifically, consider a module (1a):

$$f_k(\ell_k(t), w_k(t), \theta_k, p) = 0 \tag{6}$$

where f_k is a vector, so that the expression covers several model equations. By manipulating these equations with differential algebraic tools, they can be transformed to the following set

$$A_{k,1}(w_k, p), \ldots, A_{k,n_{A_k}}(w_k, p), \tag{7a}$$

$$B_{k,1}(w_k, \theta_{k,1}, p), B_{k,2}(w, \theta_{k,1}, \theta_{k,2}, p), \tag{7b}$$

$$B_{k,n_{\theta_k}}(w_k, \theta_{k,1}, \theta_{k,2}, \ldots, \theta_{k,n_{\theta_k}}, p),$$

$$C_{k,1}(w_k, \theta_k, \ell_k, p), \ldots, C_{k,n_{l_k}}(w_k, \theta_k, \ell_k, p). \tag{7c}$$

The A-equations are then just relationships between (derivatives of) the interconnecting signals w_k, not involving the internal variables ℓ_k nor the parameters. The B-equations will reveal whether the parameters θ_k are identifiable. The main result in [4] is that the module M_k is globally identifiable if and only if the B-equations have a linear regression form:

$$B_k = P_k(w_k, p)\theta_k - Q_k(w_k, p), \tag{8}$$

The excitation condition (see the end of Section 3) on w_k is then that P_k is invertible.

Example 1. **Capacitor:** Consider a capacitor described by the voltage drop w_1, current w_2 and capacitance θ_1. It is then described by (1a) with

$$f_1 = \begin{pmatrix} \theta_1 \dot{w}_1 - w_2 \\ \dot{\theta}_1 \end{pmatrix}. \tag{9}$$

If we consider only situations where $\dot{w}_1 \neq 0$ we get the following series of equivalences.

$$\theta_1 \dot{w}_1 - w_2 = 0, \quad \dot{\theta}_1 = 0, \quad \dot{w}_1 \neq 0$$

$$\Leftrightarrow$$

$$\theta_1 \dot{w}_1 - w_2 = 0, \quad \theta_1 \ddot{w}_1 - \dot{w}_2 = 0, \quad \dot{w}_1 \neq 0$$

$$\Leftrightarrow$$

$$\theta_1 \dot{w}_1 - w_2 = 0, \quad \theta_1 \dot{w}_1 \ddot{w}_1 - \dot{w}_1 \dot{w}_2 = 0, \quad \dot{w}_1 \neq 0$$

$$\Leftrightarrow$$

$$\theta_1 \dot{w}_1 - w_2 = 0, \quad w_2 \ddot{w}_1 - \dot{w}_1 \dot{w}_2 = 0, \quad \dot{w}_1 \neq 0 \quad (10)$$

With the notation (7) we thus have

$$A_{1,1} = w_2 \ddot{w}_1 - \dot{w}_1 \dot{w}_2 \tag{11a}$$

$$B_{1,1} = \theta_1 \dot{w}_1 - w_2 \tag{11b}$$

The capacitor is thus globally identifiable, provided $\dot{w}_1 \neq 0$.

This gives an idea of how to handle condition a) in Result 1: If we work with previously analyzed modules M_k in a model library, we have already their A-equations, (7a). We can dispense with the remaining B and C-equations. For global identifiability of the whole model we need to consider the total collection of A-equations together with the connecting equations (1b) and check whether all the interconnecting signals w can be retrieved from this set of equations,

$$A_{1,1}(w_1, p), \ldots, A_{m, n_{A_m}}(w_m, p) \tag{12a}$$

$$g(z(t), w(t)) = 0 \tag{12b}$$

knowing the external variables $z(t)$. Here (12b) is a more compact way of writing (1b).

5   Identifiability Analysis

In this section we shall illustrate how to use Result 1 for identifiability analysis based on submodels in a model libraray. We will use a minimal model library consisting of a resistor model, an inductor model, and a capacitor model. Note that these components have corresponding components for example within mechanics and fluid systems. Bond graphs make full use of such analogies: it follows that our simple model library is quite representative of more general cases. A capacitor model was described in Example 1. Similarly, we have

Example 2. **Inductor:** Next consider an inductor where w_2 is the current, w_1 the voltage and θ_1 the inductance. It is described by

$$\theta_1 \dot{w}_2 = w_1, \quad \dot{\theta}_1 = 0 \tag{13}$$

Calculations similar to those of the previous example show that this is equivalent to

$$\theta_1 \dot{w}_2 = w_1, \quad \ddot{w}_2 w_1 = \dot{w}_2 \dot{w}_1 \tag{14}$$

provided $\dot{w}_2 \neq 0$.

As discussed earlier, the transformation to (7) can always be performed for polynomial DAE. To show that calculations of this type in some cases also can be done for non-polynomial models, we consider a nonlinear resistor where the voltage drop is given by an arbitrary function.

Example 3. **Nonlinear Resistor:** Consider a nonlinear resistor with the equation

$$w_1 = R(w_2, \theta_1) \tag{15}$$

where it is assumed that the parameter θ_1 can be uniquely solved from (15) if the voltage w_1 and the current w_2 are known, so that

$$\theta_1 = \phi(w_1, w_2). \tag{16}$$

Differentiating (15) once with respect to time and inserting (16) gives

$$\dot{w}_1 = R_{w_2}\big(w_2, \phi(w_1, w_2)\big)\dot{w}_2 \tag{17}$$

which is a relation between the external variables w_1 and w_2. We use the notation R_x for the partial derivative of R with respect to the variables x. In the special case with a linear resistor, where $R = \theta_1 \cdot w_2$, this reduces to

$$\dot{w}_1 = \frac{w_1}{w_2}\dot{w}_2 \tag{18a}$$

$$\Leftrightarrow w_2\dot{w}_1 = w_1\dot{w}_2 \tag{18b}$$

(assuming $w_2 \neq 0$).

We shall here examine the identifiability of different connections of the components.

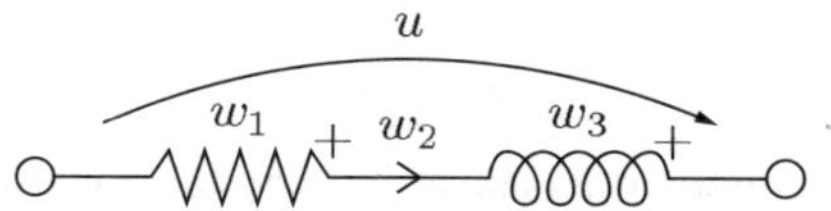

Fig. 2. A resistor and an inductor connected in series

Example 4. Consider a nonlinear resistor and an inductor connected in series where the current $w_2 = f$ and total voltage u are measured as shown in Fig. 2. Denote the voltage over the resistor with w_1 and the voltage over the inductor with w_3. Using Examples 2 and 3 we get the equations

$$\dot{w}_1 = R_{w_2}\big(w_2, \phi(w_1, w_2)\big)\dot{w}_2 \tag{19a}$$

$$\ddot{w}_2 w_3 = \dot{w}_2\dot{w}_3 \tag{19b}$$

for the components and the equation

$$w_1 + w_3 = u \tag{19c}$$

for the connections. Differentiating the last equation once gives

$$\dot{w}_1 + \dot{w}_3 = \dot{u}. \tag{19d}$$

The system of equations (19) (with w_1, $\dot{w}_1$, w_3, and $\dot{w}_3$ as unknowns) has the Jacobian

$$\begin{pmatrix} -R_{w_2,w_1}\dot{w}_2 & 1 & 0 & 0 \\ 0 & 0 & \ddot{w}_2 & -\dot{w}_2 \\ 1 & 0 & 1 & 0 \\ 0 & 1 & 0 & 1 \end{pmatrix} \tag{20}$$

where

$$R_{w_2,w_1} = \frac{\partial}{\partial w_1}\Big(R_{w_2}\big(w_2, \phi(w_1, w_2)\big) \Big). \tag{21}$$

The Jacobian has the determinant $-R_{w_2,w_1} \cdot \dot{w}_2^2 + \ddot{w}_2$, so the system of equations is solvable for most values of the external variables. This means that the system is identifiable.

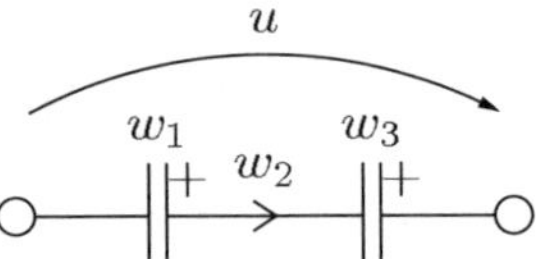

Fig. 3. Two capacitors connected in series

Example 5. Now consider two capacitors connected in series where the current $w_2 = f$ and total voltage u are measured as shown in Fig. 3. Denote the voltages over the capacitors with w_1 and w_3 respectively. Using Example 1 we get the equations

$$w_2\ddot{w}_1 = \dot{w}_1\dot{w}_2 \tag{22a}$$
$$w_2\ddot{w}_3 = \dot{w}_3\dot{w}_2 \tag{22b}$$

for the components. The connection is described by the equation

$$w_1 + w_3 = u. \tag{23}$$

These equations directly give that if

$$w_1(t) = \phi_1(t) \tag{24a}$$
$$w_3(t) = \phi_3(t) \tag{24b}$$

is a solution, then so are all functions of the form

$$w_1(t) = (1 + \lambda)\phi_1(t) \tag{25a}$$
$$w_3(t) = \phi_3(t) - \lambda\phi_1(t) \tag{25b}$$

for scalar λ. Since (11b) implies that the capacitance is an injective function (i.e. no two arguments give the same function value) of the derivative of the voltage, this shows that the system is not identifiable.

6   A Formal Theorem on Identifiability from Sub-models

We shall in this section prove a formal version of Result 1.

Consider a model structure consisting of m interconnected modules M_k, (1a). Assume that the different modules do not have common parameters, and that the differential algebraic equations (1) are polynomial in their arguments. Assume also that all modules are identifiable if their connecting variables w_i are measured. This means, that given measurements of

$$w_i \qquad i = 1, \ldots, m \tag{26}$$

the unknown parameters θ can be computed uniquely from the B-polynomials in (7b). When examining identifiability of the connected system it is not a big restriction to assume that the individual components are identifiable since information is removed when not all w_i are measured.

When the components have been connected, the only knowledge available about the w_i is the A-polynomials in (7a) and the equation
$g\big(z(t), w(t)\big) = 0$. The connected system is thus identifiable if the w_i can be computed from (12):

$$A_{ij}\big(w_i(t), p\big) = 0 \quad \begin{cases} i = 1, \ldots, m \\ j = 1, \ldots, n_{A_i} \end{cases} \tag{27a}$$

$$g\big(z(t), w(t)\big) = 0. \tag{27b}$$

Note that this means that all $w(t)$ are algebraic variables (not differential), so that no initial conditions can be specified for any component of $w(t)$. If, on the other hand, there are several solutions to the equations (27) then these different solutions can be inserted into the B polynomials, so there are also several possible parameter values. In this case the connected system is therefore not identifiable. Note again that measured inputs and outputs lead to equations of the form $w_i(t) = u(t)$, where the function u is included in the time-variability of g.

The result is formalized in the following theorems. Note that the distinction between global and local identifiability was not discussed above, but this will be done below.

6.1   Global Identifiability

Global identifiability means that there is a unique solution to the identification problem, given that the measurements are informative enough. For a subsystem (1a) that can be rewritten in the form (7) global identifiability means that the $B_{i,j}$ can be solved uniquely to give the $\theta_{i,j}$. In other words there exist functions ψ, that can in principle be calculated from the $B_{i,j}$, such that

$$\theta_i = \psi_i(w_i, p). \tag{28}$$

We then have the following formal result on identifiability.

Theorem 1. *Consider a modular model structure where the modules (1a) are globally identifiable and thus can be described in the form (28). A sufficient condition for the total model structure to be globally identifiable is that (27) can be solved uniquely for the w_i. If all the functions ψ_i of (28) are injective then this condition is also necessary.*

Proof: If (27) gives a global solution for $w(t)$, then this solution can be inserted into the B polynomials to give a global solution for θ since the components are globally identifiable. The connected system is thus globally identifiable. If there are several solutions for w_i and the functions ψ_i of (28) are injective, then there are also several solutions for θ, so the system is not globally identifiable since the identification problem has more than one solution.

6.2   Local Identifiability

Local identifiability of a model structure means that locally there is a unique solutions to the identification problem, but globally there may be more than one solution. This means that the description (28) is valid only locally. We get the following result on local identifiability.

Theorem 2. *Consider a modular model structure where the modules (1a) are locally identifiable and thus can be locally described in the form (28). A sufficient condition for the total model to be locally identifiable is that (27) can be solved locally uniquely for the w_i. If all the functions ψ_i of (28) are locally injective then this condition is also necessary.*

Proof: If (27) gives a locally unique solution for $w(t)$, then this solution can be inserted into the B polynomials to give a local solution for θ since the components are locally identifiable. The connected system is thus locally identifiable. If there locally are several solutions for w_i and the functions ψ_i of (28) are injective, then there are also several local solutions for θ, so the system is not locally identifiable since the identification problem locally has more than one solution.

7   Conclusions

This paper has shown how a modular structure of a large model can be used to simplify examination of identifiability. For modules in model libraries, the transformation to the form (7) is computed once and for all and stored with the module. This makes it possible to only consider a smaller number of equations when examining identifiability for a model composed of such modules. Although the method described in this paper may suffer from high computational complexity (depending, among other things, on the method selected for deciding the number of solutions for (12)), it can make the situation much better than when trying to use to use the differential-algebra approach described in [4] on a complete model.

The technique could be included in tools for object-oriented modeling such as DY-MOLA and OPENMODELICA. Preferably, this could be part of a complete set of system identification routines linked to the modeling software. The identification routines could either be included directly in the modeling software, or as external software that interacts with the modeling software.

Future work could include to examine if it is possible to make the method fully automatic, so that it can be included in modeling tools and to examine if other system analysis or design methods can benefit from the modularized structure in object-oriented models. It could also be interesting to examine the case when several components share the same parameter. This could occur for example if the different parts of the system are affected by environmental parameters such as temperature and fluid constants.

References

1. Peter Fritzson. *Priciples of Object-Oriented Modeling and Simulation with Modelica 2.1.* Wiley IEEE, New York, 2004.
2. Markus Gerdin. *Identification and Estimation for Models Described by Differential-Algebraic Equations.* PhD thesis, Linköping University, Linköping, Sweden, 2006. Dissertation No 1046.
3. L. Ljung. *System Identification - Theory for the User.* Prentice-Hall, Upper Saddle River, N.J., 2nd edition, 1999.
4. Lennart Ljung and Torkel Glad. On global identifiability of arbitrary model parameterizations. *Automatica*, 30(2):pp 265–276, Feb 1994.
5. P. Parrilo and Lennart Ljung. Initialization of physical parameter estimates. In P. van der Hof, B. Wahlberg, and S. Weiland, editors, *Proc. 13th IFAC Symposium on System Identification*, pages 1524 – 1529, Rotterdam, The Netherlands, Aug 2003.
6. J.F. Ritt. *Differential Algebra.* American Mathematical Society, Providence, R.I., 1950.
7. E. Walter. *Identification of State Space Models.* Springer Verlag, Berlin, 1982.
8. E. Walter, editor. *Identifiability of Parametric Models.* Pergamon Press, Oxford, 1987.

Identification of Hidden Markov Models
- Uniform LLN-s

László Gerencsér[1] and Gábor Molnár-Sáska[2]

[1] MTA SZTAKI (Computer and Automation Institute,
Hungarian Academy of Sciences), 13-17 Kende u., Budapest 1111, Hungary
Tel.: (36-1)-279-6138, (36-1)-279-6190; Fax: (36-1)-466-7503
`gerencser@sztaki.hu`

[2] Morgan Stanley Hungary Analytics, Budapest, 15 Deák Ferenc u., Budapest 1052, Hungary
Tel.: (36-1)-880-45-11
`gabor.molnar-saska@morganstanley.com`

Summary. We consider hidden Markov processes in discrete time with a finite state space $\mathcal{X}$ and a general observation or read-out space $\mathcal{Y}$. The identification of the unknown dynamics is carried out by the conditional maximum-likelihood method. The normalized log-likelihood function is shown to satisfy a uniform law of large numbers over certain compact subsets of the parameter space. Two cases are covered: first, when the running value of the transition probability matrix, denoted by Q is positive, second, when Q is primitive, but the read-out densities are strictly positive.

Keywords: hidden Markov models, maximum-likelihood estimation, predictive filters, gradient filters, L-mixing, uniform laws of large numbers.

1 Introduction

We consider Hidden Markov Models (HMM) or hidden Markov processes with a finite state space $\mathcal{X}$, $|\mathcal{X}| = N$, and a general observation or read-out space $\mathcal{Y}$, which is assumed to be a Polish space, i.e. a complete, separable metric space, or a measurable subset of it. We will identify $\mathcal{X}$ with $\{1, \ldots, N\}$.

We consider the problem of estimating the unknown true transition probability matrix and the unknown true read-out probability densities, which may or may not belong to a parametric family. The true transition probability matrix, denoted by Q^*, is assumed to be primitive, i.e $(Q^*)^r > 0$ for some positive integer r.

The first results for finite state-space $\mathcal{X}$ and finite read-out space $\mathcal{Y}$ are due to Baum and Petrie, [2]. The case of binary readouts has been studied by Arapasthotis and Marcus, [1], using different techniques. The extension of these results to continuous read-out space requires new techniques.

The identification of the unknown dynamics is carried out by the conditional maximum-likelihood method, the condition being the unknown initial state. The first step in proving consistency of the maximum likelihood method would be to show the validity of the strong law of large numbers for the log-likelihood function. This has been investigated in the literature using basically three different methods in our context. Leroux [14] used the subadditive ergodic theorem, LeGland and Mevel used the

A. Chiuso et al. (Eds.): Modeling, Estimation and Control, LNCIS 364, pp. 135–150, 2007.
springerlink.com

theory of geometric ergodicity, see [13], [12], and [15], finally in [10] and [9] the theory of L-mixing processes has been used. The advantage of the latter approach is that it enables us to adapt a variety of techniques developed for the statistical theory of linear stochastic systems for HMM-s.

The main advance of the present paper is the outline of the derivation of a uniform law of large numbers (ULLN), with rate of convergence in L_q, for the normalized log-likelihood function over compact subsets of the parameter space, satisfying two fairly weak technical conditions. These results will be derived under the condition that the running or assumed value of the transition probability matrix, denoted by Q is primitive, but the assumed read-out densities are strictly positive. It is not required that the true system belongs to the model class, see [15].

The main technical tools have been developed in [9] and [8], see also [7]. In addition we provide a simple derivation of the exponential stability of the gradient filter as opposed to the rather complicated derivation given in [13]. The basic idea for proving ULLN-s with rate of convergence has been borrowed from [4]. The results of the present paper can be directly applied to extend a basic strong approximation theorem given for ARMA-processes in [5] to HMM-s with primitive transition probability matrices and positive read-out probabilities.

2 Hidden Markov Models

Definition 2.1. *The stochastic process* $(X_n, Y_n), n \geq 0$ *taking its values in* $\mathcal{X} \times \mathcal{Y}$ *will be called a hidden Markov process, if the state process* (X_n) *is a homogenous, strictly stationary Markov process with state space* $\mathcal{X}$, *and the observations or read-outs* (Y_n) *are conditionally independent and identically distributed given* (X_n). *It is assumed that the observations* Y_k *have a conditional density given* $X_k = x$: *for any Borel-set A we have*

$$P(Y_k \in A | X_k = x) = \int_A b^{*x}(y)\lambda(dy),$$

where λ *is a fixed nonnegative, σ-finite measure on* $\mathcal{Y}$.

The upper index $*$ indicates that we take the true value of the corresponding, unknown read-out densities, as opposed to some assumed value that shows up in the estimation problem.

Example 1. Gaussian read-outs. In this case the observations are of the form

$$Y_n = h(X_n) + \sigma(X_n)\,\epsilon_n, \qquad n \geq 0$$

where (ϵ_n) is a Gaussian i.i.d. sequence in $\mathbf{R}^m$, independent of (X_n), and $h : \mathcal{X} \to \mathbf{R}^m$ and $\sigma : \mathcal{X} \to \mathbf{R}^{m \times m}$ are arbitrary mappings. If (X_n) itself is i.i.d., then we get a Gaussian *mixture*.

Define the vector $b^*(y)$ with its components being the read-out probabilities, and the diagonal matrix $B^*(y)$ its diagonal elements being $(b^{*i}(y)), i = 1, \ldots, N$:

$$b^*(y) = (b^{*1}(y), \ldots, b^{*N}(y))^T, \qquad B^*(y) = \mathrm{diag}(b^{*i}(y)).$$

The transition probability matrix of the Markov process (X_n) will be denoted by Q^*, i.e.

$$Q^*_{ij} = P(X_{n+1} = j | X_n = i).$$

We write $Q^* > 0$ if all the elements of the transition probability matrix are strictly positive.

We consider the problem of estimating the unknown transition probability matrix Q^* and the unknown read-out probability densities $b^{*x}(y)$ from a sequence of observations $y_0, \ldots y_n$ using a given class of parametric models. For this consider a parametric family of transition probability matrices $Q(\theta)$, and a parametric family of read-out probability densities $b^x(y; \theta)$, where $\theta \in D \subset \mathbf{R}^r$, with D being a measurable set of feasible parameters. The set D, to be specified later, can be arbitrary at this point. It is not required that the true system belongs to the model class. If it does, the true parameter will be denoted by $\theta^* \in D$, thus we have

$$b^{*x}(y) = b^x(y; \theta^*) \quad \text{and} \quad Q^* = Q(\theta^*).$$

To develop a maximum-likelihood method let $\theta \in D$ be any parameter-value, and let us write

$$b^x(y) = b^x(y; \theta) \quad \text{and} \quad Q = Q(\theta).$$

Let the set of probability vectors $\mathbf{q} = (q^1, \ldots, q^N)$ be denoted by $\mathcal{P}$, and let $\mathcal{P}(\epsilon) \subset \mathcal{P}$ be the set of $\mathbf{q}$-s such that $q^i \geq \epsilon > 0$ for all i. Write the conditional log-likelihood function, with condition $P(X_0 = i) = q^i$, with $\mathbf{q} \in \mathcal{P}$ as

$$\log p(y_0, \ldots y_n; \theta, \mathbf{q}) = \sum_{k=1}^{n-1} \log p(y_k | y_{k-1}, \ldots y_0; \theta, \mathbf{q}) + \log p(y_0; \theta, \mathbf{q}). \tag{1}$$

Express the k-th term as

$$\log p(y_k | y_{k-1}, \ldots y_0; \theta, \mathbf{q}) = \log \sum_i b^i(y_k; \theta) P(X_k = i | y_{k-1}, \ldots, y_0; \theta, \mathbf{q}).$$

It is seen that a basic entity is the *predictive filter*, defined as

$$p^j_{n+1} = p^j_{n+1}(\theta, \mathbf{q}) = P(X_{n+1} = j | y_n, \ldots, y_0; \theta, \mathbf{q}).$$

Writing $p_{n+1} = (p^1_{n+1}, \ldots, p^N_{n+1})^T$ it is known that the filter process satisfies the so-called *Baum-equation*, see [2], which is a discrete time version of the celebrated Duncan-Mortensen-Zakai equation, see [3, 16, 22]:

$$p_{n+1} = \pi(Q^T B(y_n) p_n), \tag{2}$$

with initial condition $p_0 = \mathbf{q}$, where π is the normalizing operator: for $x \geq 0$, $x \neq 0$ set $\pi(x)^i = x^i / \sum_j x^j$, see [2]. A shorthand notation for the Baum-equation will be

$$p_{n+1} = f(y_n, p_n), \tag{3}$$

or, in a more explicit form:

$$p_{n+1}(\theta, \mathbf{q}) = f(y_n, p_n(\theta, \mathbf{q}); \theta). \tag{4}$$

With this notation the k-th term in (1) can be written as

$$\log p(y_k | y_{k-1}, \ldots y_0; \theta, \mathbf{q}) = \log \sum_i b^i(y_k; \theta) p_k^i(\theta, \mathbf{q}).$$

Introduce the function of the pair of variables (y, p) with $p = (p^1, \ldots, p^N)$, parameterized by θ:

$$g(y, p \, ; \theta) = \log \sum_i b^i(y; \theta) p^i. \tag{5}$$

Then we can write, with $p_k = p_k(\theta, \mathbf{q})$ and $p_0 = \mathbf{q}$,

$$\log p(y_0, \ldots, y_n; \theta, \mathbf{q}) = \sum_{k=0}^{n} g(y_k, p_k \, ; \theta). \tag{6}$$

A standard step in proving consistency of the maximum likelihood is to show that

$$\lim_{n \to \infty} \frac{1}{n} \log p(Y_0, \ldots Y_n; \theta, \mathbf{q}) = W(\theta) \tag{7}$$

exists almost surely, with the limit being independent of $\mathbf{q}$. The existence of the limit in (7) has been investigated in the literature using basically three different methods as described in the introduction. In this paper we focus on [10] and the yet unpublished paper [9], where we used the theory of L-mixing processes. The advantage of the latter approach is that it enables us to adapt a variety of techniques developed for the statistical theory of linear stochastic systems for the present context. The basic definitions will be given in the next section.

3 L-Mixing Processes

In this short section we summarize the basic definitions related to L-mixing, as developed in [4]. L-mixing has been used extensively in the statistical analysis of linear stochastic systems, see [4]. It is a concept which, in its motivation, strongly exploits the stability and the linear algebraic structure of the underlying stochastic system. A major advance of [9] is that the applicability of this concept in the framework of HMM-s have been shown. First of all we need the definition of M-boundedness.

Definition 3.1. *A stochastic process* (X_n) *($n \geq 0$) taking its values in an Euclidean space is* M*-bounded if for all* $q \geq 1$

$$M_q(X) = \sup_{n \geq 0} E^{1/q} \|X_n\|^q < \infty. \tag{8}$$

If (X_n) is M-bounded we shall also write $X_n = O_M(1)$. Similarly if c_n is a positive sequence we write $X_n = O_M(c_n)$ if $X_n/c_n = O_M(1)$.

Let $(\mathcal{F}_n)$ and $(\mathcal{F}_n^+)$ be two sequences of monotone increasing and monotone decreasing σ-algebras, respectively, such that $\mathcal{F}_n$ and $\mathcal{F}_n^+$ are independent for all n.

Definition 3.2. *A stochastic process* (X_n) *taking its values in a finite-dimensional Euclidean space is L-mixing with respect to* $(\mathcal{F}_n, \mathcal{F}_n^+)$, *if it is M-bounded,* $\mathcal{F}_n$-*adapted, and with*

$$\gamma_q(\tau) = \sup_{n \geq \tau} E^{1/q} \| X_n - E(X_n | \mathcal{F}_{n-\tau}^+) \|^q \tag{9}$$

we have

$$\Gamma_q = \sum_{\tau=0}^{\infty} \gamma_q(\tau) < \infty. \tag{10}$$

The usefulness of L-mixing is that its verification is often much easier than the verification of other notions of mixing.

The above definitions extend to parameter-dependent processes as follows: let $\theta \in D \subset \mathbf{R}^r$, with D being a measurable set of feasible parameters, and let $(X_n(\theta))$, with $n \geq 0, \theta \in D$, be a measurable stochastic process defined over $\mathbf{Z} \times D$. Then in the definitions above we take

$$M_q(X) = \sup_{n \geq 0, \theta \in D} E^{1/q} \| X_n \|^q, \qquad \gamma_q(\tau) = \sup_{n \geq \tau, \theta \in D} E^{1/q} \| X_n - E(X_n | \mathcal{F}_{n-\tau}^+) \|^q,$$

and we say that $(X_n(\theta))$ is L-mixing with respect to $(\mathcal{F}_n, \mathcal{F}_n^+)$, *uniformly in* θ.

The relevance of L-mixing in the context of Markov-processes has been established in [9] and [10]. For this a convenient visualization of a Markov process is used, namely the dynamics of the process is described by

$$X_i = f(X_{i-1}, U_i),$$

where (U_i) is a sequence of i.i.d. random variables over some probability space with values in a measurable space $\mathcal{U}$, which is independent also of X_0, and f is a measurable mapping. The existence of such a representation is well-known, see Kifer [11].

In particular we can assume that U_i is distributed uniformly over $[0, 1]$. With this representation at hand let

$$\mathcal{F}_i = \sigma\{X_0, U_s : 0 \leq s \leq i\}, \quad \text{and} \quad \mathcal{F}_i^+ = \sigma\{U_s : s \geq i+1\}.$$

L-mixing in the rest of the paper will always mean L-mixing with respect to $(\mathcal{F}_n, \mathcal{F}_n^+)$ given above.

4   Asymptotic Properties of the Log-Likelihood Function

In this section we summarize some recent results on the asymptotic properties of the log-likelihood function. First we consider the case when Q is *positive*, i.e. $Q > 0$, irrespective of whether Q^* itself is positive. The essential ingredients of the following result of [9] have been given already in [10].

Proposition 4.1. *Consider a hidden Markov process* (X_n, Y_n) *as specified in Definition 2.1, and consider the Baum-equation (2). Assume that the transition probability matrix* Q^* *is primitive, and that the assumed transition probability matrix* Q *is* *positive, $Q > 0$. Furthermore, suppose that the assumed read-out probabilities are all positive: for all $x \in \mathcal{X}$ and λ almost all $y \in \mathcal{Y}$ we have*

$$b^x(y) > 0. \tag{11}$$

Furthermore assume that for all $q \geq 1$ and for all $i, j \in \mathcal{X}$

$$\int |\log b^j(y)|^q \, b^{*i}(y)\lambda(dy) < \infty. \tag{12}$$

Let $\mathbf{q} \in \mathcal{P}$ be any initial distribution, and let $p_n = p_n(\theta, \mathbf{q})$ denote the solution of the Baum-equation (2) with $y_n = Y_n$. Then the process

$$g(Y_n, p_n) = \log p(Y_n | Y_{n-1}, \ldots Y_0; \theta, \mathbf{q})$$

is L-mixing. Furthermore, the limit

$$\lim_{n \to \infty} Eg(Y_n, p_n) = W(\theta)$$

exists, and is independent of the initial value $\mathbf{q}$.

Note that condition (12) is satisfied for Gaussian read-outs with state-dependent covariance.

Corollary 4.1. *Under the conditions of Theorem 4.1 we have*

$$\frac{1}{n} \log p(Y_0, \ldots Y_n; \theta, \mathbf{q}) - W(\theta) = O_M(n^{-1/2}). \tag{13}$$

The corollary is obtained by a direct application of Theorem 1.1. of [4]. From here we get by a Borel-Cantelli argument that

$$\lim_{n \to \infty} \frac{1}{n} \log p(Y_0, \ldots Y_n; \theta, \mathbf{q}) = W(\theta) \quad \text{a.s.} \tag{14}$$

A key step in establishing Theorem 4.1 is the verification of the exponential stability of the Baum equation with respect to its initial condition $\mathbf{q}$. This has been proven by LeGland and Mevel, [13] for continuous read-outs. A similar result for hidden Markov processes with binary read-outs were given by Araposthatis and Marcus, [1]. We formulate an improved version of this result for the case of *positive Q-s*, which is obtained by a slight modification of the methods of [13], using simple inequalities for projective products in a different way:

Proposition 4.2. *Consider a hidden Markov process (X_n, Y_n) satisfying the conditions of Theorem 4.1, except (12). Then there exists a constant δ with $0 < \delta < 1$ depending only on Q, and for any $\epsilon > 0$ there exists a constant $C > 0$ depending only on Q and ϵ, such that for any $\mathbf{q}, \mathbf{q}' \in \mathcal{P}(\epsilon)$, for all $n \geq 0$, and for any sequence of observations $(y_0, \ldots, y_{n-1})$ we have*

$$\|p_n(\mathbf{q}) - p_n(\mathbf{q}')\|_{TV} \leq C(1 - \delta)^n \|\mathbf{q} - \mathbf{q}'\|_{TV}, \tag{15}$$

where $\| \cdot \|_{TV}$ denotes the total variation norm.

A key point of the above result is that it is completely independent of the underlying probabilistic structure, it is in fact a result of linear algebra, moreover δ and C do not depend on the readout densities.

An extension: an important corollary is that Proposition 4.1 remains true if the positivity condition on the read-outs, (11), is removed, and $q, q' > 0$. This extension is obtained by a simple limiting argument. Thus important examples, such as read-outs obtained by *quantization* are also covered.

5   The Case of Primitive Q-s

In extending the results of the previous section to primitive Q-s the first step is to extend Proposition 4.2 to the case when the assumed value Q is primitive. First we restate Theorem 2.1 of [13]. Let $b^x(y)$ denote the read-out density and let $Q(x, x')$ denote the transition probability matrix used in the Baum-equation (2), and let

$$\delta(y) = \frac{\max_x b^x(y)}{\min_x b^x(y)} \tag{16}$$

and

$$\epsilon = \min_{x,x'}{}^{+} Q(x, x'), \tag{17}$$

where $\min^{+}$ denotes the minimum taken over positive elements only.

Proposition 5.1. *Consider a hidden Markov process (X_n, Y_n) as specified in Definition 2.1, and consider the Baum-equation (2). Assume that the transition probability matrices Q^*, Q are primitive. Furthermore, assume that the assumed read-out probabilities are all positive, see (11). Let $q, q' \in \mathcal{P}$ be any two initializations. Then for any sequence of observations $(y_0, \ldots, y_{n-1}) \in \mathcal{Y}^n$ with $n \geq r$ we have*

$$\|p_n(q) - p_n(q')\|_{TV} \leq \epsilon^{-r}\delta(y_0)\ldots\delta(y_{r-1})$$

$$\cdot \prod_{k=1}^{\lfloor n/r \rfloor} (1 - \epsilon^r\delta^{-1}(y_{kr-r+1})\ldots\delta^{-1}(y_{kr-1}))\|q - q'\|_{TV},$$

where $\|\cdot\|_{TV}$ denotes the total variation norm.

In [8] we have proven the following significant addition to the above result:

Proposition 5.2. *Consider a hidden Markov process (X_n, Y_n) such that it satisfies the conditions of Proposition 5.1. In addition assume that for all $q \geq 1$*

$$\int |\delta(y)|^q b^{*i}(y)\lambda(dy) < \infty. \tag{18}$$

Then for any $s \geq 1$ there exist constants $0 < \delta < 1$, and a random variable $C(\omega) > 0$, with

$$\mathbb{E}C^s(\omega) < \infty, \tag{19}$$

so that for any two initial distributions q, q' we have

$$\|p_n(q) - p_n(q')\|_{TV} \leq C(\omega)(1 - \delta)^n\|q - q'\|_{TV}. \tag{20}$$

If $b^x(y)$ is bounded, then condition (18) can be formulated in the following equivalent form: for any i, j and any $q \geq 1$ we have with

$$\delta^j(y) = 1/(b^j(y)) \tag{21}$$

the inequality

$$\int |\delta^j(y)|^q b^{*i}(y)\lambda(dy) < \infty. \tag{22}$$

Consider now again a parametric family of transition and read-out probabilities $Q = Q(\theta)$ and $b^x(y) = b^x(y; \theta)$, respectively, with $\theta \in D$.

Condition 5.1. *We assume that $Q(\theta)$ is primitive for all $\theta \in D$. Furthermore, for any $x \in \mathcal{X}$ and $\theta \in D$ we have*

$$b^x(y; \theta) > 0$$

for λ-almost all $y \in \mathcal{Y}$.

Condition 5.2. *Assume that the transition probability matrices $Q(\theta)$ are continuous in θ for $\theta \in D$. Similarly, assume that that for all fixed $x \in \mathcal{X}$ the read-out probability $b^x(y; \theta)$ is measurable in (y, θ), and it is continuous in θ for $\theta \in D$ for λ-almost all y. In addition, for any $x \in \mathcal{X}$ and for any compact set $D_0 \subset D$, and $\theta \in D_0$ the densities $b^x(y; \theta)$ are uniformly bounded in y. More precisely: there exists a finite K such that for any $x \in \mathcal{X}$, and $\theta \in D_0$ we have*

$$|b^x(y; \theta)| \leq K$$

for λ-almost all $y \in \mathcal{Y}$.

A minor modification of the proof given in [8] yields the following *uniform version* of the Proposition 5.2:

Theorem 5.1. *Consider a hidden Markov process (X_n, Y_n) as specified in Definition 2.1, and consider the Baum-equation (2) with $y_n = Y_n$. Assume that the transition probability matrix Q^* is primitive. Consider a parametric family of transition and read-out probabilities $Q = Q(\theta)$ and $b^x(y) = b^x(y; \theta)$, respectively, with $\theta \in D$ satisfying Conditions 5.1 and 5.2. Finally, assume that for any compact set $D_0 \subset D$, any i, j, and any $q \geq 1$ we have with*

$$\overline{\delta}^j(y) = 1/(\min_{\theta \in D_0} b^j(y; \theta)) \tag{23}$$

the inequality

$$\int |\overline{\delta}^j(y)|^q b^{*i}(y)\lambda(dy) < \infty. \tag{24}$$

Then (20) holds with a constant $0 < \delta < 1$ that is independent of θ, and a random variable $C(\omega; \theta) > 0$ such that

$$\sup_{\theta \in D_0} \mathrm{E}C^s(\omega; \theta) < \infty. \tag{25}$$

For the proof we need two remarks. For the first remark, we quote the following result of [8], given as Theorem V.3: Let (ξ_k) be a sequence of i.i.d random variables such that $\xi_k \leq 0$, and $P(\xi_k < 0) > 0$. Then for any $s > 1$ there exist an $\alpha > 0$ such that with

$$\sigma^* = \sup_n \sum_{k=1}^n (\xi_k + \alpha) \tag{26}$$

we have $\mathrm{E}(e^{s\sigma^*}) < \infty$. See also see e.g. [17, 20]. Now it is easy to see that if a family of random variables $(\xi_k(\theta))$ is such that $(\xi_k(\theta)) \leq_s \xi^*$ for all θ, where the symbol $\leq_s$ denotes stochastically less than or equal, and ξ^* satisfies the conditions of the theorem, then there is a single, common $\alpha > 0$ with which the theorem is true.

The second remark is that if a family of transition probabilities $P(x, A; \theta)$ is such that they satisfy a minorization condition uniformly in θ in the sense that there exists a probability measure ν on $\mathcal{X}$, a $\delta > 0$ such that

$$P(x, A; \theta) \geq \delta \nu(A)$$

is valid for all $x \in \mathcal{X}$, $A \in \mathcal{B}(\mathcal{X})$ and all θ, then $P(x, A; \theta)$ can be realized on a single probability space as a mixture of a single i.i.d. sequence and some Markov processes. More precisely:

$$X_n(\theta) = \delta_n \xi_n + (1 - \delta_n) X_n'(\theta),$$

with (δ_n) being an i.i.d Bernoulli sequence taking the value 1 with probability δ, and (ξ_n) being an i.i.d. sequence with distribution ν.

A major application of the above result is given in the following extension of Proposition 4.1.

Theorem 5.2. *Consider a hidden Markov process (X_n, Y_n) satisfying the conditions of Theorem 5.1. Let $\mathbf{q} \in \mathcal{P}$ be any initial distribution and let $p_n = p_n(\theta, \mathbf{q})$ denote the solution of the Baum-equation (2) with $y_n = Y_n$. Then the process*

$$g(Y_n, p_n; \theta) = \log p(Y_n | Y_{n-1}, \dots Y_0; \theta, \mathbf{q})$$

is L-mixing uniformly in θ for $\theta \in D_0$. Furthermore, the limit

$$\lim_{n \to \infty} \mathrm{E}g(Y_n, p_n; \theta) = W(\theta)$$

exists, uniformly in θ for $\theta \in D_0$, and is independent $\mathbf{q}$.

Discussion. Condition (18) is used with $q = 1$ in Theorem of [13]. This condition is quite restrictive, though, even in the case of $q = 1$, in particular it is not satisfied for Gaussian read-outs. However, the validity of the condition can be enforced at the price of a small loss in information. A simple way of doing this is to replace our observations $Y_n \in \mathbf{R}^m$ by $Y_n' = g(Y_n)$ where

$$g(y) = \frac{y}{|y|} h(|y|), \qquad h(r) = \min\left(\frac{r}{R}, 1\right), \qquad r \geq 0,$$

where R is a large positive number. Then g maps the open ball of radius R, denoted by S_R into S_1, and it is a continuously differentiable homeomorphism there, while it maps $\mathbf{R}^m \backslash S_R$, onto the boundary of S_1, say C_1. Defining λ inside S_1 as the Lebesgue-measure in $\mathbf{R}^m$, and on C_1 as the $(m-1)$ dimensional Lebesgue-measure on C_1, it is easy to see that condition (18) will be satisfied.

A final remark: note that condition (18) with $q = 1$ implies condition (12) with any q, under the assumed boundedness condition on $b^x(y; \theta)$.

6 The Derivative of the Predictive Filter

In order to derive uniform laws of large numbers we need to analyze the derivatives of the likelihood function. Our attention will be restricted to the analysis of first derivatives, but results for higher order derivatives will also be stated. For the sake of simplicity assume that D is *open*. We need to strengthen Condition 5.2 so that continuity is replaced by first order smoothness:

Condition 6.1. *Assume that the transition probability matrices $Q(\theta)$ are continuously differentiable in θ for $\theta \in D$. Similarly, assume that for all fixed $x \in \mathcal{X}$ the read-out probability $b^x(y; \theta)$ is measurable in (y, θ), and continuously differentiable in θ for $\theta \in D$ for λ-almost all y. In addition, for any $x \in \mathcal{X}$ and for any compact set $D_0 \subset D$, and $\theta \in D_0$ the densities $b^x(y; \theta)$ are uniformly bounded in y, together with their first derivatives (see Condition 5.2).*

Taking into account the recursion given by the Baum-equation it is easy to see that $p(y_n|y_{n-1}, \ldots, y_0; \theta, \mathbf{q})$ is continuously differentiable in θ for $\theta \in D$. Differentiating

$$\log p(y_n|y_{n-1}, \ldots, y_0; \theta, \mathbf{q}) = \log b^T(y_n, \theta) p_n,$$

with respect to θ we get with

$$W_n = W_n(\theta, \mathbf{q}) = \frac{\partial p_n^T(\theta, \mathbf{q})}{\partial \theta} \quad \text{and} \quad \beta(y_n) = \beta(y_n; \theta) = \frac{\partial b^T(y_n; \theta)}{\partial \theta}$$

the expression

$$\frac{\partial}{\partial \theta} \log p(y_n|y_{n-1}, \ldots, y_0, ; \theta, \mathbf{q}) = \frac{\beta(y_n)p_n + W_n b(y_n)}{b(y_n)^T p_n}. \tag{27}$$

Introducing the function of three variables

$$g(y, p, W) = \frac{\beta(y)p + W b(y)}{b^T(y)p} \tag{28}$$

we can write the gradient of the log-likelihood function with $p_k = p_k(\theta, \mathbf{q})$ and $W_k = W_k(\theta, \mathbf{q})$ as

$$\frac{\partial}{\partial \theta} \log p(y_0, \ldots, y_n; \theta, \mathbf{q}) = \sum_{k=1}^{n} g(y_k, p_k, W_k ; \theta). \tag{29}$$

Note that since $p_0 = p_0(\theta, \mathbf{q})$ does not depend on θ, the summation starts with $k = 1$.

In order to prove a strong law of large numbers for the gradient we follow the general theory given in [10] and [9]. Thus the first step is to prove exponential stability of the gradient process with respect to initial perturbations. Let θ be fixed, and write $p_n(\mathbf{q}) = p_n(\theta, \mathbf{q})$. Obviously, $p_n(\mathbf{q})$ is a continuously differentiable function of $\mathbf{q}$. Thus from Proposition 4.2 we get by elementary arguments the following result:

Theorem 6.1. *Under the condition of Proposition 4.2 there exists a constant δ with $0 < \delta < 1$ depending only on Q, and for any $\epsilon > 0$ there exists a constant $C > 0$, depending only on Q and ϵ, such that for any $\mathbf{q} \in \mathcal{P}(\epsilon)$, for all $n \geq 0$, and for any sequence of observations $(y_0, \ldots, y_{n-1})$ we have*

$$\|\frac{\partial}{\partial \mathbf{q}} p_n(\mathbf{q})\|_{op} \leq C(1-\delta)^n$$

with $\|.\|_{op}$ denoting the operator norm.

This theorem can be reinterpreted if we consider the dynamics of

$$V_n = V_n(\mathbf{q}) = \frac{\partial}{\partial \mathbf{q}} p_n(\mathbf{q}).$$

Differentiating the Baum-equation (4) with respect to $\mathbf{q}$ it is easy to see that V_n satisfies, with $p_n = p_n(\mathbf{q})$, the matrix-equation

$$V_{n+1} = \frac{\partial}{\partial p} f(y_n, p_n) V_n, \tag{30}$$

with initial condition $V_0 = I$. Setting

$$A_n = \frac{\partial}{\partial p} f(y_n, p_n)$$

and taking into account that $Q > 0$ implies that $p_n(\mathbf{q}) \in \mathcal{P}(\epsilon)$ for all $n \geq 1$ for some $\epsilon > 0$, and the resulting shift-invariance of Proposition 4.2 we get the following corollary:

Theorem 6.2. *Under the condition of Proposition 4.2 we have for any $0 \leq m \leq n$*

$$\|A_n...A_m\| \leq C(1-\delta)^{(n-m)}.$$

Let us now return to the parametric case. To avoid the complexity of tensor calculations, let us now assume that $\theta(t) \in D, 0 \leq t < t_0$ with $\theta = \theta(0)$ is a continuously differentiable curve with tangent η at $t = 0$. Then $b^x(y; \theta(t))$ and $Q(\theta(t))$ are continuously differentiable functions of t, and thus $p_n(\theta(t); \mathbf{q})$ is a continuously differentiable function of $(t, \mathbf{q})$. Set

$$w_n = w_n(\theta, \mathbf{q}) = \frac{\partial}{\partial t} p_n(\theta(t), \mathbf{q})|_{t=0}.$$

The dynamics of (w_n) is obtained by differentiating the Baum-equation (4) with respect to t, to get, with $p_n = p_n(\theta, \mathbf{q})$,

$$w_{n+1} = \frac{\partial}{\partial p} f(y_n, p_n; \theta) w_n + \frac{\partial}{\partial \theta} f(y_n, p_n; \theta) \eta, \tag{31}$$

with initial condition $w_0 = 0$. We need to strengthen Condition 6.1 so that first order smoothness is replaced by second order smoothness:

Condition 6.2. *Assume that the transition probability matrices $Q(\theta)$ are twice continuously differentiable in θ for $\theta \in D$. Similarly, assume that that for all fixed $x \in \mathcal{X}$ the read-out probability $b^x(y; \theta)$ is measurable in (y, θ), and twice continuously differentiable in θ for $\theta \in D$, for λ-almost all y. In addition, for any $x \in \mathcal{X}$ and for any compact set $D_0 \subset D$, and $\theta \in D_0$ the densities $b^x(y; \theta)$ are uniformly bounded in y, together with their first and second derivatives, (see Condition 5.2).*

Differentiating with respect to $\mathbf{q}$, solving it for $\frac{\partial}{\partial \mathbf{q}} w_n(\mathbf{q})$, and taking into account Theorems 6.1 and 6.2 we get the following result:

Theorem 6.3. *Assume that the conditions of Proposition 4.2, and Condition 6.2 are satisfied. Then there exists a constant δ with $0 < \delta < 1$ depending only on Q, and for any $\epsilon > 0$ there exists a constant $C > 0$, depending only on Q and ϵ, such that for any $\mathbf{q} \in \mathcal{P}(\epsilon)$, and for all $n \geq 0$ we have*

$$\left\| \frac{\partial}{\partial \mathbf{q}} w_n(\mathbf{q}) \right\|_{op} \leq C(1 - \delta)^n.$$

Corollary 6.1. *Assume that the conditions of Proposition 4.2, and Condition 6.2 are satisfied. Then with the notations of Theorem 6.3 we have for any $\mathbf{q}, \mathbf{q}' \in \mathcal{P}(\epsilon)$, and for all $n \geq 0$*

$$\|w_n(\mathbf{q}) - w_n(\mathbf{q}')\|_{TV} \leq C(1 - \delta)^n \|\mathbf{q} - \mathbf{q}'\|_{TV}. \tag{32}$$

In short, the derivative process forgets its initial condition exponentially fast.

To ensure the exponential stability of the joint dynamics of $(p_n(\mathbf{q}), w_n(\mathbf{q}))$ we have to allow the initial condition of $w_n(\mathbf{q})$, as defined by in (31), vary, say write $w_0 = u$ with $1^T u = 0$, and get a process $w_n(\mathbf{q}, u)$. It is easily seen that the above two results extend to partial derivatives and finite variations with respect to u.

An extension: the above corollary remains true if the positivity condition on the read-outs, (11), is removed. This extension is obtained by a simple limiting argument, just as in the remark following Proposition 4.2. Thus, we reiterate, that important examples, such as read-outs obtained by quantization are also covered.

Note that the above argument is significantly simpler than the one given in [13]. True, their result is given for the general case, when Q is *primitive*. Can we extend the above argument to the general case? The answer is yes: the key point in such an extension is to control the constants in the inequalities stating exponential stability, written as

$$\|p_n(q) - p_n(q')\|_{TV} \leq C(\omega)(1 - \delta)^n \|q - q'\|_{TV}.$$

This is exactly the content of Theorem 5.2. It is easy to see that the results of this section can be extended to the general case. In the extension of Theorem 6.2 the constant C' on the right hand side will be replaced by a strictly stationary sequence $C'_m(\omega)$ satisfying what is stated in Proposition 5.2, i.e. (19). Altogether finally we get the following result:

Theorem 6.4. *Consider a hidden Markov process* (X_n, Y_n) *satisfying the conditions of Theorem 5.1 and Condition 6.2. Then for any* $s \geq 1$ *there exist a constant* $0 < \delta < 1$, *and a random variable* $C(\omega, \theta) > 0$, *satisfying*

$$\sup_{\theta \in D_0} \mathrm{E} C^s(\omega; \theta) < \infty, \tag{33}$$

such that for any two initial distributions $\mathbf{q}, \mathbf{q}' \in \mathcal{P}$, *and for all* $n \geq 0$ *we have*

$$\|w_n(\theta, \mathbf{q}) - w_n(\theta, \mathbf{q}')\|_{TV} \leq C(\omega; \theta)(1 - \delta)^n \|q - q'\|_{TV}. \tag{34}$$

A non-trivial result, that follows directly from the arguments given in [9] is that the above exponential stability implies the existence of a unique stationary distribution for the process (X_n, Y_n, p_n, W_n).

7   Uniform Laws of Large Numbers

In view of Theorem 6.4 we can now extend Proposition 4.1 to the gradient process. To prove that the process $g(Y_n, p_n, W_n)$ is L-mixing, with g given as

$$g(y, p, W) = \frac{\beta(y)p + Wb(y)}{b^T(y)p},$$

see (28), we follow the proof given in [9]. To ensure an appropriate integrability condition on g we impose a lower bound for the denominator. Noting that

$$b^T(y)p \geq \min_x b^x(y)$$

it is not surprising that condition (22) will play a role in the theorem below, even for the case of positive Q-s.

Theorem 7.1. *Consider a hidden Markov process* (X_n, Y_n) *as specified in Definition 2.1, such that* Q^* *is primitive. Consider the associated Baum equation (2) with* $y_n = Y_n$. *Assume that the conditions of Theorem 6.4 are satisfied. Let* $\mathbf{q} \in \mathcal{P}$ *be any initial distribution. Then*

$$g(Y_n, p_n, W_n) = \frac{\partial}{\partial \theta} \log p(Y_n | Y_{n-1}, \dots Y_0; \theta, \mathbf{q})$$

is L-*mixing, uniformly in* θ *for* $\theta \in D_0$, *where* $D_0 \subset D$ *is an arbitrary compact domain. Moreover the limit*

$$\lim_{n \to \infty} E \frac{\partial}{\partial \theta} \log p(Y_n | Y_{n-1}, \dots Y_0; \theta, \mathbf{q})$$

exists w.p.1, and is equal to $\frac{\partial}{\partial \theta} W(\theta)$.

Conditions (18) or (22) with $q = 2$ are equivalent to condition used in Theorem 4.6 of [13] to establish exponential stability of the gradient process with respect to initial conditions. See also Example 3.3 and 4.4 in [13]. Mevel and Finesso use essentially (18) with "sufficiently large q", see Assumption C on page 1124 in [15].

The above theorem remains true if D is not necessarily an open set, but (31) makes sense, and W_n is replaced by a directional derivative w_n, see the discussion preceding (31). To derive a uniform law of large numbers we follow [4]. In particular we need to extend Theorem 3.4 of [4]. For this we need the following condition:

Condition 7.1. *Let $D_0 \subset D$ be a restricted compact set of feasible parameters, such that there exist a $h_0 > 0$, and a compact set $D_1 \subset D$ such that for any $\theta, \theta' \in D_0$ with $|\theta - \theta'| \leq h \leq h_0$ there exist a smooth arc $\theta_t \in D_1$, $0 \leq t \leq 1$, connecting θ and θ' such that its length is at most ch, with some constant c.*

Define the process $(\Delta(n, \theta, \theta'))$ for $\theta \neq \theta' \in D$:

$$\Delta(n, \theta, \theta') = (\log p(Y_n | p_{n-1}, \ldots p_0; \theta', \mathbf{q}) - \log p(Y_n | p_{n-1}, \ldots p_0; \theta, \mathbf{q})) / |\theta' - \theta|.$$

Theorem 7.1 and Condition 7.1 imply that $\Delta(n, \theta, \theta')$ is L-mixing. Thus, after summation and by the application of Theorem 1.1 of [4] we arrive at the following result:

Corollary 7.1. *Assume that the conditions of Theorem 7.1 and Condition 7.1 are satisfied. Then we have that*

$$\frac{1}{n}(\log p(Y_0, \ldots Y_n; \theta, \mathbf{q}) - \log p(Y_0, \ldots Y_n; \theta', \mathbf{q})) - \frac{1}{|\theta' - \theta|}(W(\theta) - W(\theta'))$$

is of the order $O_M(n^{-1/2})$ uniformly for $\theta \neq \theta' \in D_0$ with $|\theta - \theta'| \leq h \leq h_0$.

To derive a uniform law of large numbers we need the following condition:

Condition 7.2. *The restricted set of feasible parameters D_0 is such that there exists a $0 < h_1 \leq h_0$ such that for all $\theta \in D_0$ we have*

$$\mu S(\theta, h_1) \cap D_1 > 0,$$

where $S(\theta, h_1)$ denotes the sphere with center at θ and radius h_1.

The heuristic content of this condition is that we exclude cusps and hyper-surfaces from our set of parameters D_0. It is easily seen that under Condition 7.2 the arguments of Section 3 [4] carry over, and thus we get the following *main result*:

Theorem 7.2. *Assume that the conditions of Theorem 7.1 are satisfied, and in addition assume the validity of Conditions 7.1 and 7.2. Then we have*

$$\sup_{\theta \in D_0} |\frac{1}{n} \log p(Y_0, \ldots Y_n; \theta, \mathbf{q}) - W(\theta)| = O_M(n^{-1/2}). \tag{35}$$

We note in passing that in this theorem D is not necessarily open, and then Condition 6.1 and 6.2 have to be reformulated in terms of directional derivatives.

In order to extend the techniques if [5] to HMM-s we need a uniform law of large numbers even for the second order derivatives of the log-likelihood function. For this Condition 6.2 will have to be strengthened to requiring that $Q(\theta)$ and $b^x(y; \theta)$ are *four-times* continuously differentiable in θ, and that the densities $b^x(y; \theta)$ are uniformly bounded in y, together with their first four derivatives. In our setup the same integrability condition (24) is required for all higher order derivatives. (This is in contrast with [13] or [15], where the exponents q for which integrability is required depends on the order of the derivatives).

8   Estimation of Hidden Markov Models

As pointed out in [5] uniform laws of large number with rate of convergence for the moments play a key role in deriving strong approximation theorems for off-line estimators ARMA-processes. The purpose of the present paper was to develop the tools that are needed for the extension of these results from ARMA to HMM. The usefulness of strong approximation theorems for HMM-s in adaptive encoding has been demonstrated in [6].

A major difference between the two model classes is that for ARMA-processes standard parametrizations are available, and the set of feasible parameters is typically an open set (assuming no zeros on the unit circle). A similar assumption for general HMM-s is much too restrictive. However, the tools for handling these cases as well have been developed. Applications will be described in a subsequent paper.

An important issue that has not been discussed here is *parametrization* of HMM-s. In the extreme case when the readouts are known, a natural parametrization of Q would be to take any $N-1$ entries of each row as coordinates of the parameter vector. In this case the parameter-set is a closed, convex, bounded polyhedron. If it is known a priori that $Q^* > 0$, then the parameter-set is an *open*, convex, bounded polyhedron. An intermediate case is when some elements $Q^* > 0$ are known to be positive or zero. Obviously this is a very simplistic parametrization.

A more conceptual approach is to relate parametrization issues with realization theory of HMM-s, in which fundamental contributions are due to Picci and Van Schuppen, see [18], [19]. A more recent significant advance is due to Vidyasagar, see [21]. In spite of all these basic contributions the parametrization problem of HMM-s is far from being completely settled, and we look forward to further progress.

Acknowledgment

This research was supported by the National Research Foundation of Hungary, OTKA, under grant no. T 047193.

References

1. A. Arapostathis and S.I. Marcus. Analysis of an Identification Algorithm Arising in the Adaptive Estimation of Markov Chains. *Math. Control Signals Systems*, 3:1–29., 1990.
2. L.E. Baum and T. Petrie. Statistical inference for probabilistic functions of finite state Markov chains. *Ann. Math. Stat.*, 37:1559–1563, 1966.
3. T. E. Duncan. Probability densities for diffusion processes with applications to nonlinear filtering theory. Technical report, PhD thesis, Standford, 1967.
4. L. Gerencsér. On a class of Mixing Processes. *Stochastics*, 26:165–191, 1989.
5. L. Gerencsér. On the martingale approximation of the estimation error of ARMA parameters. *Systems & Control Letters*, 15:417–423, 1990.
6. L. Gerencsér and G. Molnár-Sáska. Adaptive encoding and prediction of Hidden Markov processes. In *Proceedings of the European Control Conference, ECC 2003, Cambridge*.
7. L. Gerencsér, G. Molnár-Sáska, and Gy. Michaletzky. Improved estimation of the exponential stability of the predictive filter in Hidden Markov models. In *Proceedings of the 2006 American Control Conference, Minneapolis*, pages 5177–5182, 2006.

8. L. Gerencsér, G. Molnár-Sáska, and Gy. Michaletzky. An improved bound for the exponential stability of predictive filters of Hidden Markov models. *Communications in Information and Systems. Special Volume Dedicated to the 65th Birthday of Tyrone Duncan (guest eds.: A. Bensoussan, S. Mitter and B. Pasik-Duncan)*, 7, 2007. Accepted for publication.

9. L. Gerencsér, G. Molnár-Sáska, Gy. Michaletzky, and G. Tusnády. A new approach for the statistical analysis of Hidden Markov Models. *IEEE Transactions on Automatic Control*, under revision.

10. L. Gerencsér, G. Molnár-Sáska, Gy. Michaletzky, and G. Tusnády. New methods for the statistical analysis of Hidden Markov Models. In *Proceedings of the 41th IEEE Conference on Decision & Control, Las Vegas*, pages WeP09–6 2272–2277, 2002.

11. Y. Kifer. Ergodic Theory of Random Transformation. *Progress in Probability and Statistics*, 10, 1986.

12. F. LeGland and L. Mevel. Basic Properties of the Projective Product with Application to Products of Column-Allowable Nonnegative Matrices. *Mathematics of Control, Signals and Systems*, 13:41–62, 2000.

13. F. LeGland and L. Mevel. Exponential Forgetting and Geometric Ergodicity in Hidden Markov Models. *Mathematics of Control, Signals and Systems*, 13:63–93, 2000.

14. B.G. Leroux. Maximum-likelihood estimation for Hidden Markov-models. *Stochastic Processes and their Applications*, 40:127–143, 1992.

15. L. Mevel and L. Finesso. Asymptotical statistics of misspecified hidden Markov models. *IEEE Transactions on Automatic Control*, 49:1123 – 1132, 2004.

16. R. E. Mortensen. Optimal control of continuous time stochastic systems. Technical report, PhD thesis, University of California, Berkeley, CA, USA, 1966.

17. H.H. Panjer and G.E. Willmot. *Insurance Risk Models*. Society of Actuaries, 1992.

18. G. Picci. On the internal structure of finite state stochastic processes, recent developments in variable structure systems, Economics and Biology. In *Lecture notes in Economics and Mathematical Systems (editors R.R. Mohler and Ruberti)*, pages 288–304. Springer-Verlag, 1978.

19. G. Picci and J. H. van Schuppen. On the weak finite stochastic realization problem. In *Lecture Notes in Control and Information Sciences*, pages 237–242. Springer, New York, 1984.

20. E. Sparre Andersen. On the collective theory of risk in the case of contagation between the claims. In *Transactions of the XV-th International Congress of Actuaries*, pages 219–229, 1957.

21. M. Vidyasagar. The realization problem for Hidden Markov Models: The complete realization problem. In *Proceedings of the 44th IEEE Conference on Decision and Control*, pages 6632–6637, 2005.

22. M. Zakai. On the optimal filtering of diffusion processes. *Zeitschrift für Wahrscheinlichkeitstheorie und verwandte Gebiete*, 11:230–243, 1969.

Identifiability and Informative Experiments in Open and Closed-Loop Identification[*]

Michel Gevers[1], Alexandre Sanfelice Bazanella[1,2], and Ljubiša Mišković[1]

[1] CESAME, Université Catholique de Louvain, Louvain-la-Neuve, Belgium
`{bazanela,gevers,miskovic}@csam.ucl.ac.be`
[2] On leave from Electrical Engineering Department, Universidade Federal do Rio Grande do Sul, Porto Alegre-RS, Brazil
`bazanela@ece.ufrgs.br`

> *To Giorgio, who to our great surprise has not solved the rather fundamental problems raised in this chapter, we hope to offer some food for thought.*

1 Introduction

This chapter takes a new look at the concept of identifiability and of informative experiments for linear time-invariant systems, both in open-loop and in closed-loop identification. Some readers might think that everything has been said and written about these concepts, which were much studied all through the 1970's. We shared the same view ... until recently. The motivation for our renewed interest into these very fundamental questions is the recent surge of interest in the question of experiment design, itself triggered by the new concept of *least costly identification experiment for robust control* [3, 4, 5, 6]. Briefly speaking, least costly experiment design for robust control refers to achieving a prescribed accuracy at the lowest possible price, which is typically measured in terms of the duration of the identification experiment, the perturbation induced by the excitation signal, or any combination of these. In this context, questions like the following become relevant:

1. what is the smallest amount of external excitation that is required to achieve identifiability (or to achieve a given accuracy level)?
2. assuming that the system operates in closed-loop, is noise excitation sufficient to guarantee identifiability?
3. if noise excitation is not sufficient to guarantee identifiability in a closed-loop experiment, then how much additional reference excitation is required?
4. assuming that excitation can be applied at different entry points of a multi-input system operating in closed loop, is it necessary to excite each input to achieve identifiability (or to achieve a given accuracy level)?

Sufficient conditions for identifiability using noise excitation only (question 2) have been given, under different sets of assumptions, in [4, 11, 12]. The key condition for

---

[*] This chapter presents research results of the Belgian Programme on Interuniversity Attraction Poles, initiated by the Belgian Federal Science Policy Office. The second author is also partially supported by the Brazilian Ministry of Education through CAPES.

A. Chiuso et al. (Eds.): Modeling, Estimation and Control, LNCIS 364, pp. 151–170, 2007.
springerlink.com

identifiability using noise excitation only is in terms of the complexity of the feedback controller; this complexity condition relates the controllability (or observability) indices of the controller to the controllability (or observability) indices of the plant. Question 4 has been addressed in [2] where it is shown that, when identifiability cannot be achieved using noise excitation only, this does not imply that all reference inputs must be excited.

In attempting to address questions 1 and 3 above, we discovered to our surprise that these questions do not seem to have been addressed (or at least solved) before. As is well-known, besides the choice of an identifiable model structure, the key ingredient to achieve identifiability is the informativity of the experiment. In open-loop identification, and in all closed-loop identification experiments where the noise excitation by itself does not make the experiment informative, the informativity is achieved by applying a sufficiently rich input signal. The degree of richness of a signal is a concept that is precisely defined; a signal is said to be sufficiently rich of degree n if its spectral density is nonzero in at least n distinct frequency points in the interval $(-\pi, \pi]$. But whereas the scientific literature abounds with sufficient conditions on input signal richness, there appear to be no result on the smallest possible degree of richness that delivers informative data in a given identification setup. In other words, necessary conditions on input richness that will guarantee an informative experiment are strangely lacking.

The purpose of this contribution is to attempt to find the smallest possible degree of richness of the excitation signal that makes an experiment informative with respect to a chosen model structure, both in open-loop and in closed-loop identification. More precisely, we address the following two questions:

- assuming open-loop identification, what is the smallest degree of input signal richness that is necessary to achieve an informative experiment with respect to a chosen model structure?
- assuming closed-loop identification with a controller that is not sufficiently complex to yield identifiability using noise excitation only, what is then the smallest degree of reference signal excitation that is necessary to achieve an informative experiment with respect to a chosen model structure?

In addressing these questions, we shall introduce a new framework that allows one to handle in the same way the range space spanned by stationary stochastic vectors and that spanned by vectors of rational transfer functions. We believe this new framework to be a convenient tool to establish results on the transfer of excitation from input signals to regression vectors through linear time-invariant filters. Our analysis and results will be established for single-input single-output (SISO) systems, but the framework we develop lends itself easily to extensions to multi-input multi-output (MIMO) systems.

The chapter is organized as follows. In Section 2 we set up the notations and the key tools of the prediction error identification framework. In Section 3, we recall the basic concepts of identifiability and informative experiments. The body of our results are in Section 4 which focuses on the key role of the information matrix. Our main results concern the derivation of necessary and sufficient conditions on the input signal that make a regressor persistently exciting. This allows us to formulate necessary and sufficient conditions on input signal richness that makes the information matrix have full rank. Finally, in Section 5 we apply these results to some widely utilized model structures.

2   The Prediction Error Identification Setup

Consider the identification of a linear time-invariant discrete-time single-input-single-output process

$$\mathcal{S} : y(t) = G_0(z)u(t) + H_0(z)e(t) \tag{1}$$

In (1) z is the forward-shift operator, $G_0(z)$ and $H_0(z)$ are the process transfer functions, $u(t)$ is the control input and $e(t)$ is white noise with variance σ_e^2. Both transfer functions, $G_0(z)$ and $H_0(z)$, are rational and causal (proper); furthermore, $H_0(\infty) = 1$, that is the impulse response $h(t)$ of the filter $H_0(z)$ satisfies $h(0) = 1$.

This true system may be under feedback control with a causal rational stabilizing controller $K(z)$:

$$u(t) = K(z)[r(t) - y(t)]. \tag{2}$$

The system (1) is identified using a model structure parametrized by a vector $\theta \in \mathcal{R}^d$:

$$M(\theta) : y(t) = G(z, \theta)u(t) + H(z, \theta)\varepsilon(t) \tag{3}$$

The set of models $M(\theta)$, for all θ in some set $D_\theta \in \mathcal{R}^d$, defines the model set $\mathcal{M}$: $\mathcal{M} \triangleq \{M(\theta) \mid \theta \in D_\theta\}$. The true system is said to belong to this model set, $\mathcal{S} \in \mathcal{M}$, if there is a θ_0 such that $M(\theta_0) = \mathcal{S}$. In a prediction error identification framework, a model $[G(z, \theta) \ H(z, \theta)]$ uniquely defines the one-step-ahead predictor of $y(t)$ given all input/output data up to time t:

$$\hat{y}(t|t - 1, \theta) = W_u(z, \theta)u(t) + W_y(z, \theta)y(t), \tag{4}$$

where $W_u(z, \theta)$ and $W_y(z, \theta)$ are stable filters obtained from the model

$$[G(z, \theta) \ H(z, \theta)]$$

as follows:

$$W_u(z, \theta) = H^{-1}(z, \theta)G(z, \theta), \ W_y(z, \theta) = [I - H^{-1}(z, \theta)]. \tag{5}$$

Since there is a $1 - 1$ correspondance between $[G(z, \theta), \ H(z, \theta)]$ and $[W_u(z, \theta), W_y(z, \theta)]$, the model $M(\theta)$ will in the future refer indistinctly to either one of these equivalent descriptions. For later use, we introduce the following vector notations:

$$W(z, \theta) \triangleq [W_u(z, \theta) \ W_y(z, \theta)], \quad z(t) \triangleq \begin{bmatrix} u(t) \\ y(t) \end{bmatrix} \tag{6}$$

We shall also consider throughout this chapter that the vector process $z(t)$ is quasistationary [9], so that the spectral densitiy matrix $\Phi_z(\omega)$ is well defined.

The one-step-ahead prediction error is defined as:

$$\varepsilon(t, \theta) \triangleq y(t) - \hat{y}(t|t - 1, \theta) \tag{7}$$
$$= W(z, \theta)z(t) = H^{-1}(z, \theta)\left[y(t) - G(z, \theta)u(t)\right]$$

Using a set of input-output data of length N and a least squares prediction error criterion yields the estimate $\hat{\theta}_N$ [9]:

$$\hat{\theta}_N = \arg \min_{\theta \in D_\theta} \frac{1}{N} \sum_{t=1}^{N} \varepsilon^2(t, \theta). \tag{8}$$

Under reasonable conditions [9], $\hat{\theta}_N \xrightarrow{N \to \infty} \theta^* \triangleq \arg \min_{\theta \in D_\theta} \overline{V}(\theta)$, with

$$\overline{V}(\theta) \triangleq E[\varepsilon^2(t, \theta)]. \tag{9}$$

If $\mathcal{S} \in \mathcal{M}$ and if $\hat{\theta}_N \xrightarrow{N \to \infty} \theta_0$, the parameter error converges to a Gaussian random variable:

$$\sqrt{N}(\hat{\theta}_N - \theta_0) \xrightarrow{N \to \infty} N(0, P_\theta), \tag{10}$$

where

$$P_\theta = [I(\theta)]^{-1}|_{\theta = \theta_0}, \tag{11}$$

$$I(\theta) = \frac{1}{\sigma_e^2} E\left[\psi(t, \theta)\psi(t, \theta)^T\right], \tag{12}$$

$$\psi(t, \theta) = -\frac{\partial \varepsilon(t, \theta)}{\partial \theta} = \frac{\partial \hat{y}(t|t-1, \theta)}{\partial \theta} = W(z, \theta)z(t) \tag{13}$$

The matrix $I(\theta_0)$ is called the information matrix, and will be much coveted in this chapter.

3 Identifiability, Informative Data, and All That Jazz

Several concepts of identifiability have been proposed in the scientific literature, and these definitions have evolved over the years. They can be broadly classified into *consistency-oriented definitions*, which focus on whether the parameter estimate $\hat{\theta}_N$ converges to the 'true' parameter θ_0 in some stochastic sense, and *uniqueness-oriented definitions*, which deal with the question of whether the model structure is such that the identification criterion has a unique global minimum. Here we adopt a uniqueness-oriented definition proposed in [9].

Definition 1 (Identifiability). *A parametric model structure $M(\theta)$ is locally identifiable at a value θ_1 if $\exists \delta > 0$ such that, for all θ in $\| \theta - \theta_1 \| \leq \delta$:*

$$[W_u(z, \theta)\ W_y(z, \theta)] = [W_u(z, \theta_1)\ W_y(z, \theta_1)] \ \forall \omega \iff \theta = \theta_1.$$

The model structure is globally identifiable at θ_1 if the same holds for $\delta \to \infty$. Finally, a model structure is globally identifiable if it is globally identifiable at almost all θ_1. ∎

We now introduce the matrix $\Gamma(\theta) \in \mathcal{R}^{d \times d}$:

$$\Gamma(\theta) \triangleq \int_{-\pi}^{\pi} \nabla_\theta W(e^{j\omega}, \theta) \, \nabla_\theta W^H(e^{j\omega}, \theta) \, d\omega \tag{14}$$

where $\nabla_\theta W(e^{j\omega}, \theta) \triangleq \frac{\partial W(e^{j\omega}, \theta)}{\partial \theta}$, and for any $M(e^{j\omega})$, the notation $M^H(e^{j\omega})$ denotes $M^T(e^{-j\omega})$. The following result is then an alternative definition for local identifiability of a model structure; see problem 4G.4 in [9].

Proposition 1. *A parametric model structure $M(\theta)$ is locally identifiable at θ_1 if $\Gamma(\theta)$ is nonsingular at θ_1.* ∎

Most commonly used model structures (except ARX) are not globally identifiable, but they are globally identifiable at all values θ that do not cause pole-zero cancellations: see Chapter 4 in [9]. For the existence of a unique global minimum of $\overline{V}(\theta)$, it is required that the model structure is globally identifiable at θ_0. We illustrate the loss of rank of $\Gamma(\theta)$ at a pole-zero cancellation with the following example.

Example 1. Consider the OE (Output-Error) model structure: $y(t) = \frac{B(z^{-1})}{F(z^{-1})}u(t) + e(t)$, where $B(z^{-1}) = b_1 z^{-1} + b_2 z^{-2}$ and $F(z^{-1}) = 1 + f_1 z^{-1} + f_2 z^{-2}$, with $\theta = (b_1\ b_2\ f_1\ f_2)^T$. Then

$$
\nabla_\theta W(e^{j\omega}, \theta) = \frac{1}{F^2}
\begin{pmatrix}
Fz^{-1} \\
Fz^{-2} \\
-Bz^{-1} \\
-Bz^{-2}
\end{pmatrix}
= \frac{1}{F^2}
\underbrace{
\begin{pmatrix}
1 & f_1 & f_2 & 0 \\
0 & 1 & f_1 & f_2 \\
0 & -b_1 & -b_2 & 0 \\
0 & 0 & -b_1 & -b_2
\end{pmatrix}
}_{S_{BF}}
\begin{pmatrix}
z^{-1} \\
z^{-2} \\
z^{-3} \\
z^{-4}
\end{pmatrix}
\tag{15}
$$

The matrix S_{BF} is called a Sylvester matrix [7]: it is nonsingular if and only if the polynomials B and F have no common factor. Thus, $\Gamma(\theta)$ is nonsingular at all values θ except those that cause a pole-zero cancellation in $\frac{B}{F}$.

The definition of identifiability (local, or global) is a property of the parametrization of the model $[G(z, \theta),\ H(z, \theta)]$ or, equivalently, $[W_u(z, \theta),\ W_y(z, \theta)]$. It tells us that if the model structure is globally identifiable at some θ_1, then there is no other parameter value $\theta \neq \theta_1$ that yields the exact same predictor as $M(\theta_1)$. However, it does not tell us that if the true system is in the model set for some parameter value θ_0, then θ_0 will be the unique global minimum of the identification criterion. This requires, additionally, that the data set is informative enough to distinguish between different predictors, which leads us to the definition of informative data with respect to a model structure.

Definition 2 (Informative data). *[9] A quasistationary data set $z(t)$ is called informative with respect to a parametric model set $\{M(\theta), \theta \in \mathcal{D}_\theta\}$ if, for any two models $W(z, \theta_1)$ and $W(z, \theta_2)$ in that set,*

$$
E\{[W(z, \theta_1) - W(z, \theta_2)]z(t)\}^2 = 0 \tag{16}
$$

implies

$$
W(e^{j\omega}, \theta_1) = W(e^{j\omega}, \theta_2)\ \ \forall \omega \tag{17}
$$

∎

By Parseval's theorem, we can rewrite:

$$
E\{[W(z, \theta_1) - W(z, \theta_2)]z(t)\}^2 = \frac{1}{2\pi} \int_{-\pi}^{\pi} |W(e^{j\omega}, \theta_1) - W(e^{j\omega}, \theta_2)|^2 \Phi_z(\omega)\,d\omega
$$

$$
\tag{18}
$$

It is easy to see that an experiment that yields $\Phi_z(\omega) > 0$ for almost all ω is informative for all model structures, but such condition is of course unnecessarily strong.

The definition of informative data is with respect to a given model set, not with respect to the true system, which may or may not belong to the model set. In an identification experiment, one typically first selects a globally identifiable model structure; this is a user's choice. Experimental conditions must then be selected that make the data informative with respect to that structure; this is again a user's choice. However, the data are generated by the true system, in open or in closed loop. Thus, the conditions that make a data set $z(t)$ informative with respect to some model structure depend on the true system and on the possible feedback configuration. The information matrix (12) combines, as we shall see, information about the identifiability of the model structure and about the informativity of the experiments. In addition, as we have seen in (11), its inverse characterizes the precision with which we can estimate the model parameters from data. We thus rewrite the information matrix in a way that will make the connections with model structure and data much more transparent.

Combining (12) and (13) yields:

$$I(\theta) = \frac{1}{2\pi} \int_{-\pi}^{\pi} \nabla_\theta W(e^{j\omega}, \theta) \Phi_z(\omega) \nabla_\theta W^H(e^{j\omega}, \theta) d\omega$$

where $\Phi_z(\omega)$ is the power spectrum of the data $z(t)$ generated by an identification experiment. Comparing this expression with $\Gamma(\theta)$ in (14), we have the following result.

Proposition 2. *Consider an identification experiment that generates data with spectrum $\Phi_z(\omega)$ and assume that a model structure $W(z, \theta)$ is used. Then the information matrix is nonsingular at θ_1 if the following two conditions hold:*

(i) the model structure is locally identifiable at θ_1;
(ii) $\Phi_z(\omega) > 0$ for almost all ω.   ■

We introduce the following definition.

Definition 3 (Regularity). *We say that the information matrix $I(\theta)$ is regular at θ_1 if $I(\theta_1) \succ 0$.*   ■

While the identifiability of the model structure at θ_1 is a necessary condition for the regularity of the information matrix, the positivity of the joint spectrum $\Phi_z(\omega) > 0$ at almost all ω is again unnecessarily strong. A major contribution of this chapter will be to describe the weakest possible richness conditions on the input signal $u(t)$ (in open-loop identification) or $r(t)$ (in closed-loop identification) that make the information matrix full rank for a given model structure. This turns out to be a remarkably difficult problem. We shall examine this problem in the situation where the system is in the model set. Thus, we make the following assumption.

Assumption 1. *The true system (1) belongs to the model set $\mathcal{M}$, that is $M(\theta_0) = \mathcal{S}$ for some $\theta_0 \in \mathcal{D}_\theta$.*

Under Assumption 1 we have the following classical result [9].

Proposition 3. *Consider a model structure that obeys Assumption 1, let this model structure be globally identifiable at θ_0, and let the data be informative with respect to this model structure. Then θ_0 is the unique global minimum of $\overline{V}(\theta)$ defined by (9), and in addition $I(\theta_0) > 0$.* ■

4   Analysis of the Information Matrix

Convergence of an identification algorithm to the exact θ_0 when $\mathcal{S} \in \mathcal{M}$ rests on the satisfaction of two different conditions:

- the use of a model structure that is identifiable, at least at the global minimum θ_0 of the asymptotic criterion $\overline{V}(\theta)$;
- the application of experiments that are informative with respect to the model structure used.

These two conditions depend essentially on the used model structure. They depend on the true system only via the generation of the data. Indeed, the data $z(t)$ must be informative w.r.t. the model structure, but they are generated by the true system. As noted in Proposition 2, the information matrix combines information on the model structure and information on the data generated by the experiment. $I(\theta)$ can be regular only at values of θ that are (at least) locally identifiable, i.e. where $\Gamma(\theta) > 0$. At those values, the regularity of the information matrix depends additionally on the informativity of the data set, i.e. on $\Phi_z(\omega)$.

Thus the focus of our attention, from now on, will be to seek conditions under which the information matrix $I(\theta)$ is regular at all values of θ at which $\Gamma(\theta) > 0$, and in particular at the true θ_0, assuming that the system is globally identifiable at θ_0. To simplify all expressions, we shall assume that $\sigma_e = 1$. The information matrix is then defined as $I(\theta) = E[\psi(t,\theta)\psi(t,\theta)]$ where $\psi(t,\theta) = W(z,\theta)z(t)$ is the gradient of the predictor, which we shall call the *pseudoregression vector*: see (13). We first examine the expressions of this gradient.

4.1   Expressions of the Pseudoregression Vector

The pseudoregression vector can be written:

$$\psi(t,\theta) = [\nabla_\theta W_u(z,\theta) \; \nabla_\theta W_y(z,\theta)] \begin{bmatrix} u(t) \\ y(t) \end{bmatrix} = \nabla_\theta W(z,\theta)z(t) \qquad (19)$$

We rewrite this gradient in terms of the external excitation signals, u and e in the case of open-loop data, r and e in the case of closed-loop data. To improve readability, we delete the explicit dependence on the variables z and θ whenever it creates no confusion.

Open-Loop Identification Setup

In open-loop identification, the data are generated as

$$\begin{bmatrix} u(t) \\ y(t) \end{bmatrix} = \begin{bmatrix} 1 & 0 \\ G_0 & H_0 \end{bmatrix} \begin{bmatrix} u(t) \\ e(t) \end{bmatrix}$$

The pseudoregressor is then expressed in terms of the external signals as

$$\psi(t,\theta) = [\nabla_\theta W_u + G_0 \nabla_\theta W_y \quad H_0 \nabla_\theta W_y] \begin{bmatrix} u(t) \\ e(t) \end{bmatrix} \tag{20}$$

$$\triangleq V_{uol}(z,\theta)u(t) + V_{eol}(z,\theta)e(t) \tag{21}$$

Closed-Loop Identification Setup

In closed-loop identification, the data are generated as

$$\begin{bmatrix} u(t) \\ y(t) \end{bmatrix} = S \begin{bmatrix} K & -KH_0 \\ KG_0 & H_0 \end{bmatrix} \begin{bmatrix} r(t) \\ e(t) \end{bmatrix}$$

where $K = K(z)$ is the controller, and $S = S(z) = \frac{1}{1+K(z)G_0(z)}$ is the sensitivity function. The pseudoregressor is then expressed in terms of the external signals as

$$\psi(t,\theta) = [SK\left(\nabla_\theta W_u + G_0 \nabla_\theta W_y\right) \quad SH_0\left(\nabla_\theta W_y - K \nabla_\theta W_u\right)] \begin{bmatrix} r(t) \\ e(t) \end{bmatrix} \tag{22}$$

$$\triangleq V_{rcl}(z,\theta)r(t) + V_{ecl}(z,\theta)e(t) \tag{23}$$

4.2 The Range and Kernel of Rank-One Vector Processes

We observe that in both cases the pseudoregressor $\psi(t,\theta)$ that "feeds" the information matrix is made up of filtered versions of quasistationary scalar signals, where the filters are d-vectors of rational transfer functions. In order to study the rank of the matrix $I(\theta)$ that results from taking the expectation of these rank-one processes, we introduce the following definitions.

Definition 4. *Let $V(z) : \mathcal{C} \mapsto \mathcal{K}^d(z)$ be a d-vector of proper stable rational functions. The left-kernel of $V(z)$, denoted $Ker\{V(z)\}$, is the set spanned by all real-valued vectors $\alpha \in \Re^d$ such that $\alpha^T V(z) = 0 \ \forall z \in \mathcal{C}$. Its dimension is called the nullity and annotated ν_V. The rank of $V(z)$ is defined as $\rho_V = d - \nu_V$, and $V(z)$ is said to have full rank if $\rho_V = d$.* ∎

Definition 5. *Let $\psi(t) : \Re \mapsto \Re^d$ be a d-vector of quasi-stationary processes. The left-kernel of $\psi(t)$, denoted $Ker\{\psi(t)\}$, is the set spanned by all real-valued vectors $\alpha \in \Re^p$ such that $E[\alpha^T \psi(t)]^2 = 0$, or alternatively $\alpha^T \Phi_\psi(\omega)\alpha = 0 \ \ \forall \omega$ where $\Phi_\psi(\omega)$ is the spectral density matrix of $\psi(t)$. Its dimension is called the nullity and annotated ν_ψ. The rank of $\psi(t)$ is defined as $\rho_\psi = d - \nu_\psi$, and $\psi(t)$ is said to have full rank if $\rho_\psi = d$.* ∎

Observation. A d-vector $V(z)$ of proper stable rational functions has full rank if $V(z)$ is output reachable. A $d \times m$ transfer function matrix $H(z) = \sum_{k=0}^{\infty} H_k z^{-k}$ is called *output reachable* if $\nexists \alpha \in \mathcal{R}^d$ such that $\alpha^T H(z) = 0 \ \forall z \in \mathcal{C}$ or, equivalently, $\alpha^T H_k = 0 \ \forall k$: see e.g. [10].

With these definitions under our belt, we are now ready to analyze the rank of the information matrix $I(\theta)$ as a function of the signals u and e (in an open-loop setup), or r and e (in a closed-loop setup). The following result follows immediately from the definitions.

Lemma 1. *The rank of the information matrix $I(\theta_1)$ at some value θ_1 is the rank of $\psi(t, \theta_1)$. In particular, the information matrix is regular at θ_1 if and only if $Ker\{\psi(t, \theta_1)\} = \{0\}$; equivalent statements are $\nu_\psi(\theta_1) = 0$ and $\rho_\psi(\theta_1) = d$.* ∎

The analysis of the rank of $I(\theta)$ thus reduces to the analysis of the rank of $\psi(t, \theta)$ which itself is composed of the sum of two vector filters of scalar stationary stochastic processes: see (21) and (23). For the white noise driven terms, the analysis is very simple: we have the following theorem.

Theorem 1. *Let $\psi_e(t, \theta) = V_e(z, \theta)e(t)$, where $V_e(z, \theta)$ is a d-vector of stable proper rational filters and $e(t)$ is white noise. Then $Ker\{\psi_e(t, \theta)\} = Ker\{V_e(z, \theta)\}$, and hence $\rho_{\psi_e} = \rho_{V_e}$.*

Proof. The result follows immediately by observing that, for any $\alpha \in \mathcal{R}^d$:

$$
\begin{aligned}
0 &= \alpha^T E[\psi_e(t, \theta)\psi_e^T(t, \theta)]\alpha \\
&= \frac{1}{2\pi} \int_{-\pi}^{\pi} \alpha^T V_e(e^{j\omega}, \theta)V_e^H(e^{j\omega}, \theta)\alpha \; d\omega
\end{aligned}
$$
∎

This proof shows the coherence and the usefulness of our apparently disconnected definitions of kernels for vectors of stationary stochastic processes and for vectors of proper stable transfer functions.

4.3   Regularity Conditions for $I(\theta)$: A First Analysis

We now exploit the definitions we have just introduced to produce some first conditions on the regularity of the information matrix.

Theorem 2. *With the notations introduced in (21) and (23), the information matrix $I(\theta)$ is regular*

- *in open-loop identification if and only if*

$$
\begin{aligned}
Ker\{V_{uol}(z, \theta)u(t) + V_{eol}(z, \theta)e(t)\} &= Ker\{V_{uol}(z, \theta)u(t)\} \cap Ker\{V_{eol}(z, \theta)\} \\
&= \{0\} \tag{24}
\end{aligned}
$$

- *in closed-loop identification if and only if*

$$
\begin{aligned}
Ker\{V_{rcl}(z, \theta)r(t) + V_{ecl}(z, \theta)e(t)\} &= Ker\{V_{rcl}(z, \theta)r(t)\} \cap Ker\{V_{ecl}(z, \theta)\} \\
&= \{0\} \tag{25}
\end{aligned}
$$

Proof. Consider the case of open-loop identification. It follows from (21) and the independence of the signals u and e that

$$
\alpha^T E[\psi(t, \theta)\psi^T(t, \theta)]\alpha = E[\alpha^T V_{uol}(z, \theta)u(t)]^2 + E[\alpha^T V_{eol}(z, \theta)e(t)]^2. \tag{26}
$$

Therefore $\alpha \in Ker\{\psi(t, \theta)\}$ if and only if α belongs to the left-kernels of both $V_{uol}(z, \theta)u(t)$ and $V_{eol}(z, \theta)e(t)$, and hence to their intersection. Next, it follows from Theorem 1 that $Ker\{V_{eol}(z, \theta)e(t)\} = Ker\{V_{eol}(z, \theta)\}$. The proof is identical for the closed-loop case. ∎

Observe that the conditions (24) and (25) use the two distinct but compatible notions of kernel, defined in Definitions 4 and 5, respectively, in the same statement. These conditions show how the regularity of $I(\theta)$ depends on both the model structure through $V_{uol}(z,\theta)$ and $V_{eol}(z,\theta)$ (respectively, $V_{rcl}(z,\theta)$ and $V_{ecl}(z,\theta)$) and the excitation signal $u(t)$ (respectively $r(t)$). We now elaborate on these conditions, separately for the open-loop and for the closed-loop identification setup.

Open-Loop Identification

In open-loop identification, the filters $V_{uol}(z,\theta)$ and $V_{eol}(z,\theta)$ are given by (20) and (21). Simple calculations show that they are expressed in terms of the model transfer functions $G(z,\theta)$ and $H(z,\theta)$ as follows[1].

$$V_{uol}(z,\theta) = \nabla_\theta W_u + G_0 \nabla_\theta W_y = \frac{1}{H^2(\theta)}\left[H(\theta)\nabla_\theta G(\theta) + (G_0 - G(\theta))\nabla_\theta H(\theta)\right]$$

$$V_{eol}(z,\theta) = H_0 \nabla_\theta W_y = \frac{H_0}{H^2(\theta)}\nabla_\theta H(\theta)$$

We then have the following result.

Theorem 3. *Let $\mathcal{N}_H$ denote the left-kernel of $\nabla_\theta H(z,\theta)$. Then $I(\theta)$ is regular either if $\mathcal{N}_H = \{\mathbf{0}\}$ or if for each non-zero d-vector $\alpha \in \mathcal{N}_H$ we have*

$$E[\alpha^T \nabla_\theta G(z,\theta)u(t)]^2 \neq 0. \tag{27}$$

Proof. First note that the set of vectors $\{\alpha \in \mathcal{N}_H \subseteq \mathcal{R}^d\}$ spans $Ker\{\nabla_\theta W_y\} = Ker\{V_{eol}(z,\theta)\}$. Therefore, by Theorem 2, $I(\theta) > 0$ if and only if either $\mathcal{N}_H = \{\mathbf{0}\}$ or, for each nonzero $\alpha \in \mathcal{N}_H$, we have $E[\alpha^T (\nabla_\theta W_u + G_0 \nabla_\theta W_y)u(t)]^2 \neq 0$. Since $\alpha^T \nabla_\theta H(z,\theta) = \mathbf{0}$, this is equivalent with $E[\alpha^T \nabla_\theta G(z,\theta)u(t)]^2 \neq 0$. ∎

Closed-Loop Identification

In closed-loop identification, the filters $V_{rcl}(z,\theta)$ and $V_{ecl}(z,\theta)$ are given by (22) and (23). They are expressed in terms of the model transfer functions $G(z,\theta)$ and $H(z,\theta)$ as follows.

$$V_{rcl}(z,\theta) = KS(\nabla_\theta W_u + G_0 \nabla_\theta W_y)$$
$$= KS\{\frac{1}{H^2(\theta)}\left[H(\theta)\nabla_\theta G(\theta) + (G_0 - G(\theta))\nabla_\theta H(\theta)\right]\}$$
$$V_{ecl}(z,\theta) = H_0 S(\nabla_\theta W_y - K\nabla_\theta W_u)$$
$$= \frac{H_0 S}{H^2(\theta)}\{\nabla_\theta H(\theta) - K\left[H(\theta)\nabla_\theta G(\theta) - G(\theta)\nabla_\theta H(\theta)\right]\}$$

For the closed-loop identification setup we have the following result.

Theorem 4. *Let $\mathcal{N}_{V_{ecl}}$ denote the left-kernel of $V_{ecl}(z,\theta)$. Then $I(\theta)$ is regular either if $\mathcal{N}_{V_{ecl}} = \{\mathbf{0}\}$ or if for each non-zero d-vector $\alpha \in \mathcal{N}_{V_{ecl}}$ we have*

$$E[\alpha^T \nabla_\theta W_y(z,\theta)r(t)]^2 = E[\alpha^T K(z)\nabla_\theta W_u(z,\theta)r(t)]^2 \neq 0. \tag{28}$$

---

[1] We omit the argument z here for reasons of brevity.

Proof. First note that for each $\alpha \in \mathcal{N}_{V_{ecl}} \subseteq \mathcal{R}^d$ we have $\alpha^T \nabla_\theta W_y(z, \theta) = \alpha^T K(z)$ $\nabla_\theta W_u(z, \theta)$. By Theorem 2, $I(\theta) > 0$ if and only if either $\mathcal{N}_{V_{ecl}} = \{0\}$ or if, for each non-zero $\alpha \in \mathcal{N}_{V_{ecl}}$ we have $E[\alpha^T KS(\nabla_\theta W_u + G_0 \nabla_\theta W_y)r(t)]^2 \neq 0$. Now observe that $\alpha^T KS(\nabla_\theta W_u + G_0 \nabla_\theta W_y) = \alpha^T S(1 + KG_0)\nabla_\theta W_y = \alpha^T \nabla_\theta W_y = \alpha^T K \nabla_\theta W_u$. This proves the result. ∎

4.4   Rich and Exciting Signals

In Theorem 1 we have seen that a regressor $\psi(t)$ obtained by filtering a white noise signal $e(t)$ through a vector filter $V(z)$ has the same left-kernel as $V(z)$, i.e. white noise causes no drop of rank. The same is actually true for any input signal that has a continuous spectrum. For the parts of $\psi(t, \theta)$ driven by the controlled signals $u(t)$ or $r(t)$ (see (21) and (23)), we want to consider input signals ($u(t)$ or $r(t)$) that have discrete spectra, such as multisines. In order to analyze the rank properties of regressors obtained by filtering such signals with discrete spectra, we need to introduce the concept of richness of a signal. We first define a persistently exciting regression vector.

Definition 6. *A quasistationary vector signal $\psi(t)$ is called persistently exciting (denoted PE) if $E[\psi(t)\psi^T(t)] > 0$.* ∎

Whether a quasistationary vector signal $\psi(t)$ obtained as a filtered version (by a vector $V(z)$ of transfer functions) of a quasistationary scalar signal $u(t)$ is PE or not depends not only on whether $Ker\{V(z)\} = \{0\}$ but also on the degree of richness of the input $u(t)$. The richness of a scalar signal is defined as follows.

Definition 7. *A quasistationary scalar signal $u(t)$ is sufficiently rich of order n (denoted SRn) if the following regressor is PE:*

$$\phi_{1,n}(t) \triangleq \begin{bmatrix} u(t-1) \\ u(t-2) \\ \vdots \\ u(t-n) \end{bmatrix} = \begin{bmatrix} z^{-1} \\ z^{-2} \\ \vdots \\ z^{-n} \end{bmatrix} u(t) \tag{29}$$

∎

The vector $\phi_{1,n}(t)$ serves as a basis for all regression vectors that are obtained as (vector)-filtered versions of a scalar signal $u(t)$. For future use, we introduce the notation:

$$\mathcal{B}_{k,n}(z) \triangleq \begin{bmatrix} z^{-k} & z^{-k-1} & \dots & z^{-n} \end{bmatrix}^T, \quad \text{for } k \leq n. \tag{30}$$

Observe that, by our assumption of quasistationarity, $u(t)$ is SRn if $\mathcal{B}_{k+1,k+n}(z)u(t)$ is PE for any k. We denote by $\mathcal{U}_n$ the set of all SRn processes.

Definition 8. *A scalar signal $u(t)$ is sufficiently rich of order exactly n (denoted SREn) if $\phi_{1,n}(t)$ is PE, but $\phi_{1,n+1}(t)$ is not.* ∎

This definition is equivalent with many other classically used definitions, except that nowadays the most common terminology is to say that a signal is PE of order n rather than SR of order n. At the risk of being considered old-fashioned, we prefer the term *sufficiently rich* because sufficient intuitively reflects the notion of *degree of richness* while persistent does not. The following are commonly used definitions that are equivalent with Definitions 7 and 8.

Proposition 4. *A scalar quasistationary signal $u(t)$ is SRn if*

* *its spectral density is nonzero in at least n frequency points in the interval $(-\pi, \pi]$.*
* *it cannot be filtered to zero by a FIR filter of degree n: $\alpha_1 z^{-1} + \ldots \alpha_n z^{-n}$.*

A scalar signal $u(t)$ is SREn if its spectral density is nonzero in exactly n frequency points in the interval $(-\pi, \pi]$. ∎

The equivalence comes by observing that

$$\alpha^T E[\phi_{1,n}(t)\phi_{1,n}^T(t)]\alpha = \frac{1}{2\pi} \int_{-\pi}^{\pi} |\alpha_1 e^{-j\omega} + \ldots + \alpha_n e^{-jn\omega}|^2 \Phi_u(\omega)d\omega.$$

The question of interest here is how the richness of a scalar signal $u(t)$ transfers into the persistence of excitation of a regression vector $\psi(t)$ when this regression vector is obtained as a (vector)-filter of $u(t)$, i.e. $\psi(t) = V(z)u(t)$, where the components of $V(z)$ are stable proper transfer functions. More precisely, we would like to determine the smallest possible degree of richness of $u(t)$ that will make $\psi(t)$ PE. To help us in solving this problem, we have ...... not much. As it happens, the only available results, as far as we know, are sufficiency results. We briefly recall here the main available results.

Proposition 5. *[12] Let $u(t)$ be SRn and let $y(t) = G(z)u(t)$; then $y(t)$ is SRn if the filter $G(z)$ has no zeroes on the unit circle, i.e. if $G(e^{j\omega}) \neq 0$ for all ω.*

Proof. Since $\Phi_y(\omega) = |G(e^{j\omega})|^2 \Phi_u(\omega)$, it follows immediately that if $G(e^{j\omega})$ is nowhere zero on the unit circle, then the frequency points where $\Phi_y(\omega) \neq 0$ and where $\Phi_u(\omega) \neq 0$ are identical. ∎

Proposition 6. *[1] Let $\psi(t) \in \mathcal{R}^d$ be a vector that has the state-space model*

$$\psi(t+1) = A\psi(t) + Bu(t) \tag{31}$$

with $A \in \mathcal{R}^{d \times d}$. Then $\psi(t)$ is PE if $u(t)$ is SRd and the pair $[A, B]$ is completely reachable. ∎

Proposition 7. *[10] Let $\psi(t) \in \mathcal{R}^d$ be the output of a vector $V(z)$ of proper stable filters driven by $u(t)$:*

$$\psi(t) = V(z)u(t), \tag{32}$$

and let $\delta_V \triangleq$ McMillan degree of $V(z)$. Then $\psi(t)$ is PE if the following two conditions hold:

* *the system (32) is output reachable, i.e. $\nexists \alpha \in \mathcal{R}^d$, $\alpha \neq 0$, such that $\alpha^T V(z) = 0 \, \forall z$;*
* *$u(t)$ is SRn with $n \geq \delta_V + 1$.* ∎

We now attempt to relate the left-kernel of $\psi(t)$, the left-kernel of $V(z)$ and the richness of $u(t)$. We first state the trivial lemma.

Lemma 2. The trivial lemma. *Let $\psi(t) = V(z)u(t)$ with $\psi(t) \in \mathcal{R}^d$, $u(t)$ quasistationary, and all components of $V(z)$ proper and stable. Then*

$$Ker\{V(z)\} \subseteq Ker\{\psi(t)\}. \tag{33}$$

∎

The question we address now can be stated as follows.

What are the necessary and sufficient conditions on the richness of $u(t)$ such that $Ker\{\psi(t)\} = Ker\{V(z)\}$ when $\psi(t) = V(z)u(t)$?

$V(z)$ can always be written as

$$V(z) = \frac{N(z^{-1})}{d(z^{-1})} = \frac{z^{-m}}{d(z^{-1})}R\mathcal{B}_{0,k-1}(z) \tag{34}$$

where $d(z^{-1}) = 1 + d_1 z^{-1} + \ldots + d_p z^{-p}$, with $d_p \neq 0$, where $R \in \mathcal{R}^{d \times k}$ is the matrix of real coefficients of the expansion of the numerator matrix $N(z^{-1})$ into powers of z^{-1}, and m is a possible common delay in all elements of $N(z^{-1})$. A necessary condition for $V(z)$ to be output reachable (i.e. $Ker\{V(z)\} = \mathbf{0}$) is that $k \geq d$. As we shall see, in many cases of interest for the transfer of excitation from the signal u, or r, to the pseudo-regression vector ψ, it so happens that R is square, i.e. $k = d$. Thus, we first handle this important (and much easier) special case.

Theorem 5. *Let $\psi(t) = V(z)u(t)$ with $\psi(t) \in \mathcal{R}^d$, $u(t)$ quasistationary, $V(z)$ proper and stable, and let $V(z)$ be decomposed as in (34) with $d = k$. Then $\rho_\psi = d$ if and only if $\rho_V = d$ and $u(t)$ is SRd. (Stated otherwise: $\psi(t)$ has full rank if and only if $V(z)$ has full rank and $u(t)$ is SRd).*

Proof. Using the decomposition (34) with $k = d$, we can write

$$E[\alpha^T \psi(t)]^2 = \frac{1}{2\pi} \int_{-\pi}^{\pi} |\alpha^T R\mathcal{B}_{1,d}(e^{j\omega})|^2 \frac{\Phi_u(\omega)}{|d(e^{j\omega})|^2} d\omega, \tag{35}$$

where we have used the fact that $|e^{-mj\omega}\alpha^T R\mathcal{B}_{0,d-1}(e^{j\omega})|^2 = |\alpha^T R\mathcal{B}_{1,d}(e^{j\omega})|^2$. If $\rho_V = d$, and since $k = d$, it follows that R is nonsingular. Therefore $\alpha^T R \neq 0$ for all nonzero α, and $\alpha^T R$ spans the space of all vectors in $\mathcal{R}^d$. If in addition $u(t)$ is SRd, then by Proposition 4 the integral on the right hand side is nonzero for all $\alpha \neq 0$. Conversely, if $\rho_V < d$ then there exists $\alpha \neq 0$ such that $\alpha^T R = 0$, and if $u(t)$ is SREn with $n < d$, then there exists $\alpha \neq 0$ such that $|\alpha^T R\mathcal{B}_{1,d}(e^{j\omega})|^2 \Phi_u(\omega) = 0$ for all ω. ∎

Example 2. Let

$$V(z) = \frac{1}{z^3(z+0.5)} \begin{bmatrix} z+1 \\ z^2 \\ 1 \end{bmatrix} = \frac{z^{-2}}{1+0.5z^{-1}} \begin{bmatrix} z^{-1}+z^{-2} \\ 1 \\ z^{-2} \end{bmatrix}$$

$$= \frac{z^{-2}}{1+0.5z^{-1}} R\mathcal{B}_{0,2} \quad \text{with } R = \begin{bmatrix} 0 & 1 & 1 \\ 1 & 0 & 0 \\ 0 & 0 & 1 \end{bmatrix}$$

The McMillan degree of $V(z)$ is $\delta_V = 4$, but $\psi(t) \triangleq V(z)u(t)$ will be PE if $u(t)$ is SR3, since R has full rank.

In the more general situation where $k > d$, it is clear that $u(t) = $ SRk is a sufficient condition to guarantee that $Ker\{\psi(t)\} = Ker\{V(z)\}$. However, we shall now show that $Ker\{\psi(t)\} = Ker\{V(z)\}$ for almost all $u(t) \in \mathcal{U}_N$ if and only if $N \geq \rho_V$; recall that $\mathcal{U}_N$ is the set of $u(t)$ that are SRN, and ρ_V is the rank of $V(z)$. To show this, we need a preliminary lemma.

Lemma 3. *Let $\psi(t) = V(z)u(t)$ with $\psi(t) \in \mathcal{R}^d$, $u(t)$ quasistationary, $V(z)$ proper and stable, and let $V(z)$ be decomposed as in (34) with $\rho_V = c$. Let the rows of $Q \in \mathcal{R}^{c \times k}$ be a basis for the rowspace of R, and define the c-vectors $W(z) = \frac{z^{-m}}{d(z^{-1})}Q\mathcal{B}_{0,k-1}(z)$ and $\phi(t) = W(z)u(t)$. Then, for any $u(t)$, $Ker\{\psi(t)\} = Ker\{V(z)\}$ if and only if $Ker\{\phi(t)\} = Ker\{W(z)\} = \mathbf{0}$.*

Proof. Since the rows of Q form a basis for the rowspace of R we can write

$$R = T \begin{bmatrix} Q \\ 0 \end{bmatrix} \tag{36}$$

for some nonsingular matrix $T \in \mathcal{R}^{d \times d}$. Then for any $\alpha \in \mathcal{R}^d$ we have:

$$\alpha^T R = \alpha^T T \begin{bmatrix} Q \\ 0 \end{bmatrix} = \beta^T Q \tag{37}$$

where β is uniquely defined by $\alpha^T T \triangleq (\beta^T \quad \gamma^T)$ with $\beta \in \mathcal{R}^c$ and $\gamma \in \mathcal{R}^{d-c}$. It follows from (37) that

$$\alpha^T \psi(t) = \frac{z^{-m}}{d(z^{-1})}\alpha^T R\mathcal{B}_{0,k-1}(z)u(t) = \frac{z^{-m}}{d(z^{-1})}\beta^T Q\mathcal{B}_{0,k-1}(z)u(t) = \beta^T \phi(t)$$

Therefore the following four statements are all equivalent:

- $Ker\{\psi(t)\} = Ker\{V(z)\}$
- $E[\alpha^T \psi(t)]^2 = 0$ if and only if $\alpha^T R \in Ker\{\mathcal{B}_{0,k-1}(z)u(t)\}$
- $E[\beta^T \phi(t)]^2 = 0$ if and only if $\beta^T Q \in Ker\{\mathcal{B}_{0,k-1}(z)u(t)\}$
- $Ker\{\phi(t)\} = Ker\{W(z)\}$

Finally, since Q has full rank, $Ker\{W(z)\} = \mathbf{0}$.   ∎

Theorem 6. *Let $\psi(t) = V(z)u(t)$ with $\psi(t) \in \mathcal{R}^d$, $u(t)$ quasistationary, $V(z)$ proper and stable, and let $V(z)$ be decomposed as in (34). Then $Ker\{\psi(t)\} = Ker\{V(z)\}$ for almost all $u(t) \in \mathcal{U}_N$ if and only if $N \geq \rho_V$.*

Proof. If $\rho_V < d$, we can replace $\psi(t) = V(z)u(t)$ by $\phi(t) = W(z)u(t)$ with $W(z)$ defined from $V(z)$ as in Lemma 3 above, where $W(z)$ has full rank. Thus, using Lemma 3, we can assume without loss of generality that $\rho_V = d$.

Using Parseval's Theorem and (34) we can write

$$E[\alpha^T \psi(t)]^2 = E[\alpha^T \frac{z^{-m}}{d(z^{-1})} R\mathcal{B}_{0,k-1}(z)u(t)]^2$$

$$= \alpha^T R \left(\frac{1}{2\pi} \int_{-\pi}^{\pi} \frac{\Phi_u(\omega)}{|\,d(e^{-j\omega})\,|^2} \mathcal{B}_{0,k-1}(e^{-j\omega})\mathcal{B}_{0,k-1}^H(e^{-j\omega})d\omega \right) R^T \alpha$$

Let $u(t)$ be SRN with finite N. Its spectrum can then be written as $\Phi_u(\omega) = \sum_{i=1}^{N} \lambda_i \Phi_u(\omega_i)$ with $\omega_i \neq \omega_j$, $i \neq j$. Define its support as the vector $\mathbf{z} = [e^{j\omega_1} e^{j\omega_2} \ldots e^{j\omega_N}] \in \Omega_N$, where $\Omega_N \subset \mathcal{C}^N$ is the set of all supports $\mathbf{z}$ which result in an SRN signal, that is those $\mathbf{z}$ such that $\omega_i \neq \omega_j \; \forall i \neq j$. Ω_N is an N-dimensional subset of $\mathcal{C}^N$ which defines the class of signals $u(t)$ that we consider. Then we can write

$$E[\alpha^T \psi(t)]^2 = \alpha^T R \left(\frac{1}{2\pi} \sum_{i=1}^{N} \lambda_i' \mathcal{B}_{0,k-1}(e^{-j\omega_i})\mathcal{B}_{0,k-1}^H(e^{-j\omega_i}) \right) R^T \alpha$$

where $\lambda_i' = \frac{\lambda_i \Phi_u(\omega_i)}{|d(e^{-j\omega_i})|^2}$. Hence

$$E[\alpha^T \psi(t)]^2 = \alpha^T R Z(\mathbf{z}) \Lambda Z^H(\mathbf{z}) R^T \alpha \tag{38}$$

with

$$Z(\mathbf{z}) = \left[\mathcal{B}_{0,k-1}(e^{-j\omega_1}) \; \mathcal{B}_{0,k-1}(e^{-j\omega_2}) \; \ldots \; \mathcal{B}_{0,k-1}(e^{-j\omega_N}) \right]$$

and $\Lambda = diag\{\lambda_1', \lambda_2', \ldots, \lambda_N'\}$; note that $\rho(Z(\mathbf{z})) = N$ whenever $N \leq k$.

But $\psi(t)$ is full-rank if and only if $P(\mathbf{z}) \overset{\Delta}{=} RZ(\mathbf{z})\Lambda Z^H(\mathbf{z})R^T$ has rank equal to d, which is equivalent to $det(P(\mathbf{z})) \neq 0$. Suppose that $N < d$; then $\rho(Z(\mathbf{z})) = N < d$ which, noting that $\rho(P(\mathbf{z})) \leq \rho(Z(\mathbf{z}))$, implies $det(P(\mathbf{z})) = 0$, thus proving the necessity of $N \geq d$.

For $N \geq d$, the determinant $det(P(\mathbf{z}))$ is a nontrivial polynomial in the vector variable $\mathbf{z}$ and $\psi(t)$ loses rank exactly at the roots of this polynomial. Since the roots of a polynomial define a set of measure zero in the space of its variable, $\psi(t)$ is full-rank for almost all $\mathbf{z} \in \Omega_N$.   ∎

Our Theorem above completely characterizes the required signal richness of $u(t)$ that keeps the range of the regressor vector $V(z)u(t)$ identical to the range of $V(z)$. Yet, it is worth specifying what happens for different levels of excitation, which is given in the following theorem.

Theorem 7. *Let $\psi(t) = V(z)u(t)$ with $\psi(t) \in \mathcal{R}^d$, $u(t)$ quasistationary, $V(z)$ proper and stable, and let $V(z)$ be decomposed as in (34) with $rank(V) = \rho_V$.*

- *If $u(t)$ is **not** SR of order ρ_V, then $Ker(\psi(t)) \subset Ker(V(z))$.*
- *If $u(t)$ is SRk then $Ker(\psi(t)) = Ker(V(z))$.*

Proof. This result follows immediately from Sylvester's inequality:

$$\rho(R) + \rho(Z) - k \leq \rho(RZ) \leq \min(\rho(R), \rho(Z))$$

which yields $\rho(RZ) < \rho(R)$ for $N < \rho_V$ and $\rho(RZ) \geq \rho(R)$ for $N \geq k$.   ∎

The following example illustrates the results of our theorems.

Example 3. Consider the regressor $\psi(t) = V(z)u(t)$, with

$$V(z) = R\left[1\ z^{-1}\ z^{-2}\ z^{-3}\ z^{-4}\right]^{T} \quad \text{where} \quad R = \begin{bmatrix} 0\ 1\ 0\ 0\ 1 \\ 0\ 0\ 1\ 0\ 0 \\ 1\ 0\ 0\ 0\ 1 \end{bmatrix}$$

Consider first $u(t) = \lambda_1 + \lambda_2 \sin(\omega t)$, which is SRE3. For such signal, RZ is a 3×3 matrix, whose determinant is $det(RZ) = -2j[3\sin(\omega) - 2\sin(2\omega) - \sin(3\omega) + \sin(4\omega)]$. Its roots in $(-\pi, \pi]$ are at $-\frac{\pi}{3}$, 0, $\frac{\pi}{3}$ and π, but $\omega = 0$ and $\omega = \pi$ do not keep $u(t) \in \mathcal{U}_3$. Thus, $\psi(t)$ will have rank 3 for all $u(t) \in \mathcal{U}_3$ except for $u(t) = \lambda_1 + \lambda_2 \sin(\frac{\pi}{3}t)$, i.e. for $\omega = \frac{\pi}{3}$.

Now let $u(t) = \lambda_1 \sin(\omega_1 t) + \lambda_2 \sin(\omega_2 t)$ which is SRE4. From Theorem 7 we know that the richness of this signal is in between the "necessary" richness (SR3) and the sufficient richness (SR5). We have

$$RZ = \begin{bmatrix} e^{j\omega_1} + e^{j4\omega_1} & e^{j\omega_2} + e^{j4\omega_2} & e^{-j\omega_1} + e^{-j4\omega_1} & e^{-j\omega_2} + e^{-j4\omega_2} \\ e^{j2\omega_1} & e^{j2\omega_2} & e^{-j2\omega_1} & e^{-j2\omega_2} \\ 1 + e^{j4\omega_1} & 1 + e^{j4\omega_2} & 1 + e^{-j4\omega_1} & 1 + e^{-j4\omega_2} \end{bmatrix}$$

It is rather easy to see that RZ will have full rank for all values of ω_1 and ω_2, $\omega_1 \neq \omega_2$ except those for which $\omega_1 + \omega_2 = \pi$.

5 Regularity of $I(\theta)$ for ARMAX and BJ Model Structures

We now combine the results of Theorem 3 with those on the transfer of sufficiently rich input signals to regression vectors in order to produce necessary and sufficient richness conditions on the input signal that guarantee regularity of the information matrix at all θ at which the model structure is identifiable, i.e. $\Gamma(\theta) > 0$. We do this for ARMAX and Box-Jenkins (BJ) model structures in an open-loop identification setup.

ARMAX Model Structure
Consider the ARMAX model structure

$$A(z^{-1})y(t) = B(z^{-1})u(t) + C(z^{-1})e(t) \tag{39}$$

where $A(z^{-1}) = 1 + a_1 z^{-1} + \ldots + a_{n_a} z^{-n_a}$, $B(z^{-1}) = b_1 z^{-1} + \ldots + b_{n_b} z^{-n_b}$, and $C(z^{-1}) = 1 + c_1 z^{-1} + \ldots + c_{n_c} z^{-n_c}$. We have the following result.

Theorem 8. *For the ARMAX model structure (39), the information matrix $I(\theta)$ is regular at a θ at which the model structure is identifiable if and only if $u(t)$ is SRk, where $k = n_b + n_u(\theta)$ and $n_u(\theta)$ is the number of common roots of the polynomials $A(z^{-1})$ and $C(z^{-1})$ at that θ. $I(\theta)$ is regular at all θ at which the model structure is identifiable if and only if $u(t)$ is SRk with $k = n_b + \min\{n_a, n_c\}$.*

Proof. We first comment that for ARMAX model structures, common roots between the polynomials A and B, as well as between A and C, must be considered, because they can generically occur. However, the three polynomials A, B and C must be co-prime at any identifiable θ. For the ARMAX model structure, we have:

$$\nabla_\theta G(z,\theta) = \frac{1}{A^2}\begin{pmatrix} -Bz^{-1} \\ \vdots \\ -Bz^{-n_a} \\ \hline Az^{-1} \\ \vdots \\ Az^{-n_b} \\ \hline 0 \\ \vdots \\ 0 \end{pmatrix}, \quad \nabla_\theta H(z,\theta) = \frac{1}{A^2}\begin{pmatrix} -Cz^{-1} \\ \vdots \\ -Cz^{-n_a} \\ \hline 0 \\ \vdots \\ 0 \\ \hline Az^{-1} \\ \vdots \\ Az^{-n_c} \end{pmatrix} \qquad (40)$$

Let $\alpha^T = (\alpha_A^T \mid \alpha_B^T \mid \alpha_C^T)$ denote any vector in the left-kernel of $\nabla_\theta H(z,\theta)$, and let $\gamma_A(z^{-1}) \triangleq \alpha_A^T \mathcal{B}_{1,n_a}$, $\gamma_B(z^{-1}) \triangleq \alpha_B^T \mathcal{B}_{1,n_b}$, and $\gamma_C(z^{-1}) \triangleq \alpha_C^T \mathcal{B}_{1,n_c}$. Then

$$\alpha^T \nabla_\theta H(z,\theta) = 0 \;\Leftrightarrow\; z\gamma_A(z^{-1})C(z^{-1}) = z\gamma_C(z^{-1})A(z^{-1}) \qquad (41)$$

At all values of θ at which the polynomials A and C are coprime, it follows from the theory of Diophantine equations (see e.g. [8]) that $\alpha_A = 0$ and $\alpha_C = 0$, because $deg(z\gamma_A(z^{-1})) < deg(A(z^{-1}))$ and $deg(z\gamma_C(z^{-1})) < deg(C(z^{-1}))$. Consider now a θ at which there are common factors between A and C and let $U(z^{-1})$ denote the Greatest Common Divisor (GCD) of A and C, with $deg(U(z^{-1})) = n_u$. Then $A = A_1 U$ and $C = C_1 U$ for some coprime polynomials A_1 and C_1. Then (41) is equivalent with $z\gamma_A(z^{-1})C_1(z^{-1}) = z\gamma_C(z^{-1})A_1(z^{-1})$ where $deg(z\gamma_A) = n_a - 1$ and $deg(z\gamma_C) = n_c - 1$. The set of all solutions of this equation is described by

$$z\gamma_A = \alpha_A^T \mathcal{B}_{0,n_a-1} = A_1 T, \quad z\gamma_C = \alpha_C^T \mathcal{B}_{0,n_c-1} = C_1 T \qquad (42)$$

where $T(z^{-1})$ is an arbitrary polynomial of degree $n_u - 1$. The left-kernel of $\nabla_\theta H(z,\theta)$ is thus defined by those vectors $\alpha^T = (\alpha_A^T \mid \alpha_B^T \mid \alpha_C^T)$ such that α_A and α_C are solution of (42), while α_B is arbitrary. As stated earlier, we consider values of θ at which $\Gamma(\theta) > 0$. At these values of θ, $\alpha^T \nabla_\theta G(z,\theta) \neq 0$ for all vectors α defined above and, by Theorem 3, $I(\theta) > 0$ if $u(t)$ is such that $E[\alpha^T \nabla_\theta G(z,\theta)u(t)]^2 \neq 0$ for all such α. For such α, we have:

$$\alpha^T \nabla_\theta G(z,\theta)u(t) = \frac{1}{A^2}[-\alpha_A^T \mathcal{B}_{1,n_a} B + \alpha_B^T \mathcal{B}_{1,n_b} A]u(t)$$

$$= \frac{1}{A^2}[-z^{-1}A_1 T B + \alpha_B^T \mathcal{B}_{1,n_b} A_1 U]u(t)$$

$$= \frac{1}{AU}[-z^{-1}T B + \alpha_B^T \mathcal{B}_{1,n_b} U]u(t) \qquad (43)$$

where the coefficients of the polynomial T, of degree $n_u - 1$, as well as the coefficients of α_B are completely free. Therefore $E[\alpha^T \nabla_\theta G(z, \theta) u(t)]^2 \neq 0$ if and only if the following pseudoregressor has full rank:

$$\psi(t) = \frac{1}{AU} \begin{pmatrix} -Bz^{-1} \\ \vdots \\ -Bz^{-n_u} \\ \overline{Uz^{-1}} \\ \vdots \\ Uz^{-n_b} \end{pmatrix} u(t) = \frac{1}{AU} R \mathcal{B}_{1,n_b+n_u} u(t) \tag{44}$$

with $R \in \mathcal{R}^{(n_b+n_u) \times (n_b+n_u)}$. Since A, B, C are coprime at all θ, and U is the common factor of A and C, it follows that B and U are coprime, and hence R in (44) is nonsingular. Therefore, by Theorem 5, $\psi(t)$ in (44) is PE (and hence $I(\theta) > 0$) if and only if $u(t)$ is sufficiently rich of degree $n_b + n_u(\theta)$, where $n_u(\theta)$ represents the number of common roots between A and C. Since the maximum number of such common roots is $\min\{n_a, n_c\}$, $I(\theta)$ is regular at all identifiable θ if and only if $u(t)$ is SRk with $k = n_b + \min\{n_a, n_c\}$. ∎

BJ Model Structure

Consider now the BJ model structure:

$$y(t) = \frac{B(z^{-1})}{F(z^{-1})} u(t) + \frac{C(z^{-1})}{D(z^{-1})} e(t) \tag{45}$$

where $B(z^{-1})$ and $C(z^{-1})$ are as above, with $F(z^{-1}) = 1 + f_1 z^{-1} + \ldots f_{n_f} z^{-n_f}$ and $D(z^{-1}) = 1 + d_1 z^{-1} + \ldots d_{n_d} z^{-n_d}$.

Theorem 9. *For the BJ model structure (45), the information matrix $I(\theta)$ is regular at a θ at which the model structure is identifiable if and only if $u(t)$ is SRk, where $k = n_b + n_f$.*

Proof. The gradient vectors $V_{uol}(z, \theta)$ and $V_{eol}(z, \theta)$ defined in (21) are now partitioned into 4 blocks corresponding, successively, to the parameters of the polynomials B, F, C, and D. It is easy to see that the left-kernel of $V_{eol}(z, \theta)$ (i.e. of $\nabla_\theta H(z, \theta)$) is spanned by the set of vectors $\alpha^T = (\alpha_B^T \mid \alpha_F^T \mid 0 \ldots 0 \mid 0 \ldots 0)$. Therefore, by Theorem 3, $I(\theta) > 0$ if and only if the following pseudoregressor is PE:

$$\psi_{B,F}(t) \triangleq \frac{1}{F^2} \begin{pmatrix} Fz^{-1} \\ \vdots \\ Fz^{-n_b} \\ \overline{-Bz^{-1}} \\ \vdots \\ -Bz^{-n_f} \end{pmatrix} u(t) \tag{46}$$

$$
= \frac{1}{F^2}
\underbrace{\begin{pmatrix}
1 & f_1 & \cdots & f_{n_f} & 0 & \cdots & 0 \\
0 & 1 & f_1 & \cdots & f_{n_f} & \cdots & 0 \\
0 & 0 & \cdots & \ddots & & \ddots & 0 \\
0 & \cdots & \cdots & 1 & f_1 & \cdots & f_{n_f} \\
0 & -b_1 & \cdots & -b_{n_b} & 0 & \cdots & 0 \\
0 & 0 & -b_1 & \cdots & -b_{n_b} & 0 & \cdots \\
0 & 0 & \cdots & \ddots & & \ddots & 0 \\
0 & \cdots & \cdots & 0 & -b_1 & \cdots & -b_{n_b}
\end{pmatrix}}_{S_{BF}}
\mathcal{B}_{1,n_b+n_f}(z)u(t) \qquad (47)
$$

S_{BF} is a Sylvester matrix, with dimensions $(n_b + n_f) \times (n_b + n_f)$. It is nonsingular for all values of θ at which the polynomials B and F are coprime. Applying Theorem 5 again, we conclude that $I(\theta) > 0$ at all values of θ at which the polynomials are coprime if and only if $u(t)$ is SRk, where $k = n_b + n_f$. ∎

Just in the same vein, one can apply the results of Theorem 4 to the identification of closed-loop systems using ARMAX or BJ model structures. The results for these two closed-loop setups, which are quite illuminating, will be presented at Giorgio Picci's next 65th birthday celebration.

6 Conclusions

The information matrix plays a fundamental role in system identification, given that it combines information about the identifiability of the model structure and about the informativity of the data set. We have illustrated these connections, and we have provided conditions on the richness of the input signals that make the information matrix full rank at all values of the parameter space where the model structure is identifiable. Our objective has been to find the smallest possible degree of richness of the input signal that delivers a nonsingular information matrix. In deriving these conditions, we have presented some new results on the degree of richness required to produce a persistently exciting regressor.

Acknowledgements

We wish to thank the organizers of the Giorgio Picci workshop for giving us the opportunity to contribute this chapter. We also thank Roland Hildebrand and Luc Haine for some useful hints for the proof of Theorem 6.

References

1. E. W. Bai and S. S. Sastry. Persistence of excitation, sufficient richness and parameter convergence in discrete time adaptive control. *Systems and Control Letters*, 6:153–163, 1985.
2. A.S. Bazanella, M. Gevers, and L. Mišković. Closed-loop identification of MIMO systems: a new look at identifiability and experiment design. In *To appear, European Control Conference*, Kos, Greece, July 2007.

3. X. Bombois, G. Scorletti, M. Gevers, R. Hildebrand, and P.M.J. Van den Hof. Cheapest open-loop identification for control. In *CD-ROM Proc. 33rd IEEE Conf on Decision and Control*, pages 382–387, The Bahamas, December 2004.

4. X. Bombois, G. Scorletti, M. Gevers, P.M.J. Van den Hof, and R. Hildebrand. Least costly identification experiment for control. *Automatica*, 42(10):1651–1662, October 2006.

5. M. Gevers, L. Mišković, D. Bonvin, and A. Karimi. Identification of multi-input systems: variance analysis and input design issues. *Automatica*, 42(4):559–572, April 2006.

6. H. Jansson and H. Hjalmarsson. Optimal experiment design in closed loop. In *16th IFAC World Congress on Automatic Control, paper 04528*, July 2005.

7. T. Kailath. *Linear Systems*. Prentice-Hall, Englewood Cliffs, New Jersey, 1980.

8. V. Kučera. *Discrete linear control: the polynomial approach*. John Wiley, 1979.

9. L. Ljung. *System Identification: Theory for the User, 2nd Edition*. Prentice-Hall, Englewood Cliffs, NJ, 1999.

10. I.M.Y. Mareels and M. Gevers. Persistence of excitation criteria for linear, multivariable, time-varying systems. *Mathematics of Control, Signals and Systems*, 1(3):203–226, 1988.

11. T. S. Ng, G. C. Goodwin, and B. D. O. Anderson. Identifiability of MIMO linear dynamic systems operating in closed loop. *Automatica*, 13:477–485, 1977.

12. T. Söderström and P. Stoica. *System Identification*. Prentice-Hall International, Hemel Hempstead, Hertfordshire, 1989.

On Interpolation and the Kimura-Georgiou Parametrization

Andrea Gombani[1] and György Michaletzky[2]

[1] ISIB-CNR, Corso Stati Uniti 4, 35127 Padova, Italy
gombani@isib.cnr.it

[2] Eötvös Loránd University, H-1111 Pázmány Péter sétány 1/C,
Computer and Automation Institute of HAS, H-1111 Kende u. 13-17, Budapest, Hungary
michgy@ludens.elte.hu

Summary. We show how the Kimura-Georgiou parametrization for interpolating a function and its derivatives at 0 is independent of the particular choice of basis of Szegö-polynomials of first and second kind, but only on the map between these two polynomial bases. This leads to a more general parametrization, which extends to different interpolation points and multivariable setup.

1 Introduction

We consider here the connection between a parametrization recently obtained by the authors in [5] and [6] in the context of constrained Schur-interpolation and the Kimura–Georgiou-parametrization (see [3, 7]) for interpolating functions of a given degree n.

Historically, the problem arises in the constrained degree Schur- or Pick-Nevanlinna-interpolation problems. The unconstrained problem has been studied extensively by means of Linear Fractional Transformations (LFT) (see e.g. [1, 2] and references therein). In this setup, the degree of the interpolating function Q is – in general – the sum of the number of interpolating conditions and the degree of a parameter function S; that is, if n is the number of interpolating conditions, we have $\deg(Q) = n + \deg(S)$. Kimura in [7] posed the problem of characterizing all interpolating functions which are Schur (or Positive Real) with the constraint that they have degree n. Although the Positive Realness was hard to characterize, the parametrization of constrained degree interpolating functions in terms of Szegö-polynomials of the first and second kind described in that paper became quite important and is known as Kimura–Georgiou-parametrization (Georgiou had independently derived the same parametrization in another context in his thesis, see [3]). We show here how the Szegö-polynomials do not play a particular role in this problem and how this parametrization can easily be generalized to the multivariable case and different interpolation points.

Let Q be a rational $p \times m$ matrix of McMillan degree n. If M is a complex matrix, Tr shall denote its trace, M^T its transpose and M^* its transpose conjugate. For a rational function Q, we set $Q^*(z) := \overline{Q(-\overline{z})^T}$. If we assume that we are given a set of interpolation points $z_1, ..., z_n$ in the plane $\mathbb{C}$ and interpolating conditions

A. Chiuso et al. (Eds.): Modeling, Estimation and Control, LNCIS 364, pp. 171–182, 2007.
springerlink.com

$$
U = \begin{bmatrix} u_1 \\ u_2 \\ \vdots \\ u_n \end{bmatrix} \qquad V = \begin{bmatrix} v_1 \\ v_2 \\ \vdots \\ v_n \end{bmatrix} \tag{1}
$$

with u_i, v_i row vectors in $\mathbb{C}^p$, and we want to find the solutions Q to the problem

$$
u_i Q(z_i)^* = v_i \qquad i = 1, ..., n. \tag{2}
$$

(we will, in fact, impose the further condition that $Q(\infty) = D$), then the above conditions can be reformulated in a more general manner: set

$$
\mathcal{A} := \mathrm{diag}\{-\bar{z}_1, -\bar{z}_2, ..., -\bar{z}_n\},
$$

we can now write the problem as finding the rational functions Q such that:

$$
\begin{cases} (Q(z)U^* - V^*)(sI + \mathcal{A}^*)^{-1} & \text{is analytic on } \sigma(-\mathcal{A}^*) \\ Q(\infty) = D . \end{cases} \tag{3}
$$

2 Interpolation Conditions as Matrix Equations

Our first result is to express the interpolation conditions in Problem 3 in terms of matrix equations for a given realization for Q. Although quite simple, the following conditions do not seem to have appeared in the literature (the first formulation presented here was derived independently for the discrete time case by Marmorat and Olivi, see [8]). This result was mainly used in [5, 6] in connection with Schur-function, but it can be stated in a quite general setting.

Proposition 1. *Let Q be a rational function analytic in $\sigma(-\mathcal{A}^*)$ admitting a minimal realization*

$$
Q = \left(\begin{array}{c|c} A_Q & B_Q \\ \hline C_Q & D \end{array} \right) .
$$

Then $(Q(z)U^* - V^*)(zI + \mathcal{A}^*)^{-1}$ is analytic in $\sigma(-\mathcal{A}^*)$ if and only if there exists a matrix Y such that

$$
DU^* - V^* = C_Q Y \tag{4}
$$

and

$$
B_Q U^* = Y\mathcal{A}^* + A_Q Y . \tag{5}
$$

Proof. First suppose that $(Q(z)U^* - V^*)(zI + \mathcal{A}^*)^{-1}$ is analytic in $\sigma(-\mathcal{A}^*)$; then, it can only have poles at the eigenvalues of A_Q, and it vanishes at ∞; thus the observability of (C_Q, A_Q) implies the existence of the matrix Y in

$$
(Q(z)U^* - V^*)(zI + \mathcal{A}^*)^{-1} = C_Q (zI - A_Q)^{-1} Y . \tag{6}
$$

We can thus evaluate the identity

$$Q(z)U^* - V^* = DU^* - V^* + C_Q (zI - A_Q)^{-1} B_Q U^*$$
$$= C_Q (zI - A_Q)^{-1} Y (zI + \mathcal{A}^*)$$

at ∞ obtaining the equation

$$DU^* - V^* = C_Q Y$$

Subtraction from the previous equation yields:

$$C_Q (zI - A_Q)^{-1} B_Q U^* = C_Q (zI - A_Q)^{-1} (Y(zI + \mathcal{A}^*) - (zI - A_Q)Y)$$
$$= C_Q (zI - A_Q)^{-1} (Y\mathcal{A}^* + A_Q Y) \, .$$

Observability of (C_Q, A_Q) implies that

$$B_Q U^* = Y \mathcal{A}^* + A_Q Y \, .$$

Conversely, if a realization of Q satisfies (4) and (5), by backtracking the above argument, we see that $(Q(z)U^* - V^*)(zI + \mathcal{A}^*)^{-1} = C_Q (zI - A_Q)^{-1} Y$ is analytic in $\sigma(-\mathcal{A}^*)$ using that the assumed minimality of the realization of Q guarantees that $\sigma(A_Q)$ and $\sigma(-\mathcal{A}^*)$ are disjoint. ∎

It is worth pointing out that for the first part of the proof only the observability of (C_Q, A_Q) was needed while for backward direction the controllability of (A_Q, B_Q) is enough.

Notice also that, to obtain the equations (4)-(5), we have exploited the fact that the function Q cannot have poles at the interpolation nodes. In fact this assumption is not necessary, as it is shown in [5]. Namely, if the realization of Q is observable and the interpolation condition (3) holds then there exists a matrix Y such that equations (4) (5) are satisfied.

We would like now to parametrize all functions of degree n satisfying (3). In the scalar case – i.e. when $p = m = 1$ – this can be formulated as solutions with McMillan-degree exactly n or in other words as solutions in the form of ratio of two coprime polynomials with degree n (see [7]). In this case exactly n parameters are needed to express all the solutions. This type of formulation can be applied in the multivariate case as well, although some care is needed. In fact, in this case, saying that a function has degree n is not enough. To see this, assume that Q_1 is an interpolant of degree k strictly less than n. Then it can be shown that, under suitable conditions, there exist functions Q_2 analytic in $\sigma(-\mathcal{A}^*)$ of degree $n - k$ such $Q_2(s)U^*(sI + \mathcal{A}^*)^{-1}$ is still analytic in $\sigma(-\mathcal{A}^*)$, so that $Q = Q_1 + Q_2$ has degree less that or equal to n. By construction it is also interpolating. On the other hand, it poses the same parametrization problems as interpolants of lower degree in the scalar case. An example of how to construct such functions is given below.

In conclusion, the functions Q we want to exclude from our parametrization are not only the functions of degree strictly less than n, as in the scalar case, but also those which can be reduced to a lower degree by subtraction of a suitable function having the same state space as Q and vanishing on the the interpolation nodes.

To this end, we make the following definition.

Definition 1. *Let Q be a solution of McMillan-degree n to problem formulated in (3) and let $Q = \left[\begin{array}{c|c} A_Q & B_Q \\ \hline C_Q & D \end{array}\right]$ be a minimal realization of Q: then we say that Q is* irreducible *if the solution Y to the Sylvester-equation:*

$$A_Q Y + Y \mathcal{A}^* - B_Q U^* = 0 \tag{7}$$

has full rank.

We will therefore restrict ourselves to the set of interpolants which are irreducible; this is not a very restrictive assumption. In fact, it is quite easy to see that the excluded functions form an algebraic set. So, we can reformulate the problem in (3) more precisely as:

Problem 1. Given the matrix $\mathcal{A}$ of size $n \times n$, matrices U, V of size $n \times p$ and a constant matrix D of size $p \times p$, parametrize all functions Q, for which

(i)
$$\begin{cases} (Q(s)U^* - V^*)\,(sI + \mathcal{A}^*)^{-1} & \text{is analytic in } \sigma(-\mathcal{A}^*) \\ Q(\infty) = D \,. \end{cases} \tag{8}$$

(ii) Q is rational of McMillan-degree exactly n ,
(iii) Q is irreducible,
(iv) Q is analytic in $\sigma(-\mathcal{A}^*)$.

For any given matrix B of size $n \times p$, denote by

$$F_B^* = \left(\begin{array}{c|c} -\mathcal{A}^* & B \\ \hline -U^* & I \end{array}\right) \,,$$

$$G_B^* = \left(\begin{array}{c|c} -\mathcal{A}^* & B \\ \hline -V^* & D \end{array}\right) \,.$$

Theorem 1. *The rational function Q of McMillan-degree n is a solution of the interpolation Problem 1 if and only if it has the realization*

$$Q = \left(\begin{array}{c|c} -\mathcal{A}^* + BU^* & B \\ \hline DU^* - V^* & D \end{array}\right) \tag{9}$$

for some matrix B.
    Especially, in this case

$$Q = G_B^* \left(F_B^*\right)^{-1} \,.$$

Proof. Assume that Q is an interpolating irreducible function of degree n; then, in view of Proposition 1, any realization $\left[\begin{array}{c|c} A & B \\ \hline C & D \end{array}\right]$ of Q satisfies equations (4) and (5) for some matrix Y. Since Q is irreducible, Y is invertible and therefore, equations (4) and (5) can be rewritten as:

$$DU^* - V^* = C_Q Y \tag{10}$$

and

$$Y^{-1} A_Q Y = -\mathcal{A}^* + Y^{-1} B_Q U^* \tag{11}$$

that is, $Q = \left[\begin{array}{c|c} -\mathcal{A}^* + Y^{-1}B_Q U^* & Y^{-1}B_Q \\ \hline DU^* - V^* & D \end{array}\right]$; setting $B := Y^{-1}B_Q$, we immediately get (9).

Conversely, if Q has the minimal realization (9), then it is an interpolating function (in view of Proposition 1); it is irreducible because $Y = I$ and thus has trivial kernel.

Finally, the calculation

$$G^*(F^*)^{-1} = \left(\begin{array}{c|c} -\mathcal{A}^* & B \\ \hline -V^* & D \end{array}\right)\left(\begin{array}{c|c} -\mathcal{A}^* + BU^* & B \\ \hline U^* & I \end{array}\right)$$

$$= \left(\begin{array}{cc|c} -\mathcal{A}^* + BU^* & 0 & B \\ BU^* & -\mathcal{A}^* & B \\ \hline DU^* & -V^* & D \end{array}\right)$$

$$= \left(\begin{array}{c|c} -\mathcal{A}^* + BU^* & B \\ \hline DU^* - V^* & D \end{array}\right) = Q$$

concludes the proof. ∎

The last form of the interpolating functions are already very similar to the one provided by the Kimura–Georgiou-parametrization, where – in the scalar case – the solutions are described as ratio of two polynomials. So it is worth pointing out the converse statement, as well.

Proposition 2. *Assume that the rational function Q has the following properties:*

(i) Q is analytic on $\sigma\left(-\mathcal{A}^*\right)$;
(ii) Q satisfies condition (8)
(iii) for some matrix B the function Q has the form

$$Q = G^* \left(F_B^*\right)^{-1} ,$$

where the set of the poles of the rational functions G^* and Q are disjoint.

Then G is uniquely determined by the matrix B, as well. Namely,

$$G^* = G_B^* .$$

Proof. The function standing on the right hand side of the following identity

$$G^* - \left(D - V^*\left(zI + \mathcal{A}^*\right)^{-1} B\right) = QF^* - \left(D - V^*\left(zI + \mathcal{A}^*\right)^{-1}B\right)$$

$$= Q - D - \left(QU^* - V^*\right)\left(zI + \mathcal{A}^*\right)^{-1} B$$

is analytic on $\sigma\left(-\mathcal{A}^*\right)$, thus its poles should form a subset of the poles of Q while the set of poles of the function on the left hand side should be disjoint from the poles of Q. Consequently, they are constant functions. Evaluating at infinity, we obtain that this constant should be zero. I.e.

$$G^* = G_B^* \, ,$$

concluding the proof of the proposition. $\blacksquare$

Notice that the form of the function does not guarantee that its McMillan-degree is exactly n, only it is no greater than n.

It is now easy to see how to construct an interpolation problem together with an interpolating function of lower degree. Assume that for some matrices the equation

$$Y\mathcal{A}^* + B_1 U^* = -A_{11} Y \tag{12}$$

holds, where A_{11} and $-\mathcal{A}^*$ have no common eigenvalues and Y is a flat full row-rank matrix of dimension $n_1 \times n$. (Its right inverse will be denoted by Y^{-R}). Then obviously $A_{11} := -Y\mathcal{A}Y^{-R} + B_1 U^* Y^{-R}$. Consider any matrix C_1 of size $p \times n_1$ for which the pair (C_1, A_{11}) is observable. Define V^* as $V^* = DU^* - CY$. Then equations (5), (4) are satisfied and the full row-rank property of Y implies that (A_{11}, B_1) is a controllable pair. Thus Proposition gives that $Q_1 = \left[\begin{array}{c|c} A_{11} & B_1 \\ \hline C_1 & D \end{array}\right]$ is an interpolating function of degree n_1.

This gives a starting point how to construct – in some cases – an interpolant of degree n which is *not irreducible*. The above condition on Y means that the subspace $\mathcal{V} := span\{Y^*\xi; \xi \in \mathbb{C}^{\,n_1}\}$ is a *controlled invariant subspace* for the pair $(\mathcal{A}, U)$ (see [10]). Set $\mathcal{U} = span\{U\xi; \xi \in \mathbb{C}^{\,m}\}$ If $\mathcal{V} \cap \mathcal{U} \neq \{0\}$, then the equation

$$B_2 U^* = -A_{21} Y \tag{13}$$

has a solution, where B_2 and A_{21} are of dimensions $(n - n_1) \times m$ and $(n - n_1) \times n_1$, respectively, and not identically 0.

Thus (12) can be extended to

$$\begin{bmatrix} Y \\ 0 \end{bmatrix} \mathcal{A}^* + \begin{bmatrix} B_1 \\ B_2 \end{bmatrix} U^* = - \begin{bmatrix} A_{11} & A_{12} \\ A_{21} & A_{22} \end{bmatrix} \begin{bmatrix} Y \\ 0 \end{bmatrix}$$

where A_{12} and A_{22} are arbitrary. In particular, we can choose $A_{12} = 0$ and A_{22} such that $\sigma(A_{22}) \cap (\sigma(A_{11}) \vee \sigma(-\mathcal{A}^*)) = \emptyset$ and (A_{22}, B_2) is controllable. But then also the pair $(A, B) := \left(\begin{bmatrix} A_{11} & 0 \\ A_{21} & A_{22} \end{bmatrix}, \begin{bmatrix} B_1 \\ B_2 \end{bmatrix} \right)$ is controllable. If we now set $C = [C_1, C_2]$, with C_2 arbitrary, we obtain that the equality

$$C \begin{bmatrix} Y \\ 0 \end{bmatrix} = -DU^* + V^*$$

is satisfied and therefore, since (C_1, A_{11}) was observable, we can always choose C_2 so that also (C, A) is observable. Proposition 1 implies that the function $Q = \left(\begin{array}{cc|c} A_{11} & A_{12} & B_1 \\ A_{21} & A_{22} & B_2 \\ \hline C_1 & C_2 & D \end{array} \right)$ satisfies (i), (ii), (iv) in the Problem 1 but it is not irreducible.

3   The Connection with the Kimura-Georgiou Parametrization

We would like to show here how the Kimura–Georgiou-parametrization (see [7, 3]) overlaps with our framework; to this end, we consider the problem of interpolating the first $n + 1$ coefficients of an expansion around 0 for *scalar valued functions*. The starting point is the form $Q = G_B^* \left(F_B^*\right)^{-1}$ of the interpolating functions provided by Theorem 1. Our setup works both in discrete and continuous time; but the original parametrization was done for discrete time systems considering an expansion around infinity, (and the connection is easier to describe in this setup): therefore, in order to make the connection precise, we consider parahermitian conjugation with respect to the unit circle. That is, $Z^*(z) = \overline{Z(1/\overline{z})}^T$. Following tradition, the interpolating functions will be denoted by Z.

To get an idea of how this works, let us first consider the following case:

$$A = \begin{bmatrix} 0 & & & & \\ -1 & 0 & & & \\ & -1 & \ddots & & \\ & & \ddots & \ddots & \\ & & & -1 & 0 \end{bmatrix} \qquad U = \begin{bmatrix} 1 \\ 0 \\ \vdots \\ \vdots \\ 0 \end{bmatrix} \qquad V = \begin{bmatrix} v_1 \\ v_2 \\ \vdots \\ \vdots \\ v_n \end{bmatrix}$$

Then, the above is a classical interpolation problem of matching the value of the function and its n derivatives at 0.

$$(zI + A^*)^{-1} = \begin{bmatrix} z & -1 & & & \\ & z & -1 & & \\ & & \ddots & \ddots & \\ & & & \ddots & -1 \\ & & & & z \end{bmatrix}^{-1} = \begin{bmatrix} z^{-1} & z^{-2} & \cdots & \cdots & z^{-n} \\ & z^{-1} & z^{-2} & & \vdots \\ & & \ddots & \ddots & \vdots \\ & & & \ddots & \\ & & & & z^{-1} \end{bmatrix}$$

Thus we have

$$V^*(zI + A^*)^{-1} = [\overline{v}_1 z^{-1}, \overline{v}_1 z^{-2} + \overline{v}_2 z^{-1}, \ldots, \overline{v}_1 z^{-n} + \overline{v}_2 z^{n-1} + \ldots + \overline{v}_n z^{-1}]$$

$$U^*(zI + A^*)^{-1} = [z^{-1}, z^{-2}, \ldots, z^{-n}]$$

Choosing for example $B = [0, 0, \ldots, 1]^T$ and $D = \overline{v}_{n+1} + \overline{v}_1$, we get

$$G^* = V^*(zI + A^*)^{-1}B + D = \overline{v}_{n+1} + \overline{v}_1 + \overline{v}_1 z^{-n} + \overline{v}_2 z^{n-1} + \ldots + \overline{v}_n z^{-1}$$

$$F^* = U^*(zI + A^*)^{-1}B + 1 = z^{-n} + 1 = z^{-n}(1 + z^n) = z^{-n}\frac{1}{1 - z^n + z^{2n} - \ldots}$$

In conclusion,

$$G^* F^{-*} = \left((\overline{v}_{n+1} + \overline{v}_1)z^n + \overline{v}_1 + \overline{v}_2 z + \ldots + \overline{v}_n z^{n-1}\right)\left(1 - z^n + z^{2n} - \ldots\right)$$

$$= \overline{v}_1 + \overline{v}_2 z + \ldots + \overline{v}_n z^{n-1} + \overline{v}_{n+1} z^n + \ldots$$

as wanted. (Note that the special choice of D made it possible to interpolate the next coefficient, as well.) This interpolant was easy to calculate because of the choice of B. Nevertheless, we are going to see that the general case is not much more complicated.

Let $\bar{c}_0, \bar{c}_1, ..., \bar{c}_n$ be given; the above parametrization describes all the functions interpolating the first n conditions at 0 and having value D at infinity. Now, a suitable choice of D will allow us to interpolate also the $(n+1)$-st coefficient $\bar{c}_n$. So, we set

$$V^* := [\bar{c}_0, \bar{c}_1, ..., \bar{c}_{n-1}]$$

and let $B = [b_1, b_2, ..., b_n]^T$, with $b_n \neq 0$ so that $-\mathcal{A}^*$ and $-\mathcal{A}^* + BU^*$ have disjoint spectra. Then, defining G, F as above, the function $Z = G^* F^{-*}$ will have expression

$$G^* F^{-*} = \bar{c}_0 + \bar{c}_1 z + \bar{c}_2 z^2 + ... + \bar{c}_{n-1} z^{n-1} + O(z^n) \tag{14}$$

independently of the choice of D. So, if we impose that also the $(n+1)$-st coefficient $\bar{c}_n$ to be matched, we will obtain an equation for D. Now, setting for notational convenience $b_0 = 1$ we have that

$$F^*(z) = 1 + U^*(zI + \mathcal{A}^*)^{-1} B$$

$$= 1 + [1, 0, ..., 0] \begin{bmatrix} z^{-1} & z^{-2} & \cdots & z^{-n} \\ & z^{-1} & & z^{n-1} \\ & & \ddots & \vdots \\ & & & z^{-1} \end{bmatrix} \begin{bmatrix} b_1 \\ b_2 \\ \vdots \\ b_n \end{bmatrix} = \sum_{i=0}^{n} b_i z^{-i}$$

Similarly,

$$G^*(z) = D + V^*(zI + \mathcal{A}^*)^{-1} B$$

$$= D + [\bar{c}_0, \bar{c}_1, ..., \bar{c}_{n-1}] \begin{bmatrix} z^{-1} & z^{-2} & \cdots & z^{-n} \\ & z^{-1} & & z^{n-1} \\ & & \ddots & \vdots \\ & & & z^{-1} \end{bmatrix} \begin{bmatrix} b_1 \\ b_2 \\ \vdots \\ b_n \end{bmatrix}$$

$$= D + \sum_{l=1}^{n} z^{-l} \sum_{i=l}^{n} b_i \bar{c}_{i-l} \tag{15}$$

On the other hand, if Z interpolates the first $(n+1)$-st coefficients,

$$Z(z)F^*(z) = \left(\sum_{j=0}^{n} \bar{c}_j z^j + O(z^{n+1}) \right) \sum_{i=0}^{n} b_i z^{-i} = \sum_{l=0}^{n} z^{-l} \sum_{i=l}^{n} b_i \bar{c}_{i-l} + O(z) . \tag{16}$$

Now, equation $G^* = ZF^*$ gives that this should be compared to (15); in order for the equality to be satisfied, $O(z)$ must vanish and the constant coefficient in (16) must be equal to D. This yields

$$D = \sum_{i=0}^{n} b_i \bar{c}_i = [\bar{c}_0, \bar{c}_1, ..., \bar{c}_{n-1}, \bar{c}_n] \begin{bmatrix} b_0 \\ b_1 \\ b_2 \\ \vdots \\ b_n \end{bmatrix} \tag{17}$$

Therefore, putting together (15) and (17), G^* can be written as:

$$G^*(z) = [\bar{c}_0, \bar{c}_1, ..., \bar{c}_{n-1}, \bar{c}_n] \begin{bmatrix} 0 & z^{-1} & z^{-2} & \cdots & z^{-n} \\ & 0 & z^{-1} & & z^{n-1} \\ & & \ddots & \ddots & \vdots \\ & & & 0 & z^{-1} \\ & & & & 0 \end{bmatrix} \begin{bmatrix} b_0 \\ b_1 \\ b_2 \\ \vdots \\ b_n \end{bmatrix}$$

$$+ [\bar{c}_0, \bar{c}_1, ..., \bar{c}_{n-1}, \bar{c}_n] \begin{bmatrix} b_0 \\ b_1 \\ b_2 \\ \vdots \\ b_n \end{bmatrix}$$

$$= [\bar{c}_0, \bar{c}_1, ..., \bar{c}_{n-1}, \bar{c}_n] \begin{bmatrix} 1 & z^{-1} & z^{-2} & \cdots & z^{-n} \\ & 1 & z^{-1} & & z^{n-1} \\ & & \ddots & \ddots & \vdots \\ & & & 1 & z^{-1} \\ & & & & 1 \end{bmatrix} \begin{bmatrix} b_0 \\ b_1 \\ b_2 \\ \vdots \\ b_n \end{bmatrix} \tag{18}$$

Similarly, since $b_0 = 1$, the representation of F^* can be extended to

$$F^*(z) = [1, 0, ..., 0, 0] \begin{bmatrix} 1 & z^{-1} & z^{-2} & \cdots & z^{-n} \\ & 1 & z^{-1} & & z^{n-1} \\ & & \ddots & \ddots & \vdots \\ & & & 1 & z^{-1} \\ & & & & 1 \end{bmatrix} \begin{bmatrix} b_0 \\ b_1 \\ b_2 \\ \vdots \\ b_n \end{bmatrix} \tag{19}$$

To make the connection with the Kimura–Georgiou-parametrization, let now $c_0, c_1, ..., c_n$ be given (we assume, w.l.o.g. that $c_0 = 1$) and suppose we want to parametrize all functions $Z(z)$ of McMillan-degree n which have the expansion at infinity:

$$Z^*(z) = c_0 + c_1 z^{-1} + c_2 z^{-2} + ... + c_n z^{-n} + O(z^{-n-1}) \tag{20}$$

The solution is given by the Kimura–Georgiou- parametrization: define recursively the Szegő-polynomials ϕ_k, ψ_k as:

$$\begin{bmatrix} \phi_0 & -\psi_0 \\ \phi_0^* & \phi_0^* \end{bmatrix} := \begin{bmatrix} 1 & -1 \\ 1 & 1 \end{bmatrix}$$

and, for $k > 0$,

$$\gamma_{k+1} := -\phi_k(0) \qquad \begin{bmatrix} \phi_{k+1} & -\psi_{k+1} \\ \phi_{k+1}^* & \phi_{k+1}^* \end{bmatrix} := \begin{bmatrix} z & -\gamma_{k+1} \\ -z\gamma_{k+1} & 1 \end{bmatrix} \begin{bmatrix} \phi_k & -\psi_k \\ \phi_k^* & \phi_k^* \end{bmatrix} \tag{21}$$

Observe that ϕ_k and ψ_k are monic polynomials. Then, the Kimura–Georgiou-parametrization reads as follows: any $Z^*(z)$ of the form (20) can be written as:

$$Z^*(z) = \frac{\psi_n + \beta_1\psi_{n-1} + \ldots + \beta_n\psi_0}{\phi_n + \beta_1\phi_{n-1} + \ldots + \beta_n\phi_0} = \frac{\Psi_\beta(z)}{\Phi_\beta(z)} \tag{22}$$

We will show that this is a special case of (9)

It is well known (see [4]) that, for $1 \le k \le n$, the function $Z_k^*(z)$

$$Z_k^*(z) = \frac{\psi_k}{\phi_k} \tag{23}$$

satisfies the relation

$$Z_k^*(z) = c_0 + c_1 z^{-1} + c_{z2} z^{-2} + \ldots + c_k z^{-k} + o(z^{-k-1}) \tag{24}$$

(it is the *maximum entropy* solution for the interpolation problem (20) with $n = k$). Therefore, writing

$$\phi_k(z) = \phi_{k,k} z^k + \phi_{k,k-1} z^{k-1} + \phi_{k,k-2} z^{k-2} + \ldots + \phi_{k,0}$$

$$\psi_k(z) = \psi_{k,k} z^k + \psi_{k,k-1} z^{k-1} + \psi_{k,k-2} z^{k-2} + \ldots + \psi_{k,0}$$

(24) becomes:

$$\psi_k(z) = \phi_k(z) Z^*(z) = \sum_{i=0}^{k} z^i \sum_{j=0}^{k-i} \phi_{k,i+j} c_j$$

which, in matrix terms can be written as:

$$[\psi_{k,0}, \psi_{k,1}, \ldots, \psi_{k,k}] \begin{bmatrix} 1 \\ z \\ \vdots \\ z^k \end{bmatrix} = [\phi_{k,0}, \phi_{k,1}, \ldots, \phi_{k,k}] \begin{bmatrix} c_0 & & & \\ c_1 & c_0 & & \\ \vdots & & \ddots & \\ c_k & c_{k-1} & \ldots & c_0 \end{bmatrix} \begin{bmatrix} 1 \\ z \\ \vdots \\ z^k \end{bmatrix} \tag{25}$$

Writing relation (25) for all $k \le n$ we have the following matrix equality connecting the coefficients of the polynomials $\phi_0, \ldots, \phi_n$ and $\psi_0, \ldots, \psi_n$:

$$\begin{bmatrix} \psi_{0,0} & & & \\ \psi_{1,0} & \psi_{1,1} & & \\ \vdots & & \ddots & \\ \psi_{n,0} & \psi_{n,1} & \ldots & \psi_{n,n} \end{bmatrix} = \begin{bmatrix} \phi_{0,0} & & & \\ \phi_{1,0} & \phi_{1,1} & & \\ \vdots & & \ddots & \\ \phi_{n,0} & \phi_{n,1} & \ldots & \phi_{n,n} \end{bmatrix} \begin{bmatrix} c_0 & & & \\ c_1 & c_0 & & \\ \vdots & & \ddots & \\ c_n & c_{n-1} & \ldots & c_0 \end{bmatrix} \tag{26}$$

It is important to point out only this equation will be needed to show that the form Ψ_β/Φ_β given by the Kimura–Georgiou-parametrization is essentially the same as G/F given in Theorem 1.

Notice now that the denominator of (22) can be written as:

$$\Phi_\beta(z) = [\beta_n, \beta_{n-1}, \ldots \beta_1, 1] \begin{bmatrix} \phi_0 \\ \phi_1 \\ \vdots \\ \phi_n \end{bmatrix}$$

$$= [\beta_n, \beta_{n-1}, \ldots \beta_1, 1] \begin{bmatrix} \phi_{0,0} & & & \\ \phi_{1,0} & \phi_{1,1} & & \\ \vdots & & \ddots & \\ \phi_{n,0} & \phi_{n,1} & \ldots & \phi_{n,n} \end{bmatrix} \begin{bmatrix} 1 \\ z \\ \vdots \\ z^n \end{bmatrix}$$

Setting

$$\hat{B}^* := [\beta_n, \beta_{n-1}, \dots \beta_1, 1] \begin{bmatrix} \phi_{0,0} & & & \\ \phi_{1,0} & \phi_{1,1} & & \\ \vdots & & \ddots & \\ \phi_{n,0} & \phi_{n,1} & \cdots & \phi_{n,n} \end{bmatrix}$$

we have that the denominator $\Phi_\beta(z)$ of (22) is $\hat{B}^*[1, z, \dots, z^n]^T$. Setting the $n+1$ dimensional vector $\xi := [1, 0, \dots, 0]^T$, this can be written as:

$$\Phi_\beta(z) = \hat{B}^* \begin{bmatrix} 1 \\ z \\ \vdots \\ z^n \end{bmatrix} = \hat{B}^* \begin{bmatrix} 1 & & & \\ z & 1 & & \\ \vdots & & \ddots & \\ z^n & z^{n-1} & \cdots & 1 \end{bmatrix} \xi \tag{27}$$

Similarly, in view of (26) and setting $\eta := [c_0, c_1, \dots, c_n]^T$, the numerator Ψ_β of (22) becomes

$$\Psi_\beta(z) = [\beta_n, \beta_{n-1}, \dots \beta_1, 1] \begin{bmatrix} \psi_0 \\ \psi_1 \\ \vdots \\ \psi_n \end{bmatrix} \tag{28}$$

$$= [\beta_n, \beta_{n-1}, \dots \beta_1, 1] \begin{bmatrix} \phi_{0,0} & & & \\ \phi_{1,0} & \phi_{1,1} & & \\ \vdots & & \ddots & \\ \phi_{n,0} & \phi_{n,1} & \cdots & \phi_{n,n} \end{bmatrix} \begin{bmatrix} c_0 & & & \\ c_1 & c_0 & & \\ \vdots & & \ddots & \\ c_n & c_{n-1} & \cdots & c_0 \end{bmatrix} \begin{bmatrix} 1 \\ z \\ \vdots \\ z^n \end{bmatrix}$$

$$= \hat{B}^* \begin{bmatrix} 1 & & & \\ z & 1 & & \\ \vdots & & \ddots & \\ z^n & z^{n-1} & \cdots & 1 \end{bmatrix} \begin{bmatrix} c_0 \\ c_1 \\ \vdots \\ c_n \end{bmatrix} = \hat{B}^* \begin{bmatrix} 1 & & & \\ z & 1 & & \\ \vdots & & \ddots & \\ z^n & z^{n-1} & \cdots & 1 \end{bmatrix} \eta$$

A simple comparison of (18) with (28) and (19) with (27) shows that $\hat{B} = B/b_n$ and thus

$$F/b_n = \Phi_\beta \qquad\qquad G/b_n = \Psi_\beta \qquad\qquad Z^* = \frac{G}{F} = \frac{\Psi_\beta}{\Phi_\beta}$$

Notice that, in principle, our parametrization does not coincide with that of Kimura and Georgiou, because we require b_0 to be 1. This means, essentially, that it leaves out the polynomials. This is not surprising, since it was designed to represent interpolants as proper rational functions and thus match interpolation conditions at any points of the disk (for this section) and the constant term at infinity. It should be noted, however, that formulas (19) and (18) do make sense also for $b_0 = 0$. This implies that there is no special reason to choose the Szegő-polynomials in the parametrization: any two basis connected by the matrix

$$\begin{bmatrix} c_0 & & & \\ c_1 & c_0 & & \\ \vdots & & \ddots & \\ c_n & c_{n-1} & \cdots & c_0 \end{bmatrix}$$

will do. A generalization of this fact to the Nevanlinna-Pick problem can be found in [9].

Acknowledgements. We would like to thank Professor Paul Fuhrmann for asking us to see the details of our claim that our parametrization and that of Kimura and Georgiou were connected. Trying to answer his question led us to write the present paper.

References

1. J. A. Ball, I. Gohberg, and L. Rodman (1988) Realization and interpolation of rational matrix functions. Operator Theory, Advances and Applications, 33:1–72.
2. H. Dym (1989) J-Contractive Matrix Functions, Reproducing Kernel Hilbert Spaces and Interpolation, CBMS Regional Conference Series in Mathematics, No. 71, Amer. Math. Soc., Providence, R.I.
3. T. Georgiou (1983) Partial realization of covariance sequences. Ph.D.. Thesis.
4. Y.L. Geronimus, (1960) Polynomials orthogonal on a circle and interval, Pergamon Press.
5. A. Gombani and Gy. Michaletzky, On the Nevanlinna-Pick interpolation problem: analysis of the McMillan-degree of the solutions, to appear on Linear Algebra and Applications.
6. A. Gombani and Gy. Michaletzky, On the parametrization of Schur and Positive Real functions of degree n with fixed interpolating conditions, submitted.
7. H. Kimura (1986) Positive partial realization of covariance sequences. In C. I. Byrnes and A. Lindquist, editors, Modelling, Identification and Robust Control, pages 499–513. Elsevier Science.
8. J.P. Marmorat and M. Olivi, Nudelman Interpolation, parametrization of lossless functions and balanced realizations, submitted.
9. Gy. Michaletzky, Stochastic approach to Nevanlinna-Pick interpolation, in preparation.
10. W.M. Wonham (1984) Linear Multivariable Control, 3rd ed., Springer Verlag, New York.

The Control of Error in Numerical Methods

Daniel Holder[1], Lin Huo[2], and Clyde F. Martin[3]

[1] Texas Tech University
   daniel.holder@ttu.edu
[2] Texas Tech University
   lin.huo@ttu.edu
[3] Texas Tech University
   clyde.f.martin@ttu.edu

1 Introduction

Differential equations are types of equations that arise from the mathematical modelling, or simulation, of physical phenomena or from engineering applications; for example: the flow of water, the decay of radioactive substances, bodies in motion, electrical circuits, chemical processes, etc. When we cannot solve differential equations analytically, we must resort to numerical methods. Unfortunately, numerical methods do not give us an exact solution, and an amount of error is introduced in the answer. It is our goal to utilize concepts from Control Theory in order to minimize this error. For our study, we shall focus on ordinary differential equations (ODEs).

Without going into details, we state that ODEs may be solved analytically by methods such as: substitution, by using an integrating factor, or by separation of variables, to name a few. However, there are times when the differential equation is too complicated to be solved analytically; it is then when we need to use numerical methods to obtain an approximation to the solution.

Numerical methods may be single-step, in which case, in order to calculate the next point of the solution function of the differential equation, it is only necessary to have information of the preceding point. Examples are Euler's method, Taylor's method, or Runge-Kutta method. Or they can be multi-step, where information of at least two preceding points is necessary in order to calculate the next point of the solution function. Examples are Adams-Bashforth method or Adams-Moulton method [3].

As mentioned earlier, numerical methods can only give us an approximation of the solution, and the difference with the exact solution is referred to as the error. This error is of varying magnitude for each method. It is the motivation of this work to apply robust and optimal control concepts to these methods to reduce the error of the solution [1]; in contrast to other control theoretical techniques which look to regulate the stepsize selection in the solution of the ordinary differential equation [4]. Figure 1 is a block diagram of the complete ordinary differential equation solver.

The Plant consists of the Predictor and Corrector blocks. The Predictor generates an approximation to the solution of the ODE, while the Corrector improves this approximation. The Model contains the exact solution of the ODE for comparison purposes, from which an error is generated from the difference between the approximation and

A. Chiuso et al. (Eds.): Modeling, Estimation and Control, LNCIS 364, pp. 183–192, 2007.
springerlink.com

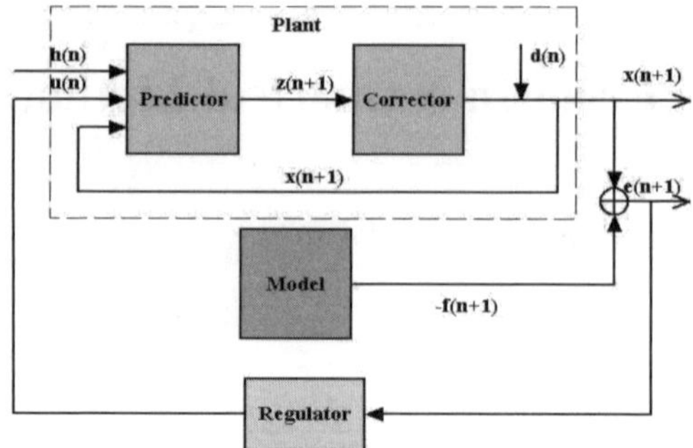

Fig. 1. Control Theoretic Model of an ODE Solver

the exact solution. Our work will focus on the error term e_{n+1} and the design of the Regulator, whose function is to minimize the error further. For the statistical analysis, we will focus on the study of error distribution for the four-step Adams-Bashforth method. We develop a model with a term which has a mean zero and a constant variance. Then we will compare these two models developed by control theory and by time series theory.

2   A Simple Example

In order to illustrate the technique. We shall begin with a simple test ODE,

$$\frac{dx}{dt} = \lambda x, \quad 0 \le t \le 1, \quad y(0) = 1,$$

with its well known solution $x(t) = e^{\lambda t}$ and we shall use Euler's method for the numerical computation of this solution. Euler's method is not sufficiently accurate to be used in real applications but it is simple enough for illustration purposes.

The Euler algorithm is given by

$$x_{n+1} = hf(t_n, x_n) \quad for \quad n = 0, 1, \ldots, N - 1,$$

where N is a positive integer and the approximate value of x, x_n, is found at given points t_n (called mesh points),

$$t_n = a + nh \quad for \quad n = 0, 1, \ldots, N,$$

where h is the step size given by $h = \dfrac{b - a}{N} = t_{n+1} - t_n$, on the interval [a,b].

Thus we have,

$$x_{n+1} = x_n + \lambda h x_n \tag{1}$$

$$x_{n+1} = (1 + \lambda h) x_n \tag{2}$$

And the error can be found from

$$e_n = x_n - e^{\lambda h n} \tag{3}$$

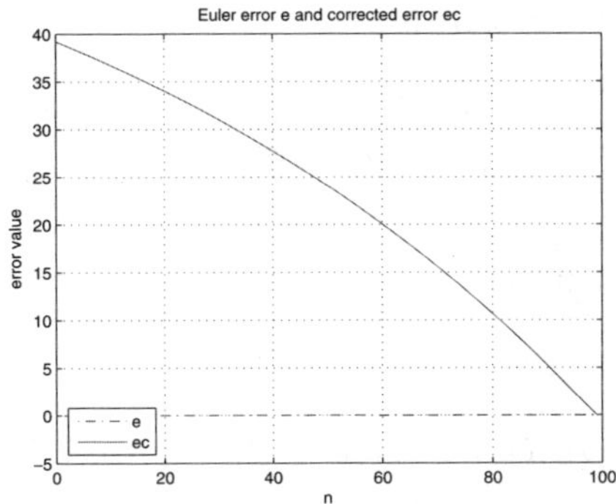

Fig. 2. Comparison of the numerical error and the minimized error over the interval [0,1] with h = 0.01

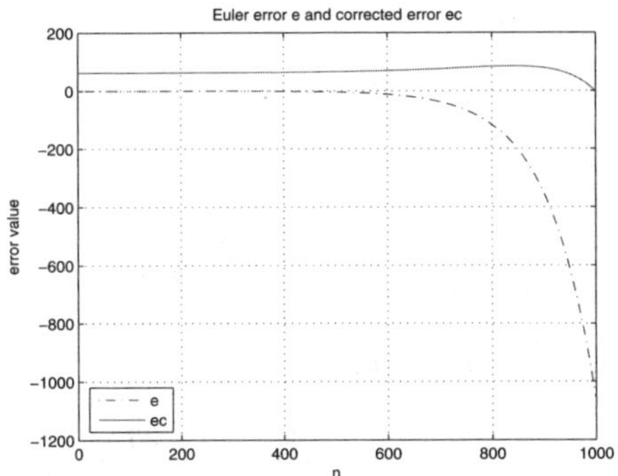

Fig. 3. Comparison of the numerical error and the minimized error over the interval [0,10] with h = 0.01

We increment (3) to n+1, insert (2) and use the expression for x_n that we get from (3) to get the following:

$$e_{n+1} = x_{n+1} - e^{\lambda h n}e^{\lambda h} \tag{4}$$

$$= (1 + \lambda h)x_n - e^{\lambda h n}e^{\lambda h} \tag{5}$$

$$e_{n+1} = (1 + \lambda h)e_n + (1 + \lambda h - e^{\lambda h})e^{\lambda h n} \tag{6}$$

We are now ready to apply the control u_n to (6) to get

$$e_{n+1} = (1 + \lambda h)e_n + u_n + s_n \tag{7}$$

where $s_n = (1 + \lambda h - e^{\lambda h})e^{\lambda h n}$ is a forcing term.

We may now use the techniques presented by Anderson and Moore in [1] to optimize the cost function

$$J(u) = \sum_{n=1}^{N+1} e_n^2 + \delta \sum_{n=1}^{N+1} u_{n-1}^2,$$

subject to (7); that is, find the control sequence $\{u_n\}$ that will minimize $J(u)$.

Thus, we first find P_n recursively from

$$P_n = A'\{Q + P_{n+1} \\ - (Q + P_{n+1})B[B'(Q + P_{n+1})B + R]^{-1} \\ B'(Q + P_{n+1})\}A,$$

with $P_N = 0,$

then we find the optimal u_n from

$$u_n = -[B'(Q + P_{n+1})B + R]^{-1}B'(Q + P_{n+1})A,$$

where Q = R = 1 for our case; and for our simple example $A = (1 + \lambda h)$ and B = 1.

We can now compute our minimized error e_{n+1} with equation (7). Figures 2 and 3 depict the behavior of the minimized error (labelled "ec" in the figures).

3   Four-Step Adams-Bashforth

We shall now study a Four-step Adams-Bashforth procedure

$$y_{n+1} = y_n + \frac{h}{24}[55f(y_n) - 59f(y_{n-1}) + 37f(y_{n-2}) - 9f(y_{n-3})]$$

as we apply it to our linear control system

$$y_{n+4} - y_{n+3} = \frac{55}{24}u_{n+3} - \frac{59}{24}u_{n+2} + \frac{37}{24}u_{n+1} - \frac{3}{8}u_n.$$

By taking z-transforms, we obtain

$$(z^4 - z^3)\hat{y}_n = (\frac{55}{24}z^3 - \frac{59}{24}z^2 + \frac{37}{24}z - \frac{3}{8})\hat{u}_n.$$

Hence, the transfer function

$$H(z) = \frac{(\frac{55}{24}z^3 - \frac{59}{24}z^2 + \frac{37}{24}z - \frac{3}{8})}{(z^4 - z^3)}$$

and the observability realization

$$x_{n+1} = Ax_n + bu_n$$

$$y_n = cx_n,$$

where

$$A = \begin{bmatrix} 0 & 0 & 0 & 0 \\ 1 & 0 & 0 & 0 \\ 0 & 1 & 0 & 0 \\ 0 & 0 & 1 & 1 \end{bmatrix}, \quad c = [0, 0, 0, 1], \quad b = [-\frac{3}{8}, \frac{37}{24}, -\frac{59}{24}, \frac{55}{24}]^T.$$

We use the test ordinary differential equation as before

$$\dot{y} = \lambda y$$

and we apply the control $u_n = \lambda h y_n + v_n$ to obtain the following:

$$x_{n+1} = (A + \lambda h b c) x_n + b v_n = \hat{A} x_n + b v_n$$

$$y_n = c x_n$$

Just as in the Two-step Adams-Bashforth method, the expression for the error between the computed solution of the ODE and its exact solution is found:

$$e_n = y_n - e^{\lambda h n} = c x_n - e^{\lambda h n} \tag{8}$$

$$
\begin{aligned}
e_{n+1} &= c x_{n+1} - e^{\lambda h} e^{\lambda h n} = c \hat{A} x_n + c b v_n \\
&\quad - e^{\lambda h} e^{\lambda h n}
\end{aligned}
\tag{9}
$$

$$e_{n+2} = c \hat{A}^2 x_n + c \hat{A} b v_n + c b v_{n+1} - e^{2\lambda h} e^{\lambda h n} \tag{10}$$

$$
\begin{aligned}
e_{n+3} &= c \hat{A}^3 x_n + c \hat{A}^2 b v_n + c \hat{A} b v_{n+1} + c b v_{n+2} \\
&\quad - e^{3\lambda h} e^{\lambda h n}
\end{aligned}
\tag{11}
$$

$$
\begin{aligned}
e_{n+4} &= c \hat{A}^4 x_n + c \hat{A}^3 b v_n + c \hat{A}^2 b v_{n+1} + c \hat{A} b v_{n+2} \\
&\quad + c b v_{n+3} - e^{4\lambda h} e^{\lambda h n}
\end{aligned}
\tag{12}
$$

now $\hat{A}$ is of the form

$$
\hat{A} = \begin{bmatrix} 0 & 0 & 0 & -\tau_0 \\ 1 & 0 & 0 & -\tau_1 \\ 0 & 1 & 0 & -\tau_2 \\ 0 & 0 & 1 & -\tau_3 \end{bmatrix}
$$

and, by the Cayley-Hamilton Theorem, it satisfies its characteristic polynomial $\hat{A}^4 + \tau_3 \hat{A}^3 + \tau_2 \hat{A}^2 + \tau_1 \hat{A} + \tau_0 I = 0$.
Hence,

$$
\begin{aligned}
\tau_0 e_n + \tau_1 e_{n+1} &+ \tau_2 e_{n+2} + \tau_3 e_{n+3} + e_{n+4} = \\
&c(\hat{A}^4 + \tau_3 \hat{A}^3 + \tau_2 \hat{A}^2 + \tau_1 \hat{A} + \tau_0 I) x_n \\
&+ (c\hat{A}^3 b + \tau_3 c \hat{A}^2 b + c \tau_2 \hat{A} b + \tau_1 c b) v_n \\
&+ (c\hat{A}^2 b + \tau_3 c \hat{A} b + \tau_2 c b) v_{n+1} \\
&+ (c\hat{A} b + \tau_3 c b) v_{n+2} \\
&- (\tau_0 + \tau_1 e^{\lambda h} + \tau_2 e^{2\lambda h} + \tau_3 e^{3\lambda h} + e^{4\lambda h}) \\
&\quad e^{\lambda h n}
\end{aligned}
\tag{13}
$$

$$
\begin{aligned}
\tau_0 e_n + \tau_1 e_{n+1} &+ \tau_2 e_{n+2} + \tau_3 e_{n+3} + e_{n+4} = \\
&(c\hat{A}^3 b + \tau_3 c \hat{A}^2 b + c \tau_2 \hat{A} b + \tau_1 c b) v_n \\
&+ (c\hat{A}^2 b + \tau_3 c \hat{A} b + \tau_2 c b) v_{n+1} \\
&+ (c\hat{A} b + \tau_3 c b) v_{n+2} \\
&- (\tau_0 + \tau_1 e^{\lambda h} + \tau_2 e^{2\lambda h} + \tau_3 e^{3\lambda h} + e^{4\lambda h}) \\
&\quad e^{\lambda h n}
\end{aligned}
\tag{14}
$$

Take the z-transform of (14) to get:

$$
\begin{aligned}
\hat{e}_n = {} & \left[\frac{\tau_1 cb + \tau_2 c\hat{A}b + \tau_3 c\hat{A}^2 b + c\hat{A}^3 b}{z^4 + \tau_3 z^3 + \tau_2 z^2 + \tau_1 z + \tau_0} \right. \\
& + \frac{(c\hat{A}^2 b + \tau_3 c\hat{A}b + \tau_2 cb)z}{z^4 + \tau_3 z^3 + \tau_2 z^2 + \tau_1 z + \tau_0} \\
& + \frac{(c\hat{A}b + \tau_3 cb)z^2}{z^4 + \tau_3 z^3 + \tau_2 z^2 + \tau_1 z + \tau_0} \\
& \left. + \frac{cb}{z^4 + \tau_3 z^3 + \tau_2 z^2 + \tau_1 z + \tau_0} \right] \hat{v}_n \\
& - \frac{1}{z^4 + \tau_3 z^3 + \tau_2 z^2 + \tau_1 z + \tau_0} \\
& Z\{(\tau_0 + \tau_1 e^{\lambda h} + \tau_2 e^{2\lambda h} + \tau_3 e^{3\lambda h} + e^{4\lambda h}) \\
& e^{\lambda h n}\}
\end{aligned}
\tag{15}
$$

Same as in the previous section, we seek a realization of the form

$$
\begin{aligned}
z_{n+1} &= F z_n + g v_n + g(\text{forcing term}) \\
e_n &= \hat{h} z_n
\end{aligned}
\tag{16}
$$

The cost function to be minimized does not change:

$$
J(v) = \sum_{n=1}^{N+1} e_n^2 + \delta \sum_{n=1}^{N+1} v_{n-1}^2 .
$$

Calculate P_n recursively as before, with $P_N = 0$ from

$$
\begin{aligned}
P_n = F'\{ & Q + P_{n+1} \\
& - (Q + P_{n+1})g[g'(Q + P_{n+1})g + R]^{-1} \\
& g'(Q + P_{n+1})\}F
\end{aligned}
$$

and find the optimal control law from

$$
K_n' = -[g'(Q + P_{n+1})g + R]^{-1}g'(Q + P_{n+1})F
\tag{17}
$$

with the associated closed-loop system being

$$
z_{n+1} = (F + gK_n')z_n
\tag{18}
$$

where $Q = \begin{bmatrix} 1 & 0 & 0 & 0 \\ 0 & 1 & 0 & 0 \\ 0 & 0 & 1 & 0 \\ 0 & 0 & 0 & 1 \end{bmatrix}$, $R = [1]$, $\hat{h} = [0, 0, 0, 1]$,

$$g = \begin{bmatrix} \tau_1 cb + \tau_2 c\hat{A}b + \tau_3 c\hat{A}^2 b + c\hat{A}^3 b \\ \tau_2 cb + \tau_3 c\hat{A}b + c\hat{A}^2 b \\ \tau_3 cb + c\hat{A}b \\ cb \end{bmatrix}, \text{ and } F = \begin{bmatrix} 0 & 0 & 0 & -\tau_0 \\ 1 & 0 & 0 & -\tau_1 \\ 0 & 1 & 0 & -\tau_2 \\ 0 & 0 & 1 & -\tau_3 \end{bmatrix}.$$

Calculate the minimized error e_n with equation (16). Given the values obtained for e_n, calculate the Four-step Adams-Bashforth error with $error = \|e_n\|_\infty$. Figure 4 shows the graph of the f Four-step Adams-Bashforth error. Figure 5 is the semilog plot of the Four-step Adams-Bashforth error (with the necessary adjustments in order to be able to take logarithms); as before, observe that the sequence is increasing for n = 0 to 100. Figure 6 is a semilog plot of the corrected error. We see a very large transient behavior but the sequence $\{ec_n\}$ converges to 0 as n goes to 100 (the value of ec_n is not plotted at n = 100 since the log 0 is undefined).

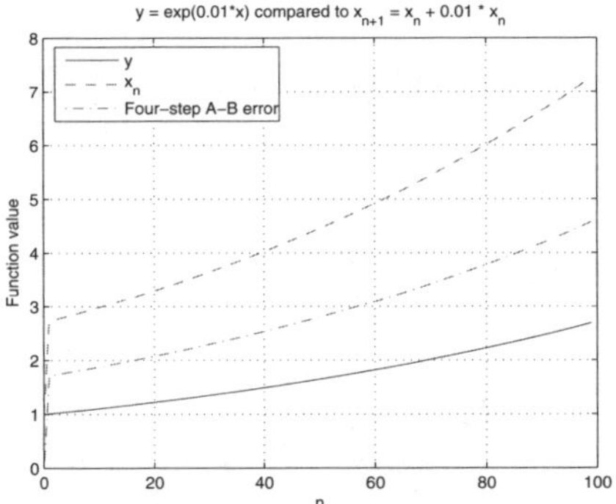

Fig. 4. Comparison of the exact solution of the ODE with the numerical approximation; with h = 0.01

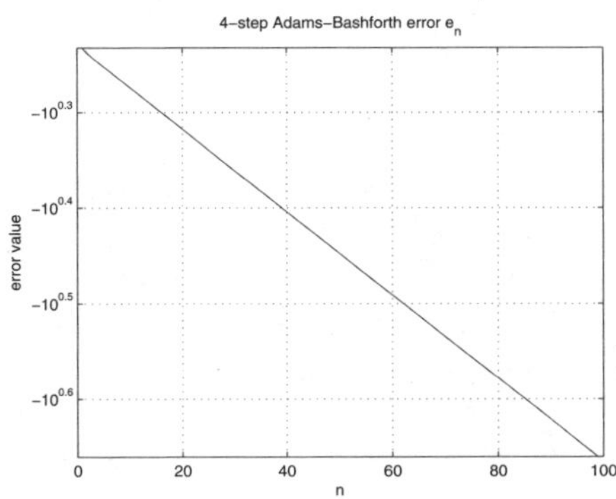

Fig. 5. Four-Step Adams-Bashforth error over the interval [0,1]; with h = 0.01

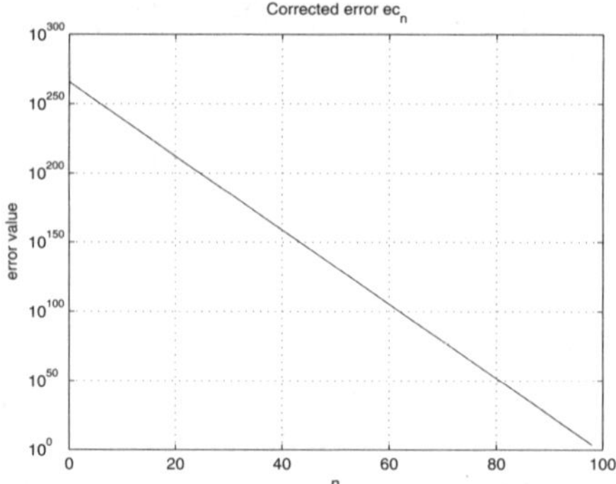

Fig. 6. Four-Step Adams-Bashforth corrected error; with h = 0.01

4   Statistical Analysis

Based on the simple example given before, we shall now study the statistical distribution of the error. For Four-Step Adams-Bashforth method, Figure 7 depicts the behavior of the error distribution.

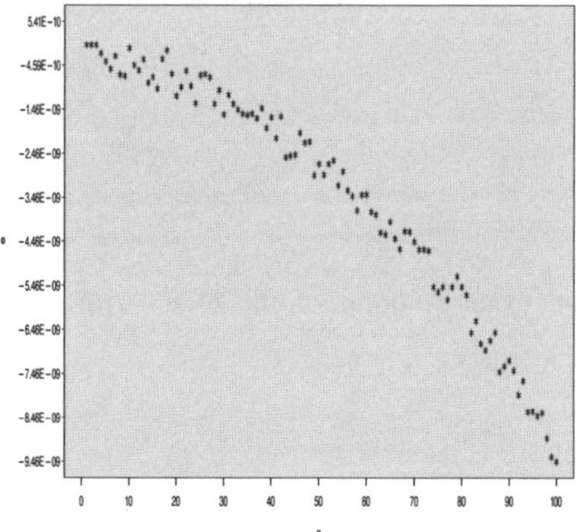

Fig. 7. Four-Step Adams-Bashforth error distribution; with $h = 0.01$ and $\lambda = 1$

The error distribution indicates a linear pattern. To reduce the linear effect, next we will fit a second-order regression model to the data then study the residuals of the model instead. The second-order regression model takes the following form:

$$e_n = \beta_0 + \beta_1 n + \beta_2 n^2 + \eta_n, \tag{19}$$

where β_0, β_1 and β_2 are the parameters we need to estimate [5]. η_n is the residual. Figure 8 shows the distribution of the residual of the regression model.

The pattern of the residual distribution indicates that the data is a stationary process. We employ the second-order autoregressive model to study the residual since this model

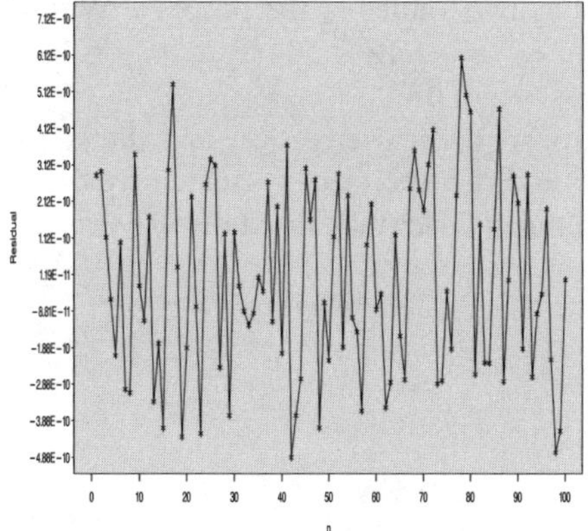

Fig. 8. Four-Step Adams-Bashforth residual distribution; with $h = 0.01$ and $\lambda = 1$

introduce a "white noise" term, which is assumed to have a normal distribution. The second-order autoregressive model [2] is given by:

$$\tilde{\eta}_n = \phi_1 \tilde{\eta}_{n-1} + \phi_2 \tilde{\eta}_{n-2} + a_n, \tag{20}$$

where $\tilde{\eta}_n$ is the deviation from μ, for example, $\tilde{\eta}_n = \eta_n - \mu$; a_k is the "white noise", usually assumed having a normal distribution; and ϕ_1 and ϕ_2 are the parameters can be estimated.

By combining the second-order regression model and second-order autoregressive, we obtain a new form of the model, which is given by:

$$e_n = \beta_0 + \beta_1 n + \beta_2 n^2 + \mu$$
$$+ \phi_1(\eta_{n-1} - \mu) + \phi_2(\eta_{n-2} - \mu) + a_n. \tag{21}$$

Then with two equations obtained from second-order regression model:

$$e_{n-1} = \beta_0 + \beta_1(n-1) + \beta_2(n-1)^2 + \eta_{n-1}$$

and

$$e_{n-2} = \beta_0 + \beta_1(n-2) + \beta_2(n-2)^2 + \eta_{n-2},$$

we can simplify the new form of the model by the following:

$$e_n = c_0 + c_1 n + c_2 n^2 + \phi_1 e_{n-1} + \phi_2 e_{n-2} + a_n, \tag{22}$$

where

$$c_0 = \beta_0 + \mu + \phi_1(-\beta_0 + \beta_1 - \beta_2 - \mu)$$
$$+ \phi_2(-\beta_0 + 2\beta_1 - 4\beta_2 - \mu),$$
$$c_1 = \beta_1 + \phi_1(-\beta_1 + 2\beta_2) + \phi_2(-\beta_1 + 4\beta_2),$$
$$c_2 = (1 - \phi_1 - \phi_2)\beta_2.$$

In equation (22), all the parameters c_0, c_1, c_2, ϕ_1 and ϕ_2 can be estimated. According to this new form of the model, the error e_n is not only related to the value of n, it is also related to the previous two values of the process. With our simple example, the estimated parameters are: $c_0 = -2.49 \times 10^{-10}, c_1 = -6.56 \times 10^{-12}, c_2 = -7.60 \times 10^{-13}, \phi_1 = 0.146$ and $\phi_2 = -0.087$.

We observe the control theoretical error, ec, and the statistical error, es, side-by-side in Figure 9 and we find that the error corrected by the control theory technique converges to a minimum much faster than the statistical error.

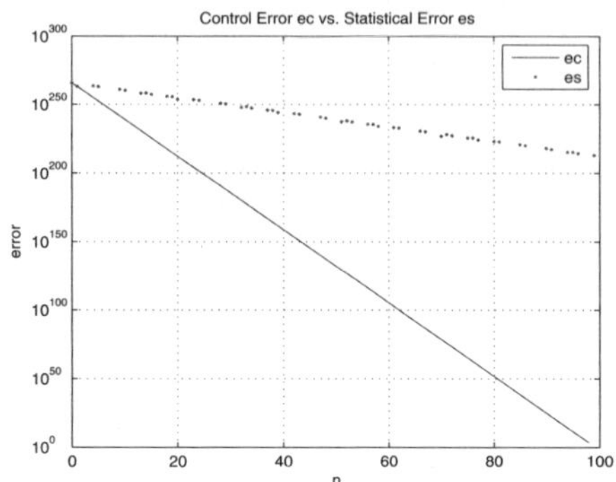

Fig. 9. Comparison of the control error with the statistical error

5   Conclusion

We approximated the solution of an Ordinary Differential Equation by using Euler's method and a Four-Step Adams-Bashforth method. It was demonstrated that the resulting error from the numerical approximations of the solution can be minimized with optimal and robust control techniques. After very high transients, the corrected error from the Adams-Bashforth method converged to the desired minimum. We developed a second-order autoregressive time series model based on the data generated from the Four-Step Adams-Bashforth method. A "white noise" term, symmetrically distributed with mean zero and constant variance, was introduced in the autoregressive model. It was shown that the minimized error converges faster than the statistical error, to the desired value.

References

1. Brian D. O. Anderson and John B. Moore. *Linear Optimal Control.* Prentice-Hall, Inc., Englewood Cliffs, New Jersey, c1971, 1971.
2. Jenkins G. M. Box, G. E. P. and G. C. Reinsel. *Time Series Analysis: Forecasting and Control.* Prentice-Hall, New Jersey, c1994, 3 edition, 1994.
3. Richard L. Burden and J. Douglas Faires. *Numerical Analysis.* Thomson, Belmont, California, c2005, 8 edition, 2005.
4. Kjell Gustafsson. Control theoretic techniques for stepsize selection in explicit runge-kutta methods. *ACM Transactions on Mathematical Software*, 17(4):533–554, 1991.
5. C. R. Rao. *Linear Statistical Inference and Its Applications.* Wiley, New York, c1973, 1973.

Contour Reconstruction and Matching Using Recursive Smoothing Splines

Maja Karasalo[1], Xiaoming Hu[1], and Clyde F. Martin[2]

[1] Optimization and Systems Theory, Royal Institute of Technology, 100 44 Stockholm, Sweden
[2] Mathematics and Statistics, Texas Tech University, Lubbock, Texas 79409-1042

Summary. In this paper a recursive smoothing spline approach is used for reconstructing a closed contour. Periodic splines are generated through minimizing a cost function subject to constraints imposed by a linear control system. The filtering effect of the smoothing splines allows for usage of noisy sensor data. An important feature of the method is that several data sets for the same closed contour can be processed recursively so that the accuracy can be improved meanwhile the current mapping can be used for planning the path for the data-collecting robot.

1 Introduction

In this paper we focus on the problem of reconstructing closed contours in a clustered environment. Our method is based on the so-called smoothing splines, both regular and periodic, constructed from sensor data. The underlying assumption is that a robot will be sent out repeatedly to collect data on the contours using range sensors. Since the data collected will be noisy, some smoothing technique has to be applied.

Data smoothing has been a classical problem in system and control history [2, 3]. It has been further shown in [8] that control theoretic smoothing splines, where the curve is found through minimizing a cost function, act as a filter and are better suited for noisy measurements. It is also noted in [16] that smoothing splines are in some sense band limited so that small changes in one data point will mainly affect the spline in a neighborhood of that point.

We note that recursive approaches to constructing splines have been investigated in [6] and [11]. In the field of robotics recursive cubic B-spline methods for path planning have been presented in [10] and [9]. Our approach differs from the previous work in this field by using control-theoretic smoothing splines that enable simultaneous planning, mapping and data filtering.

The point-to-point LQ optimal control problem has also recently been investigated in for instance [17] and [18], where [17] treats the optimal output-transition problem for linear systems while [18] considers LTI continuous-time systems with affine constraints in initial and terminal states.

This paper investigates a similar LQ problem but the important distinctions are mainly that we optimize over all periodic solutions to the LQ problem and introduce an iterative algorithm that enables the use of early contour estimates for localization and path planning while collecting new data for the refinement of the estimate.

The paper is organized as follows. In Section 2 we formulate the contour estimation problem; in Section 3 we discuss some theoretic properties of the periodic smoothing

A. Chiuso et al. (Eds.): Modeling, Estimation and Control, LNCIS 364, pp. 193–206, 2007.
springerlink.com

splines; in Section 4 we discuss how to handle the inconsistence in two consecutive data sets caused by the drifting error of the data-collecting robot. Finally, in Section 5 some experimental results are shown and discussed.

2 Problem Formulation and Motivation

We are given a data set $D = \{(t_i, z_i) : i = 1, ..., N\}$, where $t_1 = 0$ rad, $\ldots$, $t_N = T = 2\pi$ rad is the polar coordinate *angle and z_i is the radius in polar coordinates. The data in D are noise contaminated measurements from a closed continuous curve. How to find the curve $y(t)$ that best represents the data in some sense?*

The solution is found by solving the following polar second derivative L_2 smoothing problem:

Problem 1

$$\min_{u,x_0} J = \frac{1}{2} x_0^T P_0^{-1} x_0 + \frac{1}{2} \int_0^T u(t)^T Q^{-1} u(t) dt +$$

$$\sum_{i=1}^N (t_i - t_{i-1})(z_i - Cx(t_i))^T R_0^{-1}(z_i - Cx(t_i)) \quad (1)$$

$$\dot{x} = Ax + Bu$$
$$x_0 = x(0) = x(T). \quad (2)$$

The constraints (2) consist of an n-dimensional ODE with relative degree n and periodic boundary condition. The resulting smoothing spline is given by $y(t) = Cx(t)$. We refer to the system

$$\dot{x} = Ax + Bu, \quad y = Cx \quad (3)$$

as the spline generator of (1). For the purpose of finding a curve in $\mathbb{R}^2$ it suffices to choose a 2-dimensional spline generator.

As the data is noise contaminated, the resulting spline from one data set D will give a poor map of the obstacle. Thus we let the robot servo several times around the same obstacle, collecting sets of data that are used recursively to update and refine the map. We introduce a recursive smoothing spline method, where the optimal solution $(x^{k-1}(t), u^{k-1}(t))$ from the previous iteration is used in iteration k together with the new data z_i^k.

Problem 2

$$\min_{u^k,x_0^k} J_k = \frac{1}{2}(x_0^k - x_0^{k-1})^T P_0^{-1}(x_0^k - x_0^{k-1}) +$$

$$\frac{1}{2} \int_0^T (u^k(t) - u^{k-1}(t))^T Q^{-1}(u^k(t) - u^{k-1}(t)) dt +$$

$$\sum_{i=1}^N (t_i - t_{i-1})(z_i^k - Cx^k(t_i))^T R_0^{-1}(z_i^k - Cx^k(t_i)) \quad (4)$$

$$\dot{x}^k = Ax^k + Bu^k$$
$$x_0^k = x^k(0) = x^k(T). \tag{5}$$

Using a recursive method has the advantage of getting a better spline approximation for each iteration, improving the map stepwise so that the previous map can be used for path planning in each iteration. Also, if for some reason new data is obtained only for one part of the curve, we can modify that part of the spline separately by performing the next recursion using new data for that part of the curve and old data for the rest.

Due to the periodicity of the investigated curves we use polar coordinates. Thus, the interpretation of the spline generator (3) is the following:

$$x_1(t) = r(\theta)$$
$$x_2(t) = \dot{x}_1(t) = \frac{dr(\theta)}{d\theta}$$
$$\dot{x}_2(t) = \frac{d^2 r(\theta)}{d\theta^2} = u \tag{6}$$
$$y(t) = x_1(t) = r(\theta).$$

We take

$$P_0^{-1} = \begin{bmatrix} \delta_1 & 0 \\ 0 & \delta_2 \end{bmatrix}, \ Q = 1, \ R_0 = 1/\varepsilon^2, \tag{7}$$

where $\varepsilon > 0$ determines how much emphasis to put on measurement data. A large value brings the spline close to the data points while a small value yields a smoother spline.

Consequently, we refer to the spline at iteration k as $r^k(\theta)$. For this formulation we require that the origin is placed inside the closed curve. Should this not be the case we make a simple translation, compute the spline and translate back. Finally, for clarity, we state the two problems using the polar parameterization. Equations (1) and (2) become

$$\min_r J(r) = \delta_1^2 r(0)^2 + \delta_2^2 r'(0)^2 + \int_0^{2\pi} u^2(\theta)d\theta + \varepsilon^2 \sum_{i=1}^N w_i(r(t_i) - z_i)^2 \tag{8}$$
$$r(0) = r(2\pi)$$
$$r'(0) = r'(2\pi) \tag{9}$$

while Equations (4) and (5) become

$$\min_{\tilde{r}^k} J(\tilde{r}^k) = \delta_1^2 \tilde{r}^k(0)^2 + \delta_2^2 \tilde{r}^{k\prime}(0)^2 + \int_0^{2\pi} (\tilde{u}^k)^2(\theta)d\theta + \varepsilon^2 \sum_{i=1}^N w_i(\tilde{r}^k(t_i^k) - \tilde{z}_i^k)^2 \tag{10}$$
$$\tilde{r}^k(0) = \tilde{r}^k(2\pi)$$
$$\tilde{r}^{k\prime}(0) = \tilde{r}^{k\prime}(2\pi), \tag{11}$$

where

$$\tilde{z}_i^k = z_i^k - r^{k-1}(t_i^k)$$
$$\tilde{r}^k(\theta) = r^k(\theta) - r^{k-1}(\theta) \tag{12}$$
$$\tilde{u}^k = u^k - r^{k-1}(\theta)''.$$

Here (z_i^k, t_i^k) are the data points from iteration k, $r^{k-1}(t_i^k)$ is the resulting spline of iteration $k-1$ evaluated at the angle measurements of the new data set and $r^k(\theta)$ is the spline output of iteration k, found through

$$r^k(\theta) = \tilde{r}^k(\theta) + r^{k-1}(\theta). \tag{13}$$

As seen from Equations (8) and (10) the particular choice of system matrices yields a minimization over only one parameter, r. By choosing an appropriate discretization of the system both problems can be implemented as unconstrained quadratic programming problems, as is discussed in detail in [1]. Note also that Equations (10) and (11) are identical to Equation (8) and (9) except for the new variable name $\tilde{r}^k$ and input data $\tilde{z}^k$. Hence the solution methods for both problems are identical except that the input to the former is (t_i, z_i) and to the latter, at iteration k, is $(t_i, \tilde{z}_i^k)$.

In the next section we will review and discuss some theoretical properties of the optimal smoothing problems formulated in this section.

3   Some Theoretical Properties

The studied smoothing problem is a continuous time problem with discrete data and periodic boundary conditions. Such problems, without the periodic constraint, have been widely studied in the literature, see for example the books by Bryson and Ho [2] and Jazwinski [3]. However, as far as we know, it is difficult to find results concerning the periodic case. For the sake of completeness, we investigate the conditions for solving this problem. We begin by studying the proper periodicity conditions.

3.1   Proper Periodicity Conditions

In this section, we adopt the notations used in [5]. Suppose $(\overline{u}^*, \overline{x}^*)$ is optimal within the class of constant solutions of Problem 1. Namely the optimal steady state for the dynamical constraints. Let $\overline{J}^*$ denote the cost at the optimal steady state.

Definition 1. *The optimal control problem is said to be proper if there exists an admissible control $\overline{u}(\cdot)$ such that*

$$J(\overline{u}(\cdot)) < \overline{J}^*.$$

If we consider the polar coordinates (6), then it is easy to see that all steady states are circles. The problem (8) becomes proper if we can find a better curve than a circle. Here we only consider the case where δ_1 and δ_2 are set to zero, for the sake of simplicity. We believe that this problem is proper as long as the data set is nontrivial, namely, there exist at least two points $(\theta_i, r(\theta_i))$ and $(\theta_j, r(\theta_j))$, such that $r(\theta_i) \neq r(\theta_j)$. However, with the applications in mind, we will only show a weaker result here.

Theorem 1. *Consider the periodic control problem (8). Suppose $\delta_1 = \delta_2 = 0$. Then problem (8) is proper generically, namely, the data sets that make the problem not proper are contained in an algebraic set.*

Proof. By fairly straight forward but tedious calculation we can show that the optimal control for problem (8) is the solution of the following problem:

$$\min_{u} C(u) = \int_0^{2\pi} u^2 d\theta + \epsilon^2 \sum_{i=1}^{N} w_i \left(\frac{\theta_i}{2\pi} \int_0^2 \pi(2\pi - s)u(s)ds \right.$$

$$- \int_0^{\theta_i} (\theta_i - s)u(s)ds - r_i \big)^2$$

$$- \epsilon^2 \frac{\left(\sum_{i=1}^{N} w_i \left(\frac{\theta_i}{2\pi} \int_0^2 \pi(2\pi - s)u(s)ds - \int_0^{\theta_i}(\theta_i - s)u(s)ds - r_i \right) \right)^2}{\sum_{i=1}^{N} w_i}$$

$$s.t. \int_0^{2\pi} u(s)ds = 0. \tag{14}$$

In particular, $C(0) = \overline{J}^*$, the optimal steady state cost (least square solution) for (8). Now we consider only the following class of feasible control

$$u = \alpha \sin(mt + \theta_0),$$

where m is a positive integer. One can easily show that for large enough m and properly chosen θ_0, the optimal α for (14) is generically non-zero. $\qquad \square$

Next we review the optimality conditions. We first consider the continuous time, continuous data problem, then use the solution of that case to analyze the discrete data case. The final step is to study how the iterated discrete data case relates to the continuous data case.

3.2   Continuous Time, Continuous Data

The continuous time, continuous data formulation is the following: Find x_0, $u(t)$ so that J is minimized, where

$$J = \frac{1}{2}(x_0 - \hat{x}_0)^T P_0^{-1}(x_0 - \hat{x}_0) + \frac{1}{2} \int_0^T [(u(t) - \hat{u}(t))^T Q^{-1}(u(t) - \hat{u}(t)) +$$

$$(z(t) - Cx(t))^T R^{-1}(z(t) - Cx(t))]dt \tag{15}$$

$$\dot{x} = \quad Ax + Bu$$
$$x_0 = \quad x(0) = x(T). \tag{16}$$

For our setup, $\hat{x}$ and $\hat{u}$ are previous estimations of the state and control and z is the measurement data, which for now is assumed continuous. For investigating (15) our starting point will be the smoothing problem solved in [2], which is the same as above except without the periodicity constraint. The Euler-Lagrange equations **without** the periodicity constraint (16) are

$$\begin{bmatrix} \dot{x} \\ \dot{\lambda} \end{bmatrix} = \begin{bmatrix} A & -BQB^T \\ -C^T R^{-1} C & -A^T \end{bmatrix} \begin{bmatrix} x \\ \lambda \end{bmatrix} + \begin{bmatrix} B\hat{u} \\ C^T R^{-1} z \end{bmatrix} \tag{17}$$

$$x_0 = \hat{x}_0 - P_0 \lambda(0)$$
$$\lambda(T) = 0$$
$$u(t) = \hat{u} - QB^T \lambda(t). \tag{18}$$

The solution can be found using sweep methods, either starting at the terminal constraint (backward sweep) or at the initial constraint (forward sweep). The solution of our system (15) can be derived along the same lines, given that the particular system meets some specifications that will be defined in this section. We start out by deriving the Euler-Lagrange equations. They are the same as in the above, except for the boundary conditions. Obviously the terminal condition for λ must be replaced. For our system it holds that $x(0) = x(T)$. If we integrate (17) using this periodicity constraint and the unchanged initial constraint (18), we get

$$\begin{bmatrix} x(t) \\ \lambda(t) \end{bmatrix} = e^{Ht} \begin{bmatrix} x(0) \\ \lambda(0) \end{bmatrix} + \int_0^t e^{H(t-s)} g(s) ds \tag{19}$$

where H is the Hamiltonian matrix and $g(t)$ is the vector on the left hand side of (17). We rewrite this system as

$$\begin{bmatrix} x(t) \\ \lambda(t) \end{bmatrix} = E(t) \begin{bmatrix} x(0) \\ \lambda(0) \end{bmatrix} + \int_0^t E(t-s) g(s) ds. \tag{20}$$

Let $G(t) = \int_0^t E(t-s) g(s) ds$, then

$$\begin{aligned} x(t) &= E_{11}(t) x(0) + E_{12}(t) \lambda(0) + G_1(t) \\ \lambda(t) &= E_{21}(t) x(0) + E_{22}(t) \lambda(0) + G_2(t). \end{aligned} \tag{21}$$

Now we can use the fact that $x(T) = x(0)$ to derive

$$\begin{aligned} \lambda(T) = \; & E_{21}(T)(I - E_{11}(T))^{-1}(E_{12}(T)\lambda(0) + G_1(T)) + \\ & E_{22}(T)\lambda(0) + G_2(T). \end{aligned} \tag{22}$$

This is our terminal constraint on λ. As seen from (22) we need to require that the matrix $(I - E_{11}(T))$ is not singular in order to have a solution. Apparently $(I - E_{11}(t))$ is generically invertible in most cases. We will investigate this matrix for our particular system, where the matrices are defined by (7). Thus we get

$$H = \begin{bmatrix} 0 & 1 & 0 & 0 \\ 0 & 0 & 0 & -1 \\ -\varepsilon^2 & 0 & 0 & 0 \\ 0 & 0 & -1 & 0 \end{bmatrix}. \tag{23}$$

The expression for $E(T) = e^{HT}$ is tedious and will not be displayed here, but it is straight forward to show that $(I - E_{11}(2\pi))$ is invertible when ε is small. Thus we can solve the problem using the same forward sweep method but altering the terminal constraint for λ. Denote by $F(\lambda(0), T)$ the right hand side of (22), so that

$$\lambda(T) = F(\lambda(0), T). \tag{24}$$

Then our terminal constraint on x can be derived by forward sweep as

$$x(T) = \hat{x}(T) - P(T)F(\lambda(0), T). \tag{25}$$

Now recall that $\lambda(0)$ is given by $x(0)$ in (18), so (24) is really a function of $x(0) = x(T)$. Thus

$$x(T) = \hat{x}(T) - P(T)F(P_0^{-1}(\hat{x}_0 - x(T)), T)). \tag{26}$$

In order to have (26) well posed, we need

$$E_{21}(T)(I - E_{11}(T))^{-1}E_{12}(T) + E_{22}(T) - P^{-1}(T)P_0 \tag{27}$$

to be invertible. With the double integrator system, which represents a general controllable second order system, this should be true generically. Now (17) can be solved by backwards integration for the case with periodicity constraint. Note that the periodicity will only affect boundary values, the Riccati equations are the same for the two problems.

3.3 Continuous Time, Discrete Data

The continuous time, discrete data formulation is the following: Find x_0, $u(t)$ so that J is minimized, where

$$J = \frac{1}{2}(x_0 - \hat{x}_0)^T P_0^{-1}(x_0 - \hat{x}_0) +$$
$$\frac{1}{2}\int_0^T (u(t) - \hat{u}(t))^T Q^{-1}(u(t) - \hat{u}(t))dt +$$
$$\sum_{i=1}^N (t_i - t_{i-1})(z_i - Cx(t_i))^T R_0^{-1}(z_i - Cx(t_i)) \tag{28}$$

$$\begin{aligned}
\dot{x} &= \quad Ax + Bu \\
x_0 &= \quad x(0) = x(T).
\end{aligned} \tag{29}$$

Here, z_i are the sampled data at times t_i. In this section we show that the solution to (28) approaches the solution to the continuous data problem as $N \to \infty$. Since the Riccati equations do not depend on the terminal constraints, following the treatment in [2, 3] we have

$$\begin{aligned}
\dot{\hat{x}} &= \quad A\hat{x} + B\hat{u} + K_i(z_i(t_i^+) - C\hat{x})\delta(t - t_i) \\
\dot{P} &= \quad AP + PA^T + BQB^T - K_i CP(t_i^-)\delta(t - t_i)
\end{aligned} \tag{30}$$

$$\hat{x}(0) = \hat{x}_0$$
$$P(0) = P_0,$$

where $K_i = (t_i - t_{i-1})P(t_i^-)C^T[(t_i - t_{i-1})CP(t_i^-)C^T + R_0]^{-1}$. It is easy to show that as

$$\sum_{i=1}^N (t_i - t_{i-1})z_i \to \int_0^T z(t)dt, \tag{31}$$

the discrete problem also converges to the continuous problem.

3.4   Continuous Time, Discrete Data Iterated

We can write down the Riccati equations for the iterated case (4) as

$$\dot{\hat{x}}^k = \quad A\hat{x}^k + Bu^{k-1} + K_i^k(z_i^k(t_i^+) - C\hat{x}^k)\delta(t - t_i)$$
$$\dot{P}^k = \quad AP^k + P^kA^T + BQB^T - K_i^kCP^k(t_i^-)\delta(t - t_i) \tag{32}$$
$$\hat{x}^k(0) = x_0^{k-1}$$
$$P^k(0) = P_0,$$

where $K_i^k = (t_i - t_{i-1})P^k(t_i^-)C^T[(t_i - t_{i-1})CP^k(t_i^-)C^T + R_0]^{-1}$ and x_0^{k-1} is the optimal solution from iteration $k - 1$. It remains to be shown under what conditions the iterative case will converge to the continuous data case. Experimental results presented in this paper and simulations in for instance [20] do however suggest convergence for a large family of contours. In the rest of the paper, we will focus on two applications of periodic smoothing splines. First we present a method for data set reconstruction despite drifting error using the closed form spline formulation (Problem 1) and then a converging contour estimation using the recursive method (Problem 2). Both applications have been experimentally evaluated.

4   Data Set Reconstruction

The reconstruction algorithm is described for a nonholonomic platform using odometry for localization. Data points are generated by combining drifting odometry data with range measurements, which have a white noise error but no drift. The principles are however general and can be used for any system where the output is contaminated with a linearly drifting noise.

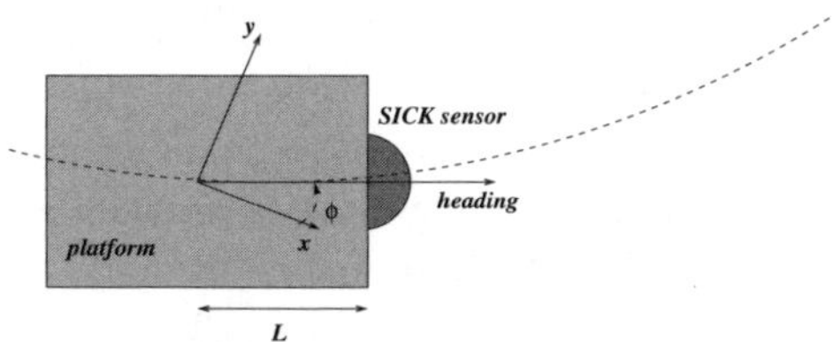

Fig. 1. Robot platform, sensor position and parameters

We call the robot state (x, y, ϕ). The agent is equipped with a range sensor located at a distances L from (x, y) as shown in Figure 1. Then a point (x_w, y_w) on the target object is obtained as follows:

$$x_w = x + L\cos\phi + S_i\cos(\phi + v_i) \tag{33}$$
$$y_w = y + L\sin\phi + S_i\sin(\phi + v_i), \tag{34}$$

Here $v_i \in [-\pi/2, \pi/2]$ is the angle of the sensor ray and S_i is the range measurement associated with it. Due to the drifting odometry error the set (x_w, y_w) yields a skew representation of the environment, see Figure 2.

To determine and compensate for the odometry drift we utilize two consecutive data sets, $\mathcal{D}_1$ and $\mathcal{D}_2$. The problem to address is to find the right translation, rotation and scaling between the two sets. We should take the following facts into consideration when developing such a method:

- We want to determine the drift in every time step as well as the rotation, translation and scaling between $\mathcal{D}_1$ and $\mathcal{D}_2$.
- We assume no knowledge of the true contour of the object.
- We cannot compare the sets pointwise since point number k of $\mathcal{D}_1$ might not correspond to point number k of $\mathcal{D}_2$. We might even have sets of different size.

Our method can be sketched as follows.

1. Translate $\mathcal{D}_1$ and $\mathcal{D}_2$ so that their respective mass centers coincide with the origin to prepare for a closed spline curve computation.
2. Transform the sets to polar coordinates in the new frames.
3. Compute smoothing splines (r_1, θ_1), (r_2, θ_2) for the sets respectively.
4. Translate $\mathcal{D}_1$ and $\mathcal{D}_2$ and the splines $z_1 = r_1 e^{\theta_1 i}$ and $z_2 = r_2 e^{\theta_2 i}$ back to the original positions of the sets.
5. Find the translation, rotation and scaling between $\mathcal{D}_1$ and $\mathcal{D}_2$ by obtaining the least square solution to $z_1 = a z_2 + b$ for complex scalars a and b. We note that the best fit might not be found by matching $z_2(k)$ to $z_1(k)$. Therefore the computation is made for every permutation of z_1 and the solution that yields the smallest residue is chosen.
6. Denote by N_j the number of data points in $\mathcal{D}_j$, $j = 1, 2$. Transform each point $z_j(t)$ in both sets by reversing the pointwise drift:

$$z_j(t) \rightarrow \left(1 - \frac{(1-a)t}{N_j}\right) z(t) + \frac{(t-1)b}{N_j} \tag{35}$$

We call the transformed sets $\tilde{\mathcal{D}}_1$ and $\tilde{\mathcal{D}}_2$,
7. The sets $\tilde{\mathcal{D}}_1$ and $\tilde{\mathcal{D}}_2$ quite accurately represent the true shape, but they differ by a translation and rotation. This transform is found again by using the same method:

$$z_2 \rightarrow a z_2 + b. \tag{36}$$

In the above procedure, if the contour cannot be well defined in polar coordinates or if they do not represent a closed contour the splines will not resemble the true curve at all. But by comparing the two splines we can still find the transformation between the two sets, the only requirement is that the investigated terrain is servoed along the same path twice and that we can assume a linear odometry drift.

The contour might not be of a kind that is ideally described in polar coordinates. This affects the discretization of the splines z_1 and z_2. For instance, a square will have more points around the center of each side than around the corners since the latter are farther away from the origin. This will affect the comparison between two curves since more emphasis will be put on matching segments with more discretization points. Should this be the case one can remove some discretization points so that the distance between two consecutive points lies within some tolerance interval for the entire vector. In Figure 2 – Figure 4 we show some results for experimental data sets from a square object.

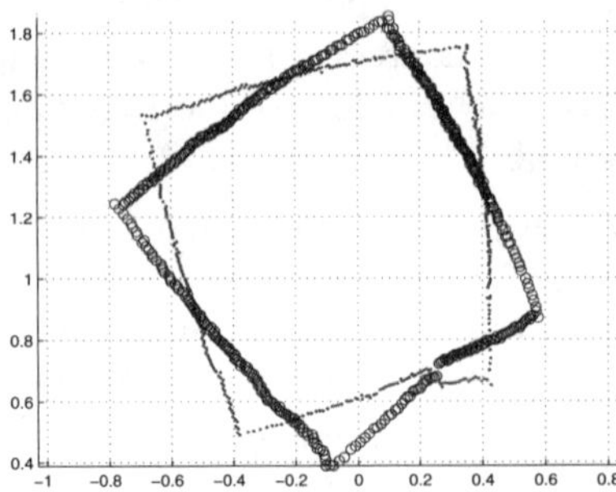

Fig. 2. *Odometry drift:* The resulting data from two consecutive revolutions along a circular path around a square object with sidelength 1 m

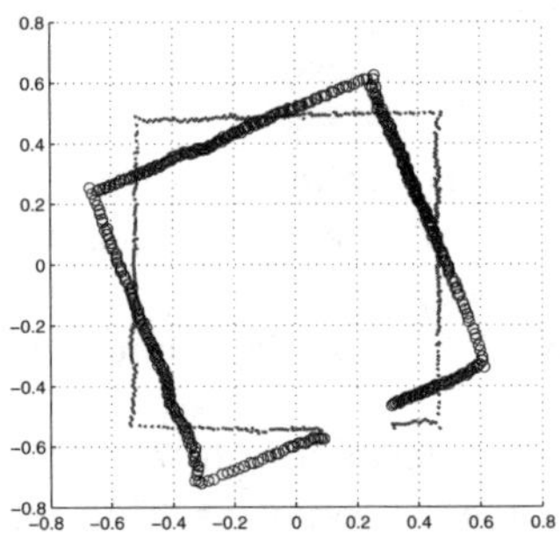

Fig. 3. *Odometry drift:* The two data sets have been pointwise transformed to compensate for the drift during one revolution

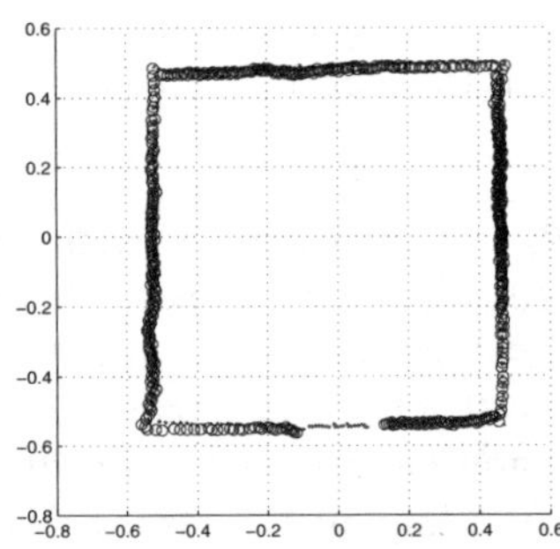

Fig. 4. *Odometry drift:* The second data set is finally transformed to coincide with the first set

5   Evaluation of Recursive Spline Method

We use the PowerBot from ActivMedia Robotics. It has unicycle dynamics and uses odometry data for localization. The robot is provided with two SICK laser scanners mounted one in front and one on the left side of the chassis. The scanners give range measurements on the interval $[-\pi/2, \pi/2]$ rad from the center of the scanner and with a resolution of 0.5^o, or 361 measurements each time step.

The robot servos the target object following a circular path. This minimizes odometry drift and allows the robot to return to the same starting position for every revolution without the use of landmarks. The algorithm was evaluated using three different target contours:

1. Single circle - the simplest possible test contour for a polar coordinate algorithm
2. Square - not ideal for polar coordinates and a quadratic smoothing cost function
3. Three circles - a non convex contour which can only be described in polar coordinates if the origin of coordinate frame is chosen carefully.

The relative error between the curve generated by the smoothing spline algorithm at iteration k and the true shape is defined as $|r^k - r_{true}|/|r_{true}|$ where r^k is the smoothing spline output and r_{true} is the real curve contour in polar coordinates. Ideally we expect the error to decrease as $1/\sqrt{k}$ due to the quadratic nature of the spline cost function.

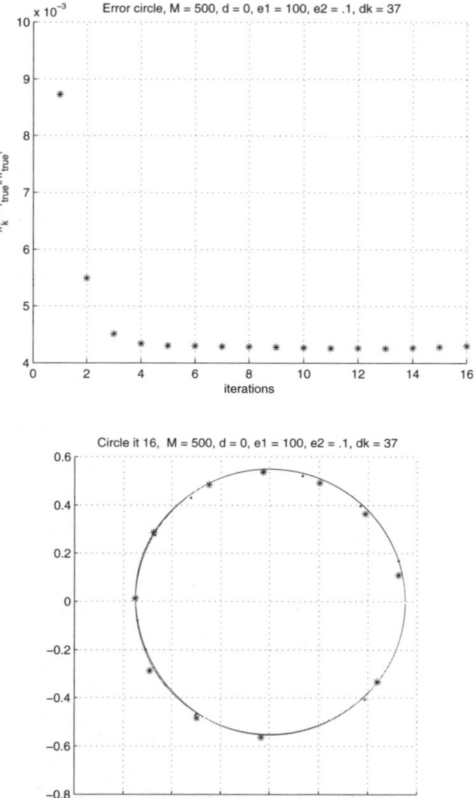

Fig. 5. Convergence of the algorithm for the circular object

This behavior is apparent for the circular shapes while the challenging square shape displays more of a linear convergence.

Single Circle

As expected, convergence is fast for the circular object and even the initial error is small. This validates the assumption that the algorithm is ideal for convex contours that are naturally defined in polar coordinates.

Square

The smoothing spline algorithm has difficulty handling sharp edges, which yields larger errors and slower convergence than for the circular shapes.

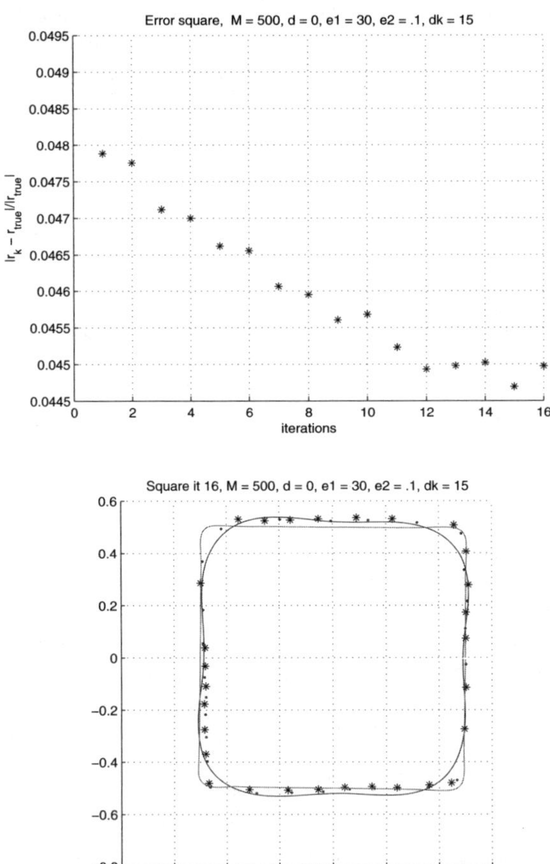

Fig. 6. Convergence of the algorithm for the square object

Three Circles

When dealing with non convex contours the performance of the algorithm varies much over the curve. For high curvature parts of the contour we would need a dense data set, or a lot of iterations, to retrieve all of the needed information. The error convergence is quadratic for this test object although it displays a noisier behavior than for the single circle.

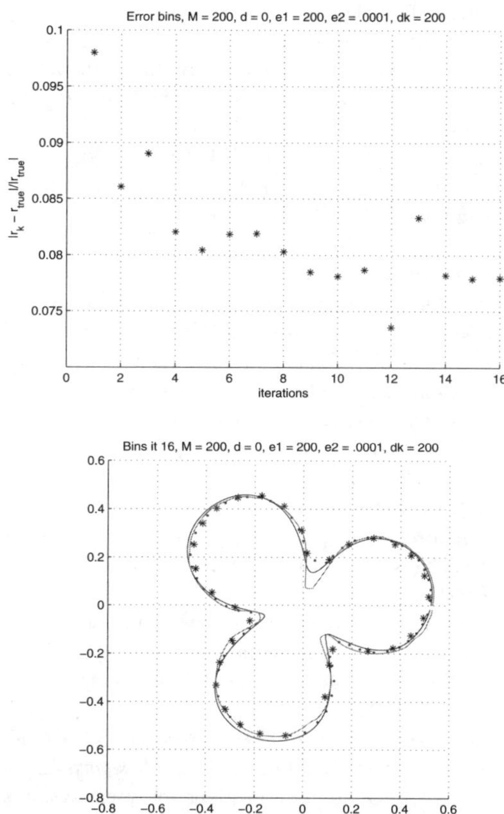

Fig. 7. Convergence of the algorithm for the non-convex object

6   Conclusions

In this paper we discussed a recursive smoothing spline approach to contour reconstruction. We derived periodic smoothing splines recursively from noisy data by solving an optimal control problem for a linear system. To complete the framework we proposed also a method for compensating the drifting error. Our experiments showed that the splines converge to the true contours we have chosen and that parts of the spline can be modified separately as new data becomes available.

References

1. G. Piccolo, M. Karasalo, D. Kragic, X. Hu Contour Reconstruction using Recursive Smoothing Splines - Experimental Validation *to appear in Proceedings of IROS 2007*, 2007 .
2. Arthur E. Bryson, Yu-Chi Ho Applied Optimal Control, Optimization estimation and control *Halsted Press*, 1975 .
3. Andrew Jazwinski. Stochastic Processes and Filtering Theory *Academic Press*, 1970.
4. C.T Leondes Advances in Control Systems, vol 16 *Academic Pr*, 1965 .
5. S. Bittanti, G. Fronza and G. Guardabssi, Periodic Control: A Frequency Domain Approach, *IEEE Trans. Aut. Control, vol. 18*, no. 1, 1973.
6. A.Z. Averbuch, A.B. Pevnyi, and V.A. Zheludev. Butterworth wavelet transforms derived from discrete interpolatory splines: recursive implementation. *Signal Processing*, 81:2363–2382, 2001.
7. M.W.M.G. Dissanayake, P. Newman, S. Clark, H.F. Durrant-Whyte, and M. Csorba. A solution to the simultaneous localization and map building (slam) problem. *IEEE Transactions on Robotics and Automation*, 17(3):229 – 241, 2001.
8. Magnus Egerstedt and Clyde F. Martin. Statistical estimates for generalized splines. *ESAIM:Control, Optimisation and Calculus of Variations*, 9:553–562, 2003.
9. R. Frezza and G. Picci. "On Line Path Following by Recursive Spline Updating". In *Proceedings of the 34th Conference of Decision and Control*, pages 367–393, 1995.
10. R. Frezza, S. Soatto, and G. Picci. "Visual Path Planning by Recursive Spline Updating". In *Proceedings of the 36th Conference of Decision and Control*, pages 367–393, 1997.
11. S. Isotani, A. de Albuquerque, M. Muratore, and N. Brasil Filho. A recursive spline-based algorithm for sensor calibration design. *Industrial Electronics, Control and Instrumentation*, 3:1952–1954, 1994.
12. H. Kano, H. Fujioka, M. Egerstedt, and C.F Martin. Optimal smoothing spline curves and contour synthesis. In *Proceedings of the 16th IFAC World Congress*, July 2005.
13. M. Karasalo, X. Hu, L. Johansson, and K. Johansson. "Multi-Robot Terrain Servoing with Proximity Sensors". *Proc. Int. Conf. Robotics and Aut., 2005*, 367–393, 2005.
14. C. F. Martin and J. Smith. Approximation, interpolation and sampling. differential geometry, the interface between pure and applied mathematics. *Contemp. Math.*, 68:227–252, 1987.
15. C. F. Martin S. Sun, M. Egerstedt. Control theoretic smoothing splines. *IEEE Transactions on Automatic Control*, 45:2271–2279, 2000.
16. Y. Zhou, W. Dayawansa, and C.F. Martin. Control theoretic smoothing splines are approximate linear filters. *Communications in Information and Systems*, 4:253–272, 2004.
17. H. Perez and S. Devasia. Optimal output-transitions for linear systems. *Automatica*, 39(2), 181–192, 2003.
18. A. Ferrante, G. Marro,and L. Ntogramatzidis. A parameterization of the solutions of the finite-horizon LQ problem with general cost and boundary condition. *Automatica*, 41(8), 1359–1366, 2005.
19. M. Egerstedt C. F. Martin, S. Sun. Optimal control, statistics and path planning. *Math. Comput. Modelling*, 33:237–253, 2001.
20. M. Karasalo, X. Hu, and C.F. Martin. Localization and Mapping using Recursive Smoothing Splines *To appear in Proc of ECC 2007*, 2007.

Role of LQ Decomposition in Subspace Identification Methods[*]

Tohru Katayama

Faculty of Culture and Information Science,
Doshisha University KyoTanabe, Kyoto 610-0394, Japan
tokataya@mail.doshisha.ac.jp

Summary. We revisit the deterministic subspace identification methods for discrete-time LTI systems, and show that each column vector of the L-matrix of the LQ decomposition in MOESP and N4SID methods is a pair of input-output vectors formed by linear combinations of given input-output data. Thus, under the assumption that the input is persistently exciting (PE) of sufficient order, we can easily compute zero-input and zero-state responses by appropriately dividing given input-output data into past and future in the LQ decomposition. This reveals the role of the LQ decomposition in subspace identification methods. Also, a related issue in stochastic realization is briefly discussed in Appendix.

1 Introduction

It is well known that the LQ decomposition (transpose of the QR decomposition), together with the singular value decomposition (SVD), has played a key role in subspace system identification methods [9, 12, 13, 14, 15]. The LQ decomposition has extensively been used as a numerical tool of pre-processing or reducing given data without loss of information. In [10], we have also employed the LQ decomposition for performing a preliminary orthogonal decomposition of the output process into deterministic and stochastic components, for which realization algorithms have been developed. In this paper, we clarify a system theoretic meaning of the LQ decomposition in subspace identification methods, motivated by the fundamental lemma due to Markovsky *et al.* [8].

Consider a discrete-time LTI system described by

$$x(t+1) = Ax(t) + Bu(t) \tag{1a}$$

$$y(t) = Cx(t) + Du(t), \quad t = 0, 1, \cdots \tag{1b}$$

where $x \in \mathbb{R}^n$ is the state vector, $u \in \mathbb{R}^m$ the control input, $y \in \mathbb{R}^p$ the output vector, and $A \in \mathbb{R}^{n \times n}$, $B \in \mathbb{R}^{n \times m}$, $C \in \mathbb{R}^{p \times n}$, $D \in \mathbb{R}^{p \times m}$ are constant matrices. In the following, we assume that (A, B) is reachable and (C, A) is observable; in this case, we say that (A, B, C) is minimal.

From (1), the transfer matrix is given by

$$G(z) = D + C(zI - A)^{-1}B \tag{2}$$

---

[*] The earlier version of this paper was presented at the International Symposium on Mathematical Theory of Networks and Systems, Kyoto, July 2006.

A. Chiuso et al. (Eds.): Modeling, Estimation and Control, LNCIS 364, pp. 207–220, 2007.
springerlink.com

and the impulse responses are

$$G_t = \begin{cases} D, & t = 0 \\ CA^{t-1}B, & t = 1, 2, \cdots \end{cases} \tag{3}$$

where $\{G_t, t = 0, 1, \cdots \}$ are also called the Markov parameters. We see that given (A, B, C, D), the transfer matrix and impulse response matrices are uniquely determined by (2) and (3), respectively; but the converse is not true.

The subspace identification problem considered is to identify the system matrices (A, B, C, D), up to similarity transformations, based on the given input-output data $\{u(t), y(t), t = 0, 1, \cdots, N - 1\}$.

2   State-Input-Output Matrix Equation

Suppose that $k > n$. According to [9], we define the stacked vectors as

$$\mathbf{y}_k(t) := \begin{bmatrix} y(t) \\ y(t+1) \\ \vdots \\ y(t+k-1) \end{bmatrix} \in \mathbb{R}^{kp}, \quad \mathbf{u}_k(t) := \begin{bmatrix} u(t) \\ u(t+1) \\ \vdots \\ u(t+k-1) \end{bmatrix} \in \mathbb{R}^{km}$$

Then, the stacked vectors satisfy the following augmented equation

$$\mathbf{y}_k(t) = \mathcal{O}_k x(t) + \mathcal{T}_k \mathbf{u}_k(t) \tag{4}$$

where $\mathcal{O}_k$ and $\mathcal{T}_k$ are respectively the extended observability matrix and the block lower triangular Toeplitz matrix defined by

$$\mathcal{O}_k = \begin{bmatrix} C \\ CA \\ \vdots \\ CA^{k-1} \end{bmatrix} \in \mathbb{R}^{kp \times n}, \quad \mathcal{T}_k = \begin{bmatrix} D & & & \mathbf{0} \\ CB & D & & \\ \vdots & & \ddots & \ddots \\ CA^{k-2}B & \cdots & CB & D \end{bmatrix} \in \mathbb{R}^{kp \times km}$$

In terms of input-output data, we further define block Hankel matrices

$$U_{0|k-1} = \begin{bmatrix} u(0) & u(1) & \cdots & u(N-1) \\ u(1) & u(2) & \cdots & u(N) \\ \vdots & \vdots & \ddots & \vdots \\ u(k-1) & u(k) & \cdots & u(k+N-2) \end{bmatrix} \in \mathbb{R}^{km \times N}$$

and

$$Y_{0|k-1} = \begin{bmatrix} y(0) & y(1) & \cdots & y(N-1) \\ y(1) & y(2) & \cdots & y(N) \\ \vdots & \vdots & \ddots & \vdots \\ y(k-1) & y(k) & \cdots & y(k+N-2) \end{bmatrix} \in \mathbb{R}^{kp \times N}$$

Since

$$U_{0|k-1} = [\mathbf{u}_k(0)\ \mathbf{u}_k(1)\ \cdots\ \mathbf{u}_k(N-1)]$$

$$Y_{0|k-1} = [\mathbf{y}_k(0)\ \mathbf{y}_k(1)\ \cdots\ \mathbf{y}_k(N-1)]$$

it follows from (4) that

$$Y_{0|k-1} = \mathcal{O}_k X_0 + \mathcal{T}_k U_{0|k-1} \tag{5}$$

where $X_0 = [x(0)\ x(1)\ \cdots\ x(N-1)] \in \mathbb{R}^{n \times N}$ is the initial state matrix. We see that (5) is the state-input-output matrix equation relating the initial state X_0 and the inputs $U_{0|k-1}$ to the output $Y_{0|k-1}$. In the following, the first term in the right-hand side of (5) is called the zero-input response, while the second term the zero-state response.

Post-multiplying (5) by a vector $\zeta \in \mathbb{R}^N$, we get

$$Y_{0|k-1}\zeta = \mathcal{O}_k X_0 \zeta + \mathcal{T}_k U_{0|k-1}\zeta$$

Thus, referring to (4), we have the following lemma [8].

Lemma 1. *(Principle of superposition) For any $\zeta \in \mathbb{R}^N$, it can be shown that $U_{0|k-1}\zeta \in \mathbb{R}^{km}$, $Y_{0|k-1}\zeta \in \mathbb{R}^{kp}$ and $X_0\zeta \in \mathbb{R}^n$, linear combinations of the input vectors, output vectors and state vectors, are a pair of input-output vectors and the unknown initial state vector satisfying (4), respectively.* $\square$

3   MOESP Method

Suppose that the LQ decomposition used in the MOESP method be given by [14, 13]

$$\begin{bmatrix} U_{0|k-1} \\ Y_{0|k-1} \end{bmatrix} = \begin{bmatrix} L_{11} & 0 \\ L_{21} & L_{22} \end{bmatrix} \begin{bmatrix} Q_1^{\mathrm{T}} \\ Q_2^{\mathrm{T}} \end{bmatrix} \tag{6}$$

where $Q_1 \in \mathbb{R}^{N \times km}$, $Q_2 \in \mathbb{R}^{N \times kp}$ are orthogonal, i.e. $Q_1^{\mathrm{T}}Q_1 = I_{km}$, $Q_2^{\mathrm{T}}Q_2 = I_{kp}$, $Q_1^{\mathrm{T}}Q_2 = 0_{km \times kp}$, and $L_{11} \in \mathbb{R}^{km \times km}$, $L_{22} \in \mathbb{R}^{kp \times kp}$ are lower triangular. Then, we have the following result, clarifying a system theoretic meaning of L-matrix in the LQ decomposition of (6).

Lemma 2. *([6]) Suppose that for the deterministic system of (1), the following conditions are satisfied.*

(i) (A, B, C) are minimal.

(ii) The input u satisfies the PE condition of order k, i.e. $\mathrm{rank}(U_{0|k-1}) = km$.

(iii) The input satisfies $\mathrm{rank}\begin{bmatrix} U_{0|k-1} \\ X_0 \end{bmatrix} = mk + n$, implying that the input-output data are obtained from an open-loop experiment.

Then it follows that

$$\mathrm{rank}\begin{bmatrix} U_{0|k-1} \\ Y_{0|k-1} \end{bmatrix} = mk + n \tag{7}$$

Moreover, each column of L-matrix is a pair of input-output vectors of length k generated from given input-output data. In particular, each column vector of L_{22} is a zero-input response of length k, so that $\mathrm{rank}(L_{22}) = n$.

Proof. Equation (7) is well known [9]. Rewriting (6), we have

$$\begin{bmatrix} L_{11} & 0 \\ L_{21} & L_{22} \end{bmatrix} = \begin{bmatrix} U_{0|k-1} \\ Y_{0|k-1} \end{bmatrix} [Q_1 \ \ Q_2] \tag{8}$$

We see from Lemma 1 that column vectors of $[L_{11} \ \ 0]$ and $[L_{21} \ \ L_{22}]$ in the left-hand side of the above equation are respectively a linear combination of column vectors of $U_{0|k-1}$ and $Y_{0|k-1}$, implying that each column vector of the L-matrix is an input-output pair generated by column vectors of given $U_{0|k-1}$ and $Y_{0|k-1}$. It also follows that each column vector of L_{22} is a zero-input response, since L_{12} is a zero matrix. It follows from (7) that the left-hand side of (8) has rank $km + n$ and $\mathrm{rank}(L_{11}) = km$, so that we have $\dim \mathrm{Im}(L_{22}) = n$, the number of independent zero-input responses. This completes a proof of lemma. $\qquad\square$

From (5) and (6), it follows that

$$L_{21}Q_1^{\mathrm{T}} + L_{22}Q_2^{\mathrm{T}} = \mathcal{O}_k X_0 + \mathcal{T}_k U_{0|k-1} \tag{9}$$

It should be noted that the two terms in the left-hand side are orthogonal, while those in the right-hand side are not. But, we see that the right-hand side is a direct sum decomposition of $Y_{0|k-1}$ by the assumption (iii) in Lemma 2. Noting that $U_{0|k-1} = L_{11}Q_1^{\mathrm{T}}$, and post-multiplying (9) by Q_1 and Q_2 respectively yield

$$L_{21} = \mathcal{O}_k X_0 Q_1 + \mathcal{T}_k L_{11} \tag{10a}$$
$$L_{22} = \mathcal{O}_k X_0 Q_2 \tag{10b}$$

where $X_0 Q_1 \in \mathbb{R}^{n \times km}$ and $X_0 Q_2 \in \mathbb{R}^{n \times kp}$ are unknown initial state matrices, so that the first term in the right hand side of (10a) is the zero-input response with the initial state $X_0 Q_1$, whereas the second term is the zero-state response, and the term in the left-hand side is the corresponding output matrix. We see that (10b) is the zero-input response with the initial state $X_0 Q_2$, so that $\mathrm{Im}(L_{22}) = \mathrm{Im}(\mathcal{O}_k)$ holds since $\mathrm{rank}(X_0 Q_2) = n$. This also shows that $\mathrm{rank}(L_{22}) = n$, since the system is assumed to be minimal.

We easily observe that the LQ decomposition gives the state-input-output equations (10) with column dimensions smaller than those of the original state-input-output equation (5). This implies that the LQ decomposition (6) plays a role of data compression as in the least-squares methods.

Let the SVD of L_{22} be given by

$$L_{22} = [U_1 \ \ U_2] \begin{bmatrix} \Sigma & 0 \\ 0 & 0 \end{bmatrix} \begin{bmatrix} V_1^{\mathrm{T}} \\ V_2^{\mathrm{T}} \end{bmatrix}, \quad \Sigma = \mathrm{diag}(\sigma_1, \cdots, \sigma_n) \tag{11}$$

where $U_1 \in \mathbb{R}^{kp \times n}$ and $U_2 \in \mathbb{R}^{kp \times (kp-n)}$. Then, we can recover the extended observability matrix as $\mathcal{O}_k = U_1 \Sigma^{1/2}$. By using the shift invariant property of $\mathcal{O}_k$, we can readily find the system matrices A and C.

Moreover, we see from (10a) that if the effect of the zero-input response therein is removed, then it will be possible to recover the zero-state response $\mathcal{T}_k L_{11}$, and hence the impulse response matrix $\mathcal{T}_k$. In order to eliminate the term of zero-input response, we pre-multiply (10a) by $U_2^{\mathrm{T}} \in \mathbb{R}^{kp \times (kp-n)}$ to get

$$U_2^{\mathrm{T}} L_{21} = U_2^{\mathrm{T}} \mathcal{T}_k(B, D) L_{11} \tag{12}$$

where $U_2^{\mathrm{T}} \mathcal{O}_k = U_2^{\mathrm{T}} U_1 \Sigma^{1/2} = 0$ is used. Since $\mathcal{T}_k(B, D)$ is linear with respect to (B, D) given $\mathcal{O}_k$, we can estimate (B, D) from (12) by using the least-squares method; see the MOESP method [14].

4 N4SID Method

We now consider the LQ decomposition employed in the N4SID method. Let $k > n$. Define $U_p := U_{0|k-1}$, $Y_p := Y_{0|k-1}$, $X_p := X_0$ and $U_f := U_{k|2k-1}$, $Y_f := Y_{k|2k-1}$, $X_f := X_k$, where the subscripts p and f denote the past and future, respectively. We also write $W_p = \begin{bmatrix} U_p \\ Y_p \end{bmatrix}$ and $W_f = \begin{bmatrix} U_f \\ Y_f \end{bmatrix}$.

We recall two state-input-output matrix equations

$$Y_p = \mathcal{O}_k X_p + \mathcal{T}_k U_p \tag{13}$$

$$Y_f = \mathcal{O}_k X_f + \mathcal{T}_k U_f \tag{14}$$

where $X_i = [x(i) \; x(i+1) \; \cdots \; x(i+N-1)]$ with $i = 0, k$.

4.1 Zero-Input Responses

Consider the LQ decomposition used in [12, 15], i.e.

$$\begin{bmatrix} U_f \\ U_p \\ Y_p \\ Y_f \end{bmatrix} = \begin{bmatrix} L_{11} & 0 & 0 & 0 \\ L_{21} & L_{22} & 0 & 0 \\ L_{31} & L_{32} & L_{33} & 0 \\ L_{41} & L_{42} & L_{43} & L_{44} \end{bmatrix} \begin{bmatrix} Q_1^{\mathrm{T}} \\ Q_2^{\mathrm{T}} \\ Q_3^{\mathrm{T}} \\ Q_4^{\mathrm{T}} \end{bmatrix} \tag{15}$$

where $Q_1, Q_2 \in \mathbb{R}^{N \times km}$, $Q_3, Q_4 \in \mathbb{R}^{N \times kp}$ are orthogonal. Since the L-matrix is block lower triangular, we see that $L_{11}, L_{22} \in \mathbb{R}^{km \times km}$, $L_{33}, L_{44} \in \mathbb{R}^{kp \times kp}$ are lower triangular, and that $L_{12} = 0, L_{13} = 0, L_{14} = 0, L_{23} = 0, L_{24} = 0, L_{34} = 0$.

From Lemma 2, each column of the L-matrix of (15) is a vector of input-output pair. Thus we have the following result, an earlier version of which is given in [6].

Lemma 3. *Suppose that for the deterministic system of (1), the following conditions are satisfied.*

(i) (A, B, C) is minimal.

(ii) The input u satisfies the PE condition of order $2k$, i.e. $\mathrm{rank}(U_{0|2k-1}) = 2km$.

(iii) The condition $\mathrm{rank} \begin{bmatrix} U_{0|2k-1} \\ X_0 \end{bmatrix} = 2mk + n$ *holds.*

Then, for the LQ decomposition of (15), it follows that

$$\mathrm{rank}(L_{33}) = n, \quad \mathrm{rank}(L_{42}) = n, \quad \mathrm{rank}(L_{43}) = n$$

$$\mathrm{rank}[L_{42}\ L_{43}] = n, \quad \mathrm{rank}\begin{bmatrix} L_{33} \\ L_{43} \end{bmatrix} = n \tag{16}$$

and that $L_{44} = 0$. Also, the following relations hold:

$$\mathrm{Ker}\begin{bmatrix} L_{22} \\ L_{32} \end{bmatrix} \subset \mathrm{Ker}L_{42}, \quad \mathrm{Ker}L_{33} \subset \mathrm{Ker}L_{43} \tag{17}$$

Proof. Since the system is minimal, and since the past input-outputs are zero ($L_{24} = 0$, $L_{34} = 0$) and the future inputs are zero ($L_{14} = 0$), we see that the future outputs are identically zero, i.e. $L_{44} = 0$. We prove the rank conditions of (16). It follows from (13) and (15) that

$$Y_p = \mathcal{O}_k X_p + \mathcal{T}_k U_p$$
$$U_p = L_{21}Q_1^{\mathrm{T}} + L_{22}Q_2^{\mathrm{T}}$$
$$Y_p = L_{31}Q_1^{\mathrm{T}} + L_{32}Q_2^{\mathrm{T}} + L_{33}Q_3^{\mathrm{T}}$$

Post-multiplying three equations above by Q_1, Q_2, Q_3 successively, we get

$$L_{31} = \mathcal{O}_k X_p Q_1 + \mathcal{T}_k L_{21}$$
$$L_{32} = \mathcal{O}_k X_p Q_2 + \mathcal{T}_k L_{22} \tag{18}$$
$$L_{33} = \mathcal{O}_k X_p Q_3$$

Since, by the assumption, $X_p Q_3$ has full rank, we see that each column vector of L_{33} is a zero-input response of length k and that $\mathrm{Im}(L_{33}) = \mathrm{Im}(\mathcal{O}_k)$ holds. Thus we have $\mathrm{rank}(L_{33}) = n$ since $\mathrm{Im}(\mathcal{O}_k) = n$ by the minimality.

Similarly, it follows from (14) and (15) that

$$Y_f = \mathcal{O}_k X_f + \mathcal{T}_k U_f$$
$$U_f = L_{11}Q_1^{\mathrm{T}}$$
$$Y_f = L_{41}Q_1^{\mathrm{T}} + L_{42}Q_2^{\mathrm{T}} + L_{43}Q_3^{\mathrm{T}}$$

Post-multiplying by three equations above by Q_1, Q_2, Q_3 successively yields

$$L_{41} = \mathcal{O}_k X_f Q_1 + \mathcal{T}_k L_{11}$$
$$L_{42} = \mathcal{O}_k X_f Q_2 \tag{19}$$
$$L_{43} = \mathcal{O}_k X_f Q_3$$

so that $[L_{42}\ L_{43}] = \mathcal{O}_k X_f [Q_2\ Q_3]$. Thus we find that each column vector of L_{42}, L_{43} is a zero-input response of length k. Moreover, the images of L_{42}, L_{43}, $[L_{42}\ L_{43}]$ coincide with $\mathrm{Im}(\mathcal{O}_k)$ if $\mathrm{rank}(X_f) = n$. To prove this latter condition, we write from (1a)

$$x(k + i) = A^k x(i) + \overline{\mathcal{C}}_k \mathbf{u}_k(i)$$

where $\overline{C}_k = [A^{k-1}B \cdots AB \ B]$ is the reversed extended reachability matrix. Thus it follows that

$$X_f = A^k X_p + \overline{C}_k U_p, \quad k > n$$

Since $\operatorname{rank}(\overline{C}_k U_p) = n$ and $\operatorname{span}(X_p) \cap \operatorname{span}(U_p) = \{0\}$, we see that $\operatorname{rank}(X_f) \geq n$. But, by definition, $\operatorname{rank}(X_f) \leq n$, so that we have $\operatorname{rank}(X_f) = n$.

Moreover, combining (13), (14) and (15), we have the following equations

$$\begin{bmatrix} Y_p \\ Y_f \end{bmatrix} = \mathcal{O}_{2k} X_p + \mathcal{T}_{2k} \begin{bmatrix} U_p \\ U_f \end{bmatrix}$$

$$\begin{bmatrix} U_p \\ U_f \end{bmatrix} = \begin{bmatrix} L_{21} \\ L_{11} \end{bmatrix} Q_1^{\mathrm{T}} + \begin{bmatrix} L_{22} \\ 0 \end{bmatrix} Q_2^{\mathrm{T}}$$

$$\begin{bmatrix} Y_p \\ Y_f \end{bmatrix} = \begin{bmatrix} L_{31} \\ L_{41} \end{bmatrix} Q_1^{\mathrm{T}} + \begin{bmatrix} L_{32} \\ L_{42} \end{bmatrix} Q_2^{\mathrm{T}} + \begin{bmatrix} L_{33} \\ L_{43} \end{bmatrix} Q_3^{\mathrm{T}}$$

Similarly to the derivation of (19),

$$\begin{bmatrix} L_{31} \\ L_{41} \end{bmatrix} = \mathcal{O}_{2k} X_p Q_1 + \mathcal{T}_{2k} \begin{bmatrix} L_{21} \\ L_{11} \end{bmatrix}$$

$$\begin{bmatrix} L_{32} \\ L_{42} \end{bmatrix} = \mathcal{O}_{2k} X_p Q_2 + \mathcal{T}_{2k} \begin{bmatrix} L_{22} \\ 0 \end{bmatrix} \tag{20}$$

$$\begin{bmatrix} L_{33} \\ L_{43} \end{bmatrix} = \mathcal{O}_{2k} X_p Q_3$$

Hence each column vector of $\begin{bmatrix} L_{33} \\ L_{43} \end{bmatrix}$ is a zero-input response of length $2k$ with $\operatorname{Im} \begin{bmatrix} L_{33} \\ L_{43} \end{bmatrix} = \operatorname{Im}(\mathcal{O}_k)$. This completes a proof of (16). Moreover, the statement in (17) is proved by noting that each column of L-matrix is a vector of input-output pair. $\square$

Since the images of L_{33}, L_{42}, L_{43}, $[L_{42} \ L_{43}]$ coincide with $\operatorname{Im}(\mathcal{O}_k)$, we can use any one of these matrices, or a combination of them, to recover $\operatorname{Im}(\mathcal{O}_k)$.

In the PO-MOESP [15], the following relations are used for obtaining the system matrices A, B, C, D:

$$\operatorname{Im}(\mathcal{O}_k) = \operatorname{Im}([L_{42} \ L_{43}])$$

$$\mathcal{O}_k^{\perp} \mathcal{T}_k(B, D)[L_{21} \ L_{22} \ L_{11}] = \mathcal{O}_k^{\perp}[L_{31} \ L_{32} \ L_{41}]$$

Though the above relations are derived via a different route, the validity of them are easily seen from (18) and (19).

Also, in [10, 6], we have derived the ORT (*ort*hogonal decomposition based) algorithm that identifies the systems subjected to both deterministic and stochastic inputs, in which the deterministic subsystem is identified by using the following relations

$$\operatorname{Im}(\mathcal{O}_k) = \operatorname{Im}(L_{42})$$

$$\mathcal{O}_k^{\perp} \mathcal{T}_k(B, D) L_{11} = \mathcal{O}_k^{\perp} L_{41}$$

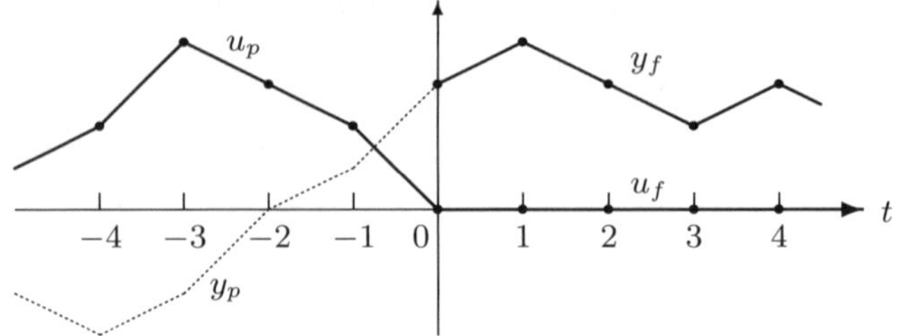

Fig. 1. Input-output response of an LTI system

4.2   Relation to Ho-Kalman's Method

It is instructive to consider the diagram of Fig. 1 to introduce the Hankel operator used in Ho-Kalman's method [5]. Let u_p be the input that assumes non-zero values up to time $t = -1$ and let u_f be the future input which is fixed to zero, i.e.

$$u = (u_p, u_f) = (\cdots, u(-3), u(-2), u(-1), 0, 0, 0, \cdots) \tag{21}$$

Applying the input u to a discrete-time LTI system, we observe the future outputs $y_f(0), y_f(1), \cdots$, where the past output y_p is shown by the dotted line. For the input sequence of (21), the future output is expressed as

$$y_f(t) = \sum_{i=-\infty}^{-1} G_{t-i} u_p(i) = CA^t x(0), \quad t = 0, 1, \cdots \tag{22}$$

where $x(0) = \sum_{i=-\infty}^{-1} A^{-i-1} B u(i)$ is the initial state excited by the past inputs. Thus we see that y_f is a zero-input response.

Thus, by the discussion above, we see that between the past and future inputs-outputs u_p, y_p, u_f, y_f in Fig. 1 and the block matrices L_{ij} of (15), there exists the following correspondences:

$$u_p \sim L_{22}, \quad y_p \sim L_{32}, \quad u_f \sim L_{12} \, (= 0), \quad y_f \sim L_{42}$$

Hence, the first k future free-responses of (22) can easily be computed by the LQ decomposition from given input-output data, if the conditions (i)-(iii) in Lemma 3 are satisfied.

4.3   State Vector and Zero-State Response

We now consider the state vector and the zero-state response of the LTI system. For convenience, we rewrite (15) as

$$\begin{bmatrix} U_f \\ W_p \\ Y_f \end{bmatrix} = \begin{bmatrix} R_{11} & 0 \\ R_{21} & R_{22} \\ R_{31} & R_{32} \end{bmatrix} \begin{bmatrix} Q_1^{\mathrm{T}} \\ Q_{2:3}^{\mathrm{T}} \end{bmatrix} \tag{23}$$

where we see that $R_{11} = L_{11}$, $R_{21} = \begin{bmatrix} L_{21} \\ L_{31} \end{bmatrix}$, $R_{22} = \begin{bmatrix} L_{22} & 0 \\ L_{32} & L_{33} \end{bmatrix}$, $R_{31} = L_{41}$,

$R_{32} = [L_{42} \ L_{43}]$ and $Q_{2:3} = [Q_2 \ Q_3]$. Since $R_{22}Q_{2:3}^{\mathrm{T}} = W_p - R_{21}Q_1^{\mathrm{T}}$, there exists a matrix $\Xi \in \mathbb{R}^{k(p+m)\times N}$ such that

$$Q_{2:3}^{\mathrm{T}} = R_{22}^{\dagger}(W_p - R_{21}Q_1^{\mathrm{T}}) + (I_{k(p+m)} - R_{22}^{\dagger}R_{22})\Xi$$

where $R_{22}^{\dagger}$ is the pseudo-inverse. It follows from (17) that $\mathrm{Ker}(R_{22}) \subset \mathrm{Ker}(R_{32})$. Also, since $\Pi := I_{k(p+m)} - R_{22}^{\dagger}R_{22}$ is the orthogonal projection onto $\mathrm{Ker}(R_{22})$, we find that $R_{32}\big(I_{k(p+m)} - R_{22}^{\dagger}R_{22}\big) = 0$. Thus, from (23),

$$Y_f = R_{32}R_{22}^{\dagger}W_p + (R_{31} - R_{32}R_{22}^{\dagger}R_{21})R_{11}^{-1}U_f \tag{24}$$

where $\mathrm{span}(U_f) \cap \mathrm{span}(W_p) = \{0\}$ since the input-output data are obtained from an open-loop experiment; see the condition (iii) of Lemma 3. It thus follows that the right-hand side of (24) is a direct sum of the oblique projections of Y_f onto $\mathrm{span}(W_p)$ along $\mathrm{span}(U_f)$ and of Y_f onto $\mathrm{span}(U_f)$ along $\mathrm{span}(W_p)$. Since $\mathrm{span}(X_f) \subset \mathrm{span}(W_p)$ [9, 6], comparing the right-hand sides of (14) and (24) yields

$$\mathcal{O}_k X_f = R_{32}R_{22}^{\dagger}W_p \tag{25}$$

$$\mathcal{T}_k(B, D)U_f = (R_{31} - R_{32}R_{22}^{\dagger}R_{21})R_{11}^{-1}U_f \tag{26}$$

We see that (25) and (26) are the zero-input and zero-state responses, respectively.

Let the SVD of $R_{32}R_{22}^{\dagger}$ in (25) be given by

$$R_{32}R_{22}^{\dagger} = [U_1 \ U_2] \begin{bmatrix} \Sigma_1 & 0 \\ 0 & 0 \end{bmatrix} \begin{bmatrix} V_1^{\mathrm{T}} \\ V_2^{\mathrm{T}} \end{bmatrix} = U_1\Sigma_1 V_1^{\mathrm{T}}$$

Let the extended observability matrix be $\mathcal{O}_k = U_1\Sigma_1^{1/2}$. Then the state vector is given by

$$X_f = \mathcal{O}_k^{\dagger}R_{32}R_{22}^{\dagger}W_p = \Sigma_1^{1/2}V_1^{\mathrm{T}}W_p$$

which can be employed in the direct N4SID method [13]. Also, since U_f has full rank, it follows from (26) that

$$\mathcal{T}_k(B, D) = (R_{31} - R_{32}R_{22}^{\dagger}R_{21})R_{11}^{-1} \tag{27}$$

so that the first k impulse responses $G_{0:k-1} := \mathrm{col}\,[G_0 \ G_1 \ \cdots \ G_{k-1}]$ are given by the first block column of the right-hand member of (27), where $\mathrm{col}(\cdot)$ denotes the stacked block matrix. In [8], the impulse responses are derived by using a slightly different idea.

4.4   Zero-State Response

We consider another method of computing zero-state responses of the system based on a related LQ decomposition. Rearranging the data matrices, we consider the block matrix

$$\begin{bmatrix} W_p \\ W_f \end{bmatrix} \in \mathbb{R}^{2k(m+p) \times N}$$

which was employed in [9] to show that the state vector X_f is a basis of the intersection $W_p \cap W_f$ of the past and future subspaces.

Consider the following LQ decomposition

$$\begin{bmatrix} W_p \\ W_f \end{bmatrix} = \begin{bmatrix} U_p \\ Y_p \\ U_f \\ Y_f \end{bmatrix} = \begin{bmatrix} \overline{L}_{11} & 0 & 0 & 0 \\ \overline{L}_{21} & \overline{L}_{22} & 0 & 0 \\ \overline{L}_{31} & \overline{L}_{32} & \overline{L}_{33} & 0 \\ \overline{L}_{41} & \overline{L}_{42} & \overline{L}_{43} & \overline{L}_{44} \end{bmatrix} \begin{bmatrix} \overline{Q}_1^{\mathrm{T}} \\ \overline{Q}_2^{\mathrm{T}} \\ \overline{Q}_3^{\mathrm{T}} \\ \overline{Q}_4^{\mathrm{T}} \end{bmatrix} \tag{28}$$

where $\overline{Q}_1, \overline{Q}_3 \in \mathbb{R}^{N \times km}$, $\overline{Q}_2, \overline{Q}_4 \in \mathrm{R}^{N \times kp}$ are orthogonal, and $\overline{L}_{11}, \overline{L}_{33} \in \mathbb{R}^{km \times km}$, $\overline{L}_{22}, \overline{L}_{44} \in \mathbb{R}^{kp \times kp}$ are lower triangular.

We see from (28) that

$$Y_f \overline{Q}_3 = \overline{L}_{43}, \quad U_f \overline{Q}_3 = \overline{L}_{33} \tag{29}$$

On the other hand, post-multiplying (14) by $\overline{Q}_3$ yields

$$Y_f \overline{Q}_3 = \mathcal{O}_k X_f \overline{Q}_3 + \mathcal{T}_k U_f \overline{Q}_3 \tag{30}$$

Since the system is minimal and since the past input-output pair is zero, i.e. $\overline{L}_{13} = 0$, $\overline{L}_{23} = 0$, we see that each column vector in $\begin{bmatrix} \overline{L}_{33} \\ \overline{L}_{43} \end{bmatrix}$ is a future input-output pair with the zero initial states, i.e. $X_f \overline{Q}_3 = 0$ in (30). Hence, from (29) and (30), we have $\overline{L}_{43} = \mathcal{T}_k L_{33}$, so that

$$\mathcal{T}_k(B, D) = \overline{L}_{43} \overline{L}_{33}^{\dagger} \tag{31}$$

This is another representation of the block Toeplitz matrix of (27).

5 Conclusions

In this paper, we have clarified the special role of LQ decomposition in subspace identification methods. In particular, we have shown that each column vector of the L-matrix of LQ decomposition is a pair of input-output vectors generated by a linear combination of given input-output data. Thus it reduces the given data so that the state-input-output matrix equation with a smaller dimension than the original one is derived.

It will be of interest to study the stochastic realization methods due to Faurre [4], Akaike [1] and the extension by Desai *et al.* [3] from the point of view of LQ decomposition. A preliminary analysis can be found in Appendix.

Acknowledgments

I would like to express my sincere thanks to Giorgio Picci who taught me several key issues in stochastic realization theory. I have enjoyed our fruitful collaboration over ten years for solving some stochastic realization problems in the presence of exogenous inputs and developing novel subspace identification methods, resulting in joint publication of several journal and conference papers. I am also grateful to Hideyuki Tanaka for his valuable comments on this paper.

References

1. H. Akaike (1975), "Markovian representation of stochastic processes by canonical variables," *SIAM J. Control*, vol. 13, no. 1, pp. 162–173.
2. K. S. Arun and S. Y. Kung (1990), "Balanced approximation of stochastic systems," *SIAM Journal on Matrix Analysis and Applications*, vol. 11, no. 1, pp. 42–68.
3. U. B. Desai, D. Pal and R. D. Kirkpatrick (1985), "A realization approach to stochastic model reduction," *Int. J. Control*, vol. 42, no. 4, pp. 821–838.
4. P. Faurre (1976), "Stochastic realization algorithms," In *System Identification: Advances and Case Studies* (R. Mehra and D. Lainiotis, eds.), Academic, pp. 1–25.
5. R. E. Kalman, P. L. Falb and M. A. Arbib (1969), *Topics in Mathematical System Theory*, McGraw-Hill.
6. T. Katayama (2005), *Subspace Methods for System Identification*, Springer.
7. A. Lindquist and G. Picci (1996), "Canonical correlation analysis, approximate covariance extension, and identification of stationary time series," *Automatica*, vol. 32, no. 5, pp. 709–733.
8. I. Markovsky, J. C. Willems, P. Rapisarda and B. L. M. De Moor (2005), "Algorithms for deterministic balanced subspace identification," *Automatica*, vol. 41, no. 5, pp. 755–766.
9. M. Moonen, B. De Moor, L. Vandenberghe and J. Vandewalle (1989), "On- and off-line identification of linear state-space models," *Int. J. Control*, vol. 49, no. 1, pp. 219–232.
10. G. Picci and T. Katayama (1996), "Stochastic realization with exogenous inputs and 'subspace methods' identification," *Signal Processing*, vol. 52, no. 2, pp. 145–160.
11. H. Tanaka and T. Katayama (2006), "A stochastic realization algorithm via block LQ decomposition in Hilbert space," *Automatica*, vol. 42, no. 5, pp. 741–746.
12. P. Van Overschee and B. De Moor (1994), "N4SID - Subspace algorithms for the identification of combined deterministic - stochastic systems," *Automatica*, vol. 30, no. 1, pp. 75–93.
13. P. Van Overschee and B. De Moor (1996), *Subspace Identification for Linear Systems*, Kluwer Academic.
14. M. Verhaegen and P. Dewilde (1992), "Subspace model identification, Part 1: The output-error state-space model identification class of algorithms & Part 2: Analysis of the elementary output-error state space model identification algorithm," *Int. J. Control*, vol. 56, no. 5, pp. 1187–1210 & pp. 1211–1241.
15. M. Verhaegen (1994), "Identification of the deterministic part of MIMO state space models given in innovations form from input-output data," *Automatica*, vol. 30, no. 1, pp. 61–74.

Appendix: LQ Decomposition in Stochastic Realization

The stochastic realization is to construct Markov models from given covariance matrices, or infinite sequence of data [1, 4]. In this appendix, we briefly discuss a role of LQ decomposition in stochastic realization based on Tanaka and Katayama [11].

A.1 Stochastic Realization

Suppose that $\{y(t),\, t = 0,\, \pm 1,\, \cdots\}$ is a regular full rank p-dimensional stationary process. We assume that the mean of y is zero and the covariance matrix is given by

$$\Lambda(l) = E\{y(t + l)y^{\mathrm{T}}(t)\}, \quad l = 0, \pm 1, \cdots \tag{32}$$

Suppose that the covariance matrices satisfy the summability condition

$$\sum_{l=-\infty}^{\infty} \|\Lambda(l)\| < \infty \tag{33}$$

Then, the spectral density matrix of y is defined by

$$\Phi(z) = \sum_{l=-\infty}^{\infty} \Lambda(l)z^{-l} \tag{34}$$

where it is assumed that the spectral density matrix is positive definite on the unit circle, i.e., $\Phi(z) > 0, |z| = 1$.

Let t be the present time. We define infinite dimensional future and past vectors

$$f(t) := \begin{bmatrix} y(t) \\ y(t+1) \\ \vdots \end{bmatrix}, \quad p(t) := \begin{bmatrix} y(t-1) \\ y(t-2) \\ \vdots \end{bmatrix}$$

Then, the cross-covariance matrix of the future and past is given by

$$H = E\{f(t)p^{\mathrm{T}}(t)\} = \begin{bmatrix} \Lambda(1) & \Lambda(2) & \Lambda(3) & \cdots \\ \Lambda(2) & \Lambda(3) & \Lambda(4) & \cdots \\ \Lambda(3) & \Lambda(4) & \Lambda(5) & \cdots \\ \vdots & \vdots & \vdots & \ddots \end{bmatrix}$$

and the covariance matrices of the future and the past are respectively given by

$$T_+ = E\{f(t)f^{\mathrm{T}}(t)\} = \begin{bmatrix} \Lambda(0) & \Lambda^{\mathrm{T}}(1) & \Lambda^{\mathrm{T}}(2) & \cdots \\ \Lambda(1) & \Lambda(0) & \Lambda^{\mathrm{T}}(1) & \cdots \\ \Lambda(2) & \Lambda(1) & \Lambda(0) & \cdots \\ \vdots & \vdots & \vdots & \ddots \end{bmatrix}$$

and

$$T_- = E\{p(t)p^{\mathrm{T}}(t)\} = \begin{bmatrix} \Lambda(0) & \Lambda(1) & \Lambda(2) & \cdots \\ \Lambda^{\mathrm{T}}(1) & \Lambda(0) & \Lambda(1) & \cdots \\ \Lambda^{\mathrm{T}}(2) & \Lambda^{\mathrm{T}}(1) & \Lambda(0) & \cdots \\ \vdots & \vdots & \vdots & \ddots \end{bmatrix}$$

It should be noted that H is an infinite block Hankel matrix, and $T_\pm$ are infinite block Toeplitz matrices, where we assume that T_+ (or equivalently T_-) is positive definite.

As in [3, 7], we compute the SVD of the normalized block Hankel matrix

$$T_+^{-1/2} H T_-^{-T/2} = U \Sigma V^T \tag{35}$$

and define the extended observability and reachability matrices as

$$\mathcal{O} = T_+^{1/2} U \Sigma^{1/2}, \quad \mathcal{C} = \Sigma^{1/2} V^T T_-^{T/2}$$

It therefore follows that the block Hankel has a canonical decomposition $H = \mathcal{O}\mathcal{C}$. Moreover, in terms of some $A \in \mathbb{R}^{n \times n}$, $C \in \mathbb{R}^{p \times n}$, $\overline{C} \in \mathbb{R}^{p \times n}$, we can express $\mathcal{O}$ and $\mathcal{C}$ as

$$\mathcal{O} = \begin{bmatrix} C \\ CA \\ \vdots \end{bmatrix}, \quad \mathcal{C} = [\overline{C}^T \ A\overline{C}^T \ \cdots]$$

so that the covariance matrix has a decomposition $\Lambda(k) = CA^{k-1}\overline{C}^T$, $k = 1, 2, \cdots$.

We now define a state vector as $\hat{x}(t) = \mathcal{C}T_-^{-1}p(t)$. Thus from [3, 7], we can show that the output process y has a stochastic realization of the form

$$\begin{bmatrix} \hat{x}(t+1) \\ y(t) \end{bmatrix} = \begin{bmatrix} A & K \\ C & I \end{bmatrix} \begin{bmatrix} \hat{x}(t) \\ e(t) \end{bmatrix} \tag{36}$$

where e is the innovation process defined by $e(t) = y(t) - C\hat{x}(t)$. We can show that the covariance matrices of $\hat{x}$ and e are respectively given by

$$E\{\hat{x}(t)\hat{x}^T(t)\} = \mathcal{C}(T_-)^{-1}\mathcal{C}^T = \Sigma$$
$$E\{e(t)e^T(t)\} = \Lambda(0) - C\Sigma C^T$$

where Σ is a stabilizing solution of the Riccati equation of the form

$$\Sigma = A\Sigma A^T + (\overline{C}^T - A\Sigma C^T)(\Lambda(0) - C\Sigma C^T)^{-1}(\overline{C} - C\Sigma A^T) \tag{37}$$

A.2 LQ Decomposition

Define the infinite matrices as

$$Y_t^- = \begin{bmatrix} \vdots & \vdots & \cdot^{\cdot^{\cdot}} \\ y(t-2) & y(t-1) & \cdots \\ y(t-1) & y(t) & \cdots \end{bmatrix}, \quad E_t^- = \begin{bmatrix} \vdots & \vdots & \cdot^{\cdot^{\cdot}} \\ e(t-2) & e(t-1) & \cdots \\ e(t-1) & e(t) & \cdots \end{bmatrix}$$

$$Y_t^+ = \begin{bmatrix} y(t) & y(t+1) & \cdots \\ y(t+1) & y(t+2) & \cdots \\ \vdots & \vdots & \ddots \end{bmatrix}, \quad E_t^+ = \begin{bmatrix} e(t) & e(t+1) & \cdots \\ e(t+1) & e(t+2) & \cdots \\ \vdots & \vdots & \ddots \end{bmatrix}$$

It has been shown in [11] that the LQ decomposition yields

$$\begin{bmatrix} Y_t^- \\ Y_t^+ \end{bmatrix} = \begin{bmatrix} \mathcal{L}^- & 0 \\ \mathcal{S} & \mathcal{L}^+ \end{bmatrix} \begin{bmatrix} E_t^- \\ E_t^+ \end{bmatrix} \tag{38}$$

where

$$\mathcal{L}^- = \begin{bmatrix} \ddots & & & \\ \cdots & L_0 & & \\ \cdots & L_1 & L_0 & \\ \cdots & L_2 & L_1 & L_0 \end{bmatrix}, \quad \mathcal{S} = \begin{bmatrix} \cdots & L_3 & L_2 & L_1 \\ \cdots & L_4 & L_3 & L_2 \\ \cdots & L_5 & L_4 & L_3 \\ \iddots & \vdots & \vdots & \vdots \end{bmatrix}, \quad \mathcal{L}^+ = \begin{bmatrix} L_0 & & & \\ L_1 & L_0 & & \\ L_2 & L_1 & L_0 & \\ \vdots & \vdots & \vdots & \ddots \end{bmatrix}$$

and where $L_0 = I_p$, $L_j = CA^{j-1}K$, $j = 1, 2, \cdots$.

Rearranging block matrices in (38) yields

$$\begin{bmatrix} E_t^+ \\ Y_t^+ \end{bmatrix} = \begin{bmatrix} \mathcal{I} & 0 \\ \mathcal{L}^+ & \mathcal{S} \end{bmatrix} \begin{bmatrix} E_t^+ \\ E_t^- \end{bmatrix} \tag{39}$$

where $\mathcal{I} = \text{block-diag}(I, I, \cdots)$. We can easily see a similarity between the decomposition of (39) and the LQ decomposition of (6). Also, it follows from (39) that

$$Y_t^+ = \mathcal{S}E_t^- + \mathcal{L}^+ E_t^+ \tag{40}$$

Since $\mathcal{S}$ in (39) is formed by zero-input responses, we see that the first term in the right-hand side of the above equation is the zero-input response and the second term the zero-state response.

Moreover, the first term of the right-hand side of (40) is expressed as [11]

$$\mathcal{S}E_t^- = \check{\mathcal{S}}\check{E}_t^- = \mathcal{O}[\hat{x}(t) \ \hat{x}(t+1) \ \cdots] = \mathcal{O}X_t$$

where

$$\check{\mathcal{S}} = \begin{bmatrix} L_1 & L_2 & L_3 & \cdots \\ L_2 & L_3 & L_4 & \cdots \\ L_3 & L_4 & L_5 & \cdots \\ \vdots & \vdots & \vdots & \ddots \end{bmatrix}, \quad \check{E}_t^- = \begin{bmatrix} e(t-1) & e(t) & e(t+1) & \cdots \\ e(t-2) & e(t-1) & e(t) & \cdots \\ \vdots & \vdots & \vdots & \ddots \end{bmatrix}$$

Thus, the extended observability matrix $\mathcal{O}$ can be determined by exploiting the fact that $\text{Im}(\mathcal{O}) = \text{Im}(\mathcal{S})$. In fact, in [11], the following decomposition is obtained:

$$\check{\mathcal{S}} = \mathcal{O}C_K, \quad C_K = [K \ \ AK \ \ \cdots]$$

so that we have $X_t = C_K\check{E}_t^-$. Thus the zero-input response of the stochastic system is expressed as $\mathcal{O}X_t = \mathcal{O}C_K\check{E}_t^-$, so that (40) is rewritten as

$$Y_t^+ = \mathcal{O}X_t + \mathcal{L}^+ E_t^+ \tag{41}$$

This is a stochastic analog of (5), since $\mathcal{L}^+$ is block Toeplitz and E_t^+ is the future stochastic input; see also Arun and Kung [2] for a scalar case.

Canonical Operators on Graphs

Matthias Kawski⋆ and Thomas J. Taylor⋆⋆

Arizona State University
kawski@asu.edu
tom.taylor@asu.edu

Dedicated to our friend and mentor Giorgio Picci on the occasion of his 65th birthday

Summary. This paper studies canonical operators on finite graphs, with the aim of characterizing the toolbox of linear feedback laws available to control networked dynamical systems.

1 Introduction

There is widespread current interest in distributed control of networked systems, e.g. [4], [5], [6], [12], [13], [15]. Much of the work to date centered on linear control laws, and has taken advantage the last twenty years of development in spectral graph theory. In particular the graph Laplacian, in various incarnations, has seen use as a stabilizing feedback. The property of the Laplacian used in these works has been essentially the fact that it is the generator of a reversible continuous time ergodic Markov chain: it has one zero eigenvalue and all others are strictly positive. The study we wish to propose is broader. We wish to ask which linear feedback laws are possible for actors which must communicate on a (possibly directed) network.

The coarse grain answer to this question is: those laws which respect the network structure. The present work, in initiating this study, precisely defines and characterizes in some detail classes of canonical (di)graph operators constructed from the incidence relations. These ideas are implicit or glossed over in a number of earlier publications; we felt that there will be those readers who, like us, benefit from the careful codification of properties. Our methods have a pronounced geometric and functorial flavor. There is a literature which has also taken this perspective; see e.g. [8], [11], [19]. We then turn to characterizing the graph Laplacian as constructed from differences of these canonical operators using basic linear algebraic operations. This section is remarkable in that it touches only tangentially the wide and profound literature of spectral graph theory. However, building on our foundation of properties of the fundamental operators and pursuing analogies with algebraic topological and differential geometric constructs, we are able to characterize operators previously little considered in the spectral graph theory literature. These include the Laplace-deRham operator on the edge space and Dirac operators. We show that these contain much the same graph theoretic information as the Laplacian.

⋆ M. Kawski was supported in part by NSF Grant DMS 05-09039.
⋆⋆ T. Taylor was supported in part by the EU project RECSYS.

A. Chiuso et al. (Eds.): Modeling, Estimation and Control, LNCIS 364, pp. 221–237, 2007.
springerlink.com   © Springer-Verlag Berlin Heidelberg 2007

The next section discusses the properties of Laplacians on weighted (di)graphs, and bears considerable relationship, and some differences in perspective, with constructions of Bensoussan and Menaldi [1] and Chung [3]. We include this section because we are able to turn this machinery to a geometric context for the Laplacian of Chung [3]. Early on this operator had mystified us and we hope that this section may provide an entry point for other readers. From these we turn to characterize the properties of operators based on other combinations of the canonical operators. In the former instance, we consider properties related to the undirected incidence operator. Here we note the prior contribution of Van Nuffelen [18] and Grossman et al [9]. Lastly we consider complex combinations of the canonical operators. There is a mathematical physics literature which touches such objects, but there the operators may be considered Laplacians on weighted graphs with complex weights [14], [16], [17]. In our situation this does not seem to be the case. In these latter sections, some of our results recapitulate the literature, and in some instances we have not yet been able to discover close analogs of our results. The literature in these areas seems to be relatively poorly developed, perhaps there are new results. We ask indulgence of the knowledgeable reader to direct us to related publications.

2   Graphs

2.1   The Geometry of Graphs and Digraphs

In this section we will make contact with graph theory, and describe our somewhat ideosyncratic perspective on the geometry of these objects. The objects of spectral graph theory tend to be described in terms of one of several matrices; as our perspective has been formed by contact with functorial constructions in differential geometry, operator theory and probability, our discussion will bear a marked resemblance to these areas of mathematics.

Definition 1

1. *A* digraph *(or* directed graph*) is a pair* $G = (V, E)$ *where V is a finite set called the* vertex set *and E is a subset of $V \times V$ called the* edge set.
2. *A* graph *is a pair* $G = (V, E)$ *where V is a finite set and E is a subset of $V \odot V$, where $\odot$ denotes the symmetric cartesian product; e.g. $V \odot V$ consists of all* unordered *pairs of elements of V, i.e. equivalence classes of the relation $(u, v) \sim (v, u)$ on $V \times V$.*
3. *A* multi-graph *is a pair* $G = (V, E)$ *where V is a finite set and E is a subset of the disjoint union of an finite number of copies of $V \odot V$.*

Note that we allow self edges unless otherwise stated. A more standard and visual set of definitions are: a digraph is a set of points in which some pairs of points are connected by arrows, while a graph is a set of points connected by lines. The *order* of a graph (resp. digraph) is the number of vertices, $|V|$. The *size* of a graph (resp. digraph) is the number of edges, $|E|$. A digraph is specified by its *transition matrix* $M(G)$, which is a $|V| \times |V|$ binary-valued matrix in which the entries $m_{i,j}(G) = 1$ iff $(v_i, v_j) \in E$. The *out-degree* of a vertex $v \in V$ of a digraph is the cardinality

$d_o(v) = |\{e \in E : \exists u \in V \, s.t. \; e = vu\}|$, the *in-degree* is the cardinality $d_i(v) = |\{e \in E : \exists u \in V \, s.t. \; e = uv\}|$ and the *degree* of v is $d(v) = d_o(v) + d_i(v)$. The degrees of the digraph are defined as $d_i(G) = \max_{v \in V} d_i(v)$, $d_o(G) = \max_{v \in V} d_o(v)$, $d(G) = \max_{v \in V} (d_i(v) + d_o(v))$. For a (multi)graph there is only one kind of edge, so only one kind of order. A symmetric transition matrix $M(G)$ specifies a digraph in which every edge is doubled by an edge in the opposite direction, the same data also specifies a graph subject to the understanding that $(u, v) \sim (v, u)$. The *forgetful morphism* Φ maps a digraph $G = (V, E)$ to the graph $\hat{G} = (V, \hat{E})$ of the same order in which each edge $e = (u, v) \in E \subset V \times V$ is mapped to its equivalence class $[u, v] \in V \odot V$.

Associated to each digraph are a canonical pair of mappings,

$$\sigma, \tau : E \to V,$$

the *source* map and the *target* map. To be precise:

Definition 2

1. *The* source *map* $\sigma : E \to V$ *is defined by* $\sigma((u, v)) = u$, *for* $(u, v) \in E$.
2. *The* target *map* $\tau : E \to V$ *is defined by* $\tau((u, v)) = v$, *for* $(u, v) \in E$.

The diagram that describes this is:

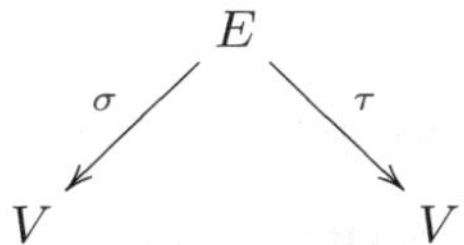

In the same way, associated to each graph is a canonical map, the incidence map:

Definition 3. *The incidence map of a graph is* $\iota : E \to 2^V$ *defined by* $\iota([u, v]) = \{u, v\} \subset V$.

Given two digraphs (resp. graphs) $G = (V, E)$ and $H = (U, F)$ a *digraph homomorphism* (resp. *graph homomorphism*) $\phi : G \to H$ is a pair of maps $\phi_V : V \to U$ and $\phi_E : E \to F$ such that the source and target maps (resp. incidence map) commute with ϕ: $\sigma \phi_E(e) = \phi_V(\sigma(e))$ and $\tau \phi_E(e) = \phi_V(\tau(e))$ (resp. $\iota \phi_E(e) = \phi_V(\iota(e))$). A digraph homomorphism is *surjective* if both ϕ_V and ϕ_E are surjective, and *injective* if both ϕ_V and ϕ_E are injective. A digraph homomorphism is *vertex surjective* (resp. *vertex injective*, resp. *edge surjective*, resp. *edge injective*) in case that ϕ_V is surjective (resp. ϕ_V is injective, resp. ϕ_E is surjective, ϕ_E is injective). A *forgetful homomorphism* of digraphs $\phi : G \to H$ is a homomorphism $\phi : \Phi G \to \Phi H$ of the associated graphs. A forgetful homomorphism maps edges to edges without specifying their direction.

We can use homomorphisms to capture properties of (di)graphs in terms of properties of simpler (di)graphs. Most important of these simpler (di)graphs are *line segments* and *circles*.

Definition 4

- *The line segment I_n is a digraph (resp. graph) in which the vertex set is the finite set of integers $\{1, 2, \cdots, n\}$, and the edges are the pairs of adjacent integers $\{(i, i+1) : i = 1, \cdots, n-1\}$ in increasing order, respectively without order.*
- *A circle is a digraph (resp. graph) in which the vertex set is $\mathbf{Z}_k$ for some k, and in which the edge set is the pairs of adjacent integers mod k in increasing order mod k, resp. the pairs of adjacent integers mod k without order.*

A *finite path* in a graph G is a forgetful homomorphism $\phi : I_k \to G$ of a line segment into G, i.e. a sequence $\{v_1, v_2, \ldots, v_k\}$ in V such that $\forall i$, $v_i v_{i+1} \in E$ or $v_{i+1} v_i \in E$; edges of the former type are called *sense* and the latter type are called *antisense*. A *directed path* is a homomorphism of I into G; in a directed path every edge is sense. The first and last vertices of a path are called the *starting vertex* and *ending vertex*, respectively. We call a path vertex which is not starting or ending is called an *interior vertex*. Recall from the theory of Markov chains that a directed graph is called *irreducible* if for every ordered pair $(u, v) \in V \times V$ there is a directed path for which u is the starting point and v is the endpoint. For digraphs irreducible implies only one connected component, but the converse is not true. We will call a vertex v such that $d_i(v) = 0$ (resp. $d_o(v) = 0$) *germinal* (resp. *terminal*) (more commonly these are called *source* and *sink*). A graph has no germinal or terminal verticies; in this case irreducible is equivalent to connected.

2.2   Operator Theory on Graphs and Digraphs

It is common to consider a pair of vector spaces associated to a graph;

Definition 5

1. *The vertex space L_V is the free linear span of V, i.e. the vector space of all real (resp. complex) valued functions defined on V.*
2. *The edge space L_E is the free linear span of E, i.e. the vector space of all real (resp. complex) valued functions defined on E.*
3. *If W is a subset of V or E we denote its free linear space by L_W.*
4. *The support of a function $f \in L_V$ (resp. $g \in L_E$) is the subset $supp(f) = \{v \in V : f(v) \neq 0\}$ (resp. $supp(g) = \{e \in E : g(e) \neq 0\}$).*

For $W \subset V$, we will regard L_W as a subset of L_V by use of the convention that functions in L_W are extended to all of V by zero. For $F \subset E$, $L_F \subset L_E$. Likewise if $supp(f) = W$ then $f \in L_W$ and if $supp(g) = F$ then $g \in L_F$.

Definition 6. *Given the constructions above, there are canonically defined linear mappings, the source and target operators, obtained as the pullback of the source and target maps. Namely, we may define linear maps $S, T : L_V \to L_E$ by $Sf = f \circ \sigma$ and $Tf = f \circ \tau$. In other words, $Sf(e) = f(e^-)$ and $Tf(e) = f(e^+)$. where we introduce the notation $e^+ = \tau(e)$ and $e^- = \sigma(e)$.*

Of course, to do computations with specific examples of these transformations it is sometimes useful to express them as matrices. (However, we will avoid doing so for the present in order to emphasize the underlying geometric structures).

We will need to bring into sharper focus certain types of localization of E over V. Each edge of E has a unique source. If v is not terminal, it is the source of an edge of E. Thus for every vertex, the set valued mapping $\sigma^{-1} : V \to 2^E$ may be regarded as assigning v to the subset $E_v^\sigma = \sigma^{-1}\{v\} \subset E$, which is the empty subset if v is a terminal vertex. This association may be viewed as a generalized localization of points of E over V (in that $E_v^\sigma \cap E_u^\sigma = \emptyset$ if $v \neq u$), which is a true localization over V_σ, the set of nonterminal vertices (in that $E_v^\sigma \neq \emptyset$). Similarly $E_v^\tau = \tau^{-1}\{v\} \subset E$ may be viewed as a (complementary) generalized localization of E over V by τ, which is a true localization over V_τ, the set of nongerminal vertices. Thus we may write E as a disjoint union:

$$E = \coprod_{v \in V_\sigma} E_v^\sigma = \coprod_{v \in V_\tau} E_v^\tau,$$

$$E = \coprod_{v \in V} E_v^\sigma = \coprod_{v \in V} E_v^\tau,$$

although in the latter equations some terms may be the empty set. For irreducible digraphs there are no germinal or terminal vertices, in this case $V_\sigma = V_\tau = V$.

Denote the free linear span of E_v^σ (resp. E_v^τ) by L_v^σ (resp. L_v^τ). Some of these vector spaces may be $\{0\}$. Nevertheless the set $\coprod_{v \in V} L_v^\sigma$ (resp. $\coprod_{v \in V} L_v^\tau$) may be regarded as a sort of generalized vector bundle over V, with the obvious projection map $\xi \mapsto v$ for $\xi \in L_v^\sigma$ (resp. $\xi \in L_v^\tau$). Then

$$L_E = \bigoplus_v L_v^\sigma = \bigoplus_v L_v^\tau,$$

so that L_E may be regarded as the space of sections both of these vector bundles.

Lemma 1. $ker(S) = L_{V-V_\sigma}$, $ker(T) = L_{V-V_\tau}$.

We may identify some distinguished subspaces in L_E.

Definition 7. $C_\sigma(E)$ *is the set of functions which are constant on each subset E_v^σ, i.e. $C_\sigma(E) = \bigoplus_v \mathbb{C}1_{E_v^\sigma}$ (we will restrict ourselves to consideration only of vector spaces over the complex numbers). Likewise $C_\tau(E)$ is the set of functions which are constant on each subset E_v^τ, i.e. $C_\tau(E) = \bigoplus_v \mathbb{C}1_{E_v^\tau}$.*

Lemma 2. $Range(S) = C_\sigma(E)$, $Range(T) = C_\tau(E)$

Now, if we give the vector spaces L_V, L_E inner products, we may consider the adjoint maps $S^*, T^* : L_E \to L_V$. Different choices of inner product give rise to different operators, which is an issue we shall consider in the following sections. Now, since L_V, L_E are free vector spaces, we consider the special cases of the dot product on these spaces, i.e. the L^2 inner products induced by the counting measures on V, respectively E. Note that for these inner products $L_v^\sigma \perp L_{v'}^\sigma$ and $L_v^\tau \perp L_{v'}^\tau$ for $v \neq v'$.

Definition 8. $M_v^\sigma \subset L_v^\sigma$ *is the subspace of functions on E_v^σ which are orthogonal to the constants. $M_v^\tau \subset L_v^\tau$ is the subspace of functions on E_v^τ which are orthogonal to the constants. $M_\sigma(E) = \bigoplus_v M_v^\sigma$. $M_\tau(E) = \bigoplus_v M_v^\tau$ (where in this context $\oplus$ denotes the orthogonal direct sum).*

Lemma 3. $L_v^\sigma = \mathbf{C}1_{E_v^\sigma} \oplus M_v^\sigma$, $L_v^\tau = \mathbf{C}1_{E_v^\tau} \oplus M_v^\tau$, $C_\sigma(E)^\perp = M_\sigma(E)$, $C_\tau(E)^\perp = M_\tau(E)$.

Lemma 4. *For an element $g \in L_E$,*

$$S^* g(v) = \sum_{e:e^- = v} g(e), \text{ and } T^* g(v) = \sum_{e:e^+ = v} g(e).$$

Proof. Let $1_v \in L_V$ denote the indicator function of the point $v \in V$. Then

$$S^* g(v) = \langle 1_v, S^* g \rangle = \langle S 1_v, g \rangle = \sum_e S 1_v(e) g(e) = \sum_e 1_v(e^-) g(e) = \sum_{e:e^- = v} g(e).$$

The case for T^* is analogous. $\qquad\qquad\qquad\qquad\qquad\qquad\qquad\qquad\qquad\qquad\square$

Lemma 5. *We have*

1. $ker(S^*) = C_\sigma(E)^\perp = M_\sigma(E)$, $ker(T^*) = C_\tau(E)^\perp = M_\tau(E)$.
2. $Range(S^*) = L_{V_\sigma}$, $Range(T^*) = L_{V_\tau}$.

We may regard the operators S, T as canonical operators associated with the digraph, as the operators S^*, T^* are also canonically associated with the digraph and our choice of inner product on L_V and L_E. Moreover, operators constructed from algebraic combinations of these operators may also be regarded as canonical. The following results will be useful.

Proposition 1. *The operators $S^* S$ and $T^* T$ satisfy:* $\forall f \in L_V, \forall v \in V$, $S^* S f(v) = d_o(v) f(v)$ *and* $T^* T f(v) = d_i(v) f(v)$.

Proof. $S f(e) = f(e^-)$, so $S^* S f(v) = \sum_{e:e^- = v} f(e^-) = f(v) \sum_{e:e^- = v} 1$. The case for $T^* T$ is analogous.

Corollary 1. $\|S\|^2 = d_o(G)$ *and* $\|T\|^2 = d_i(G)$

Proof.

$$\|S\|^2 = \sup_{f \in L_V} \frac{\langle S f, S f \rangle}{\langle f, f \rangle} = \sup_f \frac{\langle f, S^* S f \rangle}{\langle f, f \rangle} = \sup_v d_o(v),$$

since the $S^* S$ is diagonal. The situation for T is symmetric.

Proposition 2. *The operators $S^* T$ and $T^* S$ satisfy*

1. $S^* T f(v) = \sum_{e:e^- = v} f(e^+)$
2. $T^* S f(v) = \sum_{e:e^+ = v} f(e^-)$

Proposition 3. *The operators SS^* and TT^* satisfy:*

1. *$SS^*g(e) = d_o(e^-)g(e)$ for $g \in C_\sigma(E)$ and $SS^*g = 0$ for $g \in C_\sigma(E)^\perp$.*
2. *$TT^*g(e) = d_i(e^+)g(e)$ for $g \in C_\tau(E)$ and $TT^*g = 0$ for $g \in C_\tau(E)^\perp$.*

Definition 9. *the* incidence operator *of a graph $G = (V, E)$ is $\mathcal{I}_G : L_V \to L_E$ is defined by $\mathcal{I}_G f([u, v]) = f(u) + f(v)$.*

Lemma 6. *Let G be a digraph. Then $\mathcal{I}_{\Phi G} f(e) = Sf(\tilde{e}) + Tf(\tilde{e})$, where $\tilde{e}$ is any element of $\Phi^{-1}e$.*

Remarks. Unlike the source and target operator, the incidence operator is not a pull-back, e.g. of the incidence mapping. It is more common in the literature (e.g. [2], [3]) to discuss instead with the *directed incidence operator*, discussed below, for an arbitrary choice of direction to each edge. The additional ease in using D may based on a morphism with differential geometry, as has been remarked by a number of authors ([2], [3], [10]).

3 Differences, Divergences, Laplacians and Dirac Operators

One fundamental family of operators is founded on the difference operator. A fundamental reference for this section is Bollabas [2].

Definition 10. *The* difference operator *(i.e. directed incidence operator) $D : L_V \to L_E$ is defined by $Df(e) = Tf(e) - Sf(e)$.*

Definition 11. *A cut is a partition of the vertex set into two pieces $V = W \cup W^c$; equivalently a cut is an indicator function $1_W \in L_V$. The cut vector of W is the function $g \in L_E$ such that $g(e) = 1$ if $e^+ \in W$ but $e^- \notin W$, $g(e) = -1$ if $e^- \in W$ but $e^+ \notin W$ and $g(e) = 0$ otherwise. The cut space is the span of all cut vectors.*

Note that the indicator functions span L_V, and that each cut vector is equal to $D1_W$ for some subset $W \subset V$. From this it is an easy step (since $\{1_{\{v\}} : v \in V\}$ spans L_V) to

Proposition 4. *The cut space is equal to $\mathrm{Range}(D)$.*

Clearly the value of a cut vector on any self edge is zero.

Definition 12. *A cycle is a forgetful homomorphism of a circle into G, i.e. a path in which the endpoint vertices are equal, and a simple cycle is an injective forgetful homomorphism of a circle into G, i.e. a cycle in which only the endpoint vertices are repeated. By abuse of notation an edge vector $g \in L_E$ which is zero except on the edges of a simple cycle and assigns the value which assigns the value $+1$ to sense edges, -1 to antisense edges is also referred to as a simple cycle. The cycle space is the linear span of the simple cycles in L_E.*

Note that we might also consider the *self cycle space* linear span of the set of self cycles. For the following, denote the self cycle space by $K(G)$, the cycle space by $Z(G)$ and the cut space by $B(G)$.

Definition 13. *A* connected component *of a (di)graph is a maximal subset of V having the property that every two vertices are contained in a path. A function in L_V is called* locally constant *if $f(v) = f(u)$ for any v, u such that $vu \in E$ or $uv \in E$.*

Lemma 7. *Locally constant functions are constant on each path, hence are constant on each connected component of G. Locally constant functions are a vector subspace of L_V.*

Let $C(G)$ *denote the vector space of locally constant functions in L_V. Clearly the dimension of $C(G)$ is equal the number of connected components of G.*

Proposition 5. $ker(D) = C(G)$, $dim\,(B(G)) = Order(G) - dim\,(C(G))$.

Proof. Clearly $Df(e) = 0$ iff f takes the same value on both sides of e. Thus $Df = 0$ iff f is constant on every path, i.e. is locally constant.

Definition 14. *The* divergence *operator is $D^* : L_E \to L_V$, the dual operator of D.*

The following is a direct consequence of Lemma 4

Lemma 8. *The divergence operator satisfies*

$$D^*g(v) = T^*g(v) - S^*g(v) = \sum_{e:e^+=v} g(e) - \sum_{e:e^-=v} g(e)$$

Lemma 9. $ker(D^*) = B(G)^\perp$. $Range(D^*) = C(G)^\perp$.

Proof. For $g \in ker(D^*)$ and for all $f \in L_V$, $0 = \langle f, D^*g \rangle = \langle Df, g \rangle$. The result follows since $B(G) = Range(D)$. Clearly $Range(D^*) \subset C(G)^\perp$, since for $f \in C(G)$, $g \in L_E$, $\langle f, D^*g \rangle = \langle Df, g \rangle = 0$. Now suppose that $Range(D^*)$ is a proper subspace of $C(G)^\perp$. Then there exists a nonzero $f \in C(G)^\perp$ which is also in the orthogonal complement of $Range(D^*)$, so $Df \neq 0$, and for all $g \in L_E$, $0 = \langle f, D^*g \rangle = \langle Df, g \rangle$. In particular this is true for $g = Df$, which implies $0 = \langle Df, Df \rangle$, of $Df = 0$, which is a contradiction.

The proof of the following follows easily from [2], p. 53.

Proposition 6. $B(G)^\perp = Z(G) \oplus K(G)$.

Definition 15. *The* Laplacian *of G is the operator $\Delta = D^*D$ defined on L_V. The* Laplace-de Rham *operator of G is the operator $\square = D^*D \oplus DD^*$ defined on $L_V \oplus L_E$, where $\oplus$ denotes the orthogonal direct sum.*

Proposition 7

1. $\Delta = S^*S + T^*T - S^*T - T^*S$.
2. $ker(\Delta) = C(G)$ *and* $Range(\Delta) = C(G)^\perp$.
3. $0 \leq \langle f, \Delta f \rangle \leq \left(d(G) + 2\sqrt{d_o(G)d_i(G)} \right) \langle f, f \rangle \leq 2d(G) \langle f, f \rangle$.

Proof.

1. Item 1 follows from the definition of D.
2. $u \in ker(\Delta)$ iff

$$0 = \langle u, \Delta u \rangle = \langle Du, Du \rangle,$$

    i.e. iff $u \in ker(D) = C(G)$. Since Δ is self adjoint, both $ker\Delta$ and $ker(\Delta)^\perp$ are finite dimensional invariant subspaces, so $\Delta|ker(\Delta)^\perp$ is an isomorphism.
3. Note that

$$\begin{aligned}
\langle f, \Delta f \rangle &= \langle f, S^*Sf \rangle + \langle f, T^*Tf \rangle - \langle f, (S^*T + T^*S)f \rangle \\
&\leq \max_v \left(d_i(v) + d_o(v) \right) \langle f, f \rangle + 2\|Sf\|\|Tf\| \\
&= \left(d(G) + 2\sqrt{d_o(G)d_i(G)} \right) \langle f, f \rangle \\
&\leq 2d(G) \langle f, f \rangle,
\end{aligned}$$

where the inequality follows from Theorem 1, the triangle inequality and the Cauchy-Schwartz inequality, and the succeeding equality follows from Corollary 1. Note that the latter inequality in item 3 is an equality for the case of the graph which consists of a single directed cycle of even order, and f is the function which alternates between $+1$ and -1 on successive vertices; f is an eigenfunction of Δ with eigenvalue 4, $d(G) = 2$ and $d_o(G) = d_i(G) = 1$. In particular the inequalities in item 3 are tight.

Proposition 8

1. $DD^* = SS^* + TT^* - ST^* - TS^*$.
2. $ker(DD^*) = B(G)^\perp$ and $Range(DD^*) = B(G)$.
3. *The eigenvalues of DD^* are the same as those of Δ, with the same multiplicity. If $\Delta f = \lambda f$ for $\lambda \neq 0$ a constant, then Df is an eigenvector for DD^* with the same eigenvalue.*

Thus the Laplace-deRham operator contains only a little additional information about the graph geometry beyond that contained in the Laplacian. The term 'Dirac operator' refers generally to a square root of the Laplacian, although by custom not to the symmetric square root of the Laplacian. Hence:

Definition 16. *The* Dirac *operator of G is the operator* $\partial = \begin{pmatrix} 0 & D^* \\ D & 0 \end{pmatrix}$ *defined on* $L_V \oplus L_E$.

Proposition 9

1. $\partial^2 = \square$.
2. ∂ *is self adjoint.*
3. $ker(\partial) = C(G) \oplus B(G)^\perp$ *and* $Range(\partial) = C(G)^\perp \oplus B(G)$.
4. *The eigenvalues of ∂ are the (positive and negative) square roots of those of Δ, with the same multiplicity. If $\Delta f = \lambda f$ for $\lambda \neq 0$ a constant, then the eigenvector of $\sqrt{\lambda}$ (resp. $-\sqrt{\lambda}$) is $\begin{pmatrix} \sqrt{\lambda}f \\ Df \end{pmatrix}$ (resp. $\begin{pmatrix} \sqrt{\lambda}f \\ -Df \end{pmatrix}$).*

Remarks: We have taken the convention that the Laplacian is a nonnegative operator. The other common convention is, of course, that the Laplacian is nonpositive. This perspective has the net effect of replacing eigenvalues of the Laplacian and Laplace-deRham operators by their negatives, and replacing the Dirac operator by the skew-adjoint operator $\begin{pmatrix} 0 & -D^* \\ D & 0 \end{pmatrix}$, for which the eigenvalues are imaginary.

4 Operators on Weighted Graphs

Consider a function $w : E \rightarrow \mathbf{C}$ on E, and a function $\rho : V \rightarrow \mathbf{C}$ on V, which we will call *weight functions*. Although some literature considers cases in which w, rho are complex, [14], [16], [17], we will suppose that w and ρ are both positive real functions. Then the bilinear function $\langle f_1, f_2 \rangle_\rho = \sum_v \rho(v) f_1(v) f_2(v)^*$ on L_V defines an inner product. Likewise the inner product $\langle g_1, g_2 \rangle_w = \sum_e w(e) g_1(e) g_2(e)^*$ defines an inner product on L_E. Clearly there is a wide latitude of choice of weight functions, and they influence properties of the canonical operators. We may take as fundamental the definition of the operators S, T. Then the discussion of section B is valid in its entirety through Lemma 1.3 provided that orthogonality in L_E is understood to be in the weighted sense. However, Lemma 1.4 now takes the form

Lemma 10. *For a element $g \in L_E$,*

$$S^* g(v) = \frac{1}{\rho(v)} \sum_{e:e^-=v} w(e) g(e), \text{ and } T^* g(v) = \frac{1}{\rho(v)} \sum_{e:e^+=v} w(e) g(e).$$

Proof. Let $1_v \in L_V$ denote the indicator function of the point $v \in V$. Then

$$S^* g(v) = \frac{1}{\rho(v)} \langle 1_v, S^* g \rangle = \frac{1}{\rho(v)} \langle S 1_v, g \rangle = \frac{1}{\rho(v)} \sum_e w(e) S 1_v(e) g(e)$$
$$= \frac{1}{\rho(v)} \sum_e w(e) 1_v(e^-) g(e) = \frac{1}{\rho(v)} \sum_{e:e^-=v} w(e) g(e).$$

The case for T^* is analogous. $\qquad\square$

Lemma 1.5 is valid in the weighted case as stated. We will revise our definitions as follows.

Definition 17. *The* out-degree *of a vertex $v \in V$ of a digraph is the sum $d_o(v) = \frac{1}{\rho(v)} \sum_{e^-=v} w(e)$, the* in-degree *is the sum $d_i(v) = \frac{1}{\rho(v)} \sum_{e^+=v} w(e)$ and the* degree *of v is $d(v) = d_o(v) + d_i(v)$. The degrees of the digraph are defined as $d_i(G) = \max_{v \in V} d_i(v)$, $d_o(G) = \max_{v \in V} d_o(v)$, $d(G) = \max_{v \in V} (d_i(v) + d_o(v))$. For a graph there is only one kind of edge, so one kind of degree.*

Note that the degrees are positive real numbers, but need no longer be integers.

Proposition 10. *The operators $S^* S$ and $T^* T$ satisfy: $S^* S f(v) = d_o(v) f(v)$ and $T^* T f(v) = d_i(v) f(v)$.*

Proof. $Sf(e) = f(e^-)$, so

$$S^*Sf(v) = \frac{1}{\rho(v)} \sum_{e:e^-=v} w(e)f(e^-) = f(v)\frac{1}{\rho(v)} \sum_{e:e^-=v} w(e).$$

The case for T^*T is analogous. $\qquad\square$

With the above definitions of vertex degrees, the following takes the same form as in the unweighted case.

Corollary 2. $\|S\|^2 = d_o(G)$ *and* $\|T\|^2 = d_i(G)$

Proof.

$$\|S\|^2 = \sup_{f\in L_V} \frac{\langle Sf, Sf\rangle}{\langle f,f\rangle} = \sup_f \frac{\langle f, S^*Sf\rangle}{\langle f,f\rangle} = \sup_v d_o(v),$$

since the S^*S is diagonal. The situation for T is symmetric.

Proposition 11. *The operators T^*S and S^*T satisfy:*

1. $T^*Sg(v) = \frac{1}{\rho(v)} \sum\limits_{e:e^-=v} w(e)f(e^+)$
2. $S^*Tg(v) = \frac{1}{\rho(v)} \sum\limits_{e:e^+=v} w(e)f(e^-)$

Proposition 12. *The operators SS^* and TT^* satisfy:*

1. $SS^*g(e) = d_o(e^-)g(e)$ *for* $g \in C_\sigma(E)$ *and* $SS^*g = 0$ *for* $g \in C_\sigma(E)^\perp$.
2. $TT^*g(e) = d_i(e^+)g(e)$ *for* $g \in C_\tau(E)$ *and* $TT^*g = 0$ *for* $g \in C_\tau(E)^\perp$.

The definition of the difference operator D remains the same in this weighted situation, and it's range is still the space of cut vectors $B(G)$, and its kernel is still the space of locally constant functions $C(G)$. The orthogonal complement $B(G)^\perp$, and the divergence operator D^* are generally different, since they defined in terms of the inner product on L_E. Let W denote the operator on L_E of multiplication by the weight function w. The following lemma is immediate, given that $WK(G) = K(G)$

Lemma 11. $B(G)^\perp = W^{-1}Z(G) \oplus K(G)$.

Lemma 12. $ker(D^*) = B(G)^\perp$ *and* $Range(D^*) = C(G)^\perp$

The following theorem takes the same form as the non-weighted case.

Proposition 13

1. $\Delta_\rho^w f(v) = (S^*S + T^*T - S^*T - T^*S)f(v)$
2. $\Delta_\rho^w f(v) = d(v)f(v) - \frac{1}{\rho(v)}\left(\sum\limits_{e:e^-=v} w(e)f(e^+) + \sum\limits_{e:e^+=v} w(e)f(e^-)\right)$
3. $ker(\Delta_\rho^w) = C(G)$ *and* $Range(\Delta_\rho^w) = C(G)^{\perp_\rho}$
4. $0 \le \langle f, \Delta_\rho^w f\rangle \le \left(d(G) + 2\sqrt{d_o(G)d_i(G)}\right)\langle f,f\rangle$

Of particular recent interest are weighted graph Laplacians in the case that $\rho(v) = d(v)$ and $w(e) = 1$ for all edges in the graph. In this case the Laplacian has the representation $\Delta_d f(v) = f(v) - \frac{1}{d(v)} \left(\sum_{e:e^-=v} f(e^+) + \sum_{e:e^+=v} f(e^-) \right)$, and is self adjoint on L_V with respect to the inner product $\langle f, g \rangle_d = \sum_v \overline{f}(v) g(v) d(v)$. It will be a little easier to see the self adjointness of Δ_d if we express it in a unitarily equivalent form on a different inner product space. Specifically, note that the multiplication operator $U f(v) = \frac{1}{\sqrt{d(v)}} f(v)$ is a unitary map from the inner product space $(L_V, \langle \cdot, \cdot \rangle_1)$ to the inner product space $(L_V, \langle \cdot, \cdot \rangle_d)$. Thus Δ_d is unitarily equivalent to the operator $U^{-1} \Delta_d U = d(v)^{-1/2} \Delta d(v)^{-1/2}$ on $(L_V, \langle \cdot, \cdot \rangle_1)$. But on this inner product space self adjointness is just symmetry, and the symmetry of $d(v)^{-1/2} \Delta d(v)^{-1/2}$ is manifest. But the latter is just the Laplacian preferred by Chung [3] because its spectrum is so closely tied to graph geometry.

5   The Incidence Operator and Its Kin

Recall that the incidence operator $\mathcal{I} : L_V \to L_E$ is defined by $\mathcal{I} f(e) = T f(e) + S f(e)$. If a function f is in $ker(\mathcal{I})$ it must have values of equal magnitude but opposite sign at the vertices on either side of every edge. From this follows the fact that the values taken by f on a connected component of the (di)graph are determined by its value at a single vertex. Moreover, $f(v) = 0$ for v in any cycle of odd order, hence in the connected component of a cycle of odd order. This is basically everything that needs to be known about the kernel of the incidence operator.

Lemma 13. *$ker(\mathcal{I})$ is the space of functions on V which alternate sign across every edge. If G is connected, $ker(\mathcal{I})$ is zero or one dimensional according to whether G has cycles of odd order or not. In general the dimension of $ker(\mathcal{I})$ is the number of connected components without cycles of odd order.*

Remarks: Since a connected graph is bipartite iff it has no odd cycles, $dim ker(\mathcal{I})$ is the number of bipartite components. More generally the kernel contains the span of the isolated vertices. This result may be originally due to Van Nuffelen [18] in the context of graphs. A vertex with a self edge is a cycle of odd order, hence on any component containing a self edge one has $ker(\mathcal{I}) = \{0\}$.

Note that $\mathcal{I} 1_v = 1_{E_v^\sigma \cup E_v^\tau}$. The set $E_v^\sigma \cup E_v^\tau$ seems to us the shadow of v on the edge set, so we will call such a vector $1_{E_v^\sigma \cup E_v^\tau}$ a *shadow*, and call a vector in the span of such vectors a shadow vector. Denote the set of shadow vectors by $\Upsilon(G)$. Clearly $Range(\mathcal{I}) = \Upsilon(G)$.

Lemma 14. *Assume G is connected. If G has cycles of odd order, $\{1_{E_v^\sigma \cup E_v^\tau} : v \in V\}$ is basis of $Range(\mathcal{I})$. Conversely, if G has no cycles of odd order, then for any $u \in V, \{1_{E_v^\sigma \cup E_v^\tau} : v \in V - \{u\}\}$ is a basis of $Range(\mathcal{I})$.*

Lemma 15

$$\mathcal{I}^* g(v) = \sum_{e:e^-=v} g(e) + \sum_{e:e^+=v} g(e).$$

Proof. This follows directly from Lemma 4.

Another way of saying the same thing, is that $\mathcal{I}^* g(v) = \langle 1_{E_v^\sigma \cup E_v^\tau}, g \rangle$. Of course $ker(\mathcal{I}^*) = Range(\mathcal{I})^\perp$. The geometry of this statement is the following. Suppose that $\mathbf{Z}_{2k}$ is a circle of even order, and that $\hat{g} \in L_E(\mathbf{Z}_{2k})$ is the alternating function: $\hat{g}((i, i+1)) = (-1)^i$, where the addition $"i + 1"$ is interpreted as mod $2k$. Suppose that $c : \mathbf{Z}_{2k} \to G$ is a cycle in G. Define a function $g \in L_E$ with support contained in $c(\mathbf{Z}_{2k})$ by $g(e) = \sum_{i:e=c((i,i+1))} \hat{g}((i, i+1))$. Then $g \in \ker(\mathcal{I}^*)$. We will call g an *alternating cycle*. Let $A(G) \subset L_E$ denote the span of the alternating cycles. Then $A(G) \subseteq \ker(\mathcal{I}^*)$.

Definition 18. *Suppose that $g \in L_E$ and that $|supp(g) \cap (E_v^\sigma \cup E_v^\tau)| = 1$. Then we will call $e \in supp(g) \cap (E_v^\sigma \cup E_v^\tau)$ a* hanging edge.

Lemma 16. *Suppose that $\mathcal{I}^* g(v) = 0$ and that $supp(g) \cap (E_v^\sigma \cup E_v^\tau) \neq \emptyset$. Then there are at least two elements of $e, e' \in supp(g) \cap (E_v^\sigma \cup E_v^\tau)$ which satisfy $g(e)g(e') < 0$. (In other words, $supp(g)$ has no hanging edges)*

Proof. $supp(g) \cap (E_v^\sigma \cup E_v^\tau) \neq \emptyset$ implies $\sum_{e:e^-=v} |g(e)| + \sum_{e:e^+=v} |g(e)| \neq 0$. Since $\mathcal{I}^* g(v) = 0$ implies cancellation, there exists at least at least two edges $e, e' \in E_v^\sigma \cup E_v^\tau$ with $g(e) > 0$ and $g(e') < 0$. $\qquad\square$

Definition 19. *Let $\phi : I_k \to G$ be a path, let $\hat{h}((i, i+1)) = (-1)^i$ be a function $\hat{h} \in L_E(I_k)$. Define $h \in L_E(G)$ by $h(e) = \sum_{i:e=\phi((i,i+1))} \hat{h}((i, i+1))$ if $e \in \phi(I_k)$ and $h(e) = 0$ otherwise. We will call h the* alternating path *built on ϕ. If ϕ is the restriction of a path $\phi' : I_{k+n} \to G$ for $n > 0$ and h' is the alternating path built on ϕ' we shall say that h is a* restriction *of h'.*

Lemma 17. *Assume that the alternating path h has no self edges and let v be an interior vertex of h. Then $\mathcal{I}^* h(v) = 0$.*

Proof. We have:

$$\mathcal{I}^* h(v) = \sum_{e:e^-=v} h(e) + \sum_{e:e^+=v} h(e)$$
$$= \sum_{e:e^-=v} \sum_{i:e=\phi((i,i+1))} \hat{h}((i, i+1))$$
$$+ \sum_{e:e^+=v} \sum_{i:e=\phi((i,i+1))} \hat{h}((i, i+1)).$$

But, in the latter expression each summand is of magnitude one and uniquely paired with another such of opposite sign. Indeed, since v is an internal vertex, for every i such that $v = \phi(i)$, both edges $e = \phi((i-1, i))$ and $e' = \phi((i, i+1))$ are coincident with v, hence in the sum, while $\hat{h}((i, i+1)) = -\hat{h}((i-1, i))$. $\qquad\square$

Lemma 18. *An alternating path of even order cannot belong to $\ker(\mathcal{I}^*)$.*

Proof. Such a path either has a hanging edge, or in the case that the initial and terminal vertex are equal has the initial and terminal edges of equal sign, so that $\mathcal{I}^* h(v) \neq 0$ when v is the initial vertex. $\qquad\square$

Lemma 19. *Let h be an alternating path of odd order. Then $h \in \ker(\mathcal{I}^*)$ iff $v_k = \phi(2k+1) = \phi(1) = v_1$. In other words, an alternating path is in $\ker(\mathcal{I}^*)$ iff it is of odd order and an alternating cycle.*

Proof. If $v_{2k+1} \neq v_1$ then h has a hanging edge, hence cannot be in $\ker(\mathcal{I}^*)$. Conversely, if $v_{2k+1} = v_1$ then ϕ defines a cycle of even order, and h is an alternating cycle. $\square$

Theorem 1. *Suppose G is a multigraph. Then $\ker(\mathcal{I}^*) = A(G)$.*

Remark: This result seems to be first due to Grossman et al [9], and then again by ourselves some thirteen years later.

Lemma 20. *$Range(\mathcal{I}^*)$ is the orthogonal complement of the space of alternating functions in L_V. If G has no bipartite components or isolated vertices, i.e. there is a cycle of odd order in every component, then $Range(\mathcal{I}^*)$ is all of L_V.*

Lemma 21. *$\mathcal{I}^*\mathcal{I} = T^*T + S^*S + S^*T + T^*S$.*

Remark: Grossman et al [9] call $\mathcal{I}^*\mathcal{I}$ the *Unoriented Laplacian* .

Proposition 14. *Suppose that G is bipartite. Then $\mathcal{I}^*\mathcal{I}$ is unitarily equivalent to Δ, the Laplacian of G.*

Proof. Let $m \in \ker(\mathcal{I}^*\mathcal{I})$ be real valued and unimodular (which exists by Lemma 13), and let M be the operator of multiplication by m. Then M is unitary and it's own inverse. According to Bollobas [2] p.264, $(T^*S + S^*T)M = -M(T^*S + S^*T)$. Since $T^*T + S^*S$ is a multiplication operator, it commutes with M. Thus $M^{-1}\mathcal{I}^*\mathcal{I}M = \Delta$. $\square$

Remarks: This result is known to the algebraic graph theory community [7], although we are unaware of a specific reference. For a regular graph $\mathcal{I}^*\mathcal{I}$ is a linear function of the Laplacian.

The following proposition is proved in exactly as Theorem 7.

Proposition 15

1. *$\ker(\mathcal{I}^*\mathcal{I}) = \ker(\mathcal{I})$ is the space of alternating functions on V.*
2. *$Range(\mathcal{I}^*\mathcal{I}) = Range(\mathcal{I}^*)$, the orthogonal complement of the alternating functions. When G has no isolated verticies or bipartite components, $Range(\mathcal{I}^*\mathcal{I})$ is all of L_V.*
3. *$0 \leq \langle f, \mathcal{I}^*\mathcal{I}f \rangle \leq \left(d(G) + 2\sqrt{d_i(G)d_o(G)} \right) \langle f, f \rangle.$*

Proposition 16

1. *$\mathcal{I}\mathcal{I}^* = TT^* + SS^* + ST^* + TS^*$*
2. *$\ker(\mathcal{I}\mathcal{I}^*) = \ker(\mathcal{I}^*) = A(G)$, $Range(\mathcal{I}\mathcal{I}^*) = Range(\mathcal{I}) = \Upsilon(G)$.*
3. *$\mathcal{I}\mathcal{I}^*$ has the same spectrum as $\mathcal{I}^*\mathcal{I}$, and, with the possible exception of the eigenvalue 0, with the same multiplicity. If u is an eigenvector of $\mathcal{I}^*\mathcal{I}$, the $\mathcal{I}u$ is an eigenvector of $\mathcal{I}\mathcal{I}^*$.*

Remark: It would be nice to have a characterization of the spectrum in the non-bipartite case. We are unaware of progress in this arena since Grossman et al [9], 1994.

6   The Drift of a Digraph

In this section we discuss a family of fundamental operators on graphs which seems not to have been discussed in the literature. So far we have considered differences and sums of the canonical operators, S, T. At this point we will consider the operator $S + iT$, mapping L_V into L_E. The adjoint operator is $S^* - iT^*$, so the operator $(S^* - iT^*)(S + iT) = S^*S + T^*T + i(S^*T - T^*S)$ is self adjoint and positive semidefinite. We can also consider the same construction using the other square root of -1 to deduce that $(S^* + iT^*)(S - iT) = S^*S + T^*T + i(T^*S - S^*T)$ is self adjoint and positive semidefinite.

Our experience with the difference and incidence operators makes the following lemma a triviality.

Lemma 22. *In a connected digraph G, $S + iT$ (resp. $S - iT$) has a trivial kernel unless the only cycles are of order divisible by 4. In this case there is a one dimensional kernel spanned by complex functions in which the which the magnitude is locally constant and the phase rotates by a factor if i (resp. $-i$) across every edge in traversing from source to target.*

Corollary 3. *$(S^* - iT^*)(S + iT)$ (resp. $(S^* + iT^*)(S - iT)$) is invertible unless the only cycles in G have order divisible by 4. In this case there is a one dimensional kernel spanned by complex functions in which the which the magnitude is locally constant and the phase rotates by a factor if i (resp. $-i$) across every edge in traversing from source to target.*

Definition 20. *We will call the operator $\Gamma(G) = T^*S - S^*T$ the* drift operator *of the graph.*

Proposition 17. *The drift operator satisfies the following properties.*

1. *Γ is skew adjoint and real.*
2. *The eigenvalues of Γ are imaginary. For each eigenvector f, $\Gamma f = \lambda f$ implies $\Gamma \overline{f} = -\lambda \overline{f}$, i.e. the eigenvalues and eigenvectors come in complex conjugate pairs.*
3. *Γ generates a one parameter unitary group $t \mapsto e^{t\Gamma}$ on L_V (in fact a group of rotations).*
4. *$\|\Gamma f\| \leq d(G)\|f\|$.*
5. *$\Gamma f(v) = \sum_{e:e^+ = v} f(e^-) - \sum_{e:e^- = v} f(e^+)$*

Proof.

1. Since $(S^* + iT^*)(S - iT)$ and $S^*S + T^*T$ are self adjoint, so is $i(S^*T - T^*S)$, hence $S^*T - T^*S$ is skew symmetric. It also maps real functions to real functions.
2. Since $i(S^*T - T^*S)$ is self adjoint, it's eigenvalues are real, hence those of $(S^*T - T^*S)$ are imaginary. Since Γ is real taking the complex conjugate of the eigenvalue equation yields complex conjugate eigenvalues and eigenvectors.
3. This is just the Stone's theorem on generators of unitary groups; since Γ is real, this unitary group is also real, hence a group of rotations.

4. For f a complex function:

$$0 \leq \langle f, (S^* \pm iT^*)(S \mp iT)f \rangle$$
$$0 \leq \pm i \langle f, (S^*T - T^*S)f \rangle + \langle f, S^*Sf \rangle + \langle f, T^*Tf \rangle$$
$$|\langle f, (S^*T - T^*S)f \rangle| \leq \langle f, S^*Sf \rangle + \langle f, T^*Tf \rangle$$
$$|\langle f, (S^*T - T^*S)f \rangle| \leq d(G) \langle f, f \rangle.$$

But since $(S^*T - T^*S)$ is skew adjoint, its norm is the maximum of the magnitude of its numerical range.

5. This follows from Proposition 2. $\qquad\qquad\qquad\qquad\qquad\qquad\qquad\qquad\square$

Recall that a *regular graph* is a graph in which the vertex degree $d(v)$ is independent of v and $d(v) = d(G)$. The same condition is less restrictive for digraphs, since $d_i(v)$ and $d_o(G)$ may vary subject to the constraint $d_i(v) + d_o(v) = d(G)$.

Corollary 4. *If the vertex-wise degree $d(v) = d(G)$ is constant then the eigenvectors of Γ are also eigenvectors of $(S^* \mp iT^*)(S \pm iT)$. In particular, if the only cycles in G are of order divisible by 4, then Γ has eigenvalues $\pm d(G)$, and $(S^* \mp iT^*)(S \pm iT)$ has eigenvalue $2d(G)$.*

Note that while the operators Δ and $\mathcal{I}^*\mathcal{I}$ are insensitive to the choice of direction in an edge $e = uv$, changing the sense of an directed edge results in a different Γ by a change of sign in one pair of entries: $\Gamma'_{uv} = -\Gamma_{uv}$. Moreover, in general this change of a direction in a single edge also results in a change of the eigenvalues. However, Γ is insensitive to the addition or deletion of self loops $e = vv$. The following lemma follows from basic properties of rotations.

Lemma 23. $Order(G) \equiv_2 dimker(\Gamma)$. *In particular, when the order of G is odd, Γ has a nonzero kernel.*

Generally the adjacency matrix of a (weighted) graph is given by a symmetric matrix. This data may also be taken as the data of a digraph in which every edge is accompanied by an edge in the opposite direction (of the same weight). We have

Proposition 18. *When the adjacency matrix is symmetric, the drift is zero, i.e. $\Gamma = 0$.*

In this sense the drift operator measures the deviation of a digraph G from a graph.

References

1. Bensoussan A and Menaldi JL (2005) Difference Equations on Weighted Graphs, Journal of Convex Analysis (Special issue in honor of Claude LeMarechal), 12:13–44 .
2. Bollobas B (1998)Modern Graph Theory, Springer, New York.
3. Chung, F. (1997) *Spectral Graph Theory*, CBMS Lecture Notes. AMS, Philadelphia.
4. Ferrari-Trecate G , Buffa A, and Gati M (2005) Analysis of coordination in multi-agent systems through partial difference equations. Part I: The Laplacian control. 16th IFAC World Congress on Automatic Control
5. Ferrari-Trecate G , Buffa A, and Gati M (2005) Analysis of coordination in multi-agent systems through partial difference equations. Part II: Nonlinear control. 16th IFAC World Congress on Automatic Control

6. Ferrari-Trecate G , Buffa A, and Gati M (2006) Analysis of Coordination in Multi-Agent Systems through Partial Difference Equations, IEEE Trans. Automatic Control, 5(6): 1058–1063.
7. Godsil C.(2007), Private communication.
8. Gross J and Tucker T (1987) Topological Graph Theory, Wiley Interscience, New York
9. Grossman, J., Kulkarni D. and Schochetman, I. (1994) Algebraic Graph Theory Without Orientation, Linear Algebra and Its Applications 212/213: 289-307.
10. Hatcher A (2002) Algebraic Topology, Cambridge University Press Cambridge, U.K. ; New York.
11. Imrich W and Pisanski T Multiple Kronecker Covering Graphs, arXiv:math.CO/050513 v1 8 May 2005
12. Jadbabiaie A, Lin J, Morse A S (2003) Coordination of groups of mobile autonomous agents using nearest neighbor rules, IEEE Transactions on Automatic Control, 48(6): 988–1001.
13. Ji M, Egerstedt M, Ferrari-Trecate G, and Buffa A (2006) Hierarchical Containment Control in Heterogeneous Mobile Networks, Mathematical Theory of Networks and Systems, Kyoto, Japan: 2227-2231
14. Lieb E and Loss M (1993) Fluxes, Laplacians and Kesteleyn's Theorem, Duke Mathematical Journal, 71(2): 337-363
15. Muhammad A and Egerstedt M (2005) Connectivity Graphs as Models of Local Interactions, Journal of Applied Mathematics and Computation, 168(1):243-269
16. Shubin MA (1994) Discrete Magnetic Laplacian, Commun. Math. Phys. 164:259-275
17. Sunada T (1994) A Discrete Analogue of Periodic Magnetic Schrodinger Operators, Contemporary Mathematics, 173:283-299
18. Van Nuffelen C (1976) On the incidence matrix of a graph, IEEE Transactions on Circuits and Systems, 23(9):572 - 572
19. S. Vigna, The Graph Fibrations Home Page, http://vigna.dsi.unimi.it/fibrations/, (as viewed November 2006)

Prediction-Error Approximation by Convex Optimization

Anders Lindquist

Optimization and Systems Theory,
Department of Mathematics, Royal Institute of Technology. SE-10044 Stockholm, Sweden
`alq@math.kth.se`

This paper is dedicated to Giorgio Picci on the occasion of his 65th birthday. I have come to appreciate Giorgio not only as a great friend but also as a great scholar. When we first met at Brown University in 1973, he introduced me to his seminal paper [29] on splitting subspaces, which became the impetus for our joint work on the geometric theory of linear stochastic systems [23, 24, 25, 26]. This led to a life-long friendship and a book project that never seemed to converge, but now is close to being finished [27].

I have learned a lot from Giorgio. The present paper grew out of a discussion in our book project, when Giorgio taught me about the connections between prediction-error identification and the Kullback-Leibler criterion. These concepts led directly into the recent theory of analytic interpolation with complexity constraint, with which I have been deeply involved in recent times. I shall try to explain these connections in the following paper.

1 Introduction

Prediction error methods for ARMA modeling play a major role in system identification [28, 30], but in general they lead to nonconvex optimization problems for which global convergence is not guaranteed. In fact, although these algorithms are computationally simple and quite reliable, as pointed out in [32, p. 103], there is so far no theoretically satisfactory algorithm for ARMA parameter estimation. Convex optimization approaches have been proposed [7, 17] for the approximation part, but it remains to verify their practical applicability and statistical accuracy.

In this paper we identify certain classes of ARMA models in which prediction error minimization leads to convex optimization. It has been shown [2, 33] that model approximation via prediction error identification leads to an optimization problem that is related to the minimization of the Kullback-Leibler divergence criterion [18, 21]. This, in turn, leads naturally to the theory of analytic interpolation and generalized moment problems with complexity constraints developed in recent years [8, 9, 10, 11, 12, 13, 14, 16]. This has already been observed, at least in the context of covariance extension, in [4, 6].

The paper is outlined as follows. In Section 2 we review some pertinent facts on prediction error approximation and set notations. In Section 3 we define model classes

A. Chiuso et al. (Eds.): Modeling, Estimation and Control, LNCIS 364, pp. 239–249, 2007.

in terms of a finite number of, not necessarily rational, basis functions and show that the corresponding prediction-error minimizers can be obtained as the solution of a pair of dual convex optimization problems. In the rational case we can even compute the minimizer in closed form. The connections to the Kullback-Leibler criterion and maximum-likelihood identification is described in Section 4. In Section 5 we provide prediction-error approximants in model classes determined by interpolation conditions on the spectral density and its positve real part.

For simplicity this paper will only deal with the scalar case, but multivariable extensions are straightforward, given multivariable versions of the the theory of generalized moment problems with degree constraints [5, 22].

2 Prediction-Error Approximation

Let $\{y(t)\}_{\mathbb{Z}}$ be a zero-mean stationary stochastic process with a spectral density $\{\Phi(e^{i\theta}); \theta \in [-\pi, \pi]\}$ that may be rational or nonrational but is zero only in isolated points θ. Let w be a normalized minimum-phase spectral of Φ; i.e.,

$$\Phi(e^{i\theta}) = \rho|w(e^{i\theta})|^2, \quad \theta \in [-\pi, \pi],$$

where $w(0) = 1$ and $\rho > 0$ is a suitable normalizing factor. Then the process y can be modeled by passing a white noise e with covariance lags $\mathbf{E}\{e(t)e(s)\} = \rho\delta_{ts}$ through a filter with a transfer function

$$w(z) = \sum_{k=0}^{\infty} w_k z^{-k}.$$

Since $w_0 = 1$,

$$y(t) = e(t) + y(t|t-1),$$

where

$$y(t|t-1) = w_1 e(t-1) + w_2 e(t-2) + \dots$$

is the one-step ahead linear predictor of $y(t)$ given $\{y(s); \ s \le t - 1\}$. Hence $y(t|t-1)$ can be represented by passing e through a filter with transfer function $w - 1$ as shown in the block diagram

In particular,

$$y(t) - y(t|t-1) = e(t).$$

Now, let $\hat{w}$ be a normalized ($\hat{w}(0) = 1$), stable minimum-phase function belonging to some *model class* $\mathcal{W}$ to be specified later. We shall regard $\hat{w}$ as an approximation of w, from which we can form an *approximate predictor*, denoted by $\hat{y}(t|t-1)$, as in the figure

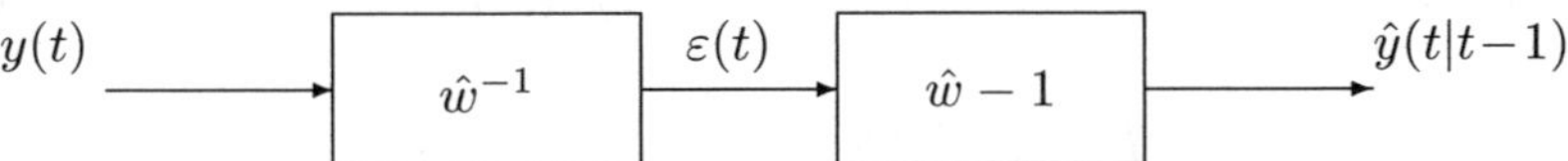

Then

$$\varepsilon(t) = y(t) - \hat{y}(t|t-1);$$

i.e., $\varepsilon(t)$ is the *prediction error*, which is not a white noise. Indeed, it is easy to see that it has the variance

$$r := \mathbf{E}\{\varepsilon(t)^2\} = \int_{-\pi}^{\pi} |\hat{w}(e^{i\theta})|^{-2}\Phi(e^{i\theta})\frac{d\theta}{2\pi}. \tag{1}$$

Since $\varepsilon(t) = e(t) + [y(t|t-1) - \hat{y}(t|t-1)]$ and $e(t)$ and $[y(t|t-1) - \hat{y}(t|t-1)]$ are uncorrelated,

$$r = \rho + \mathbf{E}\{|y(t|t-1) - \hat{y}(t|t-1)|^2\} \geq \rho.$$

The idea is now to find a $\hat{w} \in \mathcal{W}$ that minimizes the prediction error variance (1). To this end, define the class $\mathcal{F}$ of spectral densities

$$\hat{\Phi}(e^{i\theta}) = \hat{\rho}|\hat{w}(e^{i\theta})|^2, \tag{2}$$

where $\hat{w} \in \mathcal{W}$ and $\hat{\rho} > 0$. Then the prediction error takes the form

$$r := \hat{\rho} \int_{-\pi}^{\pi} \hat{\Phi}(e^{i\theta})^{-1}\Phi(e^{i\theta})\frac{d\theta}{2\pi}. \tag{3}$$

The purpose of the coefficient $\hat{\rho}$ in (3) is merely to normalize $\hat{\Phi}$. Once an optimal $\hat{\Phi} \in \mathcal{F}$ has been determined, $\hat{\rho}$ and $\hat{w} \in \mathcal{W}$ are obtained by outer spectral factorization and normalzation so that $\hat{w}(0) = 1$.

3   Prediction-Error Approximation in Restricted Model Classes

We begin by defining the model class $\mathcal{F}$. To this end, let

$$g_0, g_1, g_2, \ldots, g_n \tag{4}$$

be a linearly independent sequence of Lipschitz continuous functions on the unit circle with zeros only in isolated points, and let the model class $\mathcal{F}$ be the set of all functions $\hat{\Phi}$ such that

$$\hat{\Phi}(e^{i\theta})^{-1} = Q(e^{i\theta}) := \mathrm{Re}\left\{\sum_{k=0}^{n} q_k g_k(e^{i\theta})\right\}, \tag{5}$$

for some $q_0, q_1, \ldots, q_n \in \mathbb{C}$ such that $Q(e^{i\theta}) \geq 0$ for all $\theta \in [-\pi, \pi]$. In addition, let $\mathcal{Q}$ the class of all such functions Q.

As a simple example, consider the case $g_k = z^k$, $k = 0, 1, \ldots, n$. Then the model class $\mathcal{W}$ is the family of all $AR(n)$ models. However, more general choices of rational basis functions (4) yield model classes of $ARMA$ models. Even more generally, we may choose basis functions that are not even rational.

Theorem 1. *Let the spectral density Φ have the property that the generalized moments*

$$c_k := \int_{-\pi}^{\pi} g_k(e^{i\theta})\Phi(e^{i\theta})\frac{d\theta}{2\pi}, \quad k = 0, 1, \ldots, n, \tag{6}$$

exist, and define the functional $\mathbb{J} : \mathcal{Q} \to \overline{\mathbb{R}}$ as

$$\mathbb{J}(Q) = \int_{-\pi}^{\pi} \left[\Phi(e^{i\theta})Q(e^{i\theta}) - \log Q(e^{i\theta})\right]\frac{d\theta}{2\pi}. \tag{7}$$

Then the functional (7) has a unique minimum Q_{opt}, which is an interior point in $\mathcal{Q}$. Moreover,

$$\int_{-\pi}^{\pi} g_k(e^{i\theta})\frac{1}{Q_{opt}(e^{i\theta})}\frac{d\theta}{2\pi} = c_k, \quad k = 0, 1, \ldots, n. \tag{8}$$

Proof. Since the functions $g_0, g_1, \ldots, g_n$ are Lipschitz continuous, hypothesis **H1** in [14] is satisfied [14, Remark 1.1]. Moreover, since both $Q \in \mathcal{Q}$ and Φ are nonnegative on the unit circle with zeros only in isolated points,

$$\int_{-\pi}^{\pi} Q\Phi\frac{d\theta}{2\pi} > 0$$

for all $Q \in \mathcal{Q} \setminus \{0\}$. Hence the sequence $c = (c_1, c_2, \ldots, c_n)$ is positive in the sense prescribed in [14].

The functional $\mathbb{J} : \mathcal{Q} \to \mathbb{R}$ is strictly convex on the convex set $\mathcal{Q}$, and hence, if a minimum does exist, it must be unique. However, it is shown in [14, Theorem 1.5] that $\mathbb{J}$ has a unique minimizer, Q_{opt}, which lies in the interior of $\mathcal{Q}$, provided the sequence $c = (c_1, c_2, \ldots, c_n)$ is positive and hypthesis **H1** holds, which is what we have established above. Since the minimizer Q_{opt} is an interior point, the gradient of $\mathbb{J}$ must be zero there, and hence (8) follows.

Theorem 2. *Let Φ be an arbitrary spectral density such that the generalized moments (6) exist. Then, there is a unique spectral density $\hat{\Phi}$ in the model class $\mathcal{F}$ that minimizes the prediction error variance (3), and it is given by*

$$\hat{\Phi}_{opt} := Q_{opt}^{-1}, \tag{9}$$

where Q_{opt} is the unique minimizer in Theorem 1.

Proof. By Theorem 1, $\hat{\Phi}_{opt}$ is the unique minimizer of

$$\mathbb{J}(\hat{\Phi}^{-1}) = \int_{-\pi}^{\pi} \left[\Phi(e^{i\theta})\hat{\Phi}^{-1}(e^{i\theta}) + \log\hat{\Phi}^{-1}(e^{i\theta})\right]\frac{d\theta}{2\pi}. \tag{10}$$

However, by (3),

$$\int_{-\pi}^{\pi} \Phi(e^{i\theta})\hat{\Phi}^{-1}(e^{i\theta})\frac{d\theta}{2\pi} = \frac{r}{\hat{\rho}}.$$

Moreover, in view of (2)

$$\int_{-\pi}^{\pi} \log \hat{\Phi} \frac{d\theta}{2\pi} = \log \hat{\rho} + 2 \int_{-\pi}^{\pi} \log |\hat{w}| \frac{d\theta}{2\pi}$$

$$= \log \hat{\rho} + 2 \log \hat{w}(0) = \log \hat{\rho},$$

where we have used Jensen's formula [1, p.184] and the facts that $\hat{w}$ is outer and $\hat{w}(0) = 1$. Consequently,

$$\mathbb{J}(\hat{\Phi}^{-1}) = \frac{r}{\hat{\rho}} + \log \hat{\rho}. \tag{11}$$

Now, for any fixed $r > 0$, (11) has a unique minimum for $\hat{\rho} = r$, and hence

$$\mathbb{J}(\hat{\Phi}_{opt}^{-1}) = 1 + \min_{r} \log r.$$

Therefore $\log r$, and hence the prediction error r, takes it unique minimum value for $\hat{\Phi} = \hat{\Phi}_{opt}$, as claimed.

Now, in view of (8) and (9),

$$\int_{-\pi}^{\pi} g_k(e^{i\theta}) \hat{\Phi}_{opt}(e^{i\theta}) \frac{d\theta}{2\pi} = c_k, \quad k = 0, 1, \ldots, n. \tag{12}$$

However, $\hat{\Phi}_{opt}$ is not the only spectral density that satisfies these moment conditions. In fact, following [14, 12], we can prove that, among all such solutions, $\hat{\Phi}_{opt}$ is the one maximizing the entropy gain.

Theorem 3. *The optimal prediction-error approximation $\hat{\Phi}_{opt}$ of Theorem 2 is the unique maximizer of the entropy gain*

$$\mathbb{I}(\hat{\Phi}) := \int_{-\pi}^{\pi} \log \hat{\Phi}(e^{i\theta}) \frac{d\theta}{2\pi} \tag{13}$$

subject to the moment constraints

$$\int_{-\pi}^{\pi} g_k(e^{i\theta}) \hat{\Phi}(e^{i\theta}) \frac{d\theta}{2\pi} = c_k, \quad k = 0, 1, \ldots, n. \tag{14}$$

Let us stress again that the basis functions $g_0, g_1, \ldots, g_n$ need not be rational. Although, in general, we want the model class $\mathcal{W}$ to consist of rational functions of low degree, there may be situations when it is desirable to include nonrational components, such as, for example, exponentials.

Identification in terms of orthogonal basis functions is a well studied topic [19, 34, 35]. The most general choice is

$$g_k(z) = \frac{\sqrt{1 - |\xi_k|^2}}{z - \xi_k} \prod_{j=0}^{k-1} \frac{1 - \xi_j^* z}{z - \xi_j},$$

where $\xi_0, \xi_1, \xi_2, \ldots$ are poles to be selected by the user. The functions $g_0, g_1, g_2, \ldots$ form a complete sequence in the Hardy space $H^2(\mathbb{D}^c)$ over the complement of the unit disc $\mathbb{D}$ provided $\sum_{k=0}^{\infty}(1 - |\xi_k|) = \infty$. In [19] the problem to determine a minimum-degree rational function of the form

$$\hat{F}(z) = \frac{1}{2}c_0 g_0(z) + \sum_{k=1}^{\infty} c_k g_k(z),$$

where $c_0, c_1, \ldots, c_n$ are prescribed, was considered.

In our present setting, in order for $\hat{\Phi} := \mathrm{Re}\{\hat{F}\}$ to be a spectral density, $\hat{F}$ needs to be positive real, leading to a problem left open in [19]. Let $c_0, c_1, \ldots, c_n$ be given by (6). Then, by Theorem 3, the problem of determining the minimum prediction-error approximant of Φ in the model class defined by $g_0, g_1, \ldots, g_n$ amounts to finding the function $\hat{\Phi}$ that maximizes the entropy gain

$$\int_{-\pi}^{\pi} \log \hat{\Phi} \frac{d\theta}{2\pi},$$

subject to

$$\int_{-\pi}^{\pi} g_k \hat{\Phi} \frac{d\theta}{2\pi} = c_k, \quad k = 0, 1, \ldots, n.$$

Alternatively, we may solve the convex optimization problem of Theorem 1.

Theorem 1 enables us to determine, under general conditions, the minimum prediction-error in closed form. Here, following [16], we state such a result under the assumption that the basis functions are rational.

Proposition 1. *Suppose that the basis functions $g_0, g_1, \ldots, g_n$ are rational and analytic in the unit disc $\mathbb{D}$. Then,*

$$Q_{opt}(z) = \frac{|g^*(z)P^{-1}g(0)|^2}{g^*(0)P^{-1}g(0)}, \tag{15}$$

where

$$g(z) := \begin{bmatrix} g_1 \\ g_2 \\ \vdots \\ g_n \end{bmatrix}, \qquad P := \int_{-\pi}^{\pi} g(e^{i\theta})\Phi(e^{i\theta})g(e^{i\theta})^* \frac{d\theta}{2\pi}.$$

Proof. Clearly the basis functions $g_0, g_1, \ldots, g_n$ belong to the Hardy space $H^2(\mathbb{D})$, and $g := (g_0, g_1, \ldots, g_n)'$ has a representation

$$g(z) = (I - zA)^{-1}B,$$

where (A, B) is a reachable pair. Then

$$\varphi(z) = \frac{\det(zI - A^*)}{\det(I - zA)}$$

is an inner function, and it can be shown that the basis functions $g_0, g_1, \ldots, g_n$ span the coinvariant subspace $\mathcal{K} := H^2 \ominus \varphi H^2$. Moreover, for any $Q \in \mathcal{Q}$, there is an outer function in $a \in \mathcal{K}$ such that $Q = a^* a$ ([13, Proposition 9]). Consequently (7) can be written

$$J(a) = \int_{-\pi}^{\pi} a^* \Phi a \frac{d\theta}{2\pi} - \int_{-\pi}^{\pi} 2 \log |a| \frac{d\theta}{2\pi}.$$

Here the second term can be written $2 \log |a(0)|$ by Jensen's formula [1, p.184], and since $a \in \mathcal{K}$, there is a vector $\mathbf{a} \in \mathbb{C}^{n+1}$ such that $a(z) = g^*(z)\mathbf{a}$, so the second term be written $\mathbf{a}^* P \mathbf{a}$. Hence the optimization problem is reduced to determining the $\mathbf{a}$ that minimizes

$$\tilde{J}(\mathbf{a}) = \mathbf{a}^* P \mathbf{a} - 2 \log |\mathbf{a}^* g(0)|.$$

Setting the gradient equal to zero, we obtain $\mathbf{a} = P^{-1} g(0)/|a(0)|$ and hence $a(z) = g^*(z)P^{-1}g(0)/|a(0)|$. Then $|a(0)|^2 = g^*(0)P^{-1}g(0)$, and therefore the optimal $\mathbf{a}$ becomes

$$a(z) = \frac{g^*(z)P^{-1}g(0)}{\sqrt{g^*(0)P^{-1}g(0)}},$$

from which (15) follows.

Remark 1. The pair of dual optimization problems in Theorems 1-3 are special cases of a more general formulation [8, 9, 10, 11, 12, 13, 14, 16] where (7) is replaced by

$$\mathbb{J}_{\Psi}(Q) = \int_{-\pi}^{\pi} \left[\Phi(e^{i\theta})Q(e^{i\theta}) - \Psi(e^{i\theta}) \log Q(e^{i\theta}) \right] \frac{d\theta}{2\pi}, \tag{16}$$

with Ψ is a parahermitian function that is positive on the unit circle and available for tuning; and (13) is replaced by

$$\mathbb{I}_{\Psi}(\hat{\Phi}) := \int_{-\pi}^{\pi} \Psi(e^{i\theta}) \log \hat{\Phi}(e^{i\theta}) \frac{d\theta}{2\pi}. \tag{17}$$

The particular choice $\Psi = I$, corresponding to the minimum prediction-error approximation, is called the *central* or *maximum entropy* solution. As suggested by Blomqvist and Wahlberg [4,6] in the context of covariance extension, a nontrivial Ψ corresponds to a particular choice of prefiltering that may lead to better results; cf, page 249prefiltering.

4   The Kullback-Leibler Criterion and Maximum-Likelihood Identification

The optimization problem of Theorem 1 is intimately connected to the Kullback-Leibler divergence [21, 18]

$$D(y\|z) := \limsup_{N \to \infty} \frac{1}{N} D(p_y^N \mid p_z^N)$$

from one stationary, Gaussian stochastic processes z to another y, where p_y^N and p_z^N are the N-dimensional density functions of y and z respectively, and where

$$D(p_1 \mid p_2) := \int_{\mathbb{R}^n} p_1(x) \log \frac{p_1(x)}{p_2(x)} \, dx.$$

In fact, it was shown in [33] that, if y and z have spectral densities Φ and $\hat{\Phi}$, respectively, then

$$D(y\|z) = \frac{1}{2} \int_{-\pi}^{\pi} \left[(\Phi - \hat{\Phi})\hat{\Phi}^{-1} - \log(\Phi\hat{\Phi}^{-1}) \right] \frac{d\theta}{2\pi}. \tag{18}$$

Consequently,

$$D(y\|z) = \frac{1}{2}\mathbb{J}(\hat{\Phi}^{-1}) - \frac{1}{2} \left[1 + \int_{-\pi}^{\pi} \log \Phi \frac{d\theta}{2\pi} \right], \tag{19}$$

where the last integral is constant.

Given the process y, consider the problem to find the minimum divergence $D(y\|z)$ over all z with a spectral density $\hat{\Phi} \in \mathcal{F}$. Then we have established that this minimum is attained precisely when $\hat{\Phi}^{-1}$ is the unique minimizer of $\mathbb{J}$ in Theorem 1, which in turn is the minimum prediction-error estimate in the model class $\mathcal{F}$.

Next, suppose that we have a finite sample record

$$\{y_0, y_1, \ldots, y_N\} \tag{20}$$

of the process y and an estimate Φ_N of Φ based on (20) that is consistent in the sense that $\lim_{N \to \infty} \Phi_N(e^{i\theta}) = \Phi(e^{i\theta})$ with probability one for almost all $\theta \in [-\pi, \pi]$. The periodogram

$$\Phi_N(e^{i\theta}) = \frac{1}{N} \left| \sum_{t=0}^{N} e^{-i\theta t} y_t \right|^2.$$

is one such estimate of Φ. Then, under some mild technical assumptions,

$$J_N(\hat{\Phi}) := \frac{1}{2} \int_{-\pi}^{\pi} \left[\Phi_N(e^{i\theta})\hat{\Phi}(e^{i\theta})^{-1} + \log \hat{\Phi}(e^{i\theta}) \right] \frac{d\theta}{2\pi} \tag{21}$$

tends to $\mathbb{J}(\hat{\Phi}^{-1})$ as $N \to \infty$. The functional $J_N(\Psi)$ is known as the *Whittle log-likelihood*, and it is a widely used approximation of $-\log L_N$, where $L_N(\hat{\Phi})$ is the *likelihood function*. In fact, $J_N(\hat{\Phi})$ and $L_N(\hat{\Phi})$ tend to the same limit $\mathbb{J}(\hat{\Phi}^{-1})$ as $N \to \infty$.

5   Prediction-Error Approximation by Analytic Interpolation

Let Φ be the given (or estimated) spectral density defined as above. Then, by the Herglotz formula,

$$F(z) = \int_{-\pi}^{\pi} \frac{e^{i\theta} + z}{e^{i\theta} - z} \Phi(e^{i\theta}) \frac{d\theta}{2\pi} \tag{22}$$

is *the positive real part* of Φ. More precisely, F is the unique function in $H(\mathbb{D})$ such that $F(0)$ is real and

$$\Phi(e^{i\theta}) = \mathrm{Re}\{F(e^{i\theta})\}. \tag{23}$$

Now, let us select a number of points

$$z_0, z_1, \ldots, z_n \tag{24}$$

in the unit disc $\mathbb{D}$. Then, in view of (22),

$$F(z_k) = \int_{-\pi}^{\pi} g_k(z)\Phi(e^{i\theta})\frac{d\theta}{2\pi},$$

where

$$g_k(z) = \frac{z + z_k}{z - z_k}. \tag{25}$$

Therefore, if the points (24) are distinct, we may choose $g_0, g_1, \ldots, g_n$ as our basis functions, and then

$$F(z_k) = c_k, \quad k = 0, 1, \ldots, n,$$

where

$$c_k := \int_{-\pi}^{\pi} g_k(e^{i\theta})\Phi(e^{i\theta})\frac{d\theta}{2\pi}, \quad k = 0, 1, \ldots, n. \tag{26}$$

If (24) are not distinct, we modify $g_0, g_1, \ldots, g_n$ in the following way to make them linearly independent. If $z_k = z_{k+1} = \cdots = z_{k+m-1}$, then $g_k, \ldots, g_{k+m-1}$ are replaced by

$$g_k(z) = \frac{z + z_k}{z - z_k}, \quad g_{k+1}(z) = \frac{2z}{(z - z_k)^2}, \quad \ldots, \quad g_{k+m-1}(z) = \frac{2z}{(z - z_k)^m}. \tag{27}$$

Then, differentiating (22), we have the modified interpolation conditions

$$F(z_k) = c_k, \quad \frac{dF}{dz}(z_k) = c_{k+1}, \quad \ldots, \quad \frac{1}{(m-1)!}\frac{d^{(m-1)}F}{dz^{(m-1)}}(z_k) = c_{k+m-1}.$$

Now, given the points (24), let $\mathcal{F}(z_0, z_1, \ldots, z_n)$ be the class of all spectral densities $\hat{\Phi}$ with positive real part $\hat{F}$ of degree at most n and satisfying the interpolation conditions

$$\hat{F}(z_k) = c_k \tag{28a}$$

for distinct points and

$$\hat{F}(z_k) = c_k, \quad \frac{d\hat{F}}{dz}(z_k) = c_{k+1}, \quad \ldots, \quad \frac{1}{(m-1)!}\frac{d^{(m-1)}\hat{F}}{dz^{(m-1)}}(z_k) = c_{k+m-1}, \tag{28b}$$

if $z_k = z_{k+1} = \cdots = z_{k+m-1}$, where $c_0, c_1, \ldots, c_n$ are given by (26). In particular,

$$\hat{\Phi}(z_k) = \Phi(z_k), \quad k = 0, 1, \ldots, n \tag{29}$$

for all $\hat{\Phi} \in \mathcal{F}(z_0, z_1, \ldots, z_n)$, where some of the conditions (29) may be repeated (in case of multiple points).

With the basis (4) chosen as above, the minimum prediction-error approximation in the model class $\mathcal{F}(z_0, z_1, \ldots, z_n)$ defined by these functions is as described in the following theorem, which now is a direct consequence of Theorems 1 and 2.

Theorem 4. *The minimum prediction-error approximation of Φ in the class $\mathcal{F}(z_0, z_1, \ldots, z_n)$ is the unique $\hat{\Phi} \in \mathcal{F}(z_0, z_1, \ldots, z_n)$ that minimizes the entropy gain*

$$\int_{-\pi}^{\pi} \log \hat{\Phi}\,\frac{d\theta}{2\pi},$$

or, dually, the $\hat{\Phi}$ that minimizes (7), where $g_0, g_1, \ldots, g_n$ are given by (25), or (27) for multiple points.

It follows from Theorem 1 that all $\hat{\Phi} \in \mathcal{F}(z_0, z_1, \ldots, z_n)$, and in particular the optimal one, has (spectral) zeros that coincide with $z_0, z_1, \ldots, z_n$ and hence with the interpolation points. Recently, Sorensen [31] has developed an efficient algorithm for solving large problems of this type. In [15] we point out the connection between this approach, initiated by Antoulas [3], and our theory for analytic interpolation with degree constraints [10, 11, 12, 13, 14, 16]. We show that a better spectral fit can often be obtained by choosing a nontrivial weight P in the objective function (16). This corresponds to prefiltering; see Remark 1.

An important question in regard to the application of Theorem 4 to system identification is how to choose the interpolation points $z_0, z_1, \ldots, z_n$. Here (29) could serve as an initial guide. However, a more sofisticated procedure is proposed in [20].

6 Conclusion

In this paper we have shown that in large model classes of ARMA models, as well as in some model classes of nonrational functions, prediction-error approximation leads to convex optimization. The connections to Kullback-Leibler and maximum-likelihood criteria have been described. Model classes defined in terms of interpolation conditions have also been considered, connecting to literature in numerical linear algebra. Generalizations to the multivarable case should be straight-forward relying on mutivarable versions [5,22] of the theory of analytic interpolation and generalized moment problems with complexity constraints.

References

1. Ahlfors LV (1953) Complex Analysis. McGraw-Hill,
2. Anderson BDO, Moore JB, Hawkes RM (1978) Automatica 14: 615–622
3. Antoulas AC (2005) Systems and Control Letters 54: 361–374
4. Blomqvist, A (2005) A Convex Optimization Approach to Complexity Constrained Analytic Interpolation with Applications to ARMA Estimation and Robust Control. PhD Thesis, Royal Institute of Technology, Stockholm, Sweden
5. Blomqvist A, Lindquist A, Nagamune R (2003) IEEE Trans Autom Control 48: 2172–2190
6. Blomqvist A, Wahlberg B (2007) IEEE Trans Autom Control 55: 384–389
7. Byrnes CI, Enqvist P, Lindquist A (2002) SIAM J. Control and Optimization 41: 23–59
8. Byrnes CI, Gusev SV, Lindquist A (1998) SIAM J. Contr. and Optimiz. 37: 211–229
9. Byrnes CI, Gusev SV, Lindquist A (2001) SIAM Review 43: 645–675
10. Byrnes CI, Georgiou TT, Lindquist A (2001) IEEE Trans Autom Control 46: 822–839
11. Byrnes CI, Georgiou TT, Lindquist A (2000) IEEE Trans. on Signal Processing 49: 3189–3205

12. Byrnes CI, Lindquist A (2003) A convex optimization approach to generalized moment problems. In: Hashimoto K, Oishi Y, Yamamoto Y (eds) Control and Modeling of Complex Systems: Cybernetics in the 21st Century. Birkhäuser, Boston Basel Berlin
13. Byrnes CI, Georgiou TT, Lindquist A, Megretski (2006) Trans American Mathematical Society 358: 965–987
14. Byrnes CI, Lindquist A (2006) Integral Equations and Operator Theory 56: 163–180
15. Fanizza G, Karlsson J, Lindquist A, Nagamune R (2007) Linear Algebra and Applications. To be published
16. Georgiou TT, Lindquist A (2003) IEEE Trans. on Information Theory 49: 2910–2917
17. Georgiou TT, Lindquist A (2007) IEEE Trans Autom Control. To be published
18. Good, IJ (1963) Annals Math. Stat. 34: 911–934
19. Heuberger PSC, Van den Hof PMJ, Szabó Z (2001) Proc. 40th IEEE Conf. Decision and Control, Orlando, Florida, USA: 3673–3678
20. Karlsson J, Lindquist A (2007) Submitted for publication
21. Kullback S (1959) Information Theory and Statistics. John Wiley, New York
22. Kuroiwa Y, Lindquist A (2007) Proc 2007 Decision and Control Conference. Submitted for publication
23. Lindquist A, Picci G (1985) SIAM J Control Optim 23: 809–857
24. Lindquist A, Picci G (1995) Stochastics 15: 1–50
25. Lindquist A, Picci G (1991) J Math Systems Estim Control 1: 241–333
26. Lindquist A, Picci G (1996) Automatica 32:709–733
27. Lindquist A, Picci G (2007) Linear Stochastic Systems: A Geometric Approach to Modeling, Estimation and Identification. To appear
28. Ljung L (1987) System Identification: Theory for the User. Prentice Hall, Englewood Cliffs
29. Picci G (1976) Proc. IEEE 64: 112–122
30. Söderström T, Stoica P (1989) System Identification. Prentice Hall, New York
31. Sorensen DC (2005) Systems and Control Letters 54: 347-360
32. Stoica P, Moses R (1997) Introduction to Spectral Analysis. Prentice Hall, Upper Saddle River, NJ
33. Stoorvogel AA, van Schuppen JH (1996) System identification with information theoretic criteria. In: Bittanti S, Picci G (eds) Identification, Adaptation, Learning: The Science of learning Models from Data. Springer, Berlin Heidelberg
34. Wahlberg B (1991) IEEE Trans Autom Control 36: 551–562
35. Wahlberg B (1994) IEEE Trans Autom Control 39: 1276–1282

Patchy Solutions of Hamilton-Jacobi-Bellman Partial Differential Equations

Carmeliza Navasca[1] and Arthur J. Krener[2,⋆]

[1] ETIS Lab - UMR CNRS 8051, 6, avenue du Ponceau, 95014 Cergy-Pontoise, France
`cnavasca@gmail.com`
[2] Department of Applied Mathematics, Naval Postgraduate School,
Monterey, CA 93943-5216, USA
`ajkrener@nps.edu`

This paper is dedicated to our esteemed colleague and good friend Giorgi Picci on the occasion of his sixty fifth birthday.

1   Hamilton Jacobi Bellman PDEs

Consider the optimal control problem of minimizing the integral

$$\int_0^\infty l(x,u)\,dt \tag{1}$$

of a Lagrangian $l(x,u)$ subject to the controlled dynamics

$$\begin{aligned} \dot{x} &= f(x,u) \\ x(0) &= x^0 \end{aligned} \tag{2}$$

where f, l are smooth and l is strictly convex in $u \in I\!R^m$ for all $x \in I\!R^n$.

Suppose the dynamics and Lagrangian have Taylor series expansions about $x = 0, u = 0$ of the form

$$\dot{x} = Fx + Gu + f^{[2]}(x,u) + f^{[3]}(x,u) + \ldots \tag{3}$$

$$l(x,u) = \frac{1}{2}\left(x'Qx + u'Ru\right) + l^{[3]}(x,u) + l^{[4]}(x,u) + \ldots \tag{4}$$

where $^{[d]}$ indicates terms of degree d in the power series. We shall say that the optimal control problem is nice if F, G is stabilizable and $Q^{\frac{1}{2}}, F$ is detectable.

A special case of this optimal control problem is the linear quadratic regulator (LQR) where one seeks to minimize a quadratic cost

$$\int_0^\infty \frac{1}{2}\left(x'Qx + u'Ru\right)\,dt$$

⋆ Research supported in part by NSF grant 0505677.

A. Chiuso et al. (Eds.): Modeling, Estimation and Control, LNCIS 364, pp. 251–270, 2007.
springerlink.com

subject to linear dynamics

$$\dot{x} = Fx + Gu$$

If this is nice then there is a unique nonnegative definite solution to the algebraic Riccati equation

$$0 = F'P + PF + Q - PGR^{-1}G'P \tag{5}$$

that gives the optimal cost

$$\pi(x^0) = \frac{1}{2}(x^0)'Px^0 = \min \int_0^\infty \frac{1}{2}\left(x'Qx + u'Ru\right)\, dt \tag{6}$$

Furthermore the optimal control is given in feedback form

$$u(t) = \kappa(x(t)) = Kx(t)$$

where

$$K = -R^{-1}G'P \tag{7}$$

and the closed loop dynamics

$$\dot{x} = (F + GK)x \tag{8}$$

is exponentially stable.

Returning to the nonlinear problem, it is well-known that if it admits a smooth optimal cost $\pi(x)$ and a smooth optimal feedback $u = \kappa(x)$ locally around $x = 0$ then they must satisfy the Hamilton Jacobi Bellman (HJB) PDE

$$0 = \min_u \frac{\partial \pi}{\partial x}(x)f(x, u) + l(x, u)$$

$$\kappa(x) = \arg\min_u \frac{\partial \pi}{\partial x}(x)f(x, u) + l(x, u)$$

We shall assume that $\frac{\partial \pi}{\partial x}(x)f(x, u) + l(x, u)$ is strictly convex in u locally around $x = 0, u = 0$ then the HJB PDE can be rewritten as

$$0 = \frac{\partial \pi}{\partial x}(x)f(x, \kappa(x)) + l(x, \kappa(x))$$

$$0 = \frac{\partial \pi}{\partial x}(x)\frac{\partial f}{\partial u}(x, \kappa(x)) + \frac{\partial l}{\partial u}(x.\kappa(x)) \tag{9}$$

Al'brecht [1] has shown that for nice optimal control problems, the Hamilton Jacobi Bellman PDE can be approximately solved by Taylor series methods locally around the origin. Lukes [14] showed that under suitable conditions this series expansion converges to the true solution. The method has been implemented on examples by Garrard and Jordan [7], Yoshida and Loparo [22], Spencer, Timlin, Sain and Dyke [20] and others.   We have implemented it in the Nonlinear Systems Toolbox [11], a MATLAB based package.

Assume the dynamics and Lagrangian have power series expansions (3, 4). We assume that the unknowns, the optimal cost and optimal feedback, have similar expansions.

$$\pi(x) = \tfrac{1}{2}x'Px + \pi^{[3]}(x) + \pi^{[4]}(x) + \dots$$
$$\kappa(x) = Kx + \kappa^{[2]}(x) + \kappa^{[3]}(x) + \dots \tag{10}$$

We plug these into the HJB PDE (9) and extract terms of lowest degree to obtain the equations

$$0 = x'\left(F'P + PF + Q - K'RK\right)x$$
$$0 = x'\left(PG + K'R\right)$$

Notice the first equation is quadratic x and the second is linear in x. More importantly the first equation is linear in the unknown P but quadratic in the unknown K while the second is linear in both the unknowns. They lead to the familiar equations (5, 7) .

Having found P, K, we extract the next lowest terms from (9) and obtain

$$0 = \frac{\partial \pi^{[3]}}{\partial x}(x)(F + GK)x + x'Pf^{[2]}(x, Kx) + l^{[3]}(x, Kx)$$
$$0 = \frac{\partial \pi^{[3]}}{\partial x}(x)G + x'P\frac{\partial f^{[2]}}{\partial u}(x, Kx) + \frac{\partial l^{[3]}}{\partial u}(x, Kx) + \left(\kappa^{[2]}(x)\right)' R \tag{11}$$

Notice several things. The first equation is cubic in x and the second is quadratic. The equations involve the previously computed P, K. The unknowns $\pi^{[3]}(x)$; $\kappa^{[2]}(x)$ appear linearly in these equations. The equations are triangular, $\kappa^{[2]}(x)$ does not appear in the first one. If we can solve the first for $\pi^{[3]}(x)$ then clearly we can solve the second for $\kappa^{[2]}(x)$ as R is assumed to be invertible.

To decide the solvability of the first we study the linear operator

$$\pi^{[3]}(x) \mapsto \frac{\partial \pi^{[3]}}{\partial x}(x)(F + GK)x \tag{12}$$

from cubic polynomials to cubic polynomials.   Its eigenvalues are of the form $\lambda_i + \lambda_j + \lambda_k$ where $\lambda_i, \lambda_j, \lambda_k$ are eigenvalues of $F + GK$. A cubic resonance occurs when such a sum equals zero. But all the eigenvalues of $F + GK$ are in the open left half plane so there are no cubic resonances. Hence the linear operator (12) is invertible and (11) is solvable.

The higher degree terms are found in a similar fashion.   Suppose that $\overline{\pi}(x)$ and $\overline{\kappa}(x)$ are the expansions of the optimal cost and optimal feedback through degrees d and $d - 1$ respectively.   We wish to find the next terms $\pi^{[d+1]}(x)$ and $\kappa^{[d]}(x)$. We plug $\overline{\pi}(x) + \pi^{[d+1]}(x)$ and $\overline{\kappa}(x) + \kappa^{[d]}(x)$ into the HJB PDEs (9) and extract terms of degrees $d + 1$ and d respectively to obtain

$$
0 = \frac{\partial \pi^{[d+1]}}{\partial x}(x)\,(F+GK)\,x + \left(\frac{\partial \overline{\pi}}{\partial x}(x)f(x,\overline{\kappa}(x))\right)^{[d+1]} + x'PG\kappa^{[d]}(x)
$$

$$
+ \left(l(x,\overline{\kappa}(x))\right)^{[d+1]} + x'K'R\kappa^{[d]}(x)
$$

$$
0 = \frac{\partial \pi^{[d+1]}}{\partial x}(x)G + \left(\frac{\partial \overline{\pi}}{\partial x}(x)\frac{\partial f}{\partial u}(x,\overline{\kappa}(x))\right)^{[d]}
$$

$$
+ \left(\frac{\partial l}{\partial u}(x,\overline{\kappa}(x))\right)^{[d]} + \left(\kappa^{[d]}(x)\right)'R
$$

where $(\cdot)^{[d]}$ is the degree d part of the enclosed.

Because of (7) $\kappa^{[d]}(x)$ drops out of the first of these equations yielding

$$
0 = \frac{\partial \pi^{[d+1]}}{\partial x}(x)\,(F+GK)\,x + \left(\frac{\partial \overline{\pi}}{\partial x}(x)f(x,\overline{\kappa}(x))\right)^{[d+1]} \tag{13}
$$
$$
+ \left(l(x,\overline{\kappa}(x))\right)^{[d+1]}
$$

Consider the linear operator from degree $d+1$ polynomials to degree $d+1$ polynomials

$$
\pi^{[d+1]}(x) \mapsto \frac{\partial \pi^{[d+1]}}{\partial x}(x)\,(F+GK)\,x
$$

Its eigenvalues are of the form $\lambda_{i_1}+\ldots+\lambda_{i_{d+1}}$ where λ_j is an eigenvalue of $F+GK$. A resonance of degree $d+1$ occurs when such a sum equals zero. But all the eigenvalues of $F+GK$ are in the open left half plane so there are no resonances of degree $d+1$ and we can solve (13) for $\pi^{[d+1]}(x)$. Then the second equation can be solved for $\kappa^{[d]}(x)$

$$
\kappa^{[d]}(x) = -R^{-1}\left(\frac{\partial \pi^{[d+1]}}{\partial x}(x)G + \left(\frac{\partial \overline{\pi}}{\partial x}(x)\frac{\partial f}{\partial u}(x,\overline{\kappa}(x))\right)^{[d]} + \left(\frac{\partial l}{\partial u}(x,\overline{\kappa}(x))\right)^{[d]}\right)' \tag{14}
$$

We have developed MATLAB based software to compute the series solutions to the HJB PDE [11]. In principle the computation can be carried out to any degree in any number of variables but there are practical limitations in execution time and memory. This is the familiar curse of dimensionality. There are $n+d-1$ choose d monomials of degree d in n variables. Still the software is quite fast. For example we are able to solve an HJB PDE in six states and one control to degree six in the optimal cost and degree five in optimal feedback in less than 30 seconds on a five year old laptop (500 MHz) with limited memory (512 MB). There are 462 monomials of degree 6 in 6 variables.

The main problem with the power series approach is that is local in nature. The power series solution to the HJB PDE is very close to the true solution in some neighborhood of the origin. Increasing the degree of the approximation may increase the accuracy but does not necessarily yield a larger domain of validity of the approximation. Complicating this is the fact that in general HJB PDEs do not have globally smooth solutions. The underlying optimal control problem may have conjugate points or focal points. It is for this reason that the theory of viscosity solutions was developed [4], [5].

2   Other Approaches

There are several other approaches to solving HJB PDEs, and a large literature, for example see [3], [6], [13], [9], [10], [16], [18], [19], [21] and their references. One approach is to discretize the underlying optimal control problem and convert it into a nonlinear program in discrete time and space. But the curse of dimensionality rears its ugly head. Consider the optimal control problem generating the above mentioned HJB PDE. If each of the six states is discretized into 10 levels then there would 1,000,000 discrete states.

Other approaches involve discretizing the HJB PDE with subtle tricks so that the algorithm converges to its viscosity solution. This also suffers from the curse of dimensionality. The fast sweeping and marching method (Tsitsiklis [21], Osher et al. [16], [9], [10] and Sethian [19]) are ways to lessen this curse. It takes advantage of the fact that an HJB PDE has characteristics. These are the closed loop optimal state trajectories that converge to the origin as $t \to \infty$. The fast marching method grows the solution out from the origin discrete state by discrete state in reverse time by computing the solution at new discrete states that are on the boundary of the already computed solution.

3   New Approach

The new approach that we are proposing is a extension of the power series method of Al'brecht [1], the Cauchy-Kovalevskaya technique [8], the fast marching method [21], [19] and the patchy technique of Ancona and Bressan [2]. It is similar to that of Navasca and Krener [15]. Suppose we have computed a power series solution to some degree $d + 1$ of an HJB PDE in a neighborhood of the origin by the method of Al'brecht. We verify that this power series solution is valid in some sublevel set of the computed optimal cost function by checking how well it satisfies the HJB PDE on the level set that is its boundary. At the very least it should be a valid Lyapunov function for the dynamics with the computed optimal feedback on the sublevel set. Also the computed closed loop dynamics should point inward on the boundary of the sublevel set, in other words, the computed backward characteristics of the HJB PDE should radiate outward. This sublevel set is called the zeroth patch.

Then we pick a point on the boundary of the zeroth patch and assume the optimal cost and optimal feedback have a power series expansion around that point. We already know the partial derivatives of these in directions tangent to the boundary of the patch. Using a technique similar to that of Cauchy-Kovalevskaya, we can compute the other partial derivatives from the HJB PDE because we have assumed that the computed closed loop dynamics is not tangent to the level set, it points inward. In this way we compute the solution in a patch that overlaps the zeroth patch. Call this the first patch. Again we can estimate the size of this patch by how well the computed solution satisfies the HJB PDE.

It is not essential that the dynamics f and Lagrangian l be smooth at the boundary of zeroth patch (or other patches). If they are not smooth at the boundary we use their derivatives to the outside of the zeroth patch. This is a form of upwind differentiation We do assume that they are smooth at the origin but they can have discontinuities or corners elsewhere. If they do, we choose the patches so that these occur at patch boundaries. In this way it is an upwinding scheme because the closed loop dynamics, the characteristic curves of the PDE point inward on the boundary of the zeroth patch. When computing the solution on the second patch we use the derivative information in the backward characteristic direction.

Then we choose another point that is on the boundary of the zeroth patch but not in the first patch and repeat the process. In this way we grow a series of patches encircling the sublevel set. The validity of the computed solution on each patch is verified via how well it solves the HJB PDE. On the boundary between adjacent patches we may have two possible closed loop vector fields. If the angle between them is obtuse, the two trajectories are diverging, then there is no problem and we can choose either when on the boundary between the patches. If the angle is acute then there may be a sliding regime and another patch in between may be needed. Another possibility is to blend the computed costs across the patch boundary. This will cause a blending of the computed feedback. (These are research questions.)

After the original sublevel set has been completely encircled by new patches we have piecewise smooth approximations to the optimal cost and optimal feedback. We choose a higher sublevel set of the computed cost that is valid for all the patches and repeat the process.

The patches are ordered and the approximate solution to the problem at x is defined to be the approximate solution in the lowest ordered patch containing x.

The patches can also be defined a priori, this would simplify the method but might lead to unsatisfactory solutions if they are chosen too large or long computation times if they are chosen too small.

Of course there is the problem of shocks caused by conjugate or focal points. The assumptions that we make ensure that these do occur at the origin, the true solution is smooth around there. But that does not mean they will not occur elsewhere. When possible we will choose the patches so that they occur at patch boundaries. Not a lot is known about the types of singularities that can occur and how they affect the optimal feedback. One of the goals of our future research project is to better understand these issues.

We expect most of the time to compute the expansions to degree four for the optimal cost and degree three for the optimal feedback. But if the dynamics and/or Lagrangian is not sufficiently smooth we might compute to degrees two and one respectively.

As we noted before in many engineering problems stability of the closed loop dynamics is the principle goal. There may be considerable freedom in choosing the Lagrangian and so a smooth Lagrangian may be chosen. In many problems there are state and/or control constraints. Then the Lagrangian can be chosen so that the solution does not violate the constraints.

In the following sections we discuss the method in more detail.

4   One Dimensional HJB PDEs

For simplicity we consider an optimal control problem (1, 2) where the state dimension $n = 1$ and the control dimension $m = 1$. Occasionally to simplify the calculations we shall assume that the dynamics is affine in the control and the Lagrangian is quadratic in the control

$$f(x, u) = f(x) + g(x)u$$
$$l(x, u) = q(x) + s(x)u + \tfrac{1}{2}r(x)u^2 \tag{1}$$

with $r(x) > 0$. The method works for more general f, l but it is more complicated. In any case we shall assume that $l(x, u) = 0$ iff $x = 0, u = 0$.

We assume that the degree $d + 1$ polynomial $\pi^0(x)$ and the degree d polynomial $\kappa^0(x)$, computed by the power series method of Al'brecht described above, approximately solves this problem in a neighborhood of $x = 0$. We plug the power series expansions of π^0, κ^0 into the right side of the first HJB equation with the exact dynamics f and exact Lagrangian l and compute the local error

$$\rho^0(x) = \frac{\partial \pi^0}{\partial x}(x)f(x, \kappa^0(x)) + l(x.\kappa^0(x)) \tag{2}$$

or relative local error

$$\rho_r^0(x) = \frac{\rho^0(x)}{\pi^0(x).} \tag{3}$$

Of course the local error and some of its derivative will (nearly) vanish at $x = 0$ but it will generally be nonzero for $x \neq 0$. Suppose $\rho_r^0(x)$ is small on some interval $[0, x^1]$ then we accept the power series solution $\pi^0(x), \kappa^0(x)$, on this interval. We would like to continue the solution to the right of x^1. Let $\pi^1(x), \kappa^1(x)$ denote this continued solution. We have an approximation to the optimal cost $\pi^0(x^1)$ and optimal feedback $\kappa^0(x^1)$ at x^1, we accept the former by setting $\pi^1(x^1) = \pi^0(x^1)$ but not the latter. We shall compute $u^1 = \kappa^1(x^1)$.

We evaluate the HJB PDE (9) at x^1 using the assumption (1) to obtain

$$0 = \frac{\partial \pi^1}{\partial x}(x^1)f(x^1, u^1) + q(x^1) + s(x^1)u^1 + \frac{1}{2}r(x^1)\left(u^1\right)^2 \tag{4}$$

$$0 = \frac{\partial \pi^1}{\partial x}(x^1)g(x^1) + s(x^1) + r(x^1)u^1 \tag{5}$$

We can solve the second equation for u^1 and plug it into the first to obtain a quadratic in $\frac{\partial \pi^1}{\partial x}(x^1)$. We set u^1 to be the root nearer to $\kappa^0(x^1)$. In this way we find $\frac{\partial \pi^1}{\partial x}(x^1)$ and u^1.

If assumption (1) does not hold then we must solve a coupled pair of nonlinear equations for the unknowns $\frac{\partial \pi^1}{\partial x}(x^1)$ and u^1. This can be done by a couple of iterations of Newton's method as we already have good starting guesses, $\frac{\partial \pi^0}{\partial x}(x^1)$ and $\kappa^0(x^1)$.

Since we assumed that $l(x, u) = 0$ iff $x = 0, u = 0$ we conclude from (4) that $f(x^1, u^1) \neq 0$.

To find $\frac{\partial^2 \pi^1}{\partial x^2}(x^1)$ and $\frac{\partial \kappa^1}{\partial x}(x^1)$ we proceed as follows. Differentiate the HJB PDEs (9) with respect to x at x^1 to obtain

$$0 = \frac{\partial^2 \pi^1}{\partial x^2}(x^1)f(x^1,u^1) + \frac{\partial \pi^1}{\partial x}(x^1)\left(\frac{\partial f}{\partial x}(x^1,u^1) + \frac{\partial f}{\partial u}(x^1,u^1)\frac{\partial \kappa}{\partial x}(x^1)\right)$$
$$+\frac{\partial l}{\partial x}(x^1,u^1) + \frac{\partial l}{\partial u}(x^1,u^1)\frac{\partial \kappa}{\partial x}(x^1) \tag{6}$$

$$0 = \frac{\partial^2 \pi^1}{\partial x^2}(x^1)\frac{\partial f}{\partial u}(x^1,u^1) + \frac{\partial \pi^1}{\partial x}(x^1)\frac{\partial^2 f}{\partial x \partial u}(x^1,u^1) + \frac{\partial^2 l}{\partial x \partial u}(x^1,u^1)$$
$$+\left(\frac{\partial \pi^1}{\partial x}(x^1)\frac{\partial^2 f}{\partial u^2}(x^1,u^1) + \frac{\partial^2 l}{\partial u^2}(x^1,u^1)\right)\frac{\partial \kappa^1}{\partial x}(x^1) \tag{7}$$

Because of (4), the first equation (6) reduces to

$$0 = \frac{\partial^2 \pi^1}{\partial x^2}(x^1)f(x^1,u^1) + \frac{\partial \pi^1}{\partial x}(x^1)\frac{\partial f}{\partial x}(x^1,u^1) + \frac{\partial l}{\partial x}(x^1,u^1) \tag{8}$$

Notice the unknown $\frac{\partial \kappa^1}{\partial x}(x^1)$ does not appear in this equation so we can easily solve for the unknown $\frac{\partial^2 \pi^1}{\partial x^2}(x^1)$ since $f(x^1,u^1) \neq 0$. Because of the assumptions (1) the second equation reduces to

$$0 = \frac{\partial^2 \pi^1}{\partial x^2}(x^1)g(x^1) + \frac{\partial \pi^1}{\partial x}(x^1)\frac{\partial g}{\partial x}(x^1)$$
$$+\frac{\partial s}{\partial x}(x^1) + \frac{\partial r}{\partial x}(x^1)u^1 + r(x^1)\frac{\partial \kappa^1}{\partial x}(x^1)$$

By assumption $r(x^1) > 0$ so we can solve the second equation for other unknown $\frac{\partial \kappa^1}{\partial x}(x^1)$.

To find the next unknowns $\frac{\partial^3 \pi^1}{\partial x^3}(x^1)$ and $\frac{\partial^2 \kappa^1}{\partial x^2}(x^1)$ we proceed in a similar fashion. We differentiate HJB PDEs (9) twice with respect to x and evaluate at x^1 assuming (1) to obtain two equations,

$$0 = \frac{\partial^3 \pi^1}{\partial x^3}(x^1)f(x^1,u^1) + 2\frac{\partial^2 \pi^1}{\partial x^2}(x^1)\frac{\partial f}{\partial x}(x^1,u^1)$$
$$+\frac{\partial \pi^1}{\partial x}(x^1)\frac{\partial^2 f}{\partial x^2}(x^1,u^1) + \frac{\partial^2 l}{\partial x^2}(x^1,u^1)$$
$$+\left(\frac{\partial^2 \pi^1}{\partial x^2}(x^1)\frac{\partial f}{\partial u}(x^1,u^1) + \frac{\partial \pi^1}{\partial x}(x^1)\frac{\partial^2 f}{\partial x \partial u}(x^1,u^1) + \frac{\partial^2 l}{\partial x \partial u}(x^1,u^1)\right)\frac{\partial \kappa^1}{\partial x}(x^1)$$

$$0 = \frac{\partial^3 \pi^1}{\partial x^3}(x^1)\frac{\partial f}{\partial u}(x^1,u^1) + 2\frac{\partial^2 \pi^1}{\partial x^2}(x^1)\frac{\partial^2 f}{\partial x \partial u}(x^1,u^1)$$
$$+\frac{\partial \pi^1}{\partial x}(x^1)\frac{\partial^3 f}{\partial x^2 \partial u}(x^1,u^1) + \frac{\partial^3 l}{\partial x^2 \partial u}(x^1,u^1)$$
$$+2\left(\frac{\partial^2 \pi^1}{\partial x^2}(x^1)\frac{\partial^2 f}{\partial u^2}(x^1,u^1) + \frac{\partial \pi^1}{\partial x}(x^1)\frac{\partial^3 f}{\partial x \partial u^2}(x^1,u^1) + \frac{\partial^3 l}{\partial x \partial u^2}(x^1,u^1)\right)\frac{\partial \kappa^1}{\partial x}(x^1)$$
$$+\left(\frac{\partial \pi^1}{\partial x}(x^1)\frac{\partial^3 f}{\partial u^3}(x^1,u^1) + \frac{\partial^3 l}{\partial u^3}(x^1,u^1)\right)\left(\frac{\partial \kappa^1}{\partial x}(x^1)\right)^2$$
$$+\left(\frac{\partial \pi^1}{\partial x}(x^1)\frac{\partial^2 f}{\partial u^2}(x^1,u^1) + \frac{\partial^2 l}{\partial u^2}(x^1,u^1)\right)\frac{\partial^2 \kappa^1}{\partial x^2}(x^1)$$

The unknown $\frac{\partial^2 \kappa^1}{\partial x^2}(x^1)$ does not appear in the first equation because of (4). Since $f(x^1, u^1) \neq 0$ we can solve this equation for the unknown $\frac{\partial^3 \pi^1}{\partial x^3}(x^1)$ The second is linear in both unknowns. Under the assumptions (1) the second equation reduces to

$$
0 = \frac{\partial^3 \pi^1}{\partial x^3}(x^1)g(x^1) + 2\frac{\partial^2 \pi^1}{\partial x^2}(x^1)\frac{\partial g}{\partial x}(x^1)
$$
$$
+\frac{\partial \pi^1}{\partial x}(x^1)\frac{\partial^2 g}{\partial x^2}(x^1) + \frac{\partial^2 s}{\partial x^2}(x^1) + \frac{\partial^2 r}{\partial x^2}(x^1)u^1
$$
$$
+2\frac{\partial r}{\partial x}(x^1)\frac{\partial \kappa^1}{\partial x}(x^1) + r(x^1)\frac{\partial^2 \kappa^1}{\partial x^2}(x^1)
$$

and because $r(x^1) > 0$ it is readily solvable for the other unknown $\frac{\partial^2 \kappa^1}{\partial x^2}(x^1)$.

To find the next unknowns $\frac{\partial^4 \pi^1}{\partial x^4}(x^1)$ and $\frac{\partial^3 \kappa^1}{\partial x^3}(x^1)$ we differentiate HJB PDE (9) three times with respect to x and evaluate at x^1 assuming (1) to obtain the two equations,

$$
0 = \frac{\partial^4 \pi^1}{\partial x^4}(x^1)f(x^1, u^1)) + 3\frac{\partial^3 \pi^1}{\partial x^3}(x^1)\frac{\partial f}{\partial x}(x^1, u^1) + 3\frac{\partial^2 \pi^1}{\partial x^2}(x^1)\frac{\partial^2 f}{\partial x^2}(x^1, u^1)
$$
$$
+\frac{\partial \pi^1}{\partial x}(x^1)\frac{\partial^3 f}{\partial x^3}(x^1, u^1) + \frac{\partial^3 l}{\partial x^3}(x^1, u^1)
$$
$$
+2\left(\frac{\partial^3 \pi^1}{\partial x^3}(x^1)\frac{\partial f}{\partial u}(x^1, u^1) + 2\frac{\partial^2 \pi^1}{\partial x^2}(x^1)\frac{\partial^2 f}{\partial x \partial u}(x^1, u^1) \right.
$$
$$
\left. +\frac{\partial \pi^1}{\partial x}(x^1)\frac{\partial^3 f}{\partial x^2 \partial u}(x^1, u^1) + \frac{\partial^3 l}{\partial x^2 \partial u}(x^1, u^1) \right) \frac{\partial \kappa^1}{\partial x}(x^1)
$$
$$
+\left(\frac{\partial^2 \pi^1}{\partial x^2}(x^1)\frac{\partial^2 f}{\partial u^2}(x^1, u^1) + \frac{\partial \pi^1}{\partial x}(x^1)\frac{\partial^3 f}{\partial x \partial u^2}(x^1, u^1) \right.
$$
$$
\left. +\frac{\partial^3 l}{\partial x \partial u^2}(x^1, u^1) \right) \left(\frac{\partial \kappa}{\partial x}(x^1) \right)^2
$$
$$
+\left(\frac{\partial^2 \pi^1}{\partial x^2}(x^1)\frac{\partial f}{\partial u}(x^1, u^1) + \frac{\partial \pi^1}{\partial x}(x^1)\frac{\partial^2 f}{\partial x \partial u}(x^1, u^1) \right.
$$
$$
\left. +\frac{\partial^2 l}{\partial x \partial u}(x^1, u^1) \right) \frac{\partial^2 \kappa^1}{\partial x^2}(x^1)
$$

$$
0 = \frac{\partial^4 \pi^1}{\partial x^4}(x^1)\frac{\partial f}{\partial u}(x^1, u^1) + 3\frac{\partial^3 \pi^1}{\partial x^3}(x^1)\frac{\partial^2 f}{\partial x \partial u}(x^1, u^1) + 3\frac{\partial^2 \pi^1}{\partial x^2}(x^1)\frac{\partial^3 f}{\partial x^2 \partial u}(x^1, u^1)
$$
$$
+\frac{\partial \pi^1}{\partial x}(x^1)\frac{\partial^4 f}{\partial x^3 \partial u}(x^1, u^1) + \frac{\partial^4 l}{\partial x^3 \partial u}(x^1, u^1)
$$
$$
+3\left(\frac{\partial^3 \pi^1}{\partial x^3}(x^1)\frac{\partial^2 f}{\partial u^2}(x^1, u^1) + 2\frac{\partial^2 \pi^1}{\partial x^2}(x^1)\frac{\partial^3 f}{\partial x \partial u^2}(x^1, u^1) \right.
$$
$$
\left. +\frac{\partial \pi^1}{\partial x}(x^1)\frac{\partial^4 f}{\partial x^2 \partial u^2}(x^1, u^1) + \frac{\partial^4 l}{\partial x^2 \partial u^2}(x^1, u^1) \right) \frac{\partial \kappa^1}{\partial x}(x^1)
$$
$$
+3\left(\frac{\partial^2 \pi^1}{\partial x^2}(x^1)\frac{\partial^2 f}{\partial u^2}(x^1, u^1) + \frac{\partial \pi^1}{\partial x}(x^1)\frac{\partial^3 f}{\partial x \partial u^2}(x^1, u^1) + \frac{\partial^3 l}{\partial x \partial u^2}(x^1, u^1) \right) \frac{\partial^2 \kappa^1}{\partial x^2}(x^1)
$$

$$+3\left(\frac{\partial^2\pi^1}{\partial x^2}(x^1)\frac{\partial^3 f}{\partial u^3}(x^1,u^1)+\frac{\partial\pi^1}{\partial x}(x^1)\frac{\partial^4 f}{\partial x\partial u^3}(x^1,u^1)+\frac{\partial^4 l}{\partial x\partial u^3}(x^1,u^1)\right)\left(\frac{\partial\kappa^1}{\partial x}(x^1)\right)^2$$

$$+3\left(\frac{\partial\pi^1}{\partial x}(x^1)\frac{\partial^3 f}{\partial u^3}(x^1,u^1)+\frac{\partial^3 l}{\partial u^3}(x^1,u^1)\right)\frac{\partial\kappa^1}{\partial x}(x^1)\frac{\partial^2\kappa^1}{\partial x^2}(x^1)$$

$$+\left(\frac{\partial\pi^1}{\partial x}(x^1)\frac{\partial^4 f}{\partial u^4}(x^1,u^1)+\frac{\partial^4 l}{\partial u^4}(x^1,u^1)\right)\left(\frac{\partial\kappa^1}{\partial x}(x^1)\right)^3$$

$$+\left(\frac{\partial\pi^1}{\partial x}(x^1)\frac{\partial^2 f}{\partial u^2}(x^1,u^1)+\frac{\partial^2 l}{\partial u^2}(x^1,u^1)\right)\frac{\partial^3\kappa^1}{\partial x^3}(x^1)$$

We expect to stop at degree four most of the time, The assumptions (1) greatly simplify the last equation,

$$0=\frac{\partial^4\pi^1}{\partial x^4}(x^1)g(x^1)+3\frac{\partial^3\pi^1}{\partial x^3}(x^1)\frac{\partial g}{\partial x}(x^1)+3\frac{\partial^2\pi^1}{\partial x^2}(x^1)\frac{\partial^2 g}{\partial x^2}(x^1)$$
$$+\frac{\partial\pi^1}{\partial x}(x^1)\frac{\partial^3 g}{\partial x^3}(x^1)+\frac{\partial^3 s}{\partial x^3}(x^1)+\frac{\partial^3 r}{\partial x^3}(x^1)u^1+r(x^1)\frac{\partial^3\kappa^1}{\partial x^3}(x^1)$$

Notice the similarities with Al'brecht's method.   We successively solve for $\frac{\partial^{d+1}\pi}{\partial x^{d+1}}(x^1)$ and $\frac{\partial^d\kappa^1}{\partial x^d}(x^1)$ for $d=0,1,2,\dots$. At the lowest level the equations are coupled and if (1) holds we must solve a quadratic equation similar to a Riccati equation. At the higher levels the equations are linear and triangular in the unknowns.

Once we have computed a satisfactory approximate solution on the interval $[x^1,x^2]$ we can repeat the process and find an approximate solution to the right of x^2.

5   One Dimensional Example

Consider the simple LQR of minimizing

$$\frac{1}{2}\int_0^\infty z^2+u^2\,dt$$

subject to

$$\dot z=z+u$$

Here both z and u are one dimensional.

The Riccati equation (5) is

$$0=2P+1-P^2$$

and its unique nonnegative solution is $P=1+\sqrt{2}$. Therefore the optimal cost and optimal feedback are

$$\pi(z)=\frac{1+\sqrt{2}}{2}z^2$$
$$\kappa(z)=-(1+\sqrt{2})z$$

The optimal closed loop dynamics is

$$\dot{z} = -\sqrt{2}\,z$$

After the change of coordinates

$$z = \sin x$$

then the LQR become the nonlinear optimal control problem of minimizing

$$\frac{1}{2}\int_0^\infty \sin^2 x + u^2\,dt$$

subject to

$$\dot{x} = \frac{\sin x + u}{\cos x}$$

We know that the optimal cost and optimal feedback is

$$\pi(x) = \frac{1+\sqrt{2}}{2}\sin^2 x$$
$$\kappa(z) = -(1+\sqrt{2})\sin x$$

Notice that the optimal cost is even and the optimal feedback is odd. We can compare it with the solution computed by the method described above.

The computed solution on the interval $[0, 0.9]$ is the one of Al'brecht. As we compute the solution for larger x, the size of the patches decreases because the change of coordinates is becoming more nearly singular as we approach $\frac{\pi}{2}$. There are 15 patches. The relative error tolerance is 0.5.

6   HJB PDEs in Higher Dimensions

In this section we generalize the proposed scheme to higher dimensional state spaces $n \geq 1$. For notational simplicity we shall assume that the control is one dimensional $m = 1$, generalizing to higher control dimensions causes no conceptual difficulty. We also make the simplifying assumptions that the dynamics is affine in the control and the cost is quadratic in the control of the form

$$\dot{x} = f(x) + g(x)u$$
$$l(x, u) = q(x) + r(x)u^2/2$$

The method does not require these assumptions but they do greatly simplify it.

Suppose we have computed the Al'brecht solution $\pi^0(x)$, $\kappa^0(x)$ to the HJB PDE (9) in some neighborhood of the origin. We check the local error $\rho^0(x)$ (2) or relative local error (3) and decide that it is a reasonable solution in some sublevel set $\{x : \pi^0(x) \leq c\}$ which we call the zeroth patch $\mathcal{P}^0$. We choose x^1 on the level set $\pi(x^1) = c$ and seek to extend the solution in a patch around x^1. To do so we need to estimate the low degree partial derivatives of the optimal cost and optimal feedback at x^1.

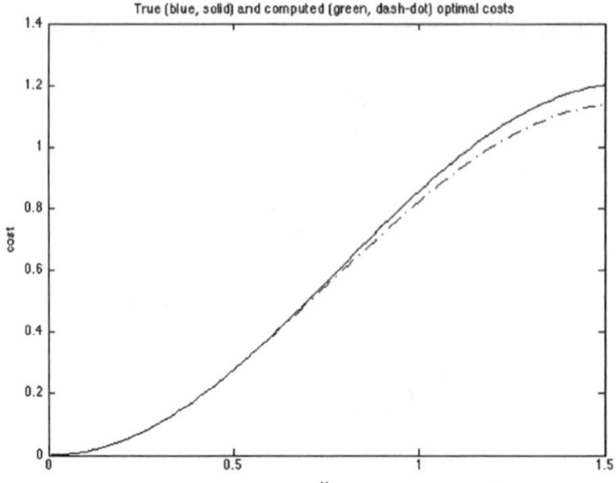

Fig. 1. True cost (solid) and the computed cost (dash-dot)

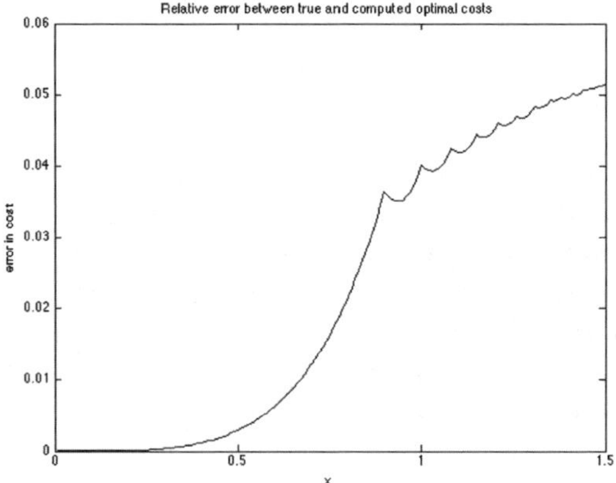

Fig. 2. Relative error between true cost and the computed cost

We assume that the Al'brecht closed loop dynamics is transverse to the boundary of the sublevel set and points inward

$$\frac{\partial \pi^0}{\partial x}(x^1)f(x^1, \kappa^0(x^1)) < 0$$

We accept that $\pi^1(x^1) = \pi^0(x^1)$ but we will compute a new $u^1 = \kappa^1(x^1)$ probably different from $\kappa^0(x^1)$.

The HJB equations become

$$0 = \frac{\partial \pi}{\partial x_\sigma}(x)\left(f_\sigma(x) + g_\sigma(x)\kappa(x)\right) + q(x) + r(x)\left(\kappa(x)\right)^2/2 \tag{1}$$

$$0 = \frac{\partial \pi}{\partial x_\sigma}(x)g_\sigma(x) + r(x)\kappa(x) \tag{2}$$

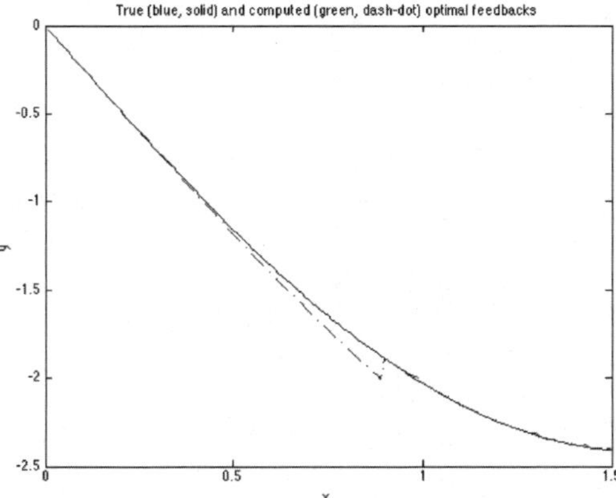

Fig. 3. True feedback (solid) and the computed feedback (dash-dot)

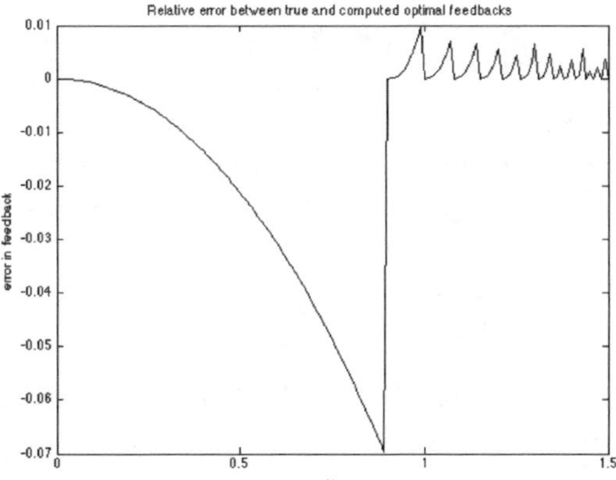

Fig. 4. Relative error between the true feedback and the computed feedback

We choose the index k that maximizes

$$|f_k(x^1) + g_k(x^1)\kappa^0(x^1)|$$

For notational convenience we assume that $k = n$.

We assume that

$$\pi^1(x^1) = \pi^0(x^1)$$
$$\frac{\partial \pi^1}{\partial x_\sigma}(x^1) = \frac{\partial \pi^0}{\partial x_\sigma}(x^1)$$

for $1 \le \sigma < n$. Then we can solve the second HJB equation for $\kappa(x^1)$ and plug it into the first to get a quadratic equation in the other unknown

$$0 = a\left(\frac{\partial \pi^1}{\partial x_n}(x^1)\right)^2 + b\frac{\partial \pi^1}{\partial x_n}(x^1) + c$$

264   C. Navasca and A.J. Krener

where

$$a = \frac{1}{2r(x^1)}(g_n(x^1))^2$$

$$b = \frac{1}{r(x^1)}g_n(x^1)\sum_{\sigma=1}^{n-1}\frac{\partial \pi^1}{\partial x_\sigma}(x^1)g_\sigma(x) - f_n(x^1)$$

$$c = \frac{1}{2r(x^1)}\sum_{\sigma=1}^{n-1}\sum_{\tau=1}^{n-1}\frac{\partial \pi^1}{\partial x_\sigma}(x^1)g_\sigma(x^1)\frac{\partial \pi^1}{\partial x_\tau}(x^1)g_\tau(x^1)$$

$$-q(x^1) - \sum_{\sigma=1}^{n-1}\frac{\partial \pi^1}{\partial x_\sigma}(x^1)f_\sigma(x^1)$$

Assuming this equation has real roots, we set $\frac{\partial \pi^1}{\partial x_n}(x^1)$ to be the root closest to $\frac{\partial \pi^0}{\partial x_n}(x^1)$ and we solve for $\kappa(x^1)$,

$$\kappa(x^1) = -\frac{1}{r(x^1)}\sum_{\sigma=1}^{n}\frac{\partial \pi^1}{\partial x_\sigma}(x^1)g_\sigma(x^1)$$

The next unknowns in a power series expansion of the optimal cost and feedback around x^1 are $\frac{\partial^2 \pi^1}{\partial x_i \partial x_j}(x^1)$ and $\frac{\partial \kappa^1}{\partial x_i}(x^1)$ for $1 \le i \le j \le n$. We assume that

$$\frac{\partial^2 \pi^1}{\partial x_i \partial x_j}(x^1) = \frac{\partial^2 \pi^0}{\partial x_i \partial x_j}(x^1)$$

for $1 \le i \le j \le n-1$ and we take the partials of (1, 2) with respect to x_i to obtain $2n$ equations

$$0 = \frac{\partial^2 \pi^1}{\partial x_i \partial x_\sigma}(x^1)\left(f_\sigma(x^1) + g_\sigma(x^1)\kappa^1(x^1)\right) \tag{3}$$

$$+\frac{\partial \pi^1}{\partial x_\sigma}(x^1)\left(\frac{\partial f_\sigma}{\partial x_i}(x^1) + \frac{\partial g_\sigma}{\partial x_i}(x^1)\kappa^1(x^1)\right)$$

$$\frac{\partial q}{\partial x_i}(x^1) + \frac{1}{2}\frac{\partial r}{\partial x_i}(x^1)(\kappa^1(x^1))^2$$

$$0 = \frac{\partial^2 \pi^1}{\partial x_i \partial x_\sigma}(x^1)g_\sigma(x^1) + \frac{\partial \pi^1}{\partial x_\sigma}(x^1)\frac{\partial g_\sigma}{\partial x_i}(x^1) \tag{4}$$

$$+\frac{\partial r}{\partial x_i}(x^1)\kappa^1(x^1) + r(x^1)\frac{\partial \kappa^1}{\partial x_i}(x^1)$$

for the remaining $2n$ unknowns. Because of the second HJB equation (2), the first n equations do not contain the unknowns $\frac{\partial \kappa^1}{\partial x_i}(x^1)$ for $1 \le i \le n$. Moreover the first n equations are decoupled and can be solved one by one

$$\frac{\partial^2 \pi^1}{\partial x_i \partial x_n}(x^1) = \frac{-1}{f_n(x^1) + g_n(x^1)\kappa^1(x^1)}$$
$$\times \left(\sum_{\sigma=1}^{n-1} \frac{\partial^2 \pi^1}{\partial x_i \partial x_\sigma}(x^1) \left(f_\sigma(x^1) + g_\sigma(x^1)\kappa^1(x^1) \right) \right.$$
$$+ \frac{\partial \pi^1}{\partial x_\sigma}(x^1) \left(\frac{\partial f_\sigma}{\partial x_i}(x^1) + \frac{\partial g_\sigma}{\partial x_i}(x^1)\kappa^1(x^1) \right)$$
$$\left. \frac{\partial q}{\partial x_i}(x^1) + \frac{1}{2}\frac{\partial r}{\partial x_i}(x^1)(\kappa^1(x^1))^2 \right)$$

We invoke the summation convention when the range of the sum is from 1 to n, otherwise we explicitly show the sum.

The remaining n equations are also solvable one by one,

$$\frac{\partial \kappa^1}{\partial x_i}(x^1) = \frac{-1}{r(x^1)} \left(\frac{\partial^2 \pi^1}{\partial x_i \partial x_\sigma}(x^1) g_\sigma(x^1) + \frac{\partial \pi^1}{\partial x_\sigma}(x^1)\frac{\partial g_\sigma}{\partial x_i}(x^1) + \frac{\partial r}{\partial x_i}(x^1)\kappa^1(x^1) \right)$$

Next we find the third partials of π^1 at x^1. We assume that

$$\frac{\partial^3 \pi^1}{\partial x_i \partial x_j \partial x_k}(x^1) = \frac{\partial^3 \pi^0}{\partial x_i \partial x_j \partial x_k}(x^1)$$

for $1 \leq i \leq j \leq k \leq n - 1$. Equations for the other third partials are obtained by differentiating the first HJB equation (1) with respect to x_i and x_j for $1 \leq i \leq j \leq n$ and evaluating at x^1 yielding

$$0 = \frac{\partial^3 \pi^1}{\partial x_i \partial x_j \partial x_\sigma}(x^1) \left(f_\sigma(x^1) + g_\sigma(x^1)\kappa^1(x^1) \right) \tag{5}$$
$$+ \frac{\partial^2 \pi^1}{\partial x_i \partial x_\sigma}(x^1) \left(\frac{\partial f_\sigma}{\partial x_j}(x^1) + \frac{\partial g_\sigma}{\partial x_j}(x^1)\kappa^1(x^1) \right)$$
$$+ \frac{\partial^2 \pi^1}{\partial x_j \partial x_\sigma}(x^1) \left(\frac{\partial f_\sigma}{\partial x_i}(x^1) + \frac{\partial g_\sigma}{\partial x_i}(x^1)\kappa^1(x^1) \right)$$
$$+ \frac{\partial \pi^1}{\partial x_\sigma}(x^1) \left(\frac{\partial^2 f_\sigma}{\partial x_i \partial x_j}(x^1) + \frac{\partial^2 g_\sigma}{\partial x_i \partial x_j}(x^1)\kappa^1(x^1) \right)$$
$$+ \frac{\partial^2 q}{\partial x_i \partial x_j}(x^1) + \frac{1}{2}\frac{\partial^2 r}{\partial x_i \partial x_j}(x^1)(\kappa^1(x^1))^2$$
$$- r(x^1)\frac{\partial \kappa^1}{\partial x_i}(x^1)\frac{\partial \kappa^1}{\partial x_j}(x^1)$$

These are $(n + 1)n/2$ equations in the $(n + 1)n/2$ unknowns $\frac{\partial^3 \pi^1}{\partial x_i \partial x_j \partial x_n}(x^1)$ for $1 \leq i \leq j \leq n$. They can be solved one by one in lexographic order. The unknowns $\frac{\partial^2 \kappa^1}{\partial x_i \partial x_j}(x^1)$ do not appear because of (2) and they are simplified by (4).

Then we differentiate the second HJB equation (3) with respect to x_i and x_j for $1 \leq i \leq j \leq n$ to obtain the $(n + 1)n/2$ equations

$$0 = \frac{\partial^3 \pi^1}{\partial x_i \partial x_j \partial x_\sigma}(x^1) g_\sigma(x^1) + \frac{\partial^2 \pi^1}{\partial x_i \partial x_\sigma}(x^1) \frac{\partial g_\sigma}{\partial x_j}(x^1) \tag{6}$$

$$+ \frac{\partial^2 \pi^1}{\partial x_j \partial x_\sigma}(x^1) \frac{\partial g_\sigma}{\partial x_i}(x^1) + \frac{\partial \pi^1}{\partial x_\sigma}(x^1) \frac{\partial^2 g_\sigma}{\partial x_i \partial x_j}(x^1)$$

$$+ \frac{\partial^2 r}{\partial x_i \partial x_j}(x^1) \kappa^1(x^1) + \frac{\partial r}{\partial x_i}(x^1) \frac{\partial \kappa^1}{\partial x_j}(x^1)$$

$$+ \frac{\partial r}{\partial x_j}(x^1) \frac{\partial \kappa^1}{\partial x_i}(x^1) + r(x^1) \frac{\partial^2 \kappa^1}{\partial x_i \partial x_j}(x^1)$$

which can be solved one by one for the $(n+1)n/2$ unknowns $\frac{\partial^2 \kappa^1}{\partial x_i \partial x_j}(x^1)$, $1 \leq i \leq j \leq n$.

To find the fourth partials of π^1 at x^1, we assume that

$$\frac{\partial^4 \pi^1}{\partial x_i \partial x_j \partial x_k \partial x_l}(x^1) = \frac{\partial^4 \pi^0}{\partial x_i \partial x_j \partial x_k \partial x_l}(x^1)$$

for $1 \leq i \leq j \leq k \leq l \leq n-1$. We differentiate the first HJB equation (1) with respect to x_i, x_j, x_k to obtain

$$0 = \frac{\partial^4 \pi^1}{\partial x_i \partial x_j \partial x_k \partial x_\sigma}(x^1) \left(f_\sigma(x^1) + g_\sigma(x^1)\kappa^1(x^1) \right) \tag{7}$$

$$+ \frac{\partial^3 \pi^1}{\partial x_i \partial x_j \partial x_\sigma}(x^1) \left(\frac{\partial f_\sigma}{\partial x_k}(x^1) + \frac{\partial g_\sigma}{\partial x_k}(x^1)\kappa^1(x^1) \right)$$

$$+ \frac{\partial^3 \pi^1}{\partial x_i \partial x_k \partial x_\sigma}(x^1) \left(\frac{\partial f_\sigma}{\partial x_j}(x^1) + \frac{\partial g_\sigma}{\partial x_j}(x^1)\kappa^1(x^1) \right)$$

$$+ \frac{\partial^3 \pi^1}{\partial x_j \partial x_k \partial x_\sigma}(x^1) \left(\frac{\partial f_\sigma}{\partial x_i}(x^1) + \frac{\partial g_\sigma}{\partial x_i}(x^1)\kappa^1(x^1) \right)$$

$$+ \frac{\partial^2 \pi^1}{\partial x_i \partial x_\sigma}(x^1) \left(\frac{\partial^2 f_\sigma}{\partial x_j \partial x_k}(x^1) + \frac{\partial^2 g_\sigma}{\partial x_j \partial x_k}(x^1)\kappa^1(x^1) \right)$$

$$+ \frac{\partial^2 \pi^1}{\partial x_j \partial x_\sigma}(x^1) \left(\frac{\partial^2 f_\sigma}{\partial x_i \partial x_k}(x^1) + \frac{\partial^2 g_\sigma}{\partial x_i \partial x_k}(x^1)\kappa^1(x^1) \right)$$

$$+ \frac{\partial^2 \pi^1}{\partial x_k \partial x_\sigma}(x^1) \left(\frac{\partial^2 f_\sigma}{\partial x_i \partial x_j}(x^1) + \frac{\partial^2 g_\sigma}{\partial x_i \partial x_j}(x^1)\kappa^1(x^1) \right)$$

$$+ \frac{\partial \pi^1}{\partial x_\sigma}(x^1) \left(\frac{\partial^3 f_\sigma}{\partial x_i \partial x_j \partial x_k}(x^1) + \frac{\partial^3 g_\sigma}{\partial x_i \partial x_j \partial x_k}(x^1)\kappa^1(x^1) \right)$$

$$+ \frac{\partial^3 q}{\partial x_i \partial x_j \partial x_k}(x^1) + \frac{1}{2} \frac{\partial^3 r}{\partial x_i \partial x_j \partial x_k}(x^1)(\kappa^1(x^1))^2$$

$$- \frac{\partial r}{\partial x_i}(x^1) \frac{\partial \kappa^1}{\partial x_j}(x^1) \frac{\partial \kappa^1}{\partial x_k}(x^1)$$

$$- \frac{\partial r}{\partial x_j}(x^1) \frac{\partial \kappa^1}{\partial x_i}(x^1) \frac{\partial \kappa^1}{\partial x_k}(x^1)$$

$$-\frac{\partial r}{\partial x_k}(x^1)\frac{\partial \kappa^1}{\partial x_i}(x^1)\frac{\partial \kappa^1}{\partial x_j}(x^1)$$

$$-r(x^1)\frac{\partial^2 \kappa^1}{\partial x_i \partial x_j}(x^1)\frac{\partial \kappa^1}{\partial x_k}(x^1)$$

$$-r(x^1)\frac{\partial^2 \kappa^1}{\partial x_i \partial x_k}(x^1)\frac{\partial \kappa^1}{\partial x_j}(x^1)$$

$$-r(x^1)\frac{\partial^2 \kappa^1}{\partial x_j \partial x_k}(x^1)\frac{\partial \kappa^1}{\partial x_i}(x^1)$$

These $(n+2)(n+1)n/6$ equations can be solved one by one in lexograhic order for the $(n+2)(n+1)n/6$ unknowns $\frac{\partial^4 \pi^1}{\partial x_i \partial x_j \partial x_k \partial x_n}(x^1)$ for $1 \le i \le j \le k \le n$. The unknowns $\frac{\partial^3 \kappa^1}{\partial x_i \partial x_j \partial x_k}(x^1)$ do not appear because of (2) and they are simplified by (4) and (7).

Then we differentiate the second HJB equation (3) with respect to x_i, x_j, x_k for $1 \le i \le j \le n$ to obtain the $(n+2)(n+1)n/6$ equations

$$0 = \frac{\partial^4 \pi^1}{\partial x_i \partial x_j \partial x_k \partial x_\sigma}(x^1)g_\sigma(x^1) \tag{8}$$

$$+\frac{\partial^3 \pi^1}{\partial x_i \partial x_j \partial x_\sigma}(x^1)\frac{\partial g_\sigma}{\partial x_k}(x^1)$$

$$+\frac{\partial^3 \pi^1}{\partial x_i \partial x_k \partial x_\sigma}(x^1)\frac{\partial g_\sigma}{\partial x_j}(x^1)$$

$$+\frac{\partial^3 \pi^1}{\partial x_j \partial x_k \partial x_\sigma}(x^1)\frac{\partial g_\sigma}{\partial x_i}(x^1)$$

$$+\frac{\partial^2 \pi^1}{\partial x_i x_\sigma}(x^1)\frac{\partial^2 g_\sigma}{\partial x_j \partial x_k}(x^1)$$

$$+\frac{\partial^2 \pi^1}{\partial x_j x_\sigma}(x^1)\frac{\partial^2 g_\sigma}{\partial x_i \partial x_k}(x^1)$$

$$+\frac{\partial^2 \pi^1}{\partial x_k x_\sigma}(x^1)\frac{\partial^2 g_\sigma}{\partial x_i \partial x_j}(x^1)$$

$$+\frac{\partial \pi^1}{\partial x_\sigma}(x^1)\frac{\partial^3 g_\sigma}{\partial x_i \partial x_j \partial x_k}(x^1)$$

$$+\frac{\partial^3 r}{\partial x_i \partial x_j \partial x_k}(x^1)\kappa^1(x^1)$$

$$+\frac{\partial^2 r}{\partial x_i \partial x_j}(x^1)\frac{\partial \kappa^1}{\partial x_k}(x^1)$$

$$+\frac{\partial^2 r}{\partial x_i \partial x_k}(x^1)\frac{\partial \kappa^1}{\partial x_j}(x^1)$$

$$+\frac{\partial^2 r}{\partial x_j \partial x_k}(x^1)\frac{\partial \kappa^1}{\partial x_i}(x^1)$$

$$+\frac{\partial r}{\partial x_i}(x^1)\frac{\partial^2\kappa^1}{\partial x_j\partial x_k}(x^1)$$

$$+\frac{\partial r}{\partial x_j}(x^1)\frac{\partial^2\kappa^1}{\partial x_i\partial x_k}(x^1)$$

$$+\frac{\partial r}{\partial x_k}(x^1)\frac{\partial^2\kappa^1}{\partial x_i\partial x_j}(x^1)$$

$$+r(x^1)\frac{\partial^3\kappa^1}{\partial x_i\partial x_j\partial x_k}(x^1)$$

which can be solved one by one for the $(n+2)(n+1)n/6$ unknowns $\frac{\partial^3\kappa^1}{\partial x_i\partial x_j\partial x_k}(x^1)$, $1\le i\le j\le k\le n$.

7   Two Dimensional Example

We consider the optimal control problem of driving a planar pendulum of length 1 and mass 1 to the upright condition by a torque u at its pivot. The dynamics is

$$\dot{x}_1 = x_2$$
$$\dot{x}_2 = \sin x_1 + u$$

We choose the Lagrangian

$$l(x,u) = \frac{1}{2}\left(|x|^2 + u^2\right)$$

We computed the Al'brecht solution around the origin to degree 4 in the cost and degree 3 in the optimal feedback. We accepted it on the sublevel set $\pi^0(x) \le 0.5$. Then using the method described above we computed the solution at four points in the eigenspaces of the quadratic part of the cost where $\pi^0(x) = 0.5$. There is one in each quadrant. These outer solutions were also computed to degree 4 in the cost and degree 3 in the feedback.

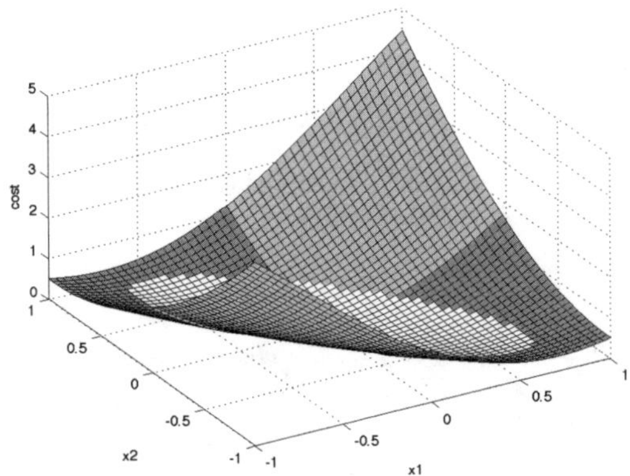

Fig. 5. Optimal cost computed on five patches. The outer patches are bounded in part by the axes.

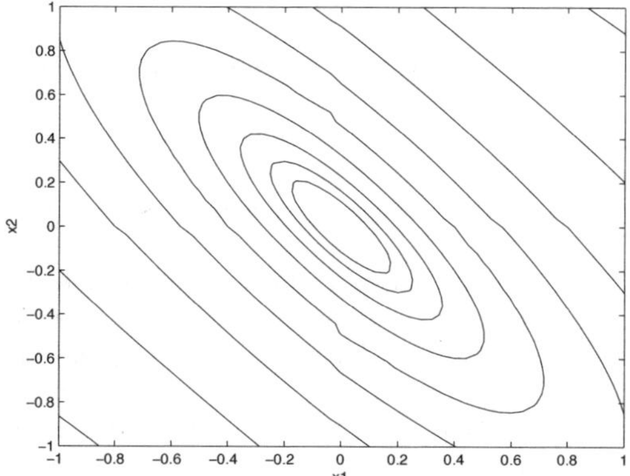

Fig. 6. Contour plot of five patch cost. The inner 4 contours are within the central patch. Notice that there is a slight mismatch of the outer contours when they meet at the axes.

8   Conclusion

We have sketched out a patchy approach to solving Hamilton Jacobi Bellman equations for nice optimal control problems and applied it to one and two dimensional examples. We were deliberately vague about some aspects of the proposed algorithm such as how to choose the boundary between outer patches. Further research is needed to clarify these issues and this can come only with extensive computation.

References

1. E. G. Al'brecht, *On the optimal stabilization of nonlinear systems*, PMM-J. Appl. Math. Mech., 25:1254-1266, 1961.
2. F. Ancona and A. Bressan, *Nearly Time Optimal Stabilizing Patchy Feedbacks*, preprint available at http://cpde.iac.rm.cnr.it/preprint.php
3. M. Bardi and I. Capuzzo-Dolcetta *Optimal Control and Viscosity Solutions of Hamilton-Jacobi-Bellman Equations*, Birkhäuser, Boston, 1997.
4. M. G. Crandall and P. L. Lions *Viscosity Solutions of Hamilton-Jacobi Equations*, Transactions of the American Mathematical Society, 227:1–42, 1983.
5. L. C. Evans, *Partial Differential Equations*. American Mathematical Society, Providence, 1998.
6. W. H. Fleming and H. M. Soner, *Controlled Markov Processes and Viscosity Solutions*. Springer-Verlag, New York, 1992.
7. W. L. Garrard and J. M. Jordan. *Design of nonlinear automatic flight control systems*, Automatica, 13:497-505, 1977.
8. F. John, *Partial Differential Equations*. Springer-Verlag, New York, 1982.
9. C. Y. Kao, S. Osher and Y. H. Tsai. *Fast Sweeping Methods for Hamilton-Jacobi Equations*, SIAM J. Numerical Analysis, 42:2612–2632, 2005.
10. C. Y. Kao, S. Osher and J. Qian. *Lax-Friedrichs Sweeping Scheme for Static Hamilton-Jacobi Equations*, J. Computational Physics, 196:367–391, 2004.

11. A. J. Krener. *Nonlinear Systems Toolbox V. 1.0, 1997*, MATLAB based toolbox available by request from ajkrener@ucdavis.edu
12. A. J. Krener. *The existence of optimal regulators*, Proc. of 1998 CDC, Tampa, FL, 3081–3086.
13. H. J. Kushner and P. G. Dupuis, *Numerical Methods for Stochastic Control Problems in Continuous Time*, Springer-Verlag, New York, 1992.
14. D. L. Lukes. *Optimal regulation of nonlinear dynamical systems*, SIAM J. Contr., 7:75–100, 1969.
15. C. L. Navasca and A. J. Krener. *Solution of Hamilton Jacobi Bellman Equations*, Proceedings of the IEEE Conference on Decision and Control, Sydney, 2000, pp. 570-574.
16. S. Osher and C. W. Shu. *High-order Essentially Nonoscillatory Schemes for Hamilton Jacobi Equations*, SIAM J. Numerical Analysis, 28:907-922, 1991.
17. H. M. Osinga and J. Hauser. *The geometry of the solution set of nonlinear optimal control problems* , to appear in Journal of Dynamics and Differential Equations.
18. W. Prager. *Numerical computation of the optimal feedback law for nonlinear infinite horizon control problems*, CALCOLO, 37:97-123, 2000.
19. J. A. Sethian. *Level Set Methods and Fast Marching Methods*, Cambridge University Press, 1999.
20. B. F. Spencer Jr., T. L. Timlin, M. K. Sain and S. J. Dyke. *Series solution of a class of nonlinear regulators*, Journal of Optimization Theory and Applications, 91:321-345, 1996.
21. J. Tsitsiklis, *Efficient algorithms for globally optimal trajectories* IEEE Trans. Auto. Con., 40:1528-1538, 1995.
22. T. Yoshida and K. A. Loparo. *Quadratic regulator theory for analytic non-linear systems with additive controls*, Automatica 25:531-544, 1989.

A Geometric Assignment Problem for Robotic Networks

Stephen L. Smith and Francesco Bullo

Department of Mechanical Engineering
Center for Control, Dynamical Systems and Computation
University of California, Santa Barbara, CA 93106-5070, USA
stephen@engineering.ucsb.edu, bullo@engineering.ucsb.edu

Summary. In this chapter we look at a geometric target assignment problem consisting of an equal number of mobile robotic agents and distinct target locations. Each agent has a fixed communication range, a maximum speed, and knowledge of every target's position. The problem is to devise a distributed algorithm that allows the agents to divide the target locations among themselves and, simultaneously, leads each agent to its unique target. We summarize two algorithms for this problem; one designed for "sparse" environments, in which communication between robots is sparse, and one for "dense" environments, where communication is more prevalent. We characterize the asymptotic performance of these algorithms as the number of agents increases and the environment grows to accommodate them.

1 Introduction

Consider a group of n mobile robotic agents, equipped with wireless transceivers for limited range communication, dispersed in an environment $\mathcal{E} \subset \mathbb{R}^2$. Suppose the environment also contains n target locations, and each agent is given a list containing their positions (these positions may be given as GPS coordinates). We would like each target location to be occupied be an agent as quickly as possible. Since no *a priori* assignment of target-agent pairs has been given, the agents must solve the problem through communication and motion. We call this the *target assignment problem*. Such a problem could arise in several applications, such as UAV's on a surveillance mission, where the targets are the centers of their desired loitering patterns.

The centralized problem of simply assigning one agent to each target is known in the combinatorial optimization literature as the *maximum matching problem* [1]. There are several polynomial time algorithms for solving this problem, the best known being [2] by Hopcroft and Karp. To efficiently assign agents to targets, we may be interested in finding a maximum matching (i.e., an assignment of one agent to each target) which minimizes a cost function. If the cost function is the sum of distances from each agent to its assigned target, then the problem is known as the *assignment problem*, or the *minimum weight maximum matching problem*, [1]. This problem can be written as an integer linear program and optimal solutions can be computed in polynomial time [3]. Another choice of cost function is to minimize the maximum distance between agents and their assigned targets. This problem is commonly referred to as the *bottleneck assignment problem* [4], and although the cost function is not linear, there still exist several

A. Chiuso et al. (Eds.): Modeling, Estimation and Control, LNCIS 364, pp. 271–284, 2007.
springerlink.com

polynomial time algorithms for its solution. There has also been work on developing algorithms for the assignment problem which can be implemented on parallel computing systems. One example is the *auction algorithm* [5], which can be implemented with one processor for each agent.

There is set of problems, commonly referred to as *decentralized task allocation*, that are closely related to our target assignment problem, see for example [6, 7, 8]. In these problems the goal is generally to assign vehicles to spatially distributed tasks while maximizing the "score" of the mission. Most works on this problem develop advanced heuristic methods, and demonstrate their effectiveness through simulation or real world implementation. In [9] the auction algorithm was adapted to solve a task allocation problem in the presence of communication delays. There has also been prior work on the target assignment problem [10, 11, 12, 13, 14]. For example, an algorithm based on hybrid systems tools is developed in [10]. The algorithm performance is characterized by a bound on the number of switches of the hybrid system; however, no analysis of the time complexity is provided.

In this chapter we summarize our recent investigations [12, 13] into the minimum-time task assignment problem and its scalability properties. We are interested in characterizing the completion time as the number of agents, n, grows, and the environment, $\mathcal{E}(n) := [0, \ell(n)]^2$, grows to accommodate them. In Section 4 we describe the ETSP ASSGMT algorithm with worst-case completion time in $O(\sqrt{n}\ell(n))$. In addition, in "sparse" environments, i.e., when $\ell(n)/\sqrt{n} \to +\infty$, the ETSP ASSGMT algorithm is asymptotically optimal among a broad class of algorithms in terms of its worst-case completion time. Then, in Section 5 we describe the GRID ASSGMT algorithm with worst-case completion time in $O(\ell(n)^2)$. We also characterize the stochastic properties of the GRID ASSGMT algorithm in "dense" environments, i.e., when $\ell(n)/\sqrt{n} \to 0$. If the agents and targets are uniformly randomly distributed, then the completion time belongs to $O(\ell(n))$ with high probability. Also, if there are n agents and only $n/\log n$ targets, then the completion time belongs to $O(1)$ with high probability.

The two algorithms are complementary: ETSP ASSGMT has better performance in sparse environments, while GRID ASSGMT has better performance in dense environments.

2 Geometric and Stochastic Preliminaries

In this section we review a few useful results on the Euclidean traveling salesperson problem, occupancy problems, and random geometric graphs. To do this, we must first briefly review some notation. We let $\mathbb{R}$ denote the set of real numbers, $\mathbb{R}_{>0}$ denote the set of positive real numbers, and $\mathbb{N}$ denote the set of positive integers. Given a finite set A, we let $|A|$ denote its cardinality. For two functions $f, g : \mathbb{N} \to \mathbb{R}_{>0}$, we write $f(n) \in O(g)$ (respectively, $f(n) \in \Omega(g)$) if there exist $N \in \mathbb{N}$ and $c \in \mathbb{R}_{>0}$ such that $f(n) \leq cg(n)$ for all $n \geq N$ (respectively, $f(n) \geq cg(n)$ for all $n \geq N$). If $f(n) \in O(g)$ and $f(n) \in \Omega(g)$ we say $f(n) \in \Theta(g)$. We say that event $A(n)$ occurs *with high probability* (w.h.p.) if the probability of $A(n)$ occurring tends to one as $n \to +\infty$.

2.1   The Euclidean Traveling Salesperson Problem

For a set of n points, $\mathcal{Q} \in \mathbb{R}^2$, we let $\mathrm{ETSP}(\mathcal{Q})$ denote the length of the shortest closed path through all points in $\mathcal{Q}$. The following result characterizes the length of this path when $\mathcal{Q} \subset [0, \ell(n)]^2$.

Theorem 1 (ETSP tour length, [15]). *For every set of n points $\mathcal{Q} \subset [0, \ell(n)]^2$, we have $\mathrm{ETSP}(\mathcal{Q}) \in O(\sqrt{n}\ell(n))$.*

The problem of computing an optimal ETSP tour is known to be NP-complete. However, there exist many efficient approximation algorithms. For example, the *Christofides' algorithm* [16], computes a tour that is no longer than $3/2$ times the optimal in $O(n^3)$ computation time.

2.2   Bins and Balls

Occupancy problems, or bins and balls problems, are concerned with randomly distributing m balls into n equally sized bins. The two results we present here will be useful in our analysis.

Theorem 2 (Bins and balls properties, [17, 18]). *Consider uniformly randomly distributing m balls into n bins and let γ_n be any function such that $\gamma_n \to +\infty$ as $n \to +\infty$. The following statements hold:*

1. *if $m = n$, then w.h.p. each bin contains $O\left(\frac{\log n}{\log \log n}\right)$ balls;*
2. *if $m = n \log n + \gamma_n n$, then w.h.p. there are no empty bins, and each bin contains $O(\log n)$ balls;*
3. *if $m = n \log n - \gamma_n n$, then w.h.p. there exists an empty bin;*
4. *if $m = Kn \log n$, where $K > 1/\log(4/e)$, then w.h.p. every bin contains $\Theta(\log n)$ balls.*

We will be interested in dividing a square environment into equally sized and openly disjoint square bins, such that the side length $\ell(B)$, of each bin is small in some sense. To do this, we require the following simple fact.

Lemma 1 (Dividing the environment). *Given $n \in \mathbb{N}$ and $r > 0$, consider an environment $\mathcal{E}(n) := [0, \ell(n)]^2$. If $\mathcal{E}(n)$ is partitioned into b^2 equally sized and openly disjoint square bins, where*

$$b := \lceil \sqrt{5}\ell(n)/r \rceil, \tag{1}$$

then $\ell(B) \leq r/\sqrt{5}$. Moreover, if $x, y \in \mathcal{E}(n)$ are in the same bin or in adjacent bins, then $\|x - y\| \leq r$.

2.3   Random Geometric Graphs

For $n \in \mathbb{N}$ and $r \in \mathbb{R}_{>0}$, a planar *geometric graph* $G(n, r)$ consists of n vertices in $\mathbb{R}^2$, and undirected edges connecting all vertex pairs $\{x, y\}$ with $\|x - y\| \leq r$. If the vertices are randomly distributed in some subset of $\mathbb{R}^2$, we call the graph a *random geometric graph*.

Theorem 3 (Connectivity of random geometric graphs, [19]). *Consider the random geometric graph $G(n, r)$ obtained by uniformly randomly distributing n points in $[0, \ell(n)]^2$. If*

$$\pi \left(\frac{r}{\ell(n)} \right)^2 = \frac{\log n + c(n)}{n},$$

then $G(n, r)$ is connected w.h.p. if and only if $c(n) \to +\infty$ as $n \to +\infty$.

This theorem will be important for understanding some of our results. If we randomly deploy n agents with communication range $r > 0$ in an environment $[0, \ell(n)]^2$, then the communication graph is connected if $\ell(n) \leq r \sqrt{n / \log n}$.

3   Network Model and Problem Statement

In this section we formalize our agent and target models and define the sparse and dense environments.

3.1   Robotic Network Model

Consider n agents in an environment $\mathcal{E}(n) := [0, \ell(n)]^2 \subset \mathbb{R}^2$, where $\ell(n) > 0$ (that is, $\mathcal{E}(n)$ is a square with side length $\ell(n)$). The environment $\mathcal{E}(n)$ is compact for each n but its size depends on n. A robotic agent, $\mathcal{A}^{[i]}$, $i \in \mathcal{I} := \{1, \ldots, n\}$, is described by the tuple

$$\mathcal{A}^{[i]} := \{\text{UID}^{[i]}, \mathbf{p}^{[i]}, r, \mathbf{u}^{[i]}, M^{[i]}\},$$

where the quantities are as follows: Its unique identifier (UID) is $\text{UID}^{[i]}$, taken from the set $I_{\text{UID}} \subset \mathbb{N}$. Note that, each agent does not know the set of UIDs being used and thus does not know the order. Its position is $\mathbf{p}^{[i]} \in \mathcal{E}(n)$. Its communication range is $r > 0$, i.e., two agents, $\mathcal{A}^{[i]}$ and $\mathcal{A}^{[k]}$, $i, k \in \mathcal{I}$, can communicate if and only if $\|\mathbf{p}^{[i]} - \mathbf{p}^{[k]}\| \leq r$. Its continuous time velocity input is $\mathbf{u}^{[i]}$, corresponding to the kinematic model $\dot{\mathbf{p}}^{[i]} = \mathbf{u}^{[i]}$, where $\|\mathbf{u}^{[i]}\| \leq v_{\max}$ for some $v_{\max} > 0$. Finally, its memory is $M^{[i]}$ and is of size $|M^{[i]}|$. From now on, we simply refer to agent $\mathcal{A}^{[i]}$ as agent i. We assume the agents move in continuous time and communicate according to a discrete time schedule $\{t_k\}_{k \in \mathbb{N}}$. We assume $|t_{k+1} - t_k| \leq t_{\max}$, for all $k \in \mathbb{N}$, where $t_{\max} \in \mathbb{R}_{>0}$. At each communication round, agents can exchange messages of length $O(\log n)$.[1]

3.2   The Target Assignment Problem

Let $\mathcal{Q} := \{\mathbf{q}_1, \ldots, \mathbf{q}_n\}$ be a set of distinct target locations, $\mathbf{q}_j \in \mathcal{E}(n)$ for each $j \in \mathcal{I}$. Agent i's memory, $M^{[i]}$, contains a copy of $\mathcal{Q}$, which we denote $\mathcal{Q}^{[i]}$. To store $\mathcal{Q}^{[i]}$ we must assume the size of each agents' memory, $|M^{[i]}|$, is in $\Omega(n)$. We refer to the assumption that each agent knows all target positions as the *full knowledge* assumption (for a more detailed discussion of this assumption see [12]). Our goal is to solve the *(full knowledge) target assignment problem*:

---

[1] $\Omega(\log n)$ bits are required to represent an ID, unique among n agents.

Determine an algorithm for $n \in \mathbb{N}$ agents, with attributes as described above, satisfying the following requirement. There exists a time $T > 0$ such that for each target $\mathbf{q}_j \in \mathcal{Q}$, there is a unique agent $i \in \mathcal{I}$, with $\mathbf{p}^{[i]}(t) = \mathbf{q}_j$ for all $t \geq T$.

3.3 Sparse and Dense Environments

We wish to study the scalability of a particular approach to the target assignment problem; that is, how the completion time increases as we increase the number of agents, n. The velocity $v_{\max}$ and communication range r of each agent are independent of n. However, we assume that the size of the environment increases with n in order to accommodate an increase in agents. Borrowing terms from the random geometric graph literature [19], we say the environment is sparse if, as we increase the number of agents, the environment grows quickly enough that the density of agents (as measured by the sum of their communication footprints) decreases; we say the environment is critical, if the density is constant, and we say the environment is dense if the density increases. Formally, we have the following definition.

Definition 1 (Dense, critical and sparse environments). *The environment $\mathcal{E}(n) := [0, \ell(n)]^2$ is* sparse *if $\ell(n)/\sqrt{n} \to +\infty$ as $n \to +\infty$,* critical *if $\ell(n)/\sqrt{n} \to C \in \mathbb{R}_{>0}$ as $n \to +\infty$, and* dense *if $\ell(n)/\sqrt{n} \to 0$, as $n \to +\infty$.*

It should be emphasized that a dense environment does not imply that the communication graph between agents is dense. On the contrary, from Theorem 3 we see that the communication graph at random agent positions in a dense environment may not even be connected.

4 Sparse Environments

We begin by studying the case when the environment is sparse, and thus there is very little communication between agents. We introduce a natural approach to the problem in the form of a class of distributed algorithms, called *assignment-based motion*. We give a worst-case lower bound on the performance of the assignment-based motion class. Next, we introduce a control and communication algorithm, called ETSP ASSGMT. In this algorithm, each agent precomputes an optimal tour through the n targets, turning the cloud of target points into an ordered ring. Agents then move along the ring, looking for the next available target. When agents communicate, they exchange information on the next available target along the ring. We show that in sparse or critical environments, the ETSP ASSGMT algorithm is an asymptotically optimal among all algorithms in the assignment-based motion class.

4.1 Assignment-Based Algorithms with Lower Bound Analysis

Here we introduce and analyze a class of deterministic algorithms for the target assignment problem. The assignment-based motion class can be described as follows.

Outline of assignment-based motion class

Initialization: In this class of algorithms agent i initially selects the closest target in $\mathcal{Q}^{[i]}$, and sets the variable $\text{curr}^{[i]}$ (agent i's current target), to the index of that target.

Motion: Agent i moves toward the target $\text{curr}^{[i]}$ at speed $v_{\max}$.

Communication: If agent i communicates with an agent k that is moving toward $\text{curr}^{[k]} = \text{curr}^{[i]}$, and if agent k is closer to $\text{curr}^{[i]}$ than agent i, then agent i "removes" $\text{curr}^{[i]}$ from $\mathcal{Q}^{[i]}$ and selects a new target.

For this class of algorithms it is convenient to adopt the following conventions: we say that agent $i \in \mathcal{I}$ is *assigned* to target $\mathbf{q}_j \in \mathcal{Q}$, when $\text{curr}^{[i]} = j$. We say that agent $i \in \mathcal{I}$ *enters a conflict* over the target $\text{curr}^{[i]}$, when agent i receives a message, $\text{msg}^{[k]}$, with $\text{curr}^{[i]} = \text{curr}^{[k]}$. Agent i *loses the conflict* if agent i is farther from $\text{curr}^{[i]}$ than agent k, and *wins the conflict* if agent i is closer to $\text{curr}^{[i]}$ than agent k, where ties are broken by comparing UIDs. Note that if an agent is assigned to the same target as another agent, it will enter a conflict in finite time.

Theorem 4 (Time complexity lower bound for target assignment). *Consider n agents, with communication range $r > 0$, in an environment $[0, \ell(n)]^2$. If $\ell(n) > r\sqrt{n}$, then for all algorithms in the assignment-based motion class, the time complexity of the target assignment problem is in $\Omega(\sqrt{n}\ell(n))$.*

In other words, the target assignment time complexity is lower bounded when the environment grows faster than some critical value, that is, when the environment is sparse or critical.

4.2   The ETSP ASSGMT Algorithm with Upper Bound Analysis

In this section we introduce the ETSP ASSGMT algorithm—an algorithm within the assignment-based motion class. We will show that when the environment is sparse or critical, this algorithm is asymptotically optimal. In the following description of ETSP ASSGMT it will be convenient to assume that the target positions are stored in each agents memory as an array, rather than as an unordered set. That is, we replace the target set $\mathcal{Q}$ with the target n-tuple $\mathbf{q} := (\mathbf{q}_1, \ldots, \mathbf{q}_n)$, and the local target set $\mathcal{Q}^{[i]}$ with the n-tuple $\mathbf{q}^{[i]}$. The algorithm can be described as follows.

For each $i \in \mathcal{I}$, agent i computes a constant factor approximation of the optimal ETSP tour of the n targets in $\mathbf{q}^{[i]}$, denoted $\text{tour}(\mathbf{q}^{[i]})$. We can think of tour as a map which reorders the indices of $\mathbf{q}^{[i]}$; $\text{tour}(\mathbf{q}^{[i]}) = (\mathbf{q}^{[i]}_{\sigma(1)}, \ldots, \mathbf{q}^{[i]}_{\sigma(n)})$, where $\sigma : \mathcal{I} \to \mathcal{I}$ is a bijection. This map is independent of i since all agents use the same method. An example is shown in Fig. 1(a). Agent i then replaces its n-tuple $\mathbf{q}^{[i]}$ with $\text{tour}(\mathbf{q}^{[i]})$. Next, agent i computes the index of the closest target in $\mathbf{q}^{[i]}$, and calls it $\text{curr}^{[i]}$. Agent i also maintains the index of the next target in the tour which may be available, $\text{next}^{[i]}$, and first target in the tour before $\text{curr}^{[i]}$ which may be available, $\text{prev}^{[i]}$. Thus, $\text{next}^{[i]}$ is initialized to $\text{curr}^{[i]} + 1 \pmod{n}$ and $\text{prev}^{[i]}$ to $\text{curr}^{[i]} - 1 \pmod{n}$. In order to "remove" assigned targets from the tuple $\mathbf{q}^{[i]}$, agent i also maintains the n-tuple, $\text{status}^{[i]}$. Letting $\text{status}^{[i]}(j)$ denote the jth entry in the n-tuple, the entries are given by

$$\text{status}^{[i]}(j) = \begin{cases} 0, & \text{if agent } i \text{ knows } \mathbf{q}_j^{[i]} \text{ is assigned to another agent,} \\ 1, & \text{otherwise.} \end{cases} \qquad (2)$$

Thus, $\text{status}^{[i]}$ is initialized as the n-tuple $(1, \ldots, 1)$. The initialization is depicted in Fig. 1(b).

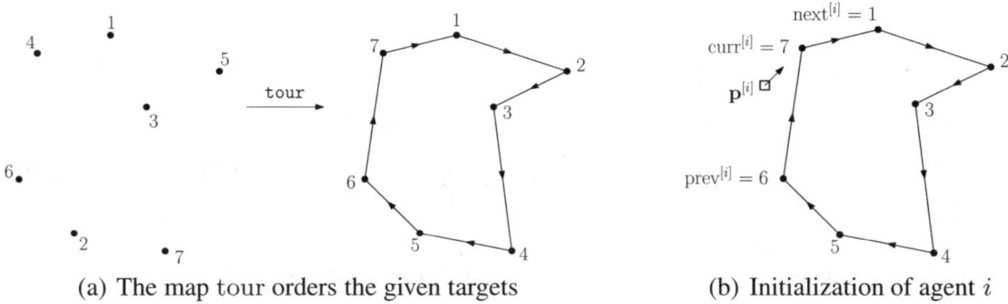

(a) The map tour orders the given targets   (b) Initialization of agent i

Fig. 1. Initialization of ETSP ASSGMT

Finally, at each communication round agent i executes the algorithm COMM-RD described below.

Outline of COMM-RD algorithm for agent i

1: Broadcast $\text{msg}^{[i]}$, consisting of the targets, $\text{prev}^{[i]}$, $\text{curr}^{[i]}$, and $\text{next}^{[i]}$, the distance to the current target $d^{[i]}$, and $\text{UID}^{[i]}$.

2: **for all** messages, $\text{msg}^{[k]}$, received **do**

3:    Set $\text{status}^{[i]}(j)$ to assigned ('0') for each target j from $\text{prev}^{[k]}+1 \pmod{n}$ to $\text{next}^{[k]} - 1 \pmod{n}$ not equal to $\text{curr}^{[i]}$.

4:    **if** $\text{prev}^{[k]} = \text{next}^{[k]} = \text{curr}^{[k]} \neq \text{curr}^{[i]}$, **then** set the status of $\text{curr}^{[k]}$ to 0 because it was missed in the previous step.

5:    **if** $\text{curr}^{[i]} = \text{curr}^{[k]}$ but agent i is farther from $\text{curr}^{[i]}$ than agent k (ties broken with UIDs) **then**

6:       Set the status of $\text{curr}^{[i]}$ to assigned ('0').

7:    **if** $\text{curr}^{[i]} = \text{curr}^{[k]}$ and agent i is closer than agent k **then**

8:       Set the status of $\text{next}^{[i]}$ and $\text{next}^{[k]}$ to assigned ('0').

9: Update $\text{curr}^{[i]}$ to the next target in the tour with status available ('1'), $\text{next}^{[i]}$ to the next available target in the tour after $\text{curr}^{[i]}$, and $\text{prev}^{[i]}$ to the first available target in the tour before $\text{curr}^{[i]}$.

In summary, the ETSP ASSGMT algorithm is the triplet consisting of the initialization of each agent, the motion law (move toward $\text{curr}^{[i]}$ at speed $v_{\max}$), and the COMM-RD algorithm executed at each communication round.

Fig. 2 gives an example of COMM-RD resolving a conflict between agents i and k, over $\text{curr}^{[i]} = \text{curr}^{[k]}$. The proposed algorithm enjoys plenty of useful properties, which are valid for any communication graph which contains the geometric graph with parameter r as a subgraph. A complete discussion is contained in [12]. Based on a careful application of Theorem 1, one can derive the following key result.

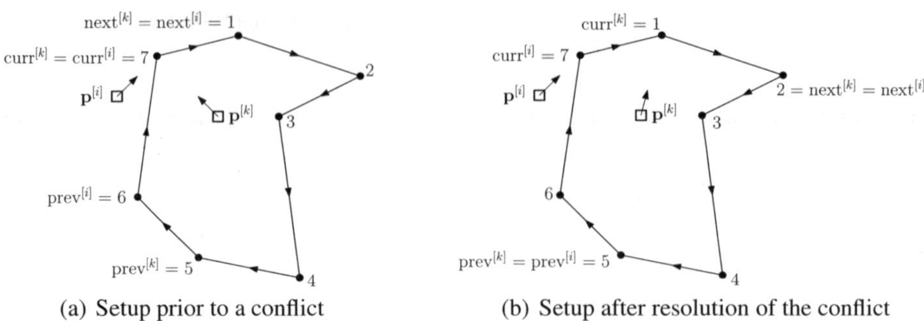

(a) Setup prior to a conflict   (b) Setup after resolution of the conflict

Fig. 2. The resolution of a conflict between agents i and k over target 7. Agent i wins the conflict since it is closer to target 7 than agent k.

Theorem 5 (Correctness and time complexity for ETSP ASSGMT). *For any $n \in \mathbb{N}$, ETSP ASSGMT solves the target assignment problem. Furthermore, consider an environment $[0, \ell(n)]^2$. If $t_{\max} < r/v_{\max}$, then ETSP ASSGMT solves the target assignment problem in $O(\sqrt{n}\ell(n) + n)$ time. If, in addition, $\ell(n) > r\sqrt{n}$, then the time complexity is in $\Theta(\sqrt{n}\ell(n))$, and ETSP ASSGMT is asymptotically optimal among algorithms in the assignment-based motion class.*

The above theorem gives a complexity bound for the case when r and $v_{\max}$ are fixed constants, and $\ell(n)$ grows with n. An equivalent setup is to consider ℓ fixed and allow the robots' attributes, r and $v_{\max}$, to vary inversely with the n, specifically, r and v proportional to $\sqrt{n}$.

Corollary 1 (Complexity with congestion). *Consider n agents moving with speed $\widetilde{v}_{\max}(n) = n^{-1/2}$ and communication radius $\widetilde{r}(n) = r_0 n^{-1/2}$, with $r_0 < 1$, in the environment $[0, 1]^2$. Then ETSP ASSGMT solves the target assignment problem with time complexity in $\Theta(n)$.*

For simplicity we have presented our time complexity results in the planar environment $[0, \ell(n)^2]$. However, in [12] we derive bounds for the more general environment

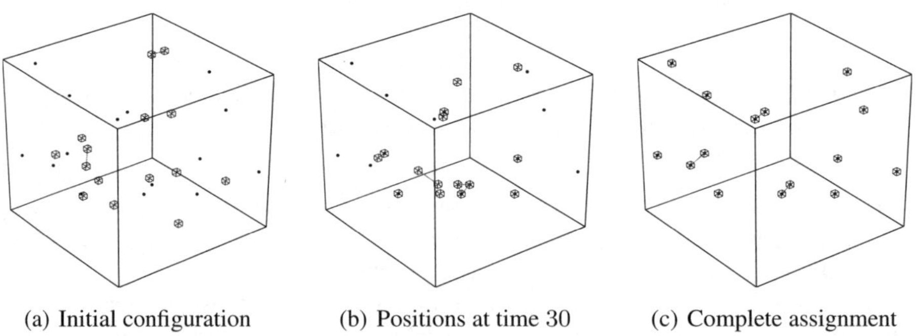

(a) Initial configuration   (b) Positions at time 30   (c) Complete assignment

Fig. 3. Simulation for 15 agents, $v_{\max} = 1$, $r = 15$ in $\mathcal{E} = [0, 100]^3$. The targets are spheres. The agents are cubes. An edge is drawn when two agents are communicating.

$[0, \ell(n)]^d$, $d \geq 1$. A simulation in $[0, 100]^3 \subset \mathbb{R}^3$ with $r = 15$ and $v = 1$ is shown in Fig. 3. To compute the ETSP tour we have used the `concorde` TSP solver.[2] The initial configuration shown in Fig. 3(a) consists of uniformly randomly generated target and agent positions.

5 Dense Environments

In the previous section we presented the ETSP ASSGMT algorithm which has provably good performance in sparse environments. In this section we introduce the GRID ASSGMT algorithm for dense environments in which communication is more prevalent. We will show that it has better worst-case performance than ETSP ASSGMT in dense environments, and that it possesses very good stochastic performance.

5.1 The GRID ASSGMT Algorithm with Complexity Analysis

In the GRID ASSGMT algorithm we assume that each agent knows the target positions, $\mathcal{Q}$, and the quantity $\ell(n)$ which describes the size of the environment. With this information, each agent partitions the environment into b^2 equally sized square cells, where $b \in \mathbb{N}$. It then labels the cells like entries in a matrix, so cell $C(w, c)$ resides in the wth row and cth column. This is shown in Fig. 4(b). Since the agents started with the same information, they all create the same partition.

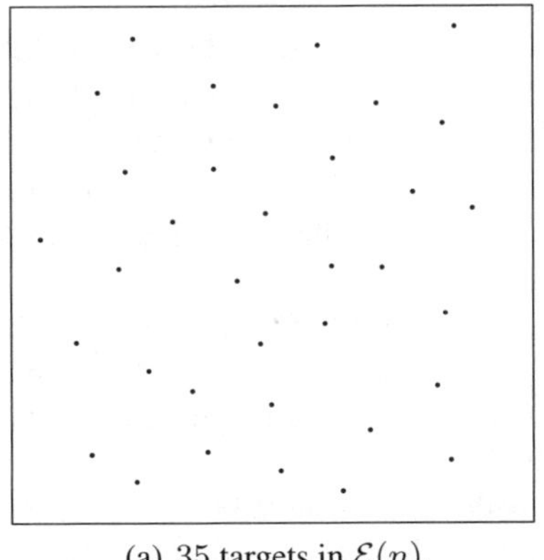

(a) 35 targets in $\mathcal{E}(n)$.

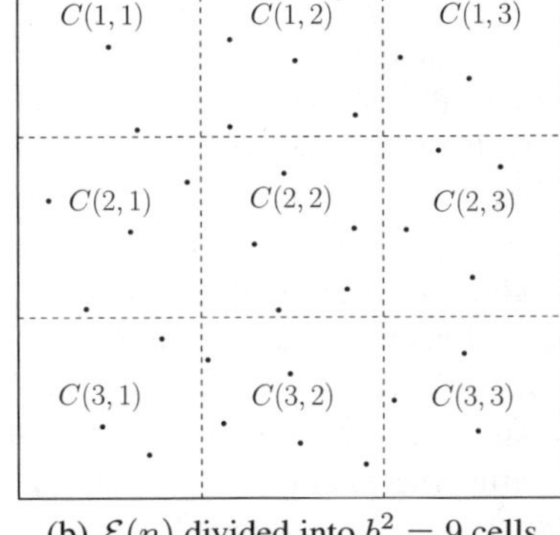

(b) $\mathcal{E}(n)$ divided into $b^2 = 9$ cells.

Fig. 4. Dividing the environment into 9 cells

In light of Lemma 1, we see that when b is given by $\lceil \sqrt{5}\ell(n)/r \rceil$, as in equation (1), the communication graph between agents in a cell is complete, and communication between agents in adjacent cells is also possible. With this in mind, an outline of the GRID ASSGMT algorithm is as follows.

---

[2] The `concorde` TSP solver is available for research use at `http://www.tsp.gatech.edu/concorde/index.html`

Outline of the GRID ASSGMT algorithm

Initialization: Each agent partitions the environment into b^2 equally sized square cells, where b is given in Lemma 1, and the cells are labeled as in Fig. 4(b).

All agents: In each cell, all agents in the cell find a maximum matching between agents and targets occupying the cell. Accordingly, agents are labeled assigned or unassigned.

Assigned agents: In each cell, all assigned agents elect a leader among them. All assigned agents, except the leaders, send their assignment information to their respective leader and then go silent.

Cell leaders: The leader in each cell communicates to the leader in the cell directly above. As a result, each leader obtains an estimate of the number of available targets in all cells below it, in its column.

Unassigned agents: First, each unassigned agent seeks a free target in its column by entering cells and querying the corresponding leader.

Second, if all targets in the unassigned agent's column are assigned, then the agent moves to the top of its column and along the top row. The agent gathers from each leader in the top row the number of available targets in the leader's column. When the agent finds a column with available targets, it travels down that column to find the free target.

To implement this algorithm agent i maintains the following variables in its memory. The variable currcell$^{[i]}$ which keeps track of the cell which agent i currently occupies. The set $\mathcal{Q}^{[i]}(w,c)$ which contains the targets in cell $C(w,c)$. The variable leader$^{[i]}$ which is set to $C(w,c)$ if agent i is the leader of $C(w,c)$, and null otherwise. The array colstatus$^{[i]}$, where colstatus$^{[i]}(c)$ is set to full if column c contains no available targets, and notfull if agent i thinks column c may contain an available target. The variable dircol$^{[i]} \in \{\text{down}, \text{up}\}$ which contains the direction of travel in a column and dirrow$^{[i]} \in \{\text{left}, \text{right}\}$ which contains the direction in the first row. Finally, the variable curr$^{[i]}$ which contains agent i's assigned target, or the entry null.

After initializing these variables, each agent runs an algorithm which allows the agents to compute a local maximum matching, and elect a leader, in each cell. Since the communication graph in each cell is complete, this can be done in one communication round by receiving the UIDs of each agent in the cell [13].

After the maximum matching and leader election the agents have been separated into three roles; assigned leader agents, assigned non-leader agents, and unassigned agents. The unassigned agents run an algorithm in which they try to find a free target. The leader of each cell runs an algorithm in which they update their estimates of available targets in various parts of the grid, and assigns unassigned targets in its cell. The leader of cell $C(w,c)$, agent i, maintains the following quantities to assign targets in its cell, and estimate the number of available targets in cells below. Agent i maintains: diff$^{[i]}(w,c)$, which records the difference between the number of targets and agents in cell $C(w,c)$; diffbelow$^{[i]}(w,c)$ which records agent i's estimate of the difference between the number of agents and targets in cells $C(w+1,c),\ldots,C(b,c)$; and taravail$^{[i]}(w,c)$ which contains the available targets in $C(w,c)$. Finally, if agent

i is the leader of $C(1, c)$ in the first row, it maintains $\text{diffright}^{[i]}(c)$ which is agent i's estimate of the number of available targets in columns $c + 1, \ldots, b$.

In summary, The GRID ASSGMT algorithm is the 4-tuple consisting of the initialization, the maximum matching and leader election algorithm, the unassigned agent algorithm, and the leader algorithm.

We can now state the main results on the GRID ASSGMT algorithm.

Theorem 6 (Correctness and worst-case upper bound). *For any initial positions of n targets and n agents in $[0, \ell(n)]^2$, GRID ASSGMT solves the target assignment problem in $O((\ell(n))^2)$ time.*

Remark 1 (GRID ASSGMT vs. ETSP ASSGMT). The worst-case bound for ETSP ASSGMT in Theorem 5 was $O(\sqrt{n}\ell(n))$. Thus, in sparse environments, when $\ell(n)$ grows faster than $\sqrt{n}$, ETSP ASSGMT performs better, and in dense environments GRID ASSGMT performs better. In critical environments, the bounds are equal. Thus, the two algorithms are complementary. In practice, if n, $\ell(n)$ and r are known, each robot in the network can determine which algorithm to run based on the following test: ETSP ASSGMT is run if $\ell(n)/\sqrt{n} > r$ and GRID ASSGMT is run if $\ell(n)/\sqrt{n} < r$.   •

In the following theorem we will see that for randomly placed targets and agents, the performance of GRID ASSGMT is considerably better than in the worst-case. The proofs of the following theorems utilize the results on bins and balls problems in Section 2.

Theorem 7 (Stochastic time complexity). *Consider n agents and n targets, uniformly randomly distributed in $[0, \ell(n)]^2$. If $\ell(n) \leq r/\sqrt{5}\sqrt{(n/K \log n)}$, where $K > 1$, then GRID ASSGMT solves the target assignment problem in $O(\ell(n))$ time with high probability.*

Remark 2 (Generalization of Theorem 7). The bound in Theorem 7 holds for any initial positions such that every cell contains at least one target and at least one agent.   •

Theorem 8 (Stochastic time complexity: More agents than targets). *Consider n agents and $n/\log n$ targets, uniformly randomly distributed in $[0, \ell(n)]^2$. If $\ell(n) \leq r/\sqrt{5}\sqrt{(n/K \log n)}$, where $K > 1/\log(4/e)$, then w.h.p., GRID ASSGMT solves the target assignment problem in $O(1)$ time.*

A representative simulation of GRID ASSGMT for 65 agents and targets uniformly randomly distributed in a dense environment is shown in Fig. 5(a)–(c). In Fig. 5(c) a dashed blue trail shows the trajectory for the final agent as it is about to reach its target in cell $C(1, 1)$. Fig. 5.1 contains a Monte Carlo simulation for uniformly randomly generated agents and targets. The side length $\ell(n)$ satisfies the bound in Theorem 7, and the agents move at unit speed. Each data point is the mean completion time of 30 trials, where each trial was performed at randomly generated agent and target positions. Error bars show plus/minus one standard deviation. The mean completion time lies between $2\ell(n)$ and $3\ell(n)$. This agrees with the $O(\ell(n))$ bound in Theorem 7 and gives some idea as to the constant in front of this bound.

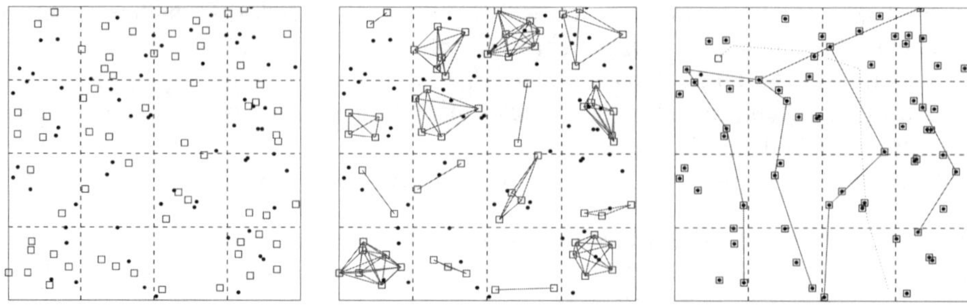

(a) Initial agent and target positions, and grid

(b) Maximum assignment and leader election

(c) Final agent reaching target

Fig. 5. A simulation of 65 agents in a dense environment. Targets are black disks and agents are blue squares. Red lines are drawn when two agents are communicating.

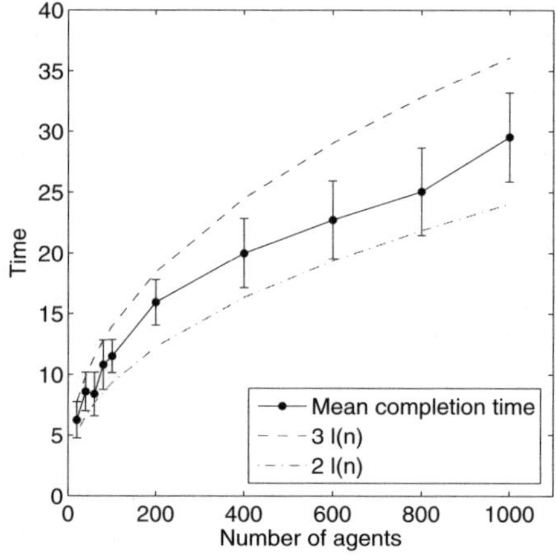

Fig. 6. A Monte Carlo simulation. Each data point is the mean of 30 trials.

5.2 A Sensor Based Version

In describing the GRID ASSGMT algorithm, we assumed that each agent knows the position of all targets. The algorithm also works when each agent does not know the position of any targets, but has a sensing range r_{sense}, with which it can sense the positions of targets in range. If each agent can partition the environment as in Fig. 4, and if $r_{\text{sense}} \geq \sqrt{2/5}r$ so that each agent can sense the position of all targets in its current cell, then GRID ASSGMT (with minor modifications) solves the target assignment problem, and the completion time results still hold.

5.3 Congestion Issues

Since wireless communication is a shared medium, simultaneous messages sent in close proximity will collide, resulting in dropped packets. In fact, clear reception of a signal requires that no other signals are present at the same point in time and space. As the density of agents increases (as measured by their communication footprints), so does wireless communication congestion. Thus, in dense environments, one would ideally account for the effects of congestion. In the design of GRID ASSGMT we have tried to limit the amount of simultaneous communication. To this end we introduced a leader in each cell, who sent messages (of size $O(\log n)$) only to its adjacent cells, and all other assigned agents were silent. However, to fully take wireless congestion into account, we would require a more sophisticated communication model than the geometric graph.

6   Conclusion and Extensions

In this chapter we have discussed two complementary algorithms for the target assignment problem, ETSP ASSGMT and GRID ASSGMT. We have shown that ETSP ASSGMT has better performance in sparse environments, where as, GRID ASSGMT has better performance in dense environments. There are many future research directions such as extensions to vehicles with motion constraints, or to the case when targets are dynamically appearing and disappearing. Another area of future research is to develop a communication framework which adequately models congestion and media access problems that are inherently present in wireless communications.

Acknowledgments: In Giorgio's Honor

The second author dedicates this work to Giorgio Picci. I am honored to have had him as my Laurea advisor where he introduced me to the exciting world of scientific research. His passion for control theory, applied mathematics and geometry was highly contagious and continues to be a part of my life today. I am honored to be able to dedicate this work to a man I consider an inspiration. Grazie di cuore!

References

1. B. Korte and J. Vygen, *Combinatorial Optimization: Theory and Algorithms*. New York: Springer Verlag, 3 ed., 2005.
2. J. E. Hopcroft and R. M. Karp, "An $n^{5/2}$ algorithm for maximum matchings in bipartite graphs," *SIAM Journal on Computing*, vol. 2, no. 4, pp. 225–231, 1973.
3. H. W. Kuhn, "The Hungarian method for the assignment problem," *Naval Research Logistics*, vol. 2, pp. 83–97, 1955.
4. R. Burkard, "Selected topics on assignment problems," *Discrete Applied Mathematics*, vol. 123, pp. 257–302, 2002.
5. D. P. Bertsekas and J. N. Tsitsiklis, *Parallel and Distributed Computation: Numerical Methods*. Belmont, MA: Athena Scientific, 1997.
6. M. F. Godwin, S. Spry, and J. K. Hedrick, "Distributed collaboration with limited communication using mission state estimates," in *American Control Conference*, (Minneapolis, MN), pp. 2040–2046, June 2006.
7. M. Alighanbari and J. P. How, "Robust decentralized task assignment for cooperative UAVs," in *AIAA Conf. on Guidance, Navigation and Control*, (Keystone, CO), Aug. 2006.

8. C. Schumacher, P. R. Chandler, S. J. Rasmussen, and D. Walker, "Task allocation for wide area search munitions with variable path length," in *American Control Conference*, (Denver, CO), pp. 3472–3477, 2003.

9. B. J. Moore and K. M. Passino, "Distributed task assignment for mobile agents," *IEEE Transactions on Automatic Control*, 2006. to appear.

10. M. Zavlanos and G. Pappas, "Dynamic assignment in distributed motion planning with local information," in *American Control Conference*, (New York), July 2007. To appear.

11. G. Arslan and J. S. Shamma, "Autonomous vehicle-target assignment: a game theoretic formulation," *IEEE Transactions on Automatic Control*, Feb. 2006. Submitted.

12. S. L. Smith and F. Bullo, "Target assignment for robotic networks: Asymptotic performance under limited communication," in *American Control Conference*, (New York), July 2007. To appear.

13. S. L. Smith and F. Bullo, "Target assignment for robotic networks: Worst-case and stochastic performance in dense environments," in *IEEE Conf. on Decision and Control*, (New Orleans, LA), Dec. 2007. Submitted.

14. D. A. Castañón and C. Wu, "Distributed algorithms for dynamic reassignment," in *IEEE Conf. on Decision and Control*, (Maui, HI), pp. 13–18, Dec. 2003.

15. K. J. Supowit, E. M. Reingold, and D. A. Plaisted, "The traveling salesman problem and minimum mathcing in the unit square," *SIAM Journal on Computing*, vol. 12, pp. 144–156, 1983.

16. N. Christofides, "Worst-case analysis of a new heuristic for the traveling salesman problem," Tech. Rep. 388, Carnegie-Mellon University, Apr. 1976.

17. R. Motwani and P. Raghavan, *Randomized Algorithms*. Cambridge, UK: Cambridge University Press, 1995.

18. F. Xue and P. R. Kumar, "The number of neighbors needed for connectivity of wireless networks," *Wireless Networks*, vol. 10, no. 2, pp. 169–181, 2004.

19. M. Penrose, *Random Geometric Graphs*. Oxford Studies in Probability, Oxford, UK: Oxford University Press, 2003.

On the Distance Between Non-stationary Time Series

Stefano Soatto

Computer Science Department
University of California, Los Angeles
soatto@ucla.edu

1 Introduction

Comparing time series is a problem of critical importance in a broad range of applications, from data mining (searching for temporal "patterns" in historical data), to speech recognition (classifying phonemes from acoustic recordings), surveillance (detecting unusual events from video and other sensory input), computer animation (concatenating and interpolating motion capture sequences), just to mention a few. The problem is difficult because the same event can manifest itself in a variety of ways, with the data subject to a large degree of variability due to *nuisance factors* in the data formation process. For instance, the presence of a person walking in a video sequence can vary based on the individual, his gait, location, orientation, speed, clothing, illumination etc. And yet, if I see Giorgio Picci, I can recognize him from one hundred yards away by the way he walks, regardless of what he is wearing, or whether it is a sunny or a cloudy day. One could conjecture that there must exist some statistics of my retinal signals that are invariant, or at least insensitive, to such nuisance factors and are instead Giorgio-Specific (GS). The information ought to be encoded in the temporal evolution of the retinal signals, for one can strip the images of their pictorial content by attaching light bulbs to one's joints and turning off the lights (or use a state-of-the-art motion capture system, for instance one made by E-motion/BTS); one can still tell a great deal from just the moving dots [6].

To be sure, searching for invariant GS statistics is not the only way to get rid of the nuisances: One can also eliminate them as part of the matching process when comparing two time series. Let us consider, for example, the nuisance of the initial observation instant, t_0. If we observe the temporal evolution of joint positions and think of them as trajectories $\{y_1(t)\}_{t \in \mathbb{R}}$ in, say, $\mathbb{L}^2$, we could compare it to a sample sequence from a database, $\{y_2(t)\}_{t \in \mathbb{R}}$, by sliding one on top of the other until the $\mathbb{L}^2$ norm of the difference is minimized: $d_0(y_1, y_2) = \min_{t_0} \int \|y_1(t) - y_2(t - t_0)\|^2 dt$. A more elegant and efficient solution is to seek for a statistic of each time series, i.e. a deterministic function $\Sigma_i \doteq \phi(y_i)$, that is invariant with respect to t_0, and then to compare such statistics directly. For the case of sequences that admit statistics that are invariant with respect to t_0, a.k.a. *stationary*, this can be done, and the resulting *realization theory* is a success story of Systems Theory, one where Giorgio Picci and his collaborators have played a key role [11]. Endowing the space of realizations with a metric and a probabilistic

A. Chiuso et al. (Eds.): Modeling, Estimation and Control, LNCIS 364, pp. 285–299, 2007.
springerlink.com

structure is the key to enabling a new level of finesse in the applications cited above, where we are not simply trying to detect the presence of a person and whether he is walking, but also to tell gender, identity, even state of mind. Some preliminary steps in this direction have shown promise in the recognition of stationary human motion [2].

In this work, I wish to *extend the formulation of the problem to more general classes of nuisances* for time series that are not stationary. In the spirit of realization theory, I will forgo the prevailing approach where sequences are compared via their likelihood. In this approach, the similarity between two sequences is measured by how well the model of one (say a realization $\Sigma_1 = \phi(y_1)$) "explains" the other, for instance quantified by the covariance of the innovation process of y_2. In this approach, the more data are available, the better the estimate of Σ_1, the worse the classification error is, an apparent paradox induced by the fact that the generalization model underlying this approach is trivial: Each realization models one sequence and noisy versions of it, without regard for the structure of the intrinsic variability that different realizations of the same process exhibit. I will also not go as far as invoking the full power of chaotic non-linear models as customary in the physics literature [7], because the processes of interest are usually observed on short time-scales and do not exhibit chaotic behavior. Instead, I will attempt to define distances between processes that are invariant, or at least insensitive, to specific classes of nuisances. Once these are available, one can attempt to construct probability distributions and therefore define priors in the space of time series, compute likelihoods, perform optimal decisions etc. in the way one would do if nuisances were absent. For instance, in $\mathbb{L}^2$ one can define a distance

$$d_0(y_1, y_2) = \int_0^T \|y_1(t) - y_2(t)\|^2 dt \tag{1}$$

and from this, with some caveats, define a probabilistic structure $dP(y)$. In our case, we can think of events of interest as equivalence classes under the action of a nuisance group, and therefore we need to define a suitable quotient space (or base) to perform the comparison. Although ideally one would want a true Riemannian metric (homogeneous spaces are in general not flat) and integrate it along geodesics to compute distances, we will limit ourselves to defining cord distances directly. We will do so in steps, first introducing nuisances in general, then describing a variety of distances. For simplicity we will assume that sequences are observed over a common finite interval $[0, T]$, although most of the considerations can be extended to the case where the initial time and the duration are also included among the nuisances.

2   Formalization

The first step to introduce nuisances is to re-write (1) in a slightly different way as

$$d_0(y_1, y_2) = \min_h \int_0^T \|y_1(t) - h(t)\|^2 + \|y_2(t) - h(t)\|^2 dt. \tag{2}$$

Note that the two expressions (1) and (2) are identical up to a factor of two, as the right hand-side of (2) is bounded from below by (1), and from above by the same quantity by choosing $\hat{h} = (y_1 + y_2)/2$. Note also that we have been deliberately vague as to the space where h lives, which we will indicate by $\mathcal{H}$; despite it being infinite-dimensional, we do not need to impose regularization on h to solve the optimization above, which is trivially done in closed form. The reason for introducing such an auxiliary variable $h \in \mathcal{H}$ will become clear shortly, but already one can see that this writing highlights the underlying data-formation model: Both time series are generated from some (deterministic but) unknown function h, corrupted by two different realizations of additive "noise" (here the word noise lumps all unmodeled phenomena, not necessarily associated to sensor errors)

$$y_i(t) = h(t) + n_i(t) \quad i = 1, 2; \ t \in [0, T] \tag{3}$$

where, for instance, $n_i(t) \overset{iid}{\sim} \mathcal{N}(0, \Sigma) \ \forall \ t \in [0, T]; i = 1, 2$. Under this model, the distance is obtained by finding the (maximum-likelihood) solution for h that minimizes

$$\phi_{data}(y_1, y_2 | h) \doteq \sum_{i=1}^{2} \int_0^T \|n_i(t)\|^2 dt \tag{4}$$

subject to (3). Note that an obvious interpretation of h is that of the *average* of the two time series. This will become handy later. Although not necessary at this stage, one could consider regularized distances, for instance

$$d_{reg}(y_1, y_2) = \min_{h \in \mathcal{H}} \phi_{data}(y_1, y_2 | h) + \phi_{reg}(h) \tag{5}$$

where, for instance, $\phi_{reg}(h) = \int_0^T \|\nabla h\| dt$. Once a distance is available, one can perform classification in a number of ways, for instance using simple k-nearest neighbors [2].

2.1   Introducing Nuisances

Let us now consider some simple nuisances of the data collection process, and how to eliminate them in computing a meaningful notion of distance between time series. We have already discussed the role of the initial condition $t_0 = \beta$, corresponding to a model $y(t) = h(t + \beta) + n(t)$. Another common accident of data collection is a different sampling frequency, which translates into an affine deformation of the temporal axis $y(t) = h(\alpha t + \beta) + n(t)$. A slightly more elaborate model is a projective transformation of the temporal axis $y(t) = h(\frac{\alpha t + \beta}{\gamma t + \delta}) + n(t)$. In order to generalize this model, we have to resort to infinite-dimensional groups of domain diffeomorphisms of the interval $[0, T]$ [14]; the data formation model (3) above then becomes

$$y_i(t) = h(x_i(t)) + n_i(t) \quad i = 1, 2. \tag{6}$$

Correspondingly, the data term of the cost functional we wish to optimize is

$$\phi_{data}(y_1, y_2 | h, x_1, x_2) \doteq \sum_{i=1}^{2} \int_0^T \|n_i(t)\|^2 dt \tag{7}$$

from which it is clear that the model is over-determined, and we must therefore impose regularization [9] in order to compute

$$d_1(y_1, y_2) = \min_{h \in \mathcal{H}, x_i \in \mathcal{U}} \phi_{data}(y_1, y_2 | h, x_1, x_2) + \phi_{reg}(h). \tag{8}$$

The functions $x_i \in \mathcal{U}$ are called *time warpings*, and in order for $\tau \doteq x(t)$ to be a viable temporal index, x must satisfy a number of properties. The first is continuity (time, alas, does not jump); in fact, it is common to assume a certain degree of smoothness, and for the sake of simplicity we will assume that x_i is infinitely differerentiable. The second is causality: The ordering of time instants has to be preserved by the time warping, which can be formalized by imposing that x_i be monotonic. Additional constraints can be imposed that either are specific to a particular application, or to make the mathematical treatment simpler. A common choice is to impose $x_i(0) = 0; x_i(T) = T$ so that the interval $[0, T]$ is fixed by the warping function. These regularization constraints on x_i are implicit in the notation $x_i \in \mathcal{U}$ and will be made explicit in Sect. 3. In the absence of additional constraints, the solution of this problem leads to a well-known technique which we describe next.

2.2   Dynamic Time Warping

The solution of (8) (or the determination of the minimizers $\hat{h}, \hat{x}_i$) is called *dynamic time warping* (DTW) and is standard practice in speech processing as well as in temporal data mining. Making the constraints more explicit, we can re-write the distance above as

$$d_2(y_1, y_2) = \min_{h \in \mathcal{H}, x_i \in \mathcal{U}} \sum_{i=1}^{2} \int_0^T \|y_i(t) - h(x_i(t))\|^2 + \lambda \|\nabla h(t)\| dt \tag{9}$$

where λ is a tuning parameter for the regularizer that is not strictly necessary for this model, and can be set equal to zero, for instance, by choosing $h(t) = y_1(x_1^{-1}(t))$ (or, similarly to what we have done earlier $h(t) = (y_1(x_1^{-1}(t)) + y_2(x_2^{-1}(t))/2)$ and $x(t) \doteq x_2(x_1^{-1}(t))$. This yields a "reduced" optimization problem with only one (functional) variable,

$$\int_0^T \|y_1(t) - y_2(x(t))\| dt + \mu \overline{\phi}_{reg}(x) \tag{10}$$

where we have assumed that $\mathcal{U}$ can be defined algebraically as $\{x \mid \overline{\phi}_{reg}(x) = 0\} \doteq \mathcal{U}$, as we will make explicit in Sect. 3. In this simplified model, x matches data-to-data, rather than each x_i matching each data y_i to a common underlying template h. This distance relates to the Skorohod topology introduced for time-of-arrival processes to account for small temporal jittering [1]. The reason for solving a seemingly more complex problem (9), rather than (10), is because, in the presence of noise in the measurements y_1, y_2, the warping x in (10) will attempt to fit the noise, causing the minimizer $\hat{x}$ to be highly irregular. This is usually addressed by enforcing heavy regularization (large $\mu \in \mathbb{R}_+$.) The advantage of the auxiliary variables x_i, h, as discussed in detail in [17], is to avoid warping the (noisy) data y_i, but instead to warp the (smooth) template h; because in general DTW is non-linear and one has to resort to gradient algorithms based on the first-order (Euler-Lagrange) optimality conditions, the presence of an explicit model h allows one to "push" the derivatives onto the model, arriving at *gradient-based* algorithms that *do not involve differentiation of the (noisy) data* [22].

Before we elucidate the structure of the space of warping functions $\mathcal{U}$, we pause to note that in general it is an infinte-dimensional group of diffeomorphisms [4]), as it is (at least locally) invertible. Under the action of this group we can now distinguish two scenarios:

- $\mathcal{U}$ acts transitively on $\mathcal{H}$: For any given y_1, y_2, there exists at least a $\hat{h} \in \mathcal{H}$ (a "template" in Grenander's nomenclature) and $\hat{x}_1, \hat{x}_2 \in \mathcal{U}$ such that the data term $\phi_{data}(y_1, y_2 | \hat{h}, \hat{x}_1, \hat{x}_2)$ is identically zero. In other words, with a group action x one can reach any y from some h. In this case, the data term of the distance is zero, and the actual distance reflects the amount of "energy" or "work" necessary to reach it. This is quantified by the regularization terms $\overline{\phi}_{reg}(x)$.
- $\mathcal{U}$ is restricted (for instance, it belongs to a parametric class of functions), in which case it can only bring h "close" to y, and their proximity is reflected in the distance.

In either case, the minimizer $\hat{h}$ can be interpreted as the "average" of the data, in the sense elucidated in [16]. As we have mentioned, the equivalence classes $[y] \doteq \{y_i \circ x_i^{-1} \,|\, x_i \in \mathcal{U}\}$ represent the objects of interest, and the quotient space can be thought of as the space where comparison is to be performed [20].

2.3 Dynamics, or Lack Thereof, in DTW

It is important to note that *there is nothing "dynamic" about dynamic time warping.*[1] There is no requirement that the warping function x be subject to physical constraints, such as the action of forces, the effects of inertia etc. However, some notion of dynamics can be coerced into the problem by characterizing the set $\mathcal{U}$ in terms of the solution of a differential equation. Following [15], as shown by [12], one can represent allowable $x \in \mathcal{U}$ in terms of a small, but otherwise unconstrained, scalar function u: $\mathcal{U} = \{x \in \mathcal{H}^2([0, \, T]) \,|\, \ddot{x} = u\dot{x}; \, u \in \mathbb{L}^2([0, T])\}$ where $\mathcal{H}^2$ denotes a Sobolev space. If we define $\rho_i \doteq \dot{x}_i$ then $\dot{\rho} = u\rho$; we can then stack the two into $\xi \doteq [x, \, \rho]^T$, and $C = [1, \, 0]$, and write the data generation model as

$$\begin{cases} \dot{\xi}_i(t) = f(\xi_i(t)) + g(\xi_i(t))u_i(t) \\ y_i(t) = h(C\xi_i(t)) + n_i(t) \end{cases} \tag{11}$$

as done by [12], where $u_i \in \mathbb{L}^2([0, T])$. Here f, g and C are given, and $h, x_i(0), u_i$ are nuisance parameters that are eliminated by minimization of the data term

$$\phi_{data}(y_1, y_2 | h, x_1(0), x_2(0), u_1, u_2) \doteq \sum_{i=1}^{2} \int_0^T \|n_i(t)\|^2 dt \tag{12}$$

---

[1] The name comes from the fact that a discretized version of this problem can be solved using dynamic programming, since the integral in (8) can be decomposed into a sum of cost-to-go terms due to the monotonicity constraint.

subject to (11), with the addition of a regularizer $\lambda\phi_{reg}(h)$ and an energy cost for u_i, for instance $\phi_{energy}(u_i) \doteq \int_0^T \|u_i\|^2 dt$. Writing explicitly all the terms, the problem of dynamic time warping can be written as

$$d_3(y_1, y_2) = \min_{h \in \mathcal{H}, u_i \in \mathbb{L}^2, x_i(0)} \sum_{i=1}^{2} \int_0^T \|y_i(t) - h(C\xi_i(t))\| + \lambda\|\nabla h(t)\| + \mu\|u_i(t)\| dt$$

$$(13)$$

subject to $\dot{\xi}_i = f(\xi_i) + g(\xi_i)u_i$. Note, however, that this differential equation does not arise out of the desire to enforce dynamic constraints exhibited in the data, but it is only an expedient to (softly) enforce causality by imposing a small "time curvature" u_i.

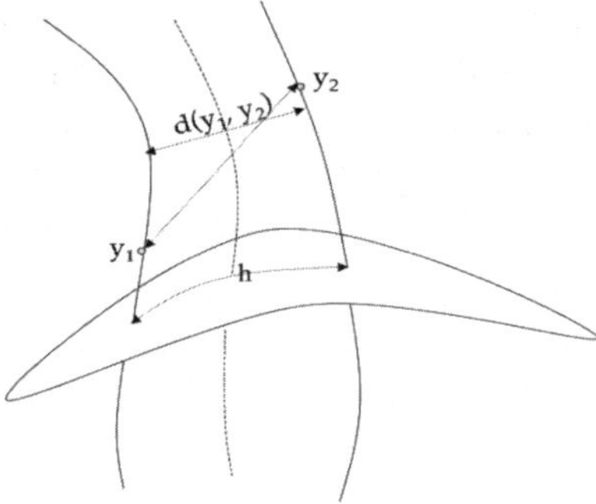

Fig. 1. Time series are points along equivalence classes (fibers), where the nuisance acts as a group that moves points along their fibers. A proper distance in the homogeneous space of equivalence classes is independent of the position of points on the fibers: This can be achieved by minimization, for instance by finding the minimum distance among the fibers (dotted line), or by canonization, by finding a base of the fiber bundle where comparisons are made (dashed lines). The distance proposed in [13] only moves one of the two points (or their average) along the fiber. One of the byproducts of the computation of the distance is the average between the data, a concept that extends to any number $N \geq 2$ time series.

Note also that the solution of the minimization above is not tantamount to a nonlinear system identification task. In fact, in system identification one is given *one* time series y, with the task of inferring the model parameters h, possibly along with the state, input and initial condition (here f, g and C are given). If that were the case, we could always choose $h = y$ for any state, input and initial condition, making the problem trivial. Here instead we are given *two* time series, and we want to jointly estimate the unknown parameters of a model h that, under suitable inputs u_i, can generate the data with minimal discrepancy, measured by ϕ_{data}.

3   Time Warping Under Dynamic Constraints

In this section we introduce a notion of time warping that respects the dynamic structure of the data. Some preliminary progress towards this goal has been made by [8], who

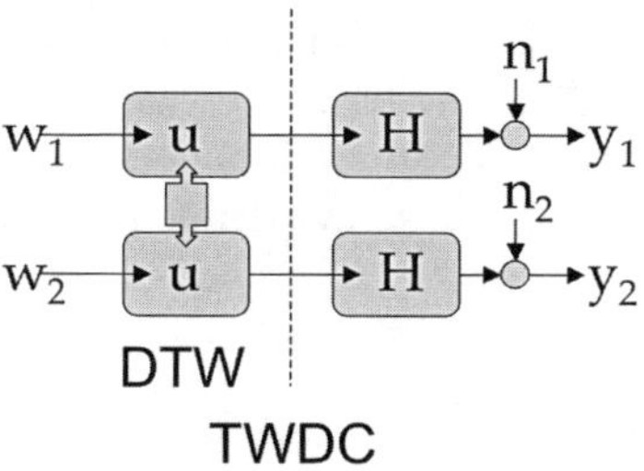

Fig. 2. Traditional dynamic time warping (DTW) assumes that the data come from a common function that is warped in different ways to yield different time series. In time warping under dynamic constraints (TWDC), the assumption is that the data are output of a dynamic model, whose inputs are warped versions of a common input function.

extended the warping function to include derivatives. However, no specific model or assumption, other than small velocity difference, is made between the two time series. Instead, we look for warpings that are compatible with the dynamics imposed by the forces and inertias of the physical processes that generated the data y_i. We will address this issue by considering a generalization of the model (11). The basic idea is illustrated in Figure 2: Rather than the data being warped versions of some common function, as in (6), we will assume that *the data are outputs of dynamical models driven by inputs that are warped versions of some common function.* In other words, given two time series y_i, $i = 1, 2$, we will assume that there exist suitable matrices A, B, C, state functions x_i of suitable dimensions, with their initial conditions, and a *common input u* such that the data are generated by the following model, *for some warping functions $w_i \in \mathcal{U}$:*

$$\begin{cases} \dot{x}_i(t) = Ax_i(t) + Bu(w_i(t)) \\ y_i(t) = Cx_i(t) + n_i(t). \end{cases} \tag{14}$$

Our goal is to find the distance between the time series by minimizing with respect to the nuisance parameters the following data discrepancy:

$$\phi_{data}(y_1, y_2 | u, w_i, x_i(0)) \doteq \sum_{i=1}^{2} \int_0^T \| n_i(t) \|^2 dt \tag{15}$$

subject to (14), together with regularizing terms $\overline{\phi}_{reg}(u)$ and with $w_i \in \mathcal{U}$. Notice that this model is considerably different from the previous one, as the state ξ earlier was used to model the temporal warping, whereas now it is used to model the data, and the warping occurs at the level of the input. It is also easy to see that the model (14), despite being linear in the state, includes (11) as a special case, because we can still model the warping functions w_i using the differential equation in (11). In order to write this *time warping under dynamic constraint* problem more explicitly, we will use the following notation:

$$y(t) = Ce^{At}x(0) + \int_0^T Ce^{A(t-\tau)}Bu(w(\tau))d\tau \doteq L_0(x(0)) + L_t(u(w)) \tag{16}$$

292   S. Soatto

in particular, notice that L_t is a convolution operator, $L_t(u) = F * u$ where F is the transfer function. We first address the problem where A, B, C (and therefore L_t) are given. For simplicity we will neglect the initial condition, although it is easy to take it into account if so desired. In this case, we define the distance between the two time series

$$d_4(y_1, y_2) = \min \sum_{i=1}^{2} \int_0^T \|y_i(t) - L_t(u_i(t))\| + \lambda \|u_i(t) - u_0(w_i(t))\| dt \qquad (17)$$

subject to $u_0 \in \mathcal{H}$ and $w_i \in \mathcal{U}$. Note that we have introduced an auxiliary variable u_0, which implies a possible discrepancy between the actual input and the warped version of the common template. This problem can be solved in two steps: A deconvolution, where u_i are chosen to minimize the first term, and a standard dynamic time warping, where w_i and u_0 are chosen to minimize the second term. Naturally the two can be solved simultaneously.

3.1   Going Blind

When the model parameters A, B, C are common to the two models, but otherwise unknown, minimization of the first term corresponds to blind system identification, which in general is ill-posed barring some assumption on the class of inputs u_i. These can be imposed in the form of generic regularizers, as common in the literature of blind deconvolution [3]. This is a general and broad problem, but beyond our scope here, so we will forgo it in favor of an approach where the input is treated as the output of an auxiliary dynamical model, also known as *exo-system* [5]. This combines standard DTW, where the monotonicity constraint is expressed in terms of a double integrator, with TWDC, where the actual stationary component of the temporal dynamics is estimated as part of the inference. The generic warping w, the output of the exo-system (see Figure 3), satisfies

$$\begin{cases} \dot{w}_i(t) = \rho_i(t), \quad i = 1, 2 \\ \dot{\rho}_i(t) = v_i(t)\rho_i(t) \end{cases} \qquad (18)$$

and $w_i(0) = 0$, $w_i(T) = T$. This is a multiplicative double integrator; one could conceivably add layers of random walks, by representing v_i are Brownian motion. Combining this with the time-invariant component of the realization yields the generative model for the time series y_i:

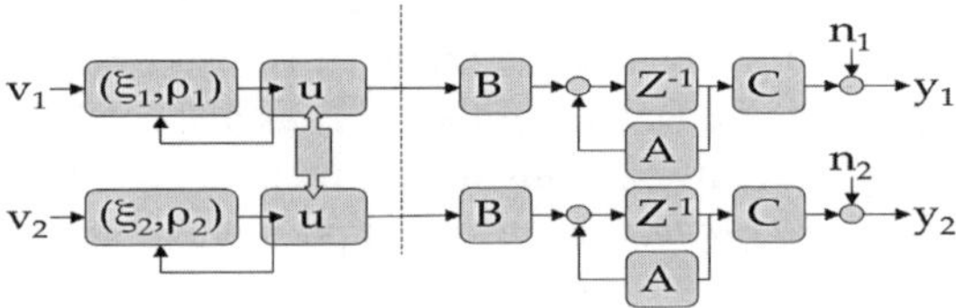

Fig. 3. TWDC can be modeled as comparison of the output of two dynamical models driven by two exo-systems that are in charge of time-warping a common input u

$$\begin{cases} \dot{w}_i(t) = \rho_i(t), \quad i = 1, 2 \\ \dot{\rho}_i(t) = v_i(t)\rho_i(t) \\ \dot{x}_i(t) = Ax_i(t) + Bu(w_i(t)) \\ y_i(t) = Cx_i(t) + n_i(t). \end{cases} \tag{19}$$

Note that the actual input function u, as well as the model parameters A, B, C, are common to the two time series. A slightly relaxed model, following the previous subsection, consists of defining $u_i(t) \doteq u(w_i(t))$, and allowing some slack between the two; correspondingly, to compute the distance one would have to minimize the data term

$$\phi_{data}(y_1, y_2 | u, w_i, A, B, C) \doteq \sum_{i=1}^{2} \int_0^T \|n_i(t)\|^2 dt \tag{20}$$

subject to (19), in addition to the regularizers

$$\overline{\phi}_{reg}(v_i, u) = \sum_{i=1}^{2} \int_0^T \|v_i(t)\|^2 + \|\nabla u(t)\|^2 dt \tag{21}$$

which yields a combined optimization problem

$$\boxed{d_5(y_1, y_2) = \min_{u, \in \mathbb{L}^2, A, B, C} \sum_{i=1}^{2} \int_0^T (\|y_i(t) - Cx_i(t)\|^2 + \|v_i(t)\|^2 + \|\nabla u(t)\|^2) dt}$$
$$\tag{22}$$

subject to (19).

3.2   Computing the Distance

In order to compute the distance for the various cases defined above we have to solve what is in effect an optimal control problem. Specifically, for the following model[2]

$$\begin{cases} \dot{x} = v \quad x(0) = 0; \ x(T) = T \\ \dot{v} = uv \end{cases} \tag{23}$$

relative to the cost function

$$J = \int_0^T \|y - h(x)\|^2 + u^2 dt \tag{24}$$

we are looking for $\min J$ subject to (23). The Hamiltonian is given by

$$H(x, u, \lambda) = \|y - h(x)\|^2 + u^2 + \lambda^T [v, uv]^T \tag{25}$$

and the value function (optimal cost) V should satisfy the Hamilton-Jacobi-Bellmann equation

---

[2] We use the simplified version where $y_1 = y$, $y_2 = h$, but all considerations can be easily extended to the more general case, where u_i, v_i, x_i have $i = 1, 2$.

$$\frac{\partial V}{\partial t}(x, t) + \min_u H\left(x, u, \frac{\partial V}{\partial x}\right) = 0 \qquad (26)$$

with suitable boundary conditions. A discretized version of this equation can be computed on a sample time sequence

$$V(x, t) = \min_u \int_t^{t+\Delta} \|y - h(x)\|^2 + u^2 d\tau + V(x(t + \Delta), t + \Delta) \qquad (27)$$

with boundary conditions for x to satisfy (23) using Dynamic Programming. If we are content with a (faster) local gradient algorithm based on the first-order (Euler-Lagrange) optimality conditions, then we simply update iteratively u, starting from an initial estimate, in the direction opposite to the gradient of the Hamiltonian.

In the case of TWDC, the additional (finite-dimensional) unknowns due to the model parameters (A, B, C) can be easily incorporated. Note that the quotient structure of the parameter space of realizations is not an issue here since all that matters is the minimum value (distance), rather than the minimizer (realization).

4   Correlation Kernels for Non-stationary Time Series

In this section we explore a distinctly different approach to defining a distance between time series. Instead of computing the data term ϕ_{data} in terms of the $\mathbb{L}^2$ distance between the time series, an alternative consists in defining an inner product between the two, via correlation, from which a cord distance can be easily computed. This has been done for the case of time-invariant models in [21]. In this section we illustrate how these concepts can be generalized to allow of time-warpings of the input. Using the notation in (16), we define the (symmetric, positive-definite) kernel

$$K(y_1, y_2 | u) \doteq \mathbb{E}_{v_1, v_2}\left[\text{trace} \int_0^T y_1(t) y_2^T(t) d\mu(t)\right]$$

$$= \mathbb{E}_{v_1, v_2}\left[\text{trace} \int_0^T L_t(u(w_1(t))) L_t^T(u(w_2(t))) d\mu(t)\right] \qquad (28)$$

where $d\mu(t) \sim \frac{e^{-\lambda t}}{t}$ includes an exponential discounting term and the expectation is computed with respect to the joint density of v_1, v_2, subject to (18). We can make the functional explicit by exploiting the calculations leading to equation (6.11) of [15], to obtain

$$L_t = C \int_0^T e^{A(t-\tau)} Bu\left(K_0 + K_1 \int_0^\tau \exp \int_0^{\tau'} v(\tau'') d\tau'' d\tau'\right) d\tau. \qquad (29)$$

This can be substituted into the previous equation and integrated against the joint density of v_1 and v_2. Several simplifying assumptions are possible for this density: One can assume, as in [21], that the two are independent, or that they are identical. One could also assume that v_1, v_2 are small an independent, an assumption implicit in the choice of regularizer in (22) (the second term in the integral). This can be enforced in practice by choosing a joint density for discretized versions of v_1, v_2 proportional to $\exp(-(\|v_1\|^2 + \|v_2\|^2))$. The two constants K_0, K_1 can be set by imposing the boundary conditions. Note that the kernel depends upon the model parameters $\{A, B, C\}$, hidden in the operator L_t, as well as on the unknown input $u \in \mathcal{U}$. Note also that the initial condition can be used to define an additive kernel, identically to what done in [21]. The kernel above satisfies Mercer's condition, and because the sum of Mercer kernels is also Mercer, this procedure yields a viable kernel in a straightforward manner.

The non-straightforward part of this program is the computation of the expectation above, for which no better strategy than general Monte Carlo is currently available. However, assuming that it can be done, one can use the kernel to define a distance via

$$\phi_{data}(y_1, y_2|u, A, B, C) \doteq K(y_1, y_1|u) + K(y_2, y_2|u) - 2K(y_1, y_2|u) \qquad (30)$$

and then optimize with respect to the unknowns $u \in \mathcal{U}, A, B, C$. As an alternative, one could marginalize the unknowns to compute

$$\boxed{d_6(y_1, y_2) = \int (\phi_{data} + \lambda \phi_{reg}(u)) dP(u)} \qquad (31)$$

as an alternative to extremization when a measure on $\mathcal{U}$ is available.

5   Invariance Via Canonization

In previous sections we have explored various alternative distances where the nuisances were eliminated by extremization (i.e. solving an optimization, or "search," problem), or by marginalization. In either cases, the computation of the distance entails the solution of a difficult computational problem. As we have pointed out at the beginning, an alternative way to endow a homogenous space with a metric structure is to reduce it to its base, that is to define for each class a *canonical representative*, and then to compute a distance in the base space that respects its geometry. For the case of stationary processes, a variety of canonical realizations has been defined. In our case, the canonical representative would have to be a diffeomorphism of the domain $[0, T]$, and "canonical" refers to the fact that given a certain time series $\{y(t)\}_{t \in [0,T]}$ and its associated equivalence class $[y] = \{y(w(t)), w \in \mathcal{U}\}$, a canonical representative $\hat{y} \doteq y(\hat{w})$ must be computed solely from $[y]$, i.e. without resorting to comparison with other fibers.

For simplicity, we illustrate the canonization process for the simple case of affine domain deformations first, although the construction can be extended to arbitrary diffeomorphisms as shown in [18] (see also Figure 4 for an illustration on this procedure). For domain transformations of the form $w(t) = \alpha t + \beta$ we have to relax the fixed boundary conditions $w(0) = 0$ (lest $\beta = 0$) and $w(T) = T$ (lest $\alpha = 1$). Assuming the boundaries of observation of the two sequences to be undetermined, we can canonize each sequence by choosing α and β that generate statistics of $\{y(w(t))\}$ that have a prescribed value. For instance, we can take some differential statistics, of the form $\phi(y) = \frac{d^k y}{dt^k}$, and impose that they take a prescribed value in uniquely identifiable positions. For instance, t_1 can be the first position where $\frac{dy}{dt} = 0$, assigned

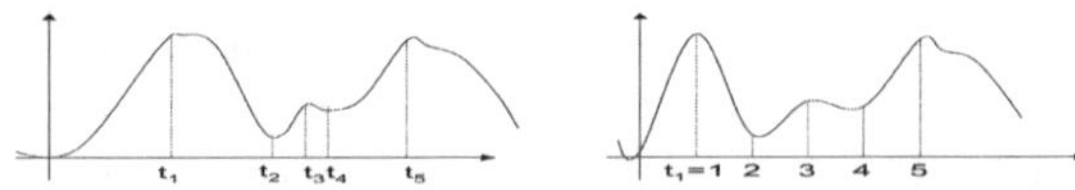

Fig. 4. Simple canonization for comparing time series: Extrema of the warped function (or a scale-space of it) are assigned to fixed value in increasing order. The back-warped time series is then, by construction, invariant to domain warpings.

to a fixed value on the axis, say $t = 0$. This fixes the translation group β. We can then take the second point where $\frac{dy}{dt} = 0$, and assign it to a fixed value, say $t = 1$. This fixes the linear scaling group α. The procedure can be extended to more general warpings, for different statistics, including integral ones (moments) of higher order. Note that if the data y are such that such distinct points do not exist (for instance if $y(t)$ is constant), then *any* α and β would do. So, in other words, where we can find statistics that depend on w, we can fix them to canonize w; where the statistics do not depend on w, these are already, by definition, invariant!

We now extend this construction more formally. Consider a set of data (time series) $\{y_i \in M\}_{i=1,\ldots,n}$ undergoing the action of a *nuisance group* $w_i \in \mathcal{U}$, to yield their *irked* versions $\tilde{y}_i = w_i y_i \doteq y_i(w_i(t))$. The ambient space, where the irked data live, is the set of orbits $N \doteq M^{\mathcal{U}}$. Now, suppose that there exists a "feature," i.e. a function $\phi : N \to \mathbb{R}^l$, where l is the dimension of $\mathcal{U}$, such that $\phi(w^{-1}\tilde{y}) = 0$ uniquely determines u up to a set of measure zero $K \subset N$.[3] Then, we can eliminate the effect of the nuisance by pre-processing each irked datum via $\hat{y} \doteq \hat{w}^{-1}\tilde{y} \mid \phi(\hat{w}^{-1}\tilde{y}) = 0$ to obtain a *canonical* element $\hat{y} \in M$. The choice of $\hat{y}$ is canonical in the sense of conforming to the rule $\phi(\hat{y}) = 0$.[4] The function ϕ is called a *pontifical feature* since it is the feature (i.e. the statistic) that determines how $\tilde{y}$ is to be canonized. If the data space M, undergoing the action of the nuisance group $\mathcal{U}$, admits a pontifical feature, it is called *sanctifiable*. We write the canonization process more succinctly as

$$\hat{y} \doteq \phi^{-1}(0|\tilde{y}) = \hat{w}^{-1}\tilde{y} \mid \phi(\hat{w}^{-1}\tilde{y}) = 0 \tag{32}$$

or, with an unholy abuse of notation, as $\hat{y} = \phi^{-1}(\tilde{y})$. It is easy to construct examples that show that not all spaces are sanctifiable [18]. Note, however, that whether a space is sanctifiable depends on the base M as well as on the nuisance $\mathcal{U}$. In general, the larger the nuisance, the more difficult it is for the space to be sanctified. Sometimes it is possible for the space to be sanctifiable, but with only one canonical element. That is, the quotient is zero-dimensional, and the entire irked population $\{\tilde{y}_i\}$ is equal under the law $\phi = 0$. In this case we say that the space N *collapses under the nuisance $\mathcal{U}$.*

---

[3] Such a set of measure zero is the set of data that is invariant under a subgroup $H \subseteq \mathcal{U}$, i.e. the *symmetry* set of H: $K \doteq \{\tilde{y} \mid \phi(w^{-1}\tilde{y}) = \phi(\tilde{y}) \,\forall\, u \in \mathcal{U}\}$.

[4] The choice of zero in the rule $\phi = 0$ is arbitrary, and any other value $\phi = k \neq 0$ would work just as well, yielding a different set of canonical representatives $\hat{y} \in N/\mathcal{U}$. Therefore, in general the base space $N/\mathcal{U}$ where the canonical representatives live is not necessarily equal to M, but it is related to it by a parallel translation $\tilde{w} \in \mathcal{U}$, so that the "true" space is given by $M = \tilde{w}N/\mathcal{U}$. Since any value $k \neq 0$ can be incorporated into the definition of ϕ, the choice of the zero level set in (32) is without loss of generality.

Example 1. In Grenander's "Deformable templates" [4] the objects of interest ("target shapes") are obtained from a common generator (the "template") under the action of an infinite-dimensional group. Because of the assumption that the group acts transitively, the entire world of objects of interest is equivalent under the action, and therefore the space collapses under the nuisance.

To construct a simple pontifical feature, consider simply the function $\phi(\tilde{y}) \doteq \frac{d\tilde{y}}{dt}$, and let $t_1, \ldots, t_N$ be the N local extrema of $\tilde{y}$:

$$t_i \doteq t \mid \phi(\tilde{y}) = 0, \ i = 1, \ldots, N \tag{33}$$

Because $\frac{d\tilde{y}}{dt} = \frac{dy}{dt}\frac{dw}{dt}$ and by assumption $w \in \mathcal{U}$ we have that $\frac{dw}{dt} > 0$, $t \in [0, T]$, we have that the values $\tilde{y}(t_i)$ are independent of w. We can then choose a canonical representative for w by imposing

$$\hat{w}(t_i) = \frac{i}{N+1}T \tag{34}$$

where we have assumed $t_0 = 0$ and $t_{N+1} = T$. The canonical representative of $\tilde{y}$ is then simply given by

$$\hat{y}(t) = \tilde{y}(\hat{w}^{-1}(t)), \quad t \in [0, T]. \tag{35}$$

Finally, the distance between canonical elements can be simply computed in $\mathbb{L}^2$:

$$d_7(y_1, y_2) = \int_0^T \|y_1(\hat{w}_1^{-1}(t)) - y_2(\hat{w}_2^{-1}(t))\|^2 dt = \int_0^T \|\hat{y}_1(t) - \hat{y}_2(t)\|^2 dt \tag{36}$$

Note that this solution is the canonization counterpart of DTW presented in Sect. 2.2. In order to extend this to TWDC as presented in Sect. 3 one would have to canonize $v(t)$, rather than $w(t)$, which can be done at the expense of additional notation, and we will therefore forgo it in this venue.

As we have already observed, the choice of canonical element is arbitrary, so that the effects of canonization in classification largely depend on the fine art of choosing a pontifical feature. Such a feature is, by design, invariant to the nuisances we have modeled explicitly: The art consists in making it also robust, or "insensitive," to other factors that we do not have explicitly modeled. In particular, canonization is sensitive to missed detections or spurious detections in the pontifical feature. For instance, in the illustrative case just discussed, in the presence of noise one would have different realization produce different numbers and location of local minima. This can be minimized by defining a scale-space of features, rather than considering the signal only at the resolution defined by the sample frequency of the sensor [10], and [19] for a more thorough discussion on this issue.

6 Discussion

The impact of the system-theoretic approach to dynamic data analysis has yet to be felt in important areas of applications such as data mining or computer vision. In order for this to happen, more general and flexible models have to be introduced. In this work we have made a modest step in this direction, by introducing Time Warping under Dynamic Constraints (TWDC), a method to compare time series that respects their dynamics. We have also introduced, albeit in a purely formal manner, a correlation kernel between processes that would respect their dynamic structure, if one could afford the time to compute the expectation with respect to the joint density of the driving noises. Finally, we have illustrated how one could, in principle, construct time-deformation-invariant statistics from time series to arrive at canonical representatives that are not affected by nuisances. These hold the promise for more efficient comparison, for each sequence can be pre-processed and comparison is performed by the simple computation of an $\mathbb{L}^2$ norm.

Acknowledgments

This work has benefited from extended discussions with Alessandro Chiuso, Andrea Vedaldi, Gregorio Guidi, Michalis Raptis, and René Vidal. Anything good there might be in it, it has been influenced – directly or indirectly – by Giorgio Picci's work. All the rest is sole responsibility of the author. The support of AFOSR and ONR is gratefully acknowledged.

References

1. P. Billingsley. *Convergence of Probability Measures*. Wiley, 1968.
2. A. Bissacco, A. Chiuso, and S. Soatto. Classification and recognition of dynamical models: the role of phase, independent components, kernels and optimal transport. *IEEE Trans. Pattern Anal. Mach. Intell.*, in press, 2007.
3. B. Giannakis and J. Mendel. Identification of nonminimum phase systems using higher order statistics. *IEEE Trans. on Acoustic, Speech and Signal Processing*, 37(3):360–377, 1989.
4. U. Grenander. *General Pattern Theory*. Oxford University Press, 1993.
5. A. Isidori. *Nonlinear Control Systems*. Springer Verlag, 1989.
6. G. Johansson. Visual perception of biological motion and a model for its analysis. *Perception and Psychophysics*, 14:201–211, 1973.
7. H. Kantz and T. Schreiber. *Nonlinear Time Series Analysis*. Cambridge University Press, 2004.
8. E. J. Keogh and M. J. Pazzani. Dynamic time warping with higher order features. In *Proceedings of the 2001 SIAM Intl. Conf. on Data Mining*, 2001.
9. A. Kirsch. An introduction to the mathematical theory of inverse problems. *Springer-Verlag, New York*, 1996.
10. T. Lindeberg. Scale space for discrete signals. *IEEE Trans. Pattern Anal. Mach. Intell.*, 12(3):234–254, 1990.
11. A. Lindquist and G. Picci. The stochastic realization problem. *SIAM J. Control Optim. 17*, pages 365–389, 1979.
12. C. F. Martin, S. Sun, and M. Egerstedt. Optimal control, statistics and path planning. 1999.
13. R. Martin. A metric for arma processes. *IEEE Trans. on Signal Processing*, 48(4):1164–1170, 2000.
14. P. J. Olver. *Equivalence, Invariants and Symmetry*. Cambridge University Press, 1995.
15. J. O. Ramsey and B. W. Silverman. *Functional Data Analysis*. Springer Verlag, 2005.

16. S. Soatto and A. Yezzi. Deformotion: deforming motion, shape average and the joint segmentation and registration of images. In *Proc. of the Eur. Conf. on Computer Vision (ECCV)*, volume 3, pages 32–47, 2002.
17. S. Soatto, A. J. Yezzi, and H. Jin. Tales of shape and radiance in multiview stereo. In *Intl. Conf. on Comp. Vision*, pages 974–981, October 2003.
18. A. Vedaldi and S. Soatto. Features for recognition: viewpoint invariance for non-planar scenes. In *Proc. of the Intl. Conf. of Comp. Vision*, October 2005.
19. A. Vedaldi and S. Soatto. Viewpoint induced deformation statitics and the design of viewpoint invariant features: singularities and occlusions. In *Eur. Conf. on Comp. Vision (ECCV)*, pages II–360–373, 2006.
20. A. Veeraraghavan, R. Chellappa, and A. K. Roy-Chowdhury. The function space of an activity. In IEEE, editor, *Proc. IEEE Conf. on Comp. Vision and Pattern Recogn.*, 2006.
21. S.V.N. Vishwanathan, R. Vidal, and A. J. Smola. Binet-cauchy kernels on dynamical systems and its application to the analysis of dynamic scenes. *International Journal of Computer Vision*, 2005.
22. A. Yezzi and S. Soatto. Stereoscopic segmentation. In *Proc. of the Intl. Conf. on Computer Vision*, pages 59–66, 2001.

Stochastic Realization for Stochastic Control with Partial Observations

Jan H. van Schuppen

CWI, P.O.Box 94079, 1090 GB Amsterdam, The Netherlands
J.H.van.Schuppen@cwi.nl

The paper is dedicated to Giorgio Picci on the occasion of his 65th birthday for his inspiring contributions to stochastic realization and to system identification.

1 Introduction

The purpose of this paper is to present a novel way to formulate control problems with partial observations of stochastic systems. The method is based on stochastic realization theory.

The contribution of the paper is the stochastic realization approach to stochastic control with partial observations. The motivation for the paper are the weaknesses of control with partial observations based on the separation principle. The separation property holds for the stochastic control problem with a Gaussian system and a quadratic cost function (LQG), but does not hold for at least one optimal stochastic control problem, LEQG, and may not hold for most other stochastic control problems. Moreover, control of decentralized stochastic systems or control of stochastic dynamic games becomes unsolvable or unnatural in the existing approaches in the literature. Therefore a novel approach may be explored.

The stochastic realization approach to stochastic control with partial observations proceeds by the following steps. (1) A stochastic realization of the input-output process is selected such that the state space is finite or finite-dimensional and at any time the state is a measurable function of the past outputs and the past inputs. (2) An optimal stochastic control problem with complete observations is solved by the existing stochastic control theory. The advantage of this approach is that the stochastic realization selected has a state set which is finite or finite-dimensional. According to the classical control theory one has to solve a filtering problem of which the state set of the filter system is not necessarily finite or finite-dimensional. This difficulty is avoided in the proposed approach.

The contents of the paper is described below. The next section contains the problem formulation. The classical approach to stochastic control with partial observations is summarized in Section 3. The stochastic realization approach to stochastic control with partial observations is presented in Section 4. Several special cases are discussed in Section 5. Concluding remarks are stated in the last section.

A. Chiuso et al. (Eds.): Modeling, Estimation and Control, LNCIS 364, pp. 301–314, 2007.
springerlink.com

2 Problem Formulation

The problem of control with partial observations of stochastic systems is formulated in this section and the approach is summarized. The approach is developed in the subsequent sections.

The notation of this paper is in accordance with the literature on control of stochastic systems. A *probability space*, denoted by (Ω, F, P), consists of a sample set Ω, a σ-algebra F of subsets of Ω, and a probability measure $P : F \to [0, 1]$. The set of the *integers* is denoted by $\mathbb{Z}$, the *natural numbers* by $\mathbb{N} = \{0, 1, 2, \ldots\}$, the *strictly positive integers* by $\mathbb{Z}_+ = \{1, 2, 3, \ldots\}$, and for $n \in \mathbb{Z}_+$, $\mathbb{Z}_n = \{1, 2, \ldots, n\}$. In this paper attention is restricted to *discrete-time stochastic control systems* hence the *time index set* is denoted by either $T = \{0, 1, \ldots\} = \mathbb{N} \subset \mathbb{Z}$, or $T = \{0, 1, \ldots, t_1\} \subset \mathbb{N}$. For a strictly positive integer $n \in \mathbb{Z}_+$, $(\mathbb{R}^n, B(\mathbb{R}^n))$ denotes the n-fold Cartesian product of the real numbers with the Borel σ-algebra generated by the open subsets of $\mathbb{R}^n$. A *Gaussian probability measure* on $\mathbb{R}^n$ is denoted by $G(m, Q)$ where the parameter corresponding to the *mean value* is $m \in \mathbb{R}^n$ and the parameter corresponding to the *variance* is $Q \in \mathbb{R}^{n \times n}$ satisfying $Q = Q^T \geq 0$. Denote by F^{x_0} the σ-algebra generated by the random variable x_0. A *Gaussian random variable* is a random variable $x : \Omega \to \mathbb{R}^n$ with a Gaussian probability measure, denoted by $x \in G(m, Q)$.

A *discrete-time Gaussian white noise process* with variance Q_v is a stochastic process $v : \Omega \times T \to \mathbb{R}^{m_v}$ such that $\{v(t), t \in T\}$ is a sequence of independent random variables and for each $t \in T$, the random variable $v(t) : \Omega \to \mathbb{R}^{m_v}$ has a Gaussian distribution with $v(t) \in G(0, Q_v)$ where $Q_v \in \mathbb{R}^{m_v \times m_v}$, $Q_v = Q_v^T \geq 0$. The *filtration* generated by a stochastic process $v : \Omega \times T \to \mathbb{R}^{m_v}$ is denoted by $\{F_t^v, t \in T \cup \{\infty\}\}$. The *open unit disc* of the complex plane is denoted by $\mathbb{D}_o = \{c \in \mathbb{C}| \ |c| < 1\}$.

The main concept of stochastic system theory is that of a *stochastic control system*. Distinguish an *input process* to be determined by the controller, an *output process* which is observed, and a *state process*. A state process of a stochastic control system is such that at any time moment the conditional distribution of future states and of future outputs conditioned on the past of the output and the past and future of the input process at that particular time depend only on the current state and the future inputs. It is a result of stochastic realization theory that there are in general several stochastic control systems which describe the same processes and any of these systems will be termed a *stochastic realization*. Attention has to be restricted to that stochastic realization of which at any time the state is measurable with respect to the past outputs and the past inputs.

The problem of control with partial observations is now formalized in terms of words, not in terms of mathematical notation. In Section 3 the problem will be formalized mathematically. Consider then a stochastic control system and a set of control laws or controllers. The interconnection of the stochastic control system and a controller is termed the corresponding *closed-loop system*. A *control objective* is a property of the closed-loop system which a control designer strives to attain. Examples of control objectives for the control problem include: stochastic stability, minimization of a performance criterion, robustness of the performance with respect to disturbance signals often the variance, and adaptation to long term changes in the stochastic control system.

Problem 1. *Stochastic control with partial observations.* Consider an input and output process. Determine a controller which specifies at each time which input to apply based on the past output and the past input values such that a set of prespecified control objective is met as well as possible.

The approach to the problem proposed in this paper is to first select a stochastic realization in the form of a stochastic system of the input and the output process of which the associated state is measurable in terms of the past outputs and the past inputs. Second, the stochastic control problem for this stochastic realization is solved. Note that the latter problem is a control problem with complete observations! For the latter problem there are many approaches including optimal control theory. The approach is formulated in Section 4, special cases are described in Section 5, but only after the classical approach is summarized in Section 3.

3 The Classical Approach

In this section the classical approach to stochastic control with partial observations is sketched briefly so as to contrast it with the to be proposed approach.

The theoretical framework for stochastic control with partial observations was mainly formulated during the 1960's. At the early 1960's there were available the theory for optimal control of deterministic systems based on the maximum principle and the dynamic programming approach. Moreover, the Kalman filter was published as an alternative to the Wiener filter theory. The control objective of suppression of disturbance signals focused attention on the control of stochastic systems later on called the problem of *stochastic control*.

The solution of the discrete-time stochastic control problem with a Gaussian stochastic control system and a quadratic cost function is credited to [13, 15]. The theoretical framework is due to C. Striebel and W.M. Wonham. C. Striebel formulated the concept of an *information system* for stochastic control, the concept of a *sufficiently informative statistic* for the control cost, and formulated the dynamic programming approach to stochastic control with partial observations, see [20]. The continuous-time framework was developed by W.M. Wonham who observed that *in general* the conditional distribution of the state depends on the control law used but that for the case of a Gaussian stochastic control system the conditional distribution does *not* depend on the control law. The stochastic control problem then exhibits the *separation property*, the problem separates into a filtering problem and a control problem with complete observations. This property has later on been used to formulate the *separation principle* of control with partial observations of nonlinear systems. Thus, according to the separation principle, one proceeds under the hypothesis that the problem of control with partial observations separates into a filtering problem and a control problem with complete observations regardless of whether the problem at hand has the separation property. The achievements of the 1960's still stand out as major contribution of control theory. Further extensions to stochastic control with partial observations for nonlinear systems with Gaussian disturbance signals were published by Ray Rishel, Pravin Varaiya, Mark H.A. Davis, Charlotte Striebel, V.E. Benes and I. Karatzas, see [2, 7, 8, 19, 21, 22]. By now the topic of stochastic control with partial observations is covered well in the text books [4], its latest edition [5, 6], and [12, 17].

The approach of separation was extended to control with partial observations in case of decentralized control and of stochastic dynamic games. Hans Witsenhausen formulated several results and conjectures on separation of filtering and control, see [25, 26, 27].

Below the classical approach to stochastic control with partial observations is sketched briefly for Gaussian stochastic control systems to contrast it later with the stochastic realization approach.

Definition 1. *Consider a Gaussian stochastic control system of the form*

$$x(t+1) = Ax(t) + Bu(t) + Mv(t), \ x(t_0) = x_0, \tag{1}$$

$$y(t) = Cx(t) + Du(t) + Nv(t). \tag{2}$$

The remaining conditions are formulated in Definition 3 except that the rank condition on the matrix N is not imposed in this definition.

A *control law* for the above system is a map $g : T \times Y^T \times U^T \to U$. Denote the *set of control laws* by (with abuse of notation)

$$G = \left\{ \begin{array}{l} g : T \times Y^T \times U^T \to U | g \text{ is a measurable function} \\ g \text{ is causal } g(t, y|_{[0,t]}, u|_{[0,t]}) = g(t, y, u) \end{array} \right\}.$$

Define the *closed-loop system* associated with the above Gaussian stochastic control system and a control law $g \in G$ as the stochastic system

$$x^g(t+1) = Ax^g(t) + Bg(t, y^g|_{[0,t)}, u^g|_{[0,t)}) + Mv(t), \ x^g(t_0) = x_0, \tag{3}$$

$$y^g(t) = Cx^g(t) + Dg(t, y^g|_{[0,t)}, u|_{[0,t)}) + Nv(t), \tag{4}$$

$$u^g(t) = g(t, y^g|_{[0,t)}, u|_{[0,t)}). \tag{5}$$

To emphasize the dependence of the state and the output process on the control law g, the control law is used as a super index. Define a *cost function* for any control law $g \in G$ as

$$J(g) = E[\sum_{s=0}^{t_1-1} b(x^g(s), u^g(s)) + b_1(x^g(t_1))], \tag{6}$$

$$J : G \to \mathbb{R}_+, \ b : X \times U \to \mathbb{R}_+, \ b_1 : X \to \mathbb{R}_+.$$

The *problem of optimal stochastic control with partial observations* for the above system, the set of control laws, and the cost function is then to solve the problem

$$\inf_{g \in G} J(g). \tag{7}$$

This amounts to determining the infimal value, establishing the existence of an optimal control law $g^* \in G$ if one exists (thus $J(g^*) = \inf_{g \in G} J(g)$), and establishing whether or not an optimal control law is unique.

The classical approach to stochastic control with partial observations for the above formulated problem as sketched below is due to W.M. Wonham. First solve the filtering problem for the Gaussian stochastic control system: determine the conditional distribution of the state conditioned on the past outputs and the past inputs

$$E[\exp(iw^T x^g(t))|F_{t-1}^{y^g} \vee F_{t-1}^{u^g}], \quad \forall t \in T, \ \forall g \in G. \tag{8}$$

As first observed by Wonham, the conditional distribution depends on the control law used. The reader may want to note that in case the control law $g \in G$ is nonlinear then the state process x^g is not even a Gaussian stochastic process in general hence the Kalman filtering approach cannot be applied. The achievement of Wonham was to prove that the conditional distribution does not depend on the control law used. Then the filtering problem can be solved by the Kalman filter theory. The resulting equations are

$$\hat{x}(t+1) = A\hat{x}(t) + Bu(t) + K(t)[y(t) - C\hat{x}(t) - Du(t)], \ \hat{x}(t_0) = 0,$$
$$K(t) = [AQ(t)C^T + MVN^T][CQ(t)C^T + NVN^T]^{-1},$$

and where the function $Q : T \to \mathbb{R}^{n \times n}$ is the solution of the forward filter Riccati difference equation. The reader will easily find that filter Riccati equation in the literature.

The second step of the classical approach is to solve a stochastic control problem with complete observations. The stochastic system is now that of the filter system but notice that now the state of the system is a function of the past outputs and the past inputs hence available to the control law. The latter problem can be solved by the existing control theory of stochastic control with complete observations. The resulting equations are

$$G_X = \{g : T \times X \to U | g \text{ measurable function}\}$$
$$g \in G_X \text{ a control law,}$$
$$J(g) = E\left[\sum_{s=t_0}^{t_1-1} b(x^g(s), u^g(s)) + b_1(x^g(t_1))\right],$$
$$b(x,u) = \begin{pmatrix} x \\ u \end{pmatrix}^T \begin{pmatrix} L_{11} & L_{12} \\ L_{12}^T & L_{22} \end{pmatrix} \begin{pmatrix} x \\ u \end{pmatrix}, \quad b_1(x) = x^T L_1 x,$$
$$L = \begin{pmatrix} L_{11} & L_{12} \\ L_{12}^T & L_{22} \end{pmatrix} = L^T \geq 0, \ L_{22} > 0, \ L_1 = L_1^T \geq 0;$$
$$\inf_{g_x \in G_X} J(g_x),$$
$$g^*(t,x) = F(t)\hat{x}^{g^*} \text{ the optimal control law,}$$
$$u^*(t) = F(t)\hat{x}^{g^*}(t) \text{ the optimal input trajectory,}$$
$$\hat{x}^{g^*}(t+1) = A\hat{x}^{g^*}(t) + BF(t)\hat{x}^{g^*}(t) +$$
$$+ K(t)[y^{g^*}(t) - C\hat{x}^{g^*}(t) - DF(t)\hat{x}^{g^*}(t)],$$
$$F(t) = -[B^T P(t+1)B + L_{22}]^{-1}[A^T P(t+1)B + L_{12}]^T,$$

where $P : T \to \mathbb{R}^{n \times n}$ is the solution of the backward control Riccati difference equation not displayed here.

4 The Stochastic Realization Approach to Stochastic Control with Partial Observations

Definition 2. The stochastic realization approach to stochastic control with partial observations. *Determine a weak stochastic realization of the input-output process in the* *form of a stochastic control system such that (1) the joint input-output process of this system equals the considered input-output process in terms of their families of finite-dimensional probability distributions (this amounts to the system being a weak stochastic realization), and (2) at any time the state of the system is measurable with respect to the σ-algebra generated by the past of the output and the past of the input process. Thus the stochastic control problem has become one with complete observations in stead of being one with partial observations.*

Then solve the stochastic control problem for the above defined stochastic realization by a method of stochastic control with complete observations.

The approach is illustrated by several special cases in Section 5.

What needs to be established and proven according to the proposed approach?

1. Formulate a stochastic system.
2. Prove that the stochastic control system is a stochastic realization with the required measurability properties.
3. Solve the stochastic control problem with complete observations.
4. Assemble the control law from the stochastic realization and the solution of the stochastic control system.

As with algebraic approaches to control, the most difficult step is the formulation of a stochastic system which has the required measurability properties. This step is a choice rather than the end product of a filtering problem.

The reader may expect a discussion on stochastic controllability of the stochastic realization. In stochastic control theory the stochastic controllability enters via the cost function, it plays a role only to produce finiteness of the cost function on an infinite horizon. For limitations of space this issue is not discussed further in this paper.

The *advantages* of the stochastic realization approach to stochastic control with partial observations are:

- The state space of the stochastic realization mentioned above is finite or finite-dimensional by formulation. This condition refers to the state space of the stochastic realization.
- The stochastic control problem in the stochastic realization approach is one with complete observations for which theory is well developed.
- No filtering problem needs to be solved in the proposed approach while it has to be solved in the classical approach. Note that for nonlinear stochastic systems the filtering problem will in many cases not admit a finite or a finite-dimensional filter system. Therefore the difficulties of deriving a finite-dimensional filter system are avoided.

For an exposition of stochastic realization see [23] or [18].

5   Special Cases

The stochastic realization approach to control with partial observations is detailed for several special cases. The first special case is that of a Gaussian stochastic control system and a quadratic cost function, often referred to as the *LQG* case. The example is rather elementary because of several particularities of this stochastic control system. In fact, the LQG case is such that the resulting control law according to the stochastic realization approach is identical to that of the classical approach. But the subsequent examples show that the situation can be completely different.

A Gaussian Stochastic Control System and a Quadratic Cost Function

Definition 3. *A* Gaussian stochastic control system. *Consider an input process and a Gaussian output process. The stochastic realization in the form of a Gaussian stochastic control system is described by the equations*

$$x(t+1) = Ax(t) + Bu(t) + Kv(t), \quad x(0) = x_0, \tag{9}$$

$$y(t) = Cx(t) + Du(t) + Nv(t), \tag{10}$$

$$n, m, m_v, p \in \mathbb{Z}_+,$$

$$x_0 : \Omega \to \mathbb{R}^n, \quad F^{x_0} \subset F_0^y \vee F_0^u, \quad x_0 \in G(0, Q_0), \tag{11}$$

$$v : \Omega \times T \to \mathbb{R}^p, \text{ is Gaussian white noise with,}$$

$$v(t) \in G(0, Q_v), \quad Q_v = Q_v^T > 0,$$

$$u : \Omega \times T \to \mathbb{R}^m = U, \text{ the input process,}$$

$$F^{x_0}, \ F_\infty^v \text{ are independent } \sigma\text{-algebras,}$$

$$F^{u(t)} \subseteq F_{t-1}^y, \ \forall t \in T,$$

$$x : \Omega \times T \to \mathbb{R}^n = X, \text{ the state process,}$$

$$y : \Omega \times T \to r_u(\omega) = Y, \text{ the output process,}$$

$$A \in \mathbb{R}^{n \times n}, \ B \in \mathbb{R}^{n \times m}, \ K \in \mathbb{R}^{n \times p},$$

$$C \in \mathbb{R}^{p \times n}, \ D \in \mathbb{R}^{p \times m}, \ N \in \mathbb{R}^{p \times p},$$

$$\text{rank}\,(N) = p, \ \text{spec}(A) \subset \mathbb{D}_o, \ \text{spec}(A - KN^{-1}C) \subset \mathbb{D}_o. \tag{12}$$

such that the joint input-output process (u, y) of the stochastic control system equals the considered input-output process in distribution.

Proposition 1. *The defined system is a stochastic realization of the input-output processes and has the specified measurability properties.*

Proof. The system is a stochastic realization by construction. Note further that

$$v(t) = N^{-1}[y(t) - Cx(t) - Du(t)], \text{ where (12) is used,}$$

$$x(t+1) = Ax(t) + Bu(t) + KN^{-1}[y(t) - Cx(t) - Du(t)]$$

$$= [A - KN^{-1}C]x(t) + [B - KN^{-1}N]u(t) + Ky(t), \quad x(0) = x_0,$$

$$x_0 \text{ is } F_0^y \vee F_0^u \text{ measurable where (11) is used,}$$

and by induction one can prove that for all $t \in T\backslash\{t_0\}$, $x(t)$ is $F_{t-1}^y \vee F_{t-1}^u$ measurable hence $F_t^x \subset F_{t-1}^y \vee F_{t-1}^u, \forall t \in T$.

On purpose the state process of the above stochastic realization is denoted by x and not by $\hat{x}$ so as to avoid analogy with the classical approach. Note however that the state process has the imposed measurability property.

The stochastic control problem can now be solved using the theory of stochastic control with complete observations. The resulting problem and the result are stated below for future reference.

Problem 2. The *stochastic control problem with partial observations according to the stochastic realization approach.* Consider the Gaussian stochastic system of Definition 3. Consider the cost function,

$$J(g) = E[\sum_{s=t_0}^{t_1-1} b(x(s), u(s)) + b_1(x(t_1))], \; J : G \to \mathbb{R}_+, \tag{13}$$

$$b(x, u) = \begin{pmatrix} x \\ u \end{pmatrix}^T \begin{pmatrix} L_{11} & L_{12} \\ L_{12}^T & L_{22} \end{pmatrix} \begin{pmatrix} x \\ u \end{pmatrix}, \; b_1(x) = x^T L_1 x,$$

$$L = \begin{pmatrix} L_{11} & L_{12} \\ L_{12}^T & L_{22} \end{pmatrix} = L^T \geq 0, \; L_{22} > 0, \; L_1 = L_1^T \geq 0.$$

Solve the problem

$$\inf_{g \in G} J(g).$$

Theorem 1. *Consider the stochastic control problem defined above. The solution to the optimal stochastic control problem is*

$$g^*(t, x) = F(t)x, \; \text{the optimal control law,} \tag{14}$$

$$u^{g^*}(t) = F(t)x^{g^*}(t), \; \text{the optimal input process,} \tag{15}$$

$$x^{g^*}(t+1) = Ax^{g^*}(t) + BF(t)x^{g^*}(t) +$$
$$+ K[y^{g^*}(t) - Cx^{g^*}(t) - DF(t)x^{g^*}(t)], \; x^{g^*}(t_0) = x_0, \tag{16}$$
$$F(t) = -[B^T P(t+1)B + L_{22}]^{-1}[A^T P(t+1)B + L_{12}]^T, \tag{17}$$

where $P : T \to \mathbb{R}^{n \times n}$ is the solution of the backward control Riccati difference equation not displayed in this paper.

A proof of the solution to the optimal stochastic control problem with complete observations may be found in [4, Section 3.1].

Note that the optimal control law consists of the proper control law Equation (14) and the system Equation (16). The latter system is the stochastic realization rewritten as a filter. The optimal control law is identical to that of the classical approach described in Section 3 except for the time invariance of the filter.

LEQG

The LEQG stochastic control problem with partial observations. The classical approach to solving the stochastic control problem with partial observations for a Gaussian stochastic system with an expected value of an exponential cost function is unsatisfactory. The conditional mean of the state based on past observations is not a sufficient information state as understood in stochastic control theory. Therefore the classical approach cannot proceed. A solution has been published of this problem in discrete-time and the proof is not based on the classical approach, see [24]. In continuous-time the corresponding problem has a solution, see [3], but the conditional mean is not the information state of the conditional distribution.

The stochastic realization approach to stochastic control with partial observations then proceeds according to Definition 2.

Definition 4. *Consider a Gaussian stochastic system*

$$x(t+1) = Ax(t) + Bu(t) + Kv(t), \ x(t_0) = x_0,$$
$$y(t) = Cx(t) + Du(t) + Nv(t),$$

with the conditions of Definition 3 and in addition $Q_v > 0$.

Problem 3. Consider the Gaussian stochastic system of Definition 4. Consider the optimal stochastic control problem

$$J(g) = E[c \exp(c \sum_{s=t_0}^{t_1-1} b(x(s), u(s)) + b_1(x(t_1)))], \ c \in \mathbb{R} \backslash \{0\}, \tag{18}$$

$$b(x, u) = \begin{pmatrix} x \\ u \end{pmatrix}^T \begin{pmatrix} L_{11} & L_{12} \\ L_{12}^T & L_{22} \end{pmatrix} \begin{pmatrix} x \\ u \end{pmatrix}, \ b_1(x) = x^T L_1 x, \tag{19}$$

$$L = \begin{pmatrix} L_{11} & L_{12} \\ L_{12}^T & L_{22} \end{pmatrix} = L^T \geq 0, \ L_{22} > 0, \ L_1 = L_1^T \geq 0,$$

$$\inf_{g \in G} J(g). \tag{20}$$

Theorem 2. *Consider the above formulated optimal stochastic control problem. The solution to the optimal stochastic control problem is*

$$g^*(t, x) = F(t)x^{g^*}, \ \textit{the optimal control law,} \tag{21}$$

$$u^*(t) = F(t)x^{g^*}(t), \ \textit{the optimal input trajectory,} \tag{22}$$

$$x^{g^*}(t+1) = Ax^{g^*}(t) + BF(t)x^{g^*}(t) +$$
$$+K[y^{g^*}(t) - Cx^{g^*}(t) - DF(t)x^{g^*}(t)], \ x^{g^*}(t_0) = x_0, \tag{23}$$

$$F(t) = -[B^T P_1(t)B + L_{22}]^{-1}[A^T P_1(t)B + L_{12}]^T, \tag{24}$$

$$P(t_1) = L_1, \ P, P_1 : T \rightarrow \mathbb{R}^{n \times n} \tag{25}$$

$$P_1(t) = P(t+1) +$$
$$cP(t+1)M[Q_v^{-1} - cMP(t+1)M]^{-1}MP(t+1), \tag{26}$$

$$P(t) = A^T P_1(t)A + L_{11} + \tag{27}$$
$$-[A^T P_1(t)B + L_{12}][B^T P_1(t)B + L_{22}]^{-1}[A^T P_1(t)B + L_{12}],$$

and the conditions are imposed that,

$$0 < Q_v^{-1} - cM^T P(t+1)M, \ 0 < B^T P_1(t)B + L_{22}, \ \forall t \in T. \tag{28}$$

The proof follows from the paper by D.H. Jacobson [14]. The conditions (28) represent stochastic controllability conditions because they assure finiteness of the cost function.

The corresponding result for the classical approach to stochastic control with partial observations is not in the literature. What is in the literature is a result of P. Whittle but that is for the nonstandard system representation in discrete-time of the form,

$$x(t+1) = Ax(t) + Bu(t) + Kv(t), \quad x(t_0) = x_0,$$
$$y(t+1) = Cx(t) + Du(t) + Nv(t).$$

The result for Whittle's system representation and the result for the standard form used in this paper are not comparable. The result of the classical approach for the continuous-time case is available in [3].

The conclusion for this LEQG case is that the solution is completely different from the solution to the classical approach as described in [24]. For the result of the continuous-time classical approach see [3, 16].

Finite Stochastic Systems

Stochastic control with partial observations of a finite stochastic system. A *finite stochastic system* is a stochastic system as defined earlier in the paper of which both the state set and the output set are finite sets.

First the classical approach of solving this stochastic control problem is sketched. The stochastic system is specified by the probability distribution of the initial state and by the stochastic transition function according to the formulas

$$n, m, p \in \mathbb{Z}_+,$$
$$X = \{x_1, x_2, \ldots, x_n\}, \text{ the state set,}$$
$$U = \{u_1, u_2, \ldots, u_m\}, \text{ the input set, } Y = \{y_1, y_2, \ldots, y_p\}, \text{ the output set,}$$
$$P(\{x_0 = x_i\}), \forall i \in \mathbb{Z}_n,$$
the probability distribution of the initial state,
$$P(\{x(t+1) = x_i, y(t) = y_k\}|F^y_{t-1} \vee F^u_{t-1}), \forall i \in \mathbb{Z}_n,$$
the stochastic transition function.

The representation of the finite stochastic system is such that the state, input, and output sets are finite while the state, the input, and the output process are such that the state at time t equals one of the values of the state set. An alternative representation often used in the literature is to take as state space $\mathbb{R}^n_+$ and then the state takes the value of the i-the unit vector if it equals the i-th element of the state set.

The filtering problem for such a finite stochastic system has been known since the early 1960's. The filter system has as stated vector

$$\hat{x}_i(t+1) = P(\{x(t+1) = x_i\}|F^y_{t-1} \vee F^u_{t-1}) \in [0, 1], \quad \forall i \in \mathbb{Z}_n.$$

Note that the state set of the filter system is $[0, 1]^n$ and is thus not a finite set. The stochastic control problem with partial observations can then be solved using the classical theory with as stochastic system the filter system. No analytic solutions to this optimal stochastic control problem are known to the author of this paper. References on the classical approach are [1, 9, 10, 11].

The stochastic realization approach to stochastic control with partial observations is to first take a stochastic realization of the input-output process with as state process a process which is a measurable function of the past outputs and the past inputs. This selection of a stochastic realization turns out to be a major step of the approach.

Definition 5. *The stochastic realization is taken to be,*

$$n, m, p \in \mathbb{Z}_+,$$
$$X = \mathbb{R}_+^n,\ \overline{X} = \{x_1, x_2, \ldots, x_n\},\ \text{the state set,}$$
$$U = \mathbb{R}_+^m,\ \overline{U} = \{u_1, u_2, \ldots, u_m\},\ \text{the input set,}$$
$$Y = \mathbb{R}_+^p,\ \overline{Y} = \{y_1, y_2, \ldots, y_p\},\ \text{the output set,}$$
$$e_i \in \mathbb{R}_+^n,\ \text{the } i\text{-the unit vector of } \mathbb{R}_+^n,$$
$$x_0 : \Omega \to \mathbb{R}^n,\ F^{x_0} \subseteq F_0^y,$$
$$x : \Omega \times T \to \mathbb{R}_+^n,\ u : \Omega \times T \to \overline{U},\ y : \Omega \times T \to \mathbb{R}_+^p,$$
$$\text{define for all } i \in \mathbb{Z}_n,$$

$$x(t+1) = e_i, \tag{29}$$
$$\text{if } S_i(x(t), u(t), y(t)) > S_j(x(t), u(t), y(t)),\ \forall j \in \mathbb{Z}_n, \tag{30}$$
$$\text{or if } i \text{ is the smallest element of } \mathbb{Z}_n \text{ such that}$$
$$S_i(x(t), u(t), y(t) \geq S_j(x(t), u(t), y(t)),\ \forall j \in \mathbb{Z}_n, \tag{31}$$
$$S : \mathbb{R}_+^n \times \overline{U} \times \mathbb{R}_+^p \to \mathbb{R}^n,$$
$$S(x(t), u(t), y(t)) = Ax(t) + Bu(t) + K[y(t) - Cx(t)], \tag{32}$$
$$A \in \mathbb{R}_+^{n \times n},\ B \in \mathbb{R}_+^{n \times m},\ C \in \mathbb{R}_+^{p \times n},$$
$$\text{spec}(A - KC) \subset \mathbb{D}_o. \tag{33}$$

The choice of the above defined stochastic realization requires comments. The stochastic realization has to be such that the state is a measurable function of the past outputs and the past inputs. Hence the dependence on $y(t)$ and $u(t)$ in the transition function of the stochastic system.

The particular form of taking $x(t+1) = e_i$ if $S_i(.) > S_j(.)$ is a choice. The choice is reasonable because if the output process is constant than one wants the state to approach the state component which generates this particular output. This choice is in accordance with the concept of a stochastic realization of which the state is a measurable function of the past outputs and the past inputs. There is the difficulty when there exist $i, j \in \mathbb{Z}_n$ with $i \neq j$ such that $S_i(x(t), u(t), y(t)) = S_j(x(t), u(t), y(t))$. The above choice is to take as next state the index of the lowest integer i in case two or more equal S_i values. This choice is not invariant with respect to ordering. However, one cannot take a random choice because that would not meet the required measurability property.

The matrices A and C relate directly to the probabilistic interpretation of the finite-state stochastic system. There remains the choice of the matrix $K \in \mathbb{R}_+^{n \times p}$. The suggestion is to take $K = sC^T \in \mathbb{R}_+^{n \times p}$ for a constant $s \in \mathbb{R}_+$ such that Equation (33) is met.

Further research is required into the appropriateness of the above defined stochastic realization.

Proposition 2. *The finite stochastic system defined above is such that*

$$F^{x(t)} \subseteq F_{t-1}^y \vee F_{t_1}^u,\ \forall t \in T \backslash \{t_0\},$$

hence the stochastic system has the measurability property of the stochastic realization approach.

Proof. The statement follows directly from the condition on the initial state $F^{x_0} \subset F_0^y$ and from the Equations (29,32).

Problem 4. The *optimal stochastic control problem for a finite stochastic system* where the system is defined in Definition 5. Solve the optimal stochastic control problem with complete observations

$$\inf_{g \in G} J(g), \tag{34}$$

$$G = \{g : T \times X \to U | g \text{ measurable function}\},$$

$$J(g) = E[\sum_{s=t_0}^{t_1-1} b(x^g(s), u^g(s)) + b_1(x^g(t_1))], \tag{35}$$

$$b : X \times U \to \mathbb{R}_+, \; b_1 : X \to \mathbb{R}_+.$$

The optimal stochastic control problem formulated above can now be solved numerically with the well known algorithms of value and policy iteration, see the books [4,17].

The conclusion of the stochastic realization approach to stochastic control with partial observations is that it is feasible, the computations can be done with available algorithms and software, and that the resulting control law is easily implementable. The approach avoids the road via the optimal stochastic control problem with the filter system with as state set the space $[0, 1]^n$.

6   Concluding Remarks

The paper presents the stochastic realization approach to stochastic control problems with partial observations. The approach prescribes: (1) to formulate a stochastic realization of the input-output process in which the state is measurable on the past of the input and the past of the output process; (2) to prove that the stochastic realization has the required measurability property; (3) to solve the optimal stochastic control problem with complete observations; and (4) to assemble the control law. For the special case of a Gaussian stochastic control system and a quadratic cost function, the stochastic realization approach yields the same control law as the classical approach. For the special cases of LEQG and of a finite stochastic system the resulting control laws are different from the classical approach.

Further research is required to explore the usefulness of the stochastic realization approach to stochastic control with partial observations. An extension to be formulated is to decentralized control of stochastic systems and to stochastic dynamic games. In these cases each of the players has his private system and a private system which models the controllers of the other players. That approach restricts the complexity of dynamic games problems considerably.

References

1. A. Arapostathis, V.S. Borkar, E. Fernández-Gaucherand, M.K. Ghosh, and S.I. Marcus. Discrete-time controlled Markov processes with average cost criterion: A survey. *SIAM J. Control & Opt.*, 31:282–344, 1993.
2. V.E. Benes and I. Karatzas. Filtering of diffusions controlled through their conditional measures. *Stochastics*, 13:1–23, 1984.
3. A. Bensoussan and J.H. van Schuppen. Optimal control of partially observable stochastic systems with an exponential-of-integral performance index. *SIAM J. Control Optim.*, 23:599–613, 1985.
4. D.P. Bertsekas. *Dynamic programming and stochastic control*. Academic Press, New York, 1976.
5. D.P. Bertsekas. *Dynamic programming and optimal control, Volume I*. Athena Scientific, Belmont, MA, 1995.
6. D.P. Bertsekas. *Dynamic programming and optimal control, Volume II*. Athena Scientific, Belmont, MA, 1995.
7. R.K. Boel and P.Varaiya. Optimal control of jump processes. *SIAM J. Control Optim.*, 15:92–119, 1977.
8. M.H.A. Davis and P. Varaiya. Dynamic programming conditions for partially observable stochastic systems. *SIAM J. Control*, 11:226–261, 1973.
9. E. Fernandez-Gaucherand and A. Arapostathis. On partially observable Markov decision processes with an average cost criterion. In *Proceedings of the 28th IEEE Conference on Decision and Control (CDC.1989)*, pages 1267–1272, New York, 1989. IEEE, IEEE Press.
10. E. Fernández-Gaucherand, A. Arapostathis, and S.I. Marcus. On the average cost optimality equation and the structure of optimal policies for partially observed Markov decision processes. *Ann. Oper. Res.*, 29:439–470, 1991.
11. E. Fernandez-Gaucherand and S.I. Marcus. Risk-sensitive optimal control of hidden markov models: Structural results. *IEEE Trans. Automatic Control*, 42:1418–1422, 1997.
12. W.H. Fleming and R.W. Rishel. *Deterministic and stochastic optimal control*. Springer-Verlag, Berlin, 1975.
13. T.L. Gunckel. Optimum design of sampled-data systems with random parameters. Technical Report SEL TR 2102-2, Stanford Electron. Lab., Stanford, 1961.
14. D.H. Jacobson. Optimal stochastic linear systems with exponential performance criteria and their relation to deterministic differential games. *IEEE Trans. Automatic Control*, 18:124–131, 1973.
15. P.D. Joseph and J.T. Tou. On linear control theory. *AIEE Trans. (Appl. Ind.)*, 80:193–196, 1961 (Sep.).
16. P.R. Kumar and J.H. van Schuppen. On the optimal control of stochastic systems with an exponential-of-integral performance index. *J. Math. Anal. Appl.*, 80:312–332, 1981.
17. P.R. Kumar and P. Varaiya. *Stochastic systems: Estimation, identification, and adaptive control*. Prentice Hall Inc., Englewood Cliffs, NJ, 1986.
18. A. Lindquist and G. Picci. A geometric approach to modelling and estimation of linear stochastic systems. *J. Math. Systems, Estimation, and Control*, 1:241–333, 1991.
19. R. Rishel. Necessary and sufficient dynamic programming conditions for continuous time stochastic optimal control. *SIAM J. Control & Opt.*, 8:559–571, 1970.
20. C. Striebel. Sufficient statistics in the optimum control of stochastic systems. *J. Math. Anal. Appl.*, 12:576–592, 1965.
21. C. Striebel. *Optimal control of discrete time stochastic systems*, volume 110 of *Lecture Notes in Economic and Mathematical Systems*. Springer-Verlag, Berlin, 1975.

22. C. Striebel. Martingale conditions for the optimal control of continuous time stochastic systems. *Stoc. Proc. Appl.*, 18:329–347, 1984.
23. J.H. van Schuppen. Stochastic realization problems. In J.M. Schumacher H. Nijmeijer, editor, *Three decades of mathematical system theory*, volume 135 of *Lecture Notes in Control and Information Sciences*, pages 480–523. Springer-Verlag, Berlin, 1989.
24. P. Whittle. Risk-sensitive Linear/Quadratic/Gaussian control. *Adv. Appl. Prob.*, 13:764–777, 1981.
25. H. Witsenhausen. Separation of estimation and control for discrete time systems. *Proc. IEEE*, 59:1557–1566, 1971.
26. H.S. Witsenhausen. On information structures, feedback and causality. *SIAM J. Control*, 9:149–160, 1971.
27. H.S. Witsenhausen. A standard form for sequential stochastic control. *Math. Systems Theory*, 7:5–11, 1973.

Experiences from Subspace System Identification - Comments from Process Industry Users and Researchers

Bo Wahlberg[1], Magnus Jansson[2], Ted Matsko[3], and Mats A. Molander[4]

[1] Automatic Control, KTH, Stockholm, Sweden
   bo@ee.kth.se
[2] Signal Processing, KTH, Stockholm, Sweden
   magnus.jansson@ee.kth.se
[3] ABB USA
   ted.matsko@us.abb.com
[4] ABB Corporate Research, Västerås, Sweden
   mats.a.molander@se.abb.com

Summary. Subspace System Identification is by now an established methodology for experimental modelling. The basic theory is well understood and it is more or less a standard tool in industry. The two main research problems in subspace system identification that have been studied in the recent years are closed loop system identification and performance analysis.

The aim of this contribution is quite different. We have asked an industrial expert working in process control a set of questions on how subspace system identification is used in design of model predictive control systems for process industry. As maybe expected, it turns out that a main issue is experiment/input design. Here, the difference between theory and practice is rather large mainly due to implementation constraints, but also lack of knowledge transfer. Motivated by the response from the expert, we will discuss several important user choices problems, such as optimal input design, merging of data sets and merging of models.

1 Introduction

System identification concerns the construction and validation of models of dynamical systems from experimental data. All advanced control and optimization methods, as model predictive control, require a reliable model of the process. System identification is thus a key technology for industry to benefit from powerful control methods. Modelling and system identification techniques suitable for industrial use and tailored for control design applications are of utmost importance The field of system identification is well developed with excellent textbooks, e.g. [12, 17], and software packages.

Subspace system identification is a class of methods for estimating state space models based on low rank properties of certain observability/prediction sets. The area was pioneered by researchers as Larimore, de Moor & van Overschee, Verhaegen and others, and was early adopted by industry. The theoretical foundation was established by for example Picci, Bauer and Jansson, and more recently by Chiuso. The industry was an early adopter. One of the first comercially available tools, released 1990, was ADAPTx from Adaptics Inc. It was based on the CVA method developped by Larimore.

A. Chiuso et al. (Eds.): Modeling, Estimation and Control, LNCIS 364, pp. 315–327, 2007.
springerlink.com

Nowadays, several vendors of Process Automation systems offer modeling tools based on subspace system identification. Examples are aspenOne from Aspen Technologies and Profit SensorPro from Honeywell.

ABB released its MPC product Predict & Control in 2000 (called 3dMPC at the time). It uses state-space models combined with Kalman-style state estimators for the predicions. The included system identification package is based on a two-step procedure where an initial model is obtained through subspace system identification, and a refined model can be obtained through the minimization of a Prediction Error criterion.

Let us now review some basic facts about subspace system identification. The matrices of a linear system in state space form

$$x(t+1) = Ax(t) + Bu(t)$$
$$y(t) = Cxt(t) + Du(t) \tag{1}$$

can efficiently be estimated from observations of the input signal $u(t)$ and the output signal $y(t)$ without any prespecified parametrization using subspace system identification methods. The basic idea is that if the state vectors $x(t)$, and the input $u(t)$ and the output $y(t)$ are known, least squares linear regression techniques can be used to directly estimate the state-space matrices A, B, C and D. The key observation is that the state $x(t)$ actually can be reconstructed from input output data using certain multiple-step ahead prediction formulas combined with low rank matrix factorizations. Combining these two ideas leads to a family of subspace system identification methods such as CVA, N4SID, and MOESP variations.

A great advantage of subspace system identification is that there is no difference between multi-input-multi-output system identification, where the input and output are vectors, and ordinary single-input single output identification. This is in contrast to e.g. prediction error methods, where the choice of model structures and corresponding parameterizations is most important. Other advantages of subspace identification are that the methods are based on robust numerical methods and avoid problem with optimization and possible local minima. The main disadvantages have been a potential performance loss compared to statistically efficient methods, as the prediction error method, and potential bias problems in identification of systems operated in closed loop. These two shortcomings have at least partly been eliminated by recent research [3, 10, 11, 14].

There are mainly three design parameters in a subspace system identification method. We will here use the notation in [12]. The maximum prediction horizon is denoted by r, the number of past outputs used in the predictors is denoted by s_y, and the number of past inputs used in the predictors by s_u. A common choice is $s_u = s_y = s$. The model order n is also important, but it is often estimated. It is often difficult to pre-specify r and s. One common way in practice is to evaluate a set of combinations and then use model validation to determine the best values.

Subspace system identification was early used in industry since it fits very well with model predictive control. It is by now almost a standard technology and hence it is of interest to see how it is used in industry and how this relates to the state-of-the-art in more theoretical academic research.

2   Questions and Answers from the User

It is important to understand some of the more typical modelling problems in process industry. The plants are often very complex and it could be very expensive or even dangerous to do experiments with too large excitations. The process dynamics is slow, and the time constants are often around minutes or hours. This means that experiments often take long time. The process noise has often low frequency characteristics, e.g. drifts and random walks.

Notation

The notation MV is used for Manipulated Variable, i.e. input signals which then will be used to control the systems. Control variables are denoted by CV, i.e. output signals which will be used for feedback control, while PV denotes process variables i.e. measurable output signals.

We have formulated a set of questions to Ted Matsko, who is an expert at ABB in implementation of control systems, and, in particular, design of model predictive controllers using subspace system identification as a tool for modelling.

First, we think it would be interesting if you could put down some kind of step by step list of actions describing your typical routine for the modelling part of a model predictive control commissioning. Preferably also with time estimates (relative or absolute) for the different steps. Also, notes on steps where special care has to be taken to not run into problems would be very interesting.

Data Collection: This averages one day per MV for refining and petrochemical processes, which have longer time constants. Powerhouse is typically two MVs per day. There tends to be a greater level of unmeasured disturbance in refining/petrochem than in power, except for coal fired units where coal quality can vary. A typical experiment takes around 4 hours. The rest of the time concerns design of experiments and is often done in collaboration with an operator. A typical input signal is series of steps (4-8), which is manually implemented by the operator. Some of these are shorter than others and the amplitude and the sign may change.

Data Review and Organization: This typically takes 2-4 hours per data set. Typically we try to step one MV at a time. This never happens (operator needs to move another MV), so we need to organize the data and merge data files together. I have had bad experience with full MIMO ID using old tools from about 12 years ago. With a dense model matrix, the package simply did not work that well with a PRBS input. This may have worked better if the system had been more diagonal. I typically import data into an ABB tool and do a lot of plotting. Then I clean and merge data files in Excel, add calculated data points.

Subspace System Identification, First Pass: Specify a fairly small range of possible models by setting the design parameters to $r = s$ rather small and check different model orders. This usually gives around 150 models. Pick 3 or 4 with low open loop error, usually with different n. My experience is that higher model orders tend to result in a step response with oscillation. Evaluate against data set, look at step response (and impulse response if integrating process). Use prediction error plots to confirm what your eyes see.

Subspace System Identification, 2nd, 3rd ...Passes: If best models are on edge of $n/r/s$ boundaries, expand the range. Push to low n, very low s and high r, gives best results in my experience.

Merge Models: Paste things together with the Merge and Connect tools (ABB tool). Use model order reduction to reduce the order of the overall system. Sometimes we convert to Laplace transfer functions before merging. This gives us a little more control over the shape, dead time, gain, ability to zero out certain MV/CV pairs. May also run subspace system identification on 1 CV x 3 or 4 MVs to control, then merge to control placing of null relationships.

I like to break things down to /actuator-to-process output/ and /control loop-to-actuator/ models and then merge. Going actuator-to-output provides a physical, intuitive model. Sometimes the base controls can cloud the issue and hide a bad job from visual inspection.

Modeling and Constraint design are the two most important steps in creating good MPC applications. Our subspace system identification tool is the best modeling tool I have used. Our Data Processing tools are almost good, but the latest release makes them difficult to use in some circumstances. Data manipulation and pre-processing can consume a lot of time if you make mistakes in data collection.

How do you set up the identification experiments? Do you try to disturb as many as possible of the MVs simultaneously? If not all MVs are disturbed simultaneously, what considerations control the splitting of the experiments? How do you typically decide the step amplitudes for the excitation?

We first create a large taglist. It is better to collect any data related to the process. If you find some inexplicable behavior, going through data for ancillary tags may allow you to save the data file and avoid doing additional tests. Additional tests may require another trip to site. We try to step one MV at a time. If the process is large and we can step variables that do not overlap in the model matrix, we may step two at the same time. An important part of working with any modeling package is looking at how well the model fits the data visually. In a full 5x5 MIMO ID with all MVs moving randomly, it is very hard to tell how good a model fit really is. To determine step test amplitudes, we monitor the noise level and talk to the operators about the process gains and settling time. The operator will be able to give you a reasonable feel for the gain and time constant. As a minimum, I want the step tests to move the PV twice as far as the noise does. After a few steps, we may adjust.

Are the identification experiments always performed in open-loop, or is it common to have the MV excitations as additions to the outputs of a running controller? If you commonly perform both types of experiments, what are your experience of the differences?

Before commissioning, most data is open loop. Infrequently the MPC will include a distillation column temperature that is controlled in the distributed control system (DCS) with a cascade loop. In the powerhouse, the pressure is always controlled. In these cases we may do some or all of the initial testing with that PV in closed loop. To fine tune a model during commissioning, we may use some closed loop data. I do not have a lot of experience with closed loop data and subspace system identification, but the header pressure models seem to work better when at least some of the data is under closed loop control. That may be a special case where the models are integrating (or very long time constants). I seem to recall that other model predictive control system identification packages discouraged the use of closed loop data.

Do you find the time needed for the identification experiments to be a main hurdle?

Data collection time is about 1/3 of the total time spent on building dynamic models. Sometimes we are able to get people at the plant to do some of the data collection under our guidance. There is also time for functional specification, configuring controller, configuring DCS, adding DCS graphics, FAT, commissioning, final documentation. So data collection time is significant, but not the majority of time.

Compared to other MIMO control systems you commissioned earlier, do you save a lot of time in the modelling phase by using the subspace system identification tools? If so, in what stages of the modelling, and how much?

Time is about the same. Quality is better and that saves time commissioning. Capability and flexibility is better (merge, connect). Better models make commissioning much easier. Biggest place for time savings would be improving our data preprocessing tools. We handle some big data sets, 100 tags by 17,500 records is a typical file (merged 4 days testing into one file in this example).

In the ABB modelling tool we have the two step procedure of a Subspace identification, followed by a Prediction Error Model identification. How important is the second PEM step? Is it always important, or have you identified situations where it is not important and others where it is? (Theoretically, one would suspect that it is more important if the identification experiment was performed in closed loop)?

I have not found that the PEM step makes much improvement. The feature to force a zero relationship does work that well (I have a recent example).

How about size restrictions? How large problems do you typically model, in number of model inputs/outputs? When do you normally run into problems regarding size?

I have had no problem with MV x PV size, but I do not stretch the limits. Even with a big controller, the biggest subset I work with is around 4x4. I use some very long files with breaks in time. This may make the subspace system identification slow, but the results are usually good. When I plot the model predictions when using data sets with breaks, I still get some strange initialization occasionally, but I am not very well educated on how the calculations work here so user error is not unlikely.

How large models (in number of states) do you normally end up with? Is it very coupled to the input/putput size, or do you normally find models with a reasonably small number of states even if the number of inputs/outputs are large?

I usually end up with models that have a low number of states from subspace system identification. Some of the models used for control, that are merged, are pretty large.

On the small powerhouse project we just finished, individual models were $n = 2$ to 6. final merge was $n = 44$ (5 PVs x 4 MVs). On a prior project with many MVs, I had a plant model with $n = 166$, but the order reduction tool pared that to 65.

How do you normally conclude that a model is good enough? What evaluation methods are most important?

Visual examination of prediction versus actual data. Of course if you made some mistakes that can still look good, even if the model is not. If you have problems commissioning, the first thing to re-examine are the models. There we take new closed loop data files and import them into the identification tool and look at the predictions versus actual output.

Do you consider the existing evaluation tools sufficient, or do you sometimes experience that a model that looks perfectly good, still does not work well for the MPC?

Yes, as I said above this happens. Usually it is some user error. Something that one overlooks in the data file. Recent example was a delay on one signal due to the communication system in the DCS. The delay did not exist for process control, just data collection. All my models were slightly off, but gains were good. This caused a nice limit cycle.

Do you always use separate identification and evaluation data sets, or do you sometimes/occassionally/always use the same data set for both identification and evaluation?

Always use the same in the identification step. Sometimes I will use another file for plotting predictions, usually data gathered from a later time period. Unless we start to do long PRBS runs, I don't think there is typically enough data to split.

As a last point, do you have specific wishes for improvements of the modelling tools?

Better results from the PEM step (a.) gains set to 0 (b.) conversion, Order reduction tool introduces some small gains where 0 gains may have existed before. A tool to convert to low order Laplace transfer functions from state space models. An online tool to monitor model performance.

3   Comments from the Researchers

The responses and answers from the process control expert are very interesting and open up a lot of very interesting research issues. Our objective, however, is not to provide a "solutions manual," but instead to relate the comments of the expert to the current status in research. We are well aware of the difficult trade-offs in real applications, and that academic examples often are severe simplifications. However, they can provide insight and results which then can be tailored real problems to the intended application. We will, in particular, discuss input design, merging of data sets and merging of estimated models.

4   Input Design

Input design for linear dynamic systems started out around 1960 and in the 1970's there were a lot of activities in this area, for which [6] acts as an excellent survey. The classical approach is to minimize some scalar measure function of the asymptotic parameter covariance matrix (e.g. the determinant or the trace) with constraints on input and/or output power.

Let Φ_u be the power spectral density of the input. The more recent approaches to optimal experiment design problems are typically posed in the form

$$\min_u \frac{1}{2\pi} \int_{-\pi}^{\pi} \Phi_u(\omega)d\omega \tag{2}$$

subject to model quality constraints, and signal constraints $\tag{3}$

i.e., formulated as optimization problems that include some constraints on the model quality together with signal constraints. It is also possible to minimize a weighted version of the input signal or the output signals. The signal constraints have to be included to obtain well-posed problems, e.g., to prevent the use of infinite input power. Powerful methods have been developed to solve a variety of such problems. Notice that the solution will be an input spectral density, and that the time domain realization problem then can be solved in different ways. We refer to [8, 2, 1] for overviews of recent methods.

System identification for control is a very important application of input design. Here the model quality constraints reflect robust performance and stability. It is very important to obtain input signals that then can be reliably implemented, while the optimal solution often only refers to frequency domain properties. To stress this issue [15] have introduced the concept of "plant-friendly" identification.

Optimal Input Design for Subspace Methods

Most of the input design methods make use of asymptotic variance analysis based on the Prediction Error identification Method (PEM) framework, [12]. Very few results have been presented for optimal input design for subspace system identification methods. A natural idea would be to do input design for PEM methods and then apply it for subspace system identification. Our conjecture is that optimal inputs for PEM would give good results for subspace system identifications methods since subspace methods often give results with quality properties close to those of PEM.

From theory it is known that subspace system identification requires that the input signal is persistently exciting of order $n + r$ or higher [9, 4], and that white noise input signals often give good performance. This is in contrast to the experiments described by the expert, which essentially are of step response type. The signal to noise ratio is, however, quite high. It would be of interest to do performance analysis of subspace system identification model estimates from step response data with high signal to noise ratio. Of course, a noise free step response contains all information about a linear system, but it is very sensitive to disturbances. Early subspace system identification techniques made use of the Hankel matrix constructed from impulse response data.

An interesting input design problem is the following. Assume that you are only allowed to change the input signal d times over a given interval with prespecified amplitude levels. What are the optimal choices of times to change the signal?

Example

Some insight can be obtained by the following simple example. Consider the problem of estimating the parameters of a first order ARX model

$$y(t) = -ay(t-1) + bu(t-1) + e(t), \quad a = 0.5, \; b = 1$$

under the assumption of zero initial conditions and only the choices $u = 1$ or $u = -1$ are allowed and that one only is allowed to change the input from 1 to -1 once. Assume that you have four measured samples. Possible input sequences are then $(1, -1, -1, -1)$, $(1, 1, -1, -1)$, $(1, 1, 1, -1)$ and $(1, 1, 1, 1)$. Since this concerns very small data records, we have used Monte Carlo computer simulations to study the quality of the estimated parameters. For the estimation of a, it turns out the first sequence is the best one and it gives a performance that is almost twice as good as the second best one. The accuracies of the estimates of b are approximately the same for the first and the third sequences. The fourth sequence is of course bad in terms of persistence of excitation. The example shows that it is important to design the input sequence even if one has hard constraints.

MIMO Systems

Identification of MIMO systems has recently been an active area of research. In particular variance analysis and input design issues. A main question is if one should excite the various inputs simultaneously or separately? From a pure performance perspective it is always better to excite all inputs simultaneously [5]. However, in the case where the output power is constrained, the conclusion is that it is possible to obtain nearly the same performance by exciting one input at a time, i.e. the strategy used by the expert. See page 25 in [5].

In case the process has strong interactions, i.e. ill conditioned systems, the dynamics of the high-gain directions tends to dominate the data and it is difficult to estimate a reliable model for the low-gain direction. This problem is discussed in e.g. [7]. An interesting approach for iterative input design of gene regulatory networks is described in [13]. The idea is to excite the inputs to obtain a uniform output excitation. It is a good idea to pre-compensate the plant to make it as diagonal as possible before identification. For a completely diagonal plant the MIMO identification problem of course simplifies to a number of SISO problems. It is of course difficult to do pre-compensation without a good model, and it could also create problems with large inputs if the plant is ill conditioned. A better solution is then to use partial decoupling as e.g. described in [21]. The idea is to transform the system to a triangular structure.

5   Merging Data

Assume that data have been collected from a series of SIMO experiments. The problem of merging data sets is discussed in details in Section 14.3 in [12]. One important aspect is that it is common that one has to remove parts of bad data, which means that data records have to then be merged.

For linear dynamical systems the main problem when merging data is to take the different initial conditions into account. In case of periodic input signals it is possible to use averaging as a means to merge data.

We will illustrate some of the issues by a simple example.

Example

Consider the model

$$y(t) = bu(t - k) + e(t)$$

i.e. a time delayed linear relation with unknown gain b. Assume that we have two data sets of equal length

$$\{y_1(t), u_1(t), t = 1, 2, \ldots, N/2\}, \quad \{y_2(t), u_2(t), t = 1, 2 \ldots, N/2\}$$

The least squares (PEM) estimate minimizes

$$\min_b \sum_{t=1+k}^{N/2} (y_1(t) - bu_1(t - k))^2 + \sum_{t=1+k}^{N/2} (y_2(t) - bu_2(t - k))^2$$

and thus equals

$$\hat{b}_N = \frac{\sum_{t=1+k}^{N/2}(y_1(t)u_1(t - k) + y_2(t)u_2(t - k))}{\sum_{t=1+k}^{N/2}(u_1^2(t - k) + u_2^2(t - k))} \tag{4}$$

Notice that this does not correspond to just concatenating the data

$$y_c(t) = \begin{cases} y_1(t) & t = 1, \ldots, N/2 \\ y_2(t - N/2), & t = N/2 + 1, \ldots, N \end{cases}$$

$$u_c(t = \begin{cases} u_1(t), & t = 1, \ldots, N/2 \\ u_2, (t - N/2), & t = N/2 + 1, \ldots, N \end{cases}$$

which would yield

$$\hat{b}_N^c = \frac{\sum_{t=1+k}^{N} y_c(t)u_c(t - k)}{\sum_{t=1+k}^{N} u_c^2(t - k)} \neq \hat{b}_N$$

since the transient has to be taken into account.

An observation is however that in case $u_1 = u_2$, i.e. the same input is used in both experiments, we have

$$\hat{b}_N = \frac{\sum_{t=1+k}^{N/2}(y_1(t) + y_2(t))(u_1(t - k) + u_2(t - k))}{\sum_{t=1+k}^{N/2} 0.5(u_1^2(t - k) + u_1^2(t - k))}$$

which just corresponds to adding the signals $y = y_1 + y_2$ and $u = u_1 + u_2$ and then use the ordinary PEM estimate

$$\hat{b}_N = \frac{\sum_{t=1+k}^{N/2} y(t)u(t - k))}{\sum_{t=1+k}^{N/2} u^2(t - k)}$$

The summation approach only works for repeated input signals and will help reducing noise contributions by averaging. Notice that we only work with data records of length $N/2$ instead of N when concatenating the data. This result holds for general models with repeated signals.

6   Merging of Models

It is easy to combine parameter estimates from different data sets by assuming that they can be regarded as independent. The minimum variance estimate is obtained by weighting the individual estimates with their inverse covariance matrix, see Equation (14.15) in [12]. It is also noted that the same result can be obtained by solving a combined least squares problem using the merged set of data.

Expression (4) in the example in the previous section illustrates exactly this. The least squares estimate from the individual data records equals

$$\hat{b}^1_{N/2} = \frac{\sum_{t=1+k}^{N/2}(y_1(t)u_1(t-k))}{\sum_{t=1+k}^{N/2} u_1^2(t-k)}, \quad \hat{b}^2_{N/2} = \frac{\sum_{t=1+k}^{N/2}(y_2(t)u_2(t-k))}{\sum_{t=1+k}^{N/2} u_2^2(t-k)}$$

with covariances

$$\mathrm{Cov}\,\hat{b}^1_{N/2} = \frac{\lambda_o}{\sum_{t=1+k}^{N/2} u_1^2(t-k)}, \quad \mathrm{Cov}\,\hat{b}^2_{N/2} = \frac{\lambda_o}{\sum_{t=1+k}^{N/2} u_2^2(t-k)}$$

Hence, the combined minimum variance estimate becomes

$$\hat{b}_N = \frac{\hat{b}^1_{N/2}/\,\mathrm{Cov}\,\hat{b}^1_{N/2} + \hat{b}^2_{N/2}/\,\mathrm{Cov}\,\hat{b}^2_{N/2}}{1/\,\mathrm{Cov}\,\hat{b}^1_{N/2} + 1/\,\mathrm{Cov}\,\hat{b}^2_{N/2}}$$

which equals (4). The observation is that it is most important to take the quality of the estimates into account when merging different models. Hence, it is not trivial to combine model estimates from subspace system identification methods, and the best idea seems to be to instead merge the data sets. One problem is that the state space realization of the estimate depends on the data, and hence it is not just to combine the different state space matrices. Instead one has to work with canonical forms such as transfer functions or frequency responses.

Consider the problem of merging two frequency response estimates $\hat{G}_1(e^{i\omega})$ and $\hat{G}_2(e^{i\omega})$, possibly of different orders, estimated by e.g. subspace methods. Assume that the corresponding variances of the two estimates are given by $W_1(e^{i\omega})$ and $W_2(e^{i\omega})$, respectively. These variance expressions can be quite complicated to obtain. However, a reasonable approximation based on the high model order theory of Ljung [12], is

$$W_i = \frac{n_i}{N_i}\frac{\widehat{\Phi}^i_v(e^{i\omega})}{\widehat{\Phi}^i_u(e^{i\omega})}, \quad i = 1, 2$$

where $\widehat{\Phi}^i_v$ is an estimate of the noise power spectrum, $\widehat{\Phi}^i_u$ is an estimate of the input power spectrum, n_i is the model order and N_i is the number of data used in experiment i.

The optimally merged estimate will then be

$$\hat{G}_3 = \frac{\hat{G}_1/W_1 + \hat{G}_2/W_2}{1/W_1 + 1/W_2}$$

with resulting variance $W_3 = 1/(1/W_1 + 1/W_2)$. In order to obtain a low order transfer function estimate the following H_2 model reduction problem can be solved

$$\min_\theta \frac{1}{2\pi} \int_{-\pi}^{\pi} \frac{|\hat{G}_3(e^{i\omega}) - G(e^{i\omega}, \theta)|^2}{W_3(e^{i\omega})} d\omega$$

where $G((e^{i\omega}, \theta)$ is a set of parameterized transfer functions. This approach was proposed in [20] and has more recently been studied in [19].

Further insight can be obtained if the noise properties in both experiments are identical and the model orders $n_1 = n_2$. Let $\hat{\Phi}_{yu}^i = \hat{G}_i \hat{\Phi}_u^i$, i.e. an estimate of the cross-spectra. Then

$$\hat{G}_3 = \frac{N_1 \hat{\Phi}_{yu}^1 + N_2 \hat{\Phi}_{yu}^2}{N_1 \hat{\Phi}_u^1 + N_2 \hat{\Phi}_u^2}$$

which can be viewed as averaging the individual spectra

$$\hat{\Phi}_{yu}^3 = \frac{N_1 \hat{\Phi}_{yu}^1 + N_2 \hat{\Phi}_{yu}^2}{N_1 + N_2}, \quad \hat{\Phi}_u^3 = \frac{N_1 \hat{\Phi}_u^1 + N_2 \hat{\Phi}_u^2}{N_1 + N_2}$$

and then taking $\hat{G}_3 = \hat{\Phi}_{yu}^3 / \hat{\Phi}_u^3$. From a non-parametric estimation point of view this makes a lot of sense. Even further insight can be obtained by the following simple example. Let

$$y_3 = y_1 + y_2, \quad u_3 = u_1 + u_2$$

and assume that u_1 and u_2 are independent. Then

$$\Phi_{y_3 u_3} = \Phi_{y_1 u_1} + \Phi_{y_2 u_2}, \quad \Phi_{u_3} = \Phi_{u_1} + \Phi_{u_2}$$

for the cross spectra and the input spectra. Hence, the merging of models corresponds in some sense to merging of data, while enforcing the independence assumption.

Combining Sub-models

Consider the following problem

$$z_1(t) = G_1(q)u(t)$$
$$z_2(t) = G_2(q)z_1(t)$$
$$y_1(t) = z_1(t) + e_1(t)$$
$$y_2(t) = z_2(t) + e_2(t)$$

Given observations of the input $u(t)$ and the outputs y_1 and y_2 we would like to estimate the transfer functions G_1 and G_2. There are at least three different ways to approach this problem. The optimal way is to use this structure in a single input two outputs (SITO) prediction error identification method. Another way is to estimate a state space model with one input and two outputs using subspace identification. The transfer functions G_1 and G_2 can then be estimated from the state space model estimates. A problem is, however, that the order of these two transfer functions will in general be of the same order as the total system. One way is to apply a model order reduction method, but then it is important to take the statistical properties of the model estimates into account.

As mentioned by the expert a common approach is to estimate individual sub-models, i.e. G_1 from observations of u and y_1 and G_2 as the transfer function from y_1 to y_2. The second problem is an error-in-variables problem, see [16], and care has to be taken in order to avoid problems caused by the noise e_1. In case the signal z_1 is of low frequency character or have a special structure due to the choice of input u this information can be used to pre-process the signal before applying the identification method.

A third way is to estimate the transfer function G_1 from u and y_1 and then $G_3 = G_1 G_2$ from u to y_2. The transfer function G_2 can then be obtained as $\hat{G}_2 = \hat{G}_3/\hat{G}_1$, but will in general be of high order and could also be unstable due to non-minimum phase zeros. A better solution is to solve

$$\min_{G_2} \frac{1}{2\pi} \int_{-\pi}^{\pi} \frac{|\hat{G}_3(e^{i\omega}) - \hat{G}_1(e^{i\omega})G_2(e^{i\omega})|^2}{W_3(e^{i\omega})}\, d\omega$$

where the weighting $W_3(e^{i\omega})$ should reflect the variance of the error. One way is to calculate

$$\mathrm{Cov}\, \frac{\hat{G}_3(e^{i\omega})}{\hat{G}_1(e^{i\omega})} = W_4(e^{i\omega})$$

and use $W_3(e^{i\omega}) = |\hat{G}_1(e^{i\omega})|^2 W_4(e^{i\omega})$. This is very closely related to indirect PEM discussed in [18].

The conclusion is that one should take the statistical properties of the model estimates into account when merging and combining sub model estimates.

7 Conclusion

The objectives of this paper have been to study how subspace system identification can be used in industrial applications and the corresponding supporting theoretical results. One very interesting area is input signal design, and in particular methods that take practical experiences and limitations and into account. Based on the input from an industrial user, we have discussed some theoretical results that can be useful for merging of data and models.

References

1. M. Barenthin, H. Jansson, H. Hjalmarsson, J. Mårtensson, and B. Wahlberg. A control perspective on optimal input design in system identification. In *Forever Ljung in System Identification*, chapter 10. Forever Ljung in System Identification, Studentlitteratur, September 2006.
2. Märta Barenthin. On input design in system identification for control. Technical report, Royal Institute of Technology (KTH), June 2006. Licentiate Thesis.

3. A. Chiuso and G. Picci. Prediction error vs. subspace methods in open and closed-loop identification. In *16th IFAC World Congress*, Prague, Czech Republic, Jul. 2005.

4. N. L. C. Chui and J. M. Maciejowski. Criteria for informative experiments with subspace identification. *International Journal of Control*, 78(5):326–44, Mar. 2005.

5. M. Gevers, L. Miskovic, D. Bonvin, and A. Karimi. Identification of multi-input systems: variance analysis and input design issues. *Automatica*, 42(4):559–572, April 2006.

6. G. Goodwin and R.L. Payne. *Dynamic System Identification: Experiment Design and Data Analysis*. Academic Press, New York, 1977.

7. E. W. Jacobsen. Identification for control of strongly interactive plants. In *AIChE Annual Meeting*, January 1994.

8. H. Jansson. *Experiment design with applications in identification for control*. PhD thesis, KTH, December 2004. TRITA-S3-REG-0404.

9. M. Jansson. *On Subspace Methods in System Identification and Sensor Array Signal Processing*. PhD thesis, October 1997.

10. M. Jansson. Subspace identification and ARX modeling. In *IFAC Symp. on System Identification*, Rotterdam, The Netherlands, Aug. 2003.

11. M. Jansson. A new subspace identification method for open and closed loop data. In *16th IFAC World Congress*, Prague, Czech Republic, Jul. 2005.

12. L. Ljung. *System Identification - Theory For the User, 2nd ed.* PTR Prentice Hall, Upper Saddle River, N.J, 1999.

13. T. E. M. Nordling and E. W. Jacobsen. Experiment design for optimal excitation of gene regulatory networks, October 2006. Poster.

14. Joe S. Qin, Weilu Lin, and Lennart Ljung. A novel subspace identification approach with enforced causal models. *Automatica*, 41(12):2043–2053, December 2005.

15. D.E. Rivera, H. Lee, M.W. Braun, and H.D. Mittelmann. Plant-friendly system identification: a challenge for the process industries. In *Proceedings of the 13th IFAC Symposium on System Identification*, pages 917–922, Rotterdam, The Netherlands, 2003.

16. T. Söderström. Errors-in-variables methods in system identification. *Automatica*, 2007. Survey paper, to appear.

17. T. Söderström and P. Stoica. *System identification*. Prentice Hall International, Hertfordshire, UK, 1989.

18. T. Söderström, P. Stoica, and B. Friedlander. An indirect prediction error method. *Automatica*, 27:183–188, 1991.

19. F. Tjärnström. *Variance Expressions and Model Reduction in System Identification*. PhD thesis, Feb 2002.

20. B. Wahlberg. Model reduction of high-order estimated models: The asymptotic ML approach. *International Journal of Control*, January 1989.

21. S.R. Weller and G.C. Goodwin. Controller design for partial decoupling of linear multivariable systems. *Int. J. Control*, 63(43):535–556, 1996.

Recursive Computation of the MPUM

Jan C. Willems

ESAT-SISTA, K.U. Leuven, B-3001 Leuven, Belgium
Jan.Willems@esat.kuleuven.be
www.esat.kuleuven.be/~jwillems

Summary. An algorithm is presented for the computation of the most powerful unfalsified model associated with an observed vector time series in the class of dynamical systems described by linear constant coefficient difference equations. This algorithm computes a module basis of the left kernel the Hankel matrix of the data, and is recursive in the elements of the basis. It is readily combined with subspace identification ideas, in which a state trajectory is computed first, directly from the data, and the parameters of the identified model are derived from the state trajectory.

1 Introduction

It is a pleasure to contribute an article to this Festschrift in honor of Giorgio Picci on the occasion of his 65-th birthday. In the 35 years since our original acquaintance, I have learned to appreciate Giorgio as a deep thinker and a kind friend. As the topic of this paper, I chose a subject that has dominated Giorgio's research throughout his scientific career: *system identification.*

My paper is purely deterministic in nature, whereas the usual approach to system identification (SYSID) is stochastic. It has always baffled me that so many subjects in systems and control — and in other scientific endeavors as well — immediately pass to a stochastic setting. Motivated by the thought that in the end uncertainty will have to be dealt with, a stochastic framework is adopted *ab initio*, and the question of how the problem would look in a deterministic setting is not even addressed. Moreover, it is considered evident that uncertainty leads to stochasticity. My belief is that from a methodological point of view, it is more reasonable to travel from exact deterministic, to approximate deterministic, to stochastic, and end with approximate stochastic SYSID. See [14] for a more elaborate explanation of my misgivings for using stochastics as basis for SYSID.

This brings up the question what we should mean by 'the' exact deterministic model identified by an observed vector time series. The concept that fits this aim is the *most powerful unfalsified model* (MPUM), the model in the model class that explains the observations, but as little else as possible. The purpose of this article is to develop an algorithm to pass from an observed vector time series to the MPUM in the familiar model class of systems described by linear constant coefficient difference equations.

We start the development with the well-known state construction based on the intersection of the row spaces of a past/future partition of the Hankel matrix of the data. By scrutinizing this algorithm, using the Hankel structure, we deduce that this state

A. Chiuso et al. (Eds.): Modeling, Estimation and Control, LNCIS 364, pp. 329–344, 2007.
springerlink.com

construction can be done without the past/future partition, and requires only the computation of the left kernel of the Hankel matrix itself. However, this left kernel is infinite dimensional, but it has the structure of a finitely generated module, and the state construction can be done using a module basis. It therefore suffices to compute a finite number of elements of the left kernel.

This computation can be done recursively, as follows. Truncate the Hankel matrix at consecutive rows, until an element in the left kernel of the Hankel matrix is found. Next, consider the system defined by this element, and construct a direct complement of it. Subsequently, compute the error defined by a kernel representation of this complement, applied to the original time series. This error has lower dimension than the original one. This is recursively repeated until the error is persistently exciting. This leads to an algorithm that computes the MPUM. It readily also gives the state trajectory corresponding to the observed trajectory. The algorithm is therefore very adapted to be used in concordance with subspace identification, in which also a state model of the MPUM is computed, with all the advantages thereof, for example for model reduction.

We present only the ideas underlying this algorithm. Proofs and simulations will appear elsewhere.

A couple of words about the notation used. $\mathbb{N}$ denotes the set of natural numbers, and $\mathbb{R}$ the reals. $\mathbb{R}[\xi]$ denotes, as usual, the ring of polynomials with real coefficients, and $\mathbb{R}(\xi)$ the field of real rational functions, with ξ the indeterminate. Occasionally, we use the notation $\mathbb{R}[\xi]^{\bullet \times \mathtt{w}}$ for polynomial matrices with $\mathtt{w}$ columns but an unspecified (finite) number of rows. The backwards shift is denote by σ, and defined, for $f : \mathbb{N} \to \mathbb{F}$, by

$$\sigma f : \mathbb{N} \to \mathbb{F}, \quad \sigma f(t) := f(t+1).$$

2   Problem Statement

The problem discussed in this paper may be compactly formulated as follows.

Given an observed vector time series

$$\tilde{w} = (\tilde{w}(1), \tilde{w}(2), \ldots, \tilde{w}(t), \ldots)$$

with $\tilde{w}(t) \in \mathbb{R}^{\mathtt{w}}$ and $t \in \mathbb{N}$, find the most powerful unfalsified model associated with $\tilde{w}$ in the model class of dynamical systems consisting of the set of solutions of linear constant coefficient difference equations.

In section 8 we discuss how the ideas can be adapted to deal with more realistic situations: finite time series, missing data (due to erasures or censoring), multiple time series, approximate modeling, etc. In the present section, the terminology used in the problem statement is explained.

We consider discrete time systems with time set $\mathbb{N}$ and with signal space a finite dimensional real vector space, say $\mathbb{R}^{\mathtt{w}}$. With a dynamical model, we mean a family of trajectories from $\mathbb{N}$ to $\mathbb{R}^{\mathtt{w}}$, a *behavior* $\mathcal{B} \subseteq (\mathbb{R}^{\mathtt{w}})^{\mathbb{N}}$. $\mathcal{B}$ is

$$[\![\textit{unfalsified by } \tilde{w}]\!] :\Leftrightarrow [\![\tilde{w} \in \mathcal{B}]\!].$$

Let $\mathcal{B}_1, \mathcal{B}_2 \subseteq (\mathbb{R}^{\mathtt{w}})^{\mathbb{N}}$. Call

$$[\![\, \mathcal{B}_1 \text{ more powerful than } \mathcal{B}_2 \,]\!] :\Leftrightarrow [\![\, \mathcal{B}_1 \subset \mathcal{B}_2 \,]\!].$$

Modeling is prohibition, and the more a model forbids, the better it is. A *model class* is a set of behaviors. The *most powerful unfalsified model* (MPUM) associated with $\tilde{w}$ in a model class is a behavior that is unfalsified by $\tilde{w}$ and more powerful than any other unfalsified model in this model class. In other words, the MPUM explains the data $\tilde{w}$, but as little else as possible. Obviously, for a general model class, the MPUM may not exist. We now describe a model class for which the MPUM does exist.

This model class is a very familiar one. It consists of the behaviors that are the set of solutions of linear constant coefficient difference equations. Explicitly, each behavior $\mathcal{B}$ in this model class is defined by a real polynomial matrix $R \in \mathbb{R}\left[\xi\right]^{\bullet \times \mathtt{w}}$ as

$$\mathcal{B} = \{w : \mathbb{N} \to \mathbb{R}^{\mathtt{w}} \mid R(\sigma)w = 0\}.$$

Since $\mathcal{B} = \mathrm{kernel}\left(R(\sigma)\right)$, with $R(\sigma)$ viewed as a map from $(\mathbb{R}^{\mathtt{w}})^{\mathbb{N}}$ to $(\mathbb{R}^{\mathrm{rowdim}(\mathrm{R})})^{\mathbb{N}}$, we call

$$R(\sigma)w = 0 \tag{$\mathcal{K}$}$$

a *kernel representation* of the corresponding behavior.

We denote this model class by $\mathcal{L}^{\mathtt{w}}$. The many ways of arriving at it, and various equivalent representations of its elements are described, for example, in [14, section 3]. Perhaps the simplest, 'representation free', way of characterizing $\mathcal{L}^{\mathtt{w}}$ is as follows. $\mathcal{B} \subseteq (\mathbb{R}^{\mathtt{w}})^{\mathbb{N}}$ belongs to $\mathcal{L}^{\mathtt{w}}$ if and only if it has the following three properties:

(i) $\mathcal{B}$ is *linear*,
(ii) *shift-invariant* ($\sigma\mathcal{B} \subseteq \mathcal{B}$), and
(iii) *closed* in the topology of point-wise convergence.

Consider $\mathbb{R}\left[\xi\right]^{\mathtt{w}}$. Obviously it is a module over $\mathbb{R}\left[\xi\right]$. Let $\mathcal{M}^{\mathtt{w}}$ denote the set of $\mathbb{R}\left[\xi\right]$-submodules of $\mathbb{R}\left[\xi\right]^{\mathtt{w}}$. It is well known that each element of $\mathcal{M}^{\mathtt{w}}$ is finitely generated, meaning that for each $\mathbb{M} \in \mathcal{M}^{\mathtt{w}}$, there exist $g_1, g_2, \ldots, g_{\mathtt{p}}$ such that

$$\mathbb{M} = \{r \in \mathbb{R}\left[\xi\right]^{\mathtt{w}} \mid \exists \, \alpha_1, \alpha_2, \ldots, \alpha_{\mathtt{p}} \in \mathbb{R}\left[\xi\right] \text{ such that } r = \alpha_1 g_1 + \alpha_2 g_2 + \cdots + \alpha_{\mathtt{p}} g_{\mathtt{p}}\}.$$

There exists a $1 \leftrightarrow 1$ relation between $\mathcal{L}^{\mathtt{w}}$ and $\mathcal{M}^{\mathtt{w}}$. This may be seen as follows. Call

$$[\![\, r \in \mathbb{R}\left[\xi\right]^{\mathtt{w}} \text{ an } \textit{annihilator} \text{ for } \mathcal{B} \,]\!] :\Leftrightarrow [\![\, r^{\top}(\sigma)\mathcal{B} = 0 \,]\!].$$

Denote the set of annihilators of $\mathcal{B}$ by $\mathcal{B}^{\perp}$. Clearly $\mathcal{B}^{\perp} \in \mathcal{M}^{\mathtt{w}}$. This identifies the map $\mathcal{B} \in \mathcal{L}^{\mathtt{w}} \mapsto \mathcal{B}^{\perp} \in \mathcal{M}^{\mathtt{w}}$. It can be shown that this map is surjective (not totally trivial, but true). When $\mathcal{B}$ has kernel representation $(\mathcal{K})$, then $\mathcal{B}^{\perp}$ is the $\mathbb{R}\left[\xi\right]$-module generated by the transposes of the rows of R. To travel the reverse route and associate an element $\mathcal{B} \in \mathcal{L}^{\mathtt{w}}$ with a module $\mathbb{M} \in \mathcal{M}^{\mathtt{w}}$, take the behavior of the system with kernel representation $(\mathcal{K})$ generated by the polynomial matrix R with as rows the transposes of a set of generators of $\mathbb{M}$.

The module of annihilators is a more appropriate way of thinking about an element $\mathcal{B} \in \mathcal{L}^{\mathtt{w}}$ than the specific, but less intrinsic, difference equation ($\mathcal{K}$) which one happens to have chosen to define $\mathcal{B}$.

The special case of ($\mathcal{K}$) given by the overly familiar

$$P(\sigma)y = Q(\sigma)u, \quad w = (u, y) \tag{i/o}$$

with $P \in \mathbb{R}\left[\xi\right]^{\mathtt{p} \times \mathtt{p}}, Q \in \mathbb{R}\left[\xi\right]^{\mathtt{p} \times \mathtt{m}}, \det(P) \neq 0$, and with proper transfer function $G = P^{-1}Q \in \mathbb{R}(\xi)^{\mathtt{p} \times \mathtt{m}}$, is called an *input/output* (i/o) system, with $u : \mathbb{N} \to \mathbb{R}^{\mathtt{m}}$ the *input* and $y : \mathbb{N} \to \mathbb{R}^{\mathtt{p}}$ the *output*. The conditions imposed on P, Q ensure that u is free, and that y does not anticipate u, the usual requirements on an input/output system. Clearly (i/o) defines an element of $\mathcal{L}^{\mathtt{m}+\mathtt{p}}$. Conversely, for every $\mathcal{B} \in \mathcal{L}^{\mathtt{w}}$, there exists a system (i/o) with $\mathtt{m} + \mathtt{p} = \mathtt{w}$, that has, up to a mere reordering of the components, behavior $\mathcal{B}$. With this reordering, we mean that there exists a permutation matrix $\Pi \in \mathbb{R}^{\mathtt{w} \times \mathtt{w}}$ (depending on $\mathcal{B}$, of course), such that (i/o) has behavior $\Pi\mathcal{B}$. In the sequel, we often silently assume that the permutation that makes the inputs the leading, and the outputs the trailing components of w has been carried out already.

As in all of system theory, controllability plays an important role also in the theory surrounding the MPUM. We recall the behavioral definition of controllability.

$[\![\, \mathcal{B} \in \mathcal{L}^{\mathtt{w}} \text{ is } controllable \,]\!]$

$:\Leftrightarrow [\![\, \forall\, w_1, w_2 \in \mathcal{B} \text{ and } t_1 \in \mathbb{N}, \exists\, w \in \mathcal{B} \text{ and } t_2 \in \mathbb{N}, t_2 \geq t_1, \text{ such that}$

$wW(t) = w_1(t) \text{ for } 1 \leq t \leq t_1, \text{ and } w(t) = w_2(t - t_1 - t_2) \text{ for } t > t_1 + t_2 \,]\!]$.

Various characterizations of controllability may be found, for example, in [14, section 5].

It is easy to see that there exists an MPUM in $\mathcal{L}^{\mathtt{w}}$ associated with $\tilde{w}$. Denote it by $\tilde{\mathcal{B}}$. The most convincing proof that this MPUM exists, is by showing what it is:

$$\tilde{\mathcal{B}} = \overline{\text{linear span } (\{\tilde{w}, \sigma\tilde{w}, \ldots, \sigma^t\tilde{w}, \ldots\})},$$

where the right hand side means the closure in the topology of point-wise convergence of the linear span of the elements in the set. Obviously, this linear span is linear, it is shift invariant since it is constructed from $\tilde{w}$ and its shifts, and after taking the closure, it is closed in the topology of point-wise convergence. Consequently it belongs to $\mathcal{L}^{\mathtt{w}}$. It is also unfalsified, since $\tilde{w} \in \tilde{\mathcal{B}}$, and clearly any unfalsified element in $\mathcal{L}^{\mathtt{w}}$ must contain all the $\sigma^t\tilde{w}$'s and hence their linear span, and be closed in the topology of point-wise convergence. This proves that $\tilde{\mathcal{B}}$ is indeed the MPUM in $\mathcal{L}^{\mathtt{w}}$ associated with $\tilde{w}$.

($\mathcal{K}$) is unfalsified by $\tilde{w}$ iff $R(\sigma)\tilde{w} = 0$. It follows that among all polynomial matrices $R \in \mathbb{R}\left[\xi\right]^{\bullet \times \mathtt{w}}$ such that $R(\sigma)\tilde{w} = 0$, there is one whose behavior is more powerful than any other. And, of course, this MPUM allows an i/o representation. Our aim are algorithms to go from the observed time series $\tilde{w}$ to a representation of its associated MPUM in $\mathcal{L}^{\mathtt{w}}$. The most direct way to go about this is to compute, from $\tilde{w}$, a polynomial matrix R such that ($\mathcal{K}$) is a kernel representation of this MPUM. Equivalently, to compute a set of generators for $\mathcal{B}^{\perp}$. In [11] several such algorithms are described.

The 'consistency' problem consists of finding conditions so that the system that has generated the data is indeed the one that is identified by the system identification algorithm. In our deterministic setting this comes down to checking when the system that has generated $\tilde{w}$ is actually the MPUM in $\mathcal{L}^{\mathtt{w}}$ associated with $\tilde{w}$. Persistency of excitation, but also controllability, are the key conditions leading to consistency. The vector time series $f : \mathbb{N} \to \mathbb{R}^{\mathtt{k}}$ is

$$[\![\,persistently\ exciting\,]\!] :\Leftrightarrow [\![\,\text{the MPUM in } \mathcal{L}^{\mathtt{k}} \text{ associated with } f \text{ equals } \left(\mathbb{R}^{\mathtt{k}}\right)^{\mathbb{N}}\,]\!].$$

Consider $\mathcal{B} \in \mathcal{L}^{\mathtt{w}}$, with i/o partition $w = (u, y)$. Assume that $\tilde{w} = (\tilde{u}, \tilde{\mathbf{y}})$ is partitioned accordingly. Then $\mathcal{B}$ is the MPUM in $\mathcal{L}^{\mathtt{w}}$ associated with $\tilde{w}$ if

1. $\tilde{w} \in \mathcal{B}$,
2. $\tilde{u}$ is persistently exciting,
3. $\mathcal{B}$ is controllable.

In [13] a more general version of this consistency result is proven. Note that the first two conditions are clearly also necessary for consistency.

This result provides additional motivation for making the MPUM the aim of deterministic system identification.

3 Subspace Identification

In addition to looking for a kernel representation of this MPUM, we are even more interested in obtaining a state space representation of it. We first explain what we mean by this.

Let $\mathtt{m}, \mathtt{p}, \mathtt{n}$ be nonnegative integers, $A \in \mathbb{R}^{\mathtt{n} \times \mathtt{n}}, B \in \mathbb{R}^{\mathtt{n} \times \mathtt{m}}, C \in \mathbb{R}^{\mathtt{p} \times \mathtt{n}}, D \in \mathbb{R}^{\mathtt{p} \times \mathtt{m}}$, and consider the ubiquitous system

$$\sigma x = Ax + Bu, \quad y = Cx + Du, \quad w = (u, y). \tag{$\mathcal{S}$}$$

In this equation, $u : \mathbb{N} \to \mathbb{R}^{\mathtt{m}}$ is the *input*, $y : \mathbb{N} \to \mathbb{R}^{\mathtt{p}}$ the *output*, and $x : \mathbb{N} \to \mathbb{R}^{\mathtt{n}}$ the *state* trajectory. The behavior

$$\{(u, y) : \mathbb{N} \to \mathbb{R}^{\mathtt{m}} \times \mathbb{R}^{\mathtt{p}} \mid \exists x : \mathbb{N} \to \mathbb{R}^{\mathtt{n}} \text{ such that } (\mathcal{S}) \text{ holds}\}$$

is called the *external behavior* of $(\mathcal{S})$. It can be shown that this external behavior belongs to $\mathcal{L}^{\mathtt{m}+\mathtt{p}}$. The (u, y, x)-behavior is obviously an element of $\mathcal{L}^{\mathtt{m}+\mathtt{p}+\mathtt{n}}$. This implies that the (u, y)-behavior, what we call the external behavior, is an element of $\mathcal{L}^{\mathtt{m}+\mathtt{p}}$. This is due to the fact than the projection onto a subset of the components of a linear shift-invariant closed subspace of $\left(\mathbb{R}^{\mathtt{w}}\right)^{\mathbb{N}}$ is again linear, shift-invariant, and closed. This result is called the 'elimination theorem', and is an important element in the behavioral theory of systems. It implies, for example, that $\mathcal{L}^{\mathtt{w}}$ is closed under addition.

So, the external behavior of $(\mathcal{S})$ belongs of to $\mathcal{L}^{\mathtt{m}+\mathtt{p}}$. Conversely, for every $\mathcal{B} \in \mathcal{L}^{\mathtt{w}}$, there exists a system $(\mathcal{S})$, with $\mathtt{m} + \mathtt{p} = \mathtt{w}$, that has, up to a mere reordering of the components (u, y), external behavior $\mathcal{B}$. With this reordering, we mean that there exists a permutation matrix $\Pi \in \mathbb{R}^{\mathtt{w} \times \mathtt{w}}$ such that $(\mathcal{S})$ has external behavior $\Pi \mathcal{B}$. In the sequel, we again often silently assume that the permutation that makes the inputs the leading, and the outputs the trailing components of w has been carried out already.

$(\mathcal{S})$ is called an *input/state/output* (i/s/o) representation of its external behavior. $(\mathcal{S})$ is called *minimal* if its state has minimal dimension among all i/s/o systems with the same external behavior. It can be shown that minimal is equivalent to state observable, meaning that if (u, y, x') and (u, y, x'') both satisfy $(\mathcal{S})$, then $x' = x''$. In other words, observability means that the state trajectory x can be deduced from the input and output trajectories (u, y) jointly. As is very well-known, observability holds iff the $(np \times n)$ matrix $\mathrm{col}\left(C, CA, \cdots, CA^{n-1}\right)$ has rank n. Minimality does not imply controllability. But a minimal i/s/o representation is state controllable iff its external behavior is controllable, in the sense we have defined controllability of behaviors.

As explained in the previous section, we are looking for algorithms that pass from the observed time series $\tilde{w}$ to its MPUM in $\mathcal{L}^{\mathtt{w}}$, for example, by computing a kernel representation $(\mathcal{K})$ of this MPUM. There is, however, another way to go about this, by first computing the state trajectory corresponding to $\tilde{w}$ in the MPUM, and subsequently the system parameters (A, B, C, D) corresponding to an i/s/o representation. Explicitly, assume that we had somehow found the MPUM. We could then compute a minimal i/s/o representation for it, and obtain the (unique) state trajectory $\tilde{x}$ corresponding to $\tilde{w}$. Of course, for every $T \in \mathbb{N}$, there holds (assuming that the reordering of the components discussed before such that $\tilde{w} = (\tilde{u}, \tilde{\mathbf{y}})$ has been carried out)

$$\begin{bmatrix} \tilde{x}(2)\ \tilde{x}(3)\ \cdots\ \tilde{x}(t+1)\ \cdots \\ \tilde{\mathbf{y}}(1)\ \tilde{\mathbf{y}}(2)\ \cdots\ \ \tilde{\mathbf{y}}(t)\ \ \ \cdots \end{bmatrix} = \begin{bmatrix} A\ B \\ C\ D \end{bmatrix} \begin{bmatrix} \tilde{x}(1)\ \tilde{x}(2)\ \cdots\ \tilde{x}(t)\ \cdots \\ \tilde{u}(1)\ \tilde{u}(2)\ \cdots\ \tilde{u}(t)\ \cdots \end{bmatrix}. \qquad (\$)$$

So, if

$$\begin{bmatrix} \tilde{x}(1)\ \tilde{x}(2)\ \cdots\ \tilde{x}(T) \\ \tilde{u}(1)\ \tilde{u}(2)\ \cdots\ \tilde{u}(T) \end{bmatrix}$$

is of full row rank, (\$), truncated at column T, provides an equation for computing A, B, C, D, and yields an i/s/o representation of the MPUM.

As we explained it, this approach appears to be a vicious circle. For in order to compute $\tilde{x}$, we seem to need the MPUM to start with. But, if we could somehow compute $\tilde{x}$, directly from the data $\tilde{w}$, *without* deriving it from the MPUM, then (\$) gives a viable and (see section 8) attractive way to compute an i/s/o representation of the MPUM. In section 8 we shall explain that even when we deduce $\tilde{x}$ from a kernel representation of the MPUM, it is advantageous to return to (\$) for the purpose of system identification because of its built-in model reduction.

The SYSID methods that first compute the state trajectory from the data, and then derive the system model from the state trajectory have become known as *subspace identification* algorithms. Before the emergence of these methods, state space representations played a somewhat secondary role in system identification. The purpose of this paper is to take a closer look at (the deterministic version of) these algorithms.

4   State Construction by Past/Future Partition

The question is:

> *How can we compute the state trajectory $\tilde{x}$*
> *directly from $\tilde{w}$, without first computing the MPUM $\tilde{\mathcal{B}}$?*

The doubly infinite matrix

$$
\mathcal{H} := \begin{bmatrix}
\tilde{w}(1) & \tilde{w}(2) & \cdots & \tilde{w}(t) & \cdots \\
\tilde{w}(2) & \tilde{w}(3) & \cdots & \tilde{w}(t+1) & \cdots \\
\vdots & \vdots & \vdots\vdots\vdots & \vdots & \vdots\vdots\vdots \\
\tilde{w}(t') & \tilde{w}(t'+1) & \cdots & \tilde{w}(t+t'-1) & \cdots \\
\vdots & \vdots & \vdots\vdots\vdots & \vdots & \vdots\vdots\vdots
\end{bmatrix}
\tag{$\mathcal{H}$}
$$

is called the *Hankel matrix* of the data $\tilde{w}$. It holds the key to the state construction.

The earliest subspace algorithms are based on the intersection of the span of the rows of a past/future partition of this Hankel matrix, and deduce the state trajectory as the common linear combinations of the past and the future. This proceeds as follows. Partition a row truncation of $\mathcal{H}$ as

$$
\left[\frac{\mathcal{H}_p}{\mathcal{H}_f}\right] = \begin{bmatrix}
\tilde{w}(1) & \tilde{w}(2) & \cdots & \tilde{w}(t) & \cdots \\
\tilde{w}(2) & \tilde{w}(3) & \cdots & \tilde{w}(t+1) & \cdots \\
\vdots & \vdots & \vdots\vdots\vdots & \vdots & \vdots\vdots\vdots \\
\tilde{w}(T_p) & \tilde{w}(T_p+1) & \cdots & \tilde{w}(T_p+t-1) & \cdots \\
\hline
\tilde{w}(T_p+1) & \tilde{w}(T_p+2) & \cdots & \tilde{w}(T_p+t) & \cdots \\
\vdots & \vdots & \vdots\vdots\vdots & \vdots & \vdots\vdots\vdots \\
\tilde{w}(T_p+T_f-1) & \tilde{w}(T_p+T_f) & \cdots & \tilde{w}(T_p+T_f+t-2) & \cdots \\
\tilde{w}(T_p+T_f) & \tilde{w}(T_p+T_f+1) & \cdots & \tilde{w}(T_f+T_p+t-1) & \cdots
\end{bmatrix}
\tag{$\mathcal{H}_p/\mathcal{H}_f$}
$$

and refer to $\mathcal{H}_p$ as the 'past', and to $\mathcal{H}_f$ as the 'future' of the Hankel matrix. T_p and T_f are sufficiently large nonnegative integers. Actually, it is possible to proceed after truncating these Hankel matrices also column-wise at a sufficiently large column T. We do not enter into details about what 'sufficiently large' exactly means in these statements — those issues are glossed over here.

Consider the intersection of the linear space spanned by the rows of $\mathcal{H}_p$ and the linear space spanned by the rows of $\mathcal{H}_f$. Let n be the dimension of this intersection. This means that there are n linearly independent linear combinations of the rows of $\mathcal{H}_p$ that are equal to linear combinations of the rows of $\mathcal{H}_f$. These linear combinations can be stacked into a matrix with n rows,

$$
\tilde{X} = \begin{bmatrix} \tilde{x}(T_p+1) & \tilde{x}(T_p+2) & \cdots & \tilde{x}(T_p+t) & \cdots \end{bmatrix}
$$

It turns out that, under suitable conditions which are spelled out in [11], the dimension of this intersection equals the dimension of the state space of a minimal i/s/o representation of the MPUM in $\mathcal{L}^{\mathtt{w}}$ associated with $\tilde{w}$. Moreover, as the notation suggests, the columns of $\tilde{X}$ form the state trajectory corresponding to $\tilde{w}$ in the MPUM $\tilde{\mathcal{B}}$ (more precisely, corresponding to a minimal i/s/o representation of $\tilde{\mathcal{B}}$). This then leads, by equation (\$), to an algorithm to compute the matrices A, B, C, D and to an identification procedure for the MPUM.

In [11] this intersection algorithm is applied to a variety of situations, including classical realization theory. These algorithms have been given good numerical linear algebra based implementations in [8].

In the *purely deterministic* case the state trajectory can be obtained, as we have just seen, as the *intersection* of the linear span of the rows of the past with the linear span of the rows of the future of the Hankel matrix of the data. This is, in a sense, analogous to the fact that in the *purely stochastic* case the state trajectory can be obtained as the *orthogonal projection* of the linear span of the rows of the past onto the linear span of the rows of the future of the Hankel matrix of the data, as noted in [1]. This idea was used in [6] for the purposes of stochastic SYSID. The resulting subspace methods in the context of the purely stochastic case have been followed up by many authors, see, in particular, [5] and [4]. The combined deterministic/stochastic case is a significant generalization of the purely deterministic case and the purely stochastic case individually. It has been studied in [7, 8]. Similar algorithms have been developed in [10]. In the mean time, many articles dealing with subspace algorithms for the combined case have appeared, for instance [2, 3].

5   The Hankel Structure and the Past/Future Partition

Let us now take a closer look at the intersection of the spaces spanned by the rows of $\mathcal{H}_p$ and by the rows of $\mathcal{H}_f$. *How can we obtain this intersection?* Consider this question first for a general partitioned matrix

$$M = \left[\frac{M'}{M''} \right].$$

The common linear combinations of the row span of M' and the row span of M'' can be computed from the left kernel of M. Indeed,

$$\llbracket\, k\, M = 0 \leftrightarrow \left[k' \mid k'' \right] \left[\frac{M'}{M''} \right] = 0\, \rrbracket \Leftrightarrow \llbracket\, k'\, M' = -k''\, M''\, \rrbracket,$$

and hence the common linear combinations of the span of the rows of M' and M'' follow immediately from a set of vectors that span the left kernel of M, by truncating these vectors conformably with the partition of the matrix M, to k', and multiplying by M'. This can be applied to the partitioned Hankel matrix $(\mathcal{H}_p/\mathcal{H}_f)$, and we observe that the state construction amounts to computing the left kernel of the partitioned matrix $(\mathcal{H}_p/\mathcal{H}_f)$.

We shall now argue that, *because of the Hankel structure*, the left kernel of $(\mathcal{H}_p/\mathcal{H}_f)$ can be deduced from the left kernel of $\mathcal{H}_p$ all by itself, and so, there is no need to use the past/future partitioning in order to construct the left kernel and the state. To see this, assume that

$$\left[\, k_1\ k_2\ \cdots\ k_{T_p}\, \right]$$

is in the left kernel of $\mathcal{H}_p$, i.e. $k\,\mathcal{H}_p = 0$. Notice that, because of the Hankel structure of $\mathcal{H}$, the vectors

$$\left[\, 0 \cdots 0 \; k_1 \; k_2 \; \cdots \; k_{T_p} \; 0 \cdots 0 \,\right],$$

obtained by putting in total T_f zeros in front and in back of $\left[\, k_1 \; k_2 \; \cdots \; k_{T_p} \,\right]$, are all contained in the left kernel of $(\mathcal{H}_p/\mathcal{H}_f)$. It can be shown that, provided T_p is sufficiently large (but it need not be larger that what was required to validate the intersection argument of the row spans of $\mathcal{H}_p$ and $\mathcal{H}_f$ of the previous section), we obtain this way, from a set of vectors that span the left kernel of $\mathcal{H}_p$, a set of vectors that span the whole left kernel of $(\mathcal{H}_p/\mathcal{H}_f)$. After truncation to its first T_p elements,

$$\left[\, 0 \cdots 0 \; k_1 \; k_2 \; \cdots \; k_L \,\right],$$

this leads to a set of vectors that, when multiplied from the right with $\mathcal{H}_p$, span the intersection of the spaces spanned by the rows of $\mathcal{H}_p$ and the rows of $\mathcal{H}_f$. Note that this truncation results from applying repeatedly the shift-and-cut operator to the row vector $\left[\, k_1 \; k_2 \; \cdots \; k_{T_p} \,\right]$, i.e. putting a zero in the first element and deleting the last element of this row vector, so as to obtain a vector of length T_p. In other words, from the vector

$$\left[\, k_1 \; k_2 \; \cdots \; k_{T_p} \,\right]$$

in the left kernel of $\mathcal{H}_p$, we obtain the vectors

$$
\begin{array}{l}
\left[\; 0 \; k_1 \; k_2 \; \cdots \; k_{T_p-2} \; k_{T_p-1} \;\right] \\[4pt]
\left[\; 0 \; 0 \; k_1 \; \cdots \; k_{T_p-3} \; k_{T_p-2} \;\right] \\[4pt]
\qquad\qquad\vdots \\[4pt]
\left[\; 0 \; 0 \; 0 \; \cdots \; k_1 \quad k_2 \;\right] \\[4pt]
\left[\; 0 \; 0 \; 0 \; \cdots \; 0 \quad k_1 \;\right]
\end{array}
$$

that are truncations of elements from the left kernel of $\left[\dfrac{\mathcal{H}_f}{\mathcal{H}_p}\right]$

Using the ideas explained in the previous section, this leads to the construction of the state trajectory associated with $\tilde{w}$ in the MPUM, by computing a basis of the left kernel of $\mathcal{H}_p$, stacking these vectors as the rows of the matrix

$$\left[\, K_1 \; K_2 \; \cdots \; K_{T_p-1} \; K_{T_p} \,\right],$$

and repeatedly applying the shift-and-cut operator to obtain the state trajectory

$$\left[\, \tilde{x}(T_p+1) \; \tilde{x}(T_p+2) \; \cdots \; \tilde{x}(T_p+t) \; \cdots \,\right] =$$

$$
\begin{bmatrix}
0 & K_1 & K_2 & \cdots & K_{T_p-2} & K_{T_p-1} \\
0 & 0 & K_1 & \cdots & K_{T_p-3} & K_{T_p-2} \\
\vdots & \vdots & \vdots & \vdots\vdots\vdots & \vdots & \vdots \\
0 & 0 & 0 & \cdots & K_1 & K_2 \\
0 & 0 & 0 & \cdots & 0 & K_1
\end{bmatrix}
\begin{bmatrix}
\tilde{w}(1) & \tilde{w}(2) & \cdots & \tilde{w}(t) & \cdots \\
\tilde{w}(2) & \tilde{w}(3) & \cdots & \tilde{w}(t+1) & \cdots \\
\tilde{w}(3) & \tilde{w}(4) & \cdots & \tilde{w}(t+2) & \cdots \\
\vdots & \vdots & \vdots\vdots\vdots & \vdots & \vdots\vdots\vdots \\
\tilde{w}(T_p-1) & \tilde{w}(T_p) & \cdots & \tilde{w}(T_p+t-2) & \cdots \\
\tilde{w}(T_p) & \tilde{w}(T_p+1) & \cdots & \tilde{w}(T_p+t-3) & \cdots
\end{bmatrix}.
$$

Actually, it turns out that we can also apply the shift-and-cut backwards, leading to

$$\begin{bmatrix} \tilde{x}(1) & \tilde{x}(2) & \cdots & \tilde{x}(t) & \cdots \end{bmatrix} =$$

$$\begin{bmatrix} K_2 & K_3 & \cdots & K_{T_p-1} & K_{T_p} \\ K_3 & K_4 & \cdots & K_{T_p} & 0 \\ \vdots & \vdots & \vdots\vdots\vdots & \vdots & \vdots \\ K_{T_p-1} & K_{T_p} & \cdots & 0 & 0 \\ K_{T_p} & 0 & \cdots & 0 & 0 \end{bmatrix} \begin{bmatrix} \tilde{w}(1) & \tilde{w}(2) & \cdots & \tilde{w}(t) & \cdots \\ \tilde{w}(2) & \tilde{w}(3) & \cdots & \tilde{w}(t+1) & \cdots \\ \vdots & \vdots & \vdots\vdots\vdots & \vdots & \vdots\vdots\vdots \\ \tilde{w}(T_p-2) & \tilde{w}(T_p-1) & \cdots & \tilde{w}(T_p+t-3) & \cdots \\ \tilde{w}(T_p-1) & \tilde{w}(T_p) & \cdots & \tilde{w}(T_p+t-2) & \cdots \end{bmatrix}.$$

This then yields the desired state trajectory to which the subspace algorithm (\$) can be applied in order to obtain an i/s/o representation of the MPUM.

6 The Left Kernel of a Hankel Matrix

In the previous section, we have seen the relevance to the problem at hand of computing the left kernel of the doubly infinite Hankel matrix $\mathcal{H}$. We are interested in characterizing the infinite vectors in its left kernel that have 'compact support', i.e. the infinite vectors of the form

$$k = \begin{bmatrix} k_1 & k_2 & \cdots & k_t & \cdots & 0 & \cdots & 0 & \cdots \end{bmatrix}, \quad k_t \in \mathbb{R}^{1 \times \mathtt{w}}, t \in \mathbb{N},$$

with $k\,\mathcal{H} = 0$. Denote the set of compact support elements in the left kernel by $\mathcal{N}$. For simplicity, we call $\mathcal{N}$ the left kernel of $\mathcal{H}$.

In general, $\mathcal{N}$ is infinite dimensional. In fact, $\mathcal{N}$ equals $\{0\}$, or it is infinite dimensional. However, we shall now argue that by considering the left kernel of $\mathcal{H}$ as a module, $\mathcal{N}$ is effectively finite dimensional, of dimension $\leq \mathtt{w}$. Observe that $\mathcal{N}$ is closed under addition (obvious), scalar multiplication (obvious), and under the right shift (also obvious, using the Hankel structure):

$$\llbracket \begin{bmatrix} k_1 & k_2 & \cdots & k_t & 0 & 0 & \cdots & 0 & \cdots \end{bmatrix} \in \mathcal{N} \rrbracket \Rightarrow \llbracket \begin{bmatrix} 0 & k_1 & \cdots & k_{t-1} & k_t & 0 & \cdots & 0 & \cdots \end{bmatrix} \in \mathcal{N} \rrbracket.$$

This implies (identify elements $k \in \mathcal{N}$ with polynomial vectors $k_1 + k_2\xi + \cdots + K_t\xi^{t-1} + \cdots \in \mathbb{R}\left[\xi\right]^{1 \times \mathtt{w}}$, and the right shift with multiplication by ξ) that $\mathcal{N}$ has the structure of a module (a submodule of $\mathbb{R}\left[\xi\right]^{1 \times \mathtt{w}}$, viewed as an $\mathbb{R}\left[\xi\right]$-module). This submodule is finitely generated (all $\mathbb{R}\left[\xi\right]$-submodules of $\mathbb{R}\left[\xi\right]^{1 \times \mathtt{w}}$ are finitely generated, with at most $\mathtt{w}$ generators). This means that there exist elements $n_1, n_2, \ldots, n_{\mathtt{p}} \in \mathcal{N}$, with $\mathtt{p} \leq \mathtt{w}$, such that all other elements of $\mathcal{N}$ can be obtained as linear combinations of these elements and their repeated right shifts.

It turns out that *for the construction of the state trajectory, we need only these generators.* In other words, rather that compute the whole left kernel of $\mathcal{H}_p$, it suffices to obtain a set of generators of the left kernel of $\mathcal{H}$ in the left kernel of $\mathcal{H}_p$. We assume that L_p is sufficiently large, so that the left kernel of $\mathcal{H}_p$ contains a set of generators of the left kernel of $\mathcal{H}$.

This leads to the following state construction algorithm. Let $n_1, n_2, \cdots, n_{\mathtt{p}}$ of $\mathcal{N}$ be a set of generators of the left kernel of $\mathcal{H}$. Truncate these vectors at their last non-zero element:

$$n_i \cong \begin{bmatrix} n_{i,1} & n_{i,2} & \cdots & n_{i,L_i} \end{bmatrix}, \quad n_{i,t} \in \mathbb{R}^{1 \times \mathtt{w}}.$$

Now apply the shift-and-cut to the i-th generator. This leads to the 'partial' state trajectory

$$\begin{bmatrix} \tilde{x}_i(1) & \tilde{x}_i(2) & \cdots & \tilde{x}_i(t) & \cdots \end{bmatrix} =$$

$$\begin{bmatrix} n_{i,2} & n_{i,3} & \cdots & n_{i,L_i-1} & n_{i,L_i} \\ n_{i,3} & n_{i,4} & \cdots & n_{i,L_i} & 0 \\ \vdots & \vdots & \vdots\vdots\vdots & \vdots & \vdots \\ n_{i,L_i-1} & n_{i,L_i} & \cdots & 0 & 0 \\ n_{i,L_i} & 0 & \cdots & 0 & 0 \end{bmatrix} \begin{bmatrix} \tilde{w}(1) & \tilde{w}(2) & \cdots & \tilde{w}(t) & \cdots \\ \tilde{w}(2) & \tilde{w}(3) & \cdots & \tilde{w}(t+1) & \cdots \\ \vdots & \vdots & \vdots\vdots\vdots & \vdots & \vdots\vdots\vdots \\ \tilde{w}(L_i-2) & \tilde{w}(L_i-1) & \cdots & \tilde{w}(L_i+t-3) & \cdots \\ \tilde{w}(L_i-1) & \tilde{w}(L_i) & \cdots & \tilde{w}(L_i+t-2) & \cdots \end{bmatrix}.$$

Now, stack the $\tilde{x}_i$'s. This leads to the state trajectory

$$\begin{bmatrix} \tilde{x}_1(1) \\ \tilde{x}_2(1) \\ \vdots \\ \tilde{x}_{\mathrm{p}}(1) \end{bmatrix}, \begin{bmatrix} \tilde{x}_1(2) \\ \tilde{x}_2(2) \\ \vdots \\ \tilde{x}_{\mathrm{p}}(2) \end{bmatrix}, \ldots, \begin{bmatrix} \tilde{x}_1(t) \\ \tilde{x}_2(t) \\ \vdots \\ \tilde{x}_{\mathrm{p}}(t) \end{bmatrix}, \ldots$$

to which the subspace algorithm (\$) can be applied.

In conclusion, from a set of generators of $\mathcal{N}$ viewed as a module, we obtain, by repeatedly using the shift-and-cut, and matrix which multiplied by the Hankel matrix of the data, yields the state trajectory $\tilde{x}$. In the next section, we provide a recursive way of computing a set of generators.

7   Recursive Computation of a Module Basis

Consider the left kernel $\mathcal{N}$ of $\mathcal{H}$. Call a minimal set of elements $n_1, n_2, \ldots, n_{\mathrm{p}} \in \mathcal{N}$ that generate all of $\mathcal{N}$, through linear combinations of these elements and their repeated right shifts, a *module basis* of $\mathcal{N}$. In the present section, we set up a recursive algorithm to compute a module basis of $\mathcal{N}$ from $\tilde{w}$.

Assume henceforth that the MPUM in $\mathcal{L}^{\mathtt{w}}$ associated with $\tilde{w}$ is controllable. This assumption is made for reasons of exposition. The algorithm can be generalized without this assumption, but it becomes considerably more involved to explain.

We start with a brief digression about controllability. The following basic property characterizes controllable behaviors in $\mathcal{L}^{\mathtt{w}}$:

$$[\![\, \mathcal{B} \in \mathcal{L}^{\mathtt{w}} \text{ is controllable}\,]\!] \Leftrightarrow [\![\, \exists \mathcal{B}' \in \mathcal{L}^{\mathtt{w}} \text{ such that } \mathcal{B} \oplus \mathcal{B}' = (\mathbb{R}^{\mathtt{w}})^{\mathbb{N}}\,]\!].$$

Evidently, $\mathcal{B}'$ is also controllable. In other words, a behavior has a direct complement iff it is controllable. This property of controllable behaviors can be translated in terms of a kernel representation ($\mathcal{K}$) of $\mathcal{B}$, with R of full row rank (every $\mathcal{B} \in \mathcal{L}^{\mathtt{w}}$ allows such a full row rank representation). It states that

$$[\![\, \text{kernel}\,(R\,(\sigma)) \in \mathcal{L}^{\mathtt{w}} \text{ is controllable}\,]\!]$$

$$\Leftrightarrow [\![\, \exists\, R' \in \mathbb{R}\,[\xi]^{\bullet \times \mathtt{w}} \text{ such that } \begin{bmatrix} R \\ \hline R' \end{bmatrix} \text{ is unimodular}\,]\!].$$

In fact, $R'(\sigma)w = 0$ is a kernel representation of the direct complement $\mathcal{B}'$.

We now return to the construction of the MPUM. Start with the data $\tilde{w}$. Consider the associated Hankel matrix $\mathcal{H}$, and its consecutive truncations

$$\begin{bmatrix} \tilde{w}(1) & \tilde{w}(2) & \cdots & \tilde{w}(t) & \cdots \end{bmatrix},$$

$$\begin{bmatrix} \tilde{w}(1) & \tilde{w}(2) & \cdots & \tilde{w}(t) & \cdots \\ \tilde{w}(2) & \tilde{w}(3) & \cdots & \tilde{w}(t+1) & \cdots \end{bmatrix},$$

$$\vdots$$

$$\begin{bmatrix} \tilde{w}(1) & \tilde{w}(2) & \cdots & \tilde{w}(t) & \cdots \\ \tilde{w}(2) & \tilde{w}(3) & \cdots & \tilde{w}(t+1) & \cdots \\ \vdots & \vdots & \vdots\vdots\vdots & \vdots & \vdots\vdots\vdots \\ \tilde{w}(L) & \tilde{w}(L+1) & \cdots & \tilde{w}(t+L-1) & \cdots \end{bmatrix},$$

until a vector in the left kernel is obtained. Denote this element by

$$\begin{bmatrix} k_1 & k_2 & \cdots & k_L \end{bmatrix} \in \mathbb{R}^{1 \times \mathtt{w}L}.$$

Now consider the corresponding vector polynomial

$$n(\xi) = k_1 + k_2\xi + \cdots + k_L\xi^{L-1} \in \mathbb{R}\left[\xi\right]^{1 \times \mathtt{w}}.$$

It can be shown that, because of the assumed controllability of the MPUM, the system with kernel representation $n(\sigma)w = 0$ is also controllable. Consequently, there exists $N \in \mathbb{R}\left[\xi\right]^{(\mathtt{w}-1)\times\mathtt{w}}$ such that the polynomial matrix

$$\left[\frac{n}{N}\right]$$

is unimodular. The system described by $n(\sigma)w = 0$ is unfalsified by $\tilde{w}$, but, of course, $N(\sigma)w = 0$ need not be. Compute the 'error' vector time series

$$\tilde{e} = N(\sigma)\tilde{w} = (\tilde{e}(1), \tilde{e}(2), \ldots, \tilde{e}(t), \ldots),$$

Now apply the above algorithm again, with $\tilde{w}$ replaced by $\tilde{e}$, and proceed recursively. Note that $\tilde{e}(t) \in \mathbb{R}^{\mathtt{w}-1}$: the dimension of the time series that needs to be examined goes down by one at each step.

Recursively this leads to the algorithm

$$\tilde{w} \mapsto n_1 \mapsto N_1 \mapsto \tilde{e}_1 \mapsto n_2 \mapsto N_2 \mapsto \tilde{e}_2 \mapsto \cdots \mapsto \tilde{e}_{p-2} \mapsto n_{p-1} \mapsto N_{p-1} \mapsto \tilde{e}_{p-1} \mapsto n_p.$$

This algorithm terminates when there are no more vectors in the left kernel of the associated Hankel matrix, i.e. when the error $\tilde{e}$ is persistently exciting. If we assume that the MPUM has $\mathtt{m}$ input and $\mathtt{p}$ output components, then $\tilde{e}_{\mathtt{p}}$ will be the first persistently exciting error time series obtained.

Now consider the polynomial vectors

$$r_1 = n_1, r_2 = n_2 N_1, r_3 = n_3 N_2 N_1, \cdots, r_{\mathtt{p}} = n_{\mathtt{p}} N_{\mathtt{p}-1} N_{\mathtt{p}-2} \cdots N_2 N_1.$$

Define

$$\tilde{R} = \begin{bmatrix} r_1 \\ r_2 \\ \vdots \\ r_{\mathrm{p}} \end{bmatrix}.$$

It can be shown that

$$\tilde{R}(\sigma)w = 0$$

is a kernel representation of $\tilde{\mathcal{B}}$, the MPUM in $\mathcal{L}^{\mathtt{w}}$ associated with $\tilde{w}$. The intermediate calculations of $r_1, r_2, \ldots, r_{\mathrm{p}}$ lead to the state trajectory in a similar way as explained in section 6. Let

$$r_i(\xi) = r_{i,1} + r_{i,2}\xi + \cdots + r_{i,L_i}\xi^{L_1-1} \in \mathbb{R}[\xi]^{1\times\mathtt{w}}.$$

Form the vector

$$r_i \cong \begin{bmatrix} r_{i,1} & r_{i,2} & \cdots & r_{i,L_i} \end{bmatrix}, \quad r_{i,t} \in \mathbb{R}^{1\times\mathtt{w}}.$$

Now apply the shift-and-cut. This leads to the 'partial' state trajectory

$$\begin{bmatrix} \tilde{x}_i(1) & \tilde{x}_i(2) & \cdots & \tilde{x}_i(t) & \cdots \end{bmatrix} =$$

$$\begin{bmatrix} r_{i,2} & r_{i,3} & \cdots & r_{i,L_i-1} & r_{i,L_i} \\ r_{i,3} & r_{i,4} & \cdots & r_{i,L_i} & 0 \\ \vdots & \vdots & \vdots\vdots\vdots & \vdots & \vdots \\ r_{i,L_i-1} & r_{i,L_i} & \cdots & 0 & 0 \\ r_{i,L_i} & 0 & \cdots & 0 & 0 \end{bmatrix} \begin{bmatrix} \tilde{w}(1) & \tilde{w}(2) & \cdots & \tilde{w}(t) & \cdots \\ \tilde{w}(2) & \tilde{w}(3) & \cdots & \tilde{w}(t+1) & \cdots \\ \vdots & \vdots & \vdots\vdots\vdots & \vdots & \vdots\vdots\vdots \\ \tilde{w}(L_i-2) & \tilde{w}(L_i-1) & \cdots & \tilde{w}(L_i+t-3) & \cdots \\ \tilde{w}(L_i-1) & \tilde{w}(L_i) & \cdots & \tilde{w}(L_i+t-2) & \cdots \end{bmatrix}.$$

Now, stack the $\tilde{x}_i$'s, and obtain the state trajectory

$$\begin{bmatrix} \tilde{x}_1(1) \\ \tilde{x}_2(1) \\ \vdots \\ \tilde{x}_{\mathrm{p}}(1) \end{bmatrix}, \begin{bmatrix} \tilde{x}_1(2) \\ \tilde{x}_2(2) \\ \vdots \\ \tilde{x}_{\mathrm{p}}(2) \end{bmatrix}, \ldots, \begin{bmatrix} \tilde{x}_1(t) \\ \tilde{x}_2(t) \\ \vdots \\ \tilde{x}_{\mathrm{p}}(t) \end{bmatrix}, \ldots$$

to which the subspace algorithm ($) can be applied.

8 Concluding Remarks

8.1 Subspace ID

Solving equation ($) as the basis for system identification has many appealing features. We can begin by reducing the number of rows of

$$\tilde{X} = \begin{bmatrix} \tilde{x}(1) & \tilde{x}(2) & \cdots & \tilde{x}(t) & \cdots \end{bmatrix}$$

by numerically approximating this matrix by one with fewer rows. This leads to a reduction of the state dimension and hence of the model complexity. Of course, ($)

will then no longer be exactly solvable, even when the observed data is noise free, but since this equation is linear in the unknown matrices A, B, C, D, it is very amenable to a least-squares (LS) solution. Introducing the state in a sense linearizes the SYSID problem. Solving equation (\$) using (LS) also accommodates for noisy data and for numerical errors in the intermediate calculations.

Missing data can be dealt with by deleting columns in the equation (\$). If multiple time series are observed (this is the case in classical realization theory), then equation (\$) can readily be extended with the vectors $\tilde{u}, \tilde{y}, \tilde{x}$ replaced by matrices.

8.2   State Construction by Shift-and-Cut

The state construction that permeates sections 5, 6, and 7 is actually well-known, and our presentation of it via the past/future partition and the left kernel of the Hankel matrix $\mathcal{H}$ to some extent hides the simplicity and generality of the ideas behind it.

Indeed, in [9], a state construction algorithm based on the shift-and-cut operator has been presented as a very direct and general method for constructing state representations, starting from a variety of other system representations. Let $p \in \mathbb{R}\left[\xi\right]$ and define the *shift-and-cut* operator $\sigma^* : \mathbb{R}\left[\xi\right] \to \mathbb{R}\left[\xi\right]$ by

$$\sigma^* : \; p_0 + p_1\xi + \cdots + p_L\xi^L \mapsto p_1 + p_2\xi + \cdots + p_L\xi^{L-1}.$$

By applying the operator σ^* element-wise, it is readily extended to polynomial vectors and matrices.

We now explain how this state construction works, starting with a kernel representation $(\mathcal{K})$. Associate with

$$R(\xi) = R_0 + R_1\xi + \cdots + R_L\xi^L$$

the stacked polynomial matrix

$$X(\xi) = \begin{bmatrix} \sigma^* R \\ \sigma^{*2} R \\ \vdots \end{bmatrix}(\xi) = \begin{bmatrix} R_1 + R_2\xi + R_3\xi^3 + \cdots + R_{L-1}\xi^{L-2} + R_L\xi^{L-1} \\ R_2 + R_3\xi + \cdots + R_{L-1}\xi^{L-3} + R_L\xi^{L-2} \\ \vdots \\ R_{L-1} + R_L\xi \\ R_L \end{bmatrix}(\xi),$$

obtained by repeatedly applying σ^* to R until we get the zero matrix. It can be shown that $X(\sigma)$ is a *state map*: it associates to $w \in \mathrm{kernel}(R(\sigma))$ the corresponding state trajectory $x = X(\sigma)w$ of a (in general non-minimal) state representation of $(\mathcal{K})$. In other words, as soon as we have a kernel representation of the MPUM, the shift-and-cut operator gives us the underlying state.

This shift-and-cut state construction is precisely what is done in sections 5, 6, and 7. Starting from the MPUM as

$$\tilde{\mathcal{B}} = \overline{\mathrm{linear\ span}\ (\{\tilde{w}, \sigma\tilde{w}, \ldots, \sigma^t\tilde{w}, \ldots\})},$$

we construct annihilators for $\tilde{\mathcal{B}}$. These annihilators are, of course, exactly the elements of the left kernel of the Hankel matrix $\mathcal{H}$. By subsequently applying the shift-and-cut operator, we obtain the state trajectory $\tilde{x}$ corresponding to $\tilde{w}$.

8.3   Return to the Data

One of the attractive features of subspace methods, is that after construction of $\tilde{x}$, the model is obtained using equation ($) which involves $\tilde{w}$ and $\tilde{x}$. In other words, it allows to return to the original observed time series $\tilde{w}$ in order to fit the final parameter estimates of the identified system to the data.

The shift-and-cut state construction shows how to obtain a state trajectory from any kernel representation of the MPUM. We originally posed the question of how to construct the state trajectory by avoiding the intermediate computation of a kernel representation of the MPUM. But, we have come full circle on this. We demonstrated that the state construction based on the intersection of the row spans of $\mathcal{H}_p$ and $\mathcal{H}_f$ actually amounts to finding a module basis of a kernel representation of the MPUM. It is an interesting matter to investigate to what extent these insights can also be used in the purely stochastic or in the mixed deterministic/stochastic case.

8.4   Approximation and Balanced Reduction

The algorithm proposed in section 7 lends itself very well for approximate implementation. Checking whether the consecutive truncations of the Hankel matrix have, up to reasonable level of approximation, a non-trivial element in the left kernel, and finding the optimal element in the left kernel, are typical decisions that can be made using SVD based numerical linear algebra computations.

It is of interest to combine our recursive algorithms with model reduction. In particular, it ought to be possible to replace the state construction based on the shift-and-cut operator applied to an annihilator by an alternative set of low order polynomial vectors that lead to a balanced state model. Some ideas in this direction have been given in [12].

8.5   The Complementary System

The most original feature of this article is the recursive computation of a basis of the module of annihilators of a given behavior in section 7. In the controllable case, this may be done by complementing a kernel representation to a unimodular polynomial matrix. It is of interest to explore if the recursive computation of the module basis explained in section 7, combined with the shift-and-cut map, can also be used in the general constructions of a state map, starting from a kernel, an image, or a latent variable representation of a behavior.

The finer features and numerical aspects of this recursive computation and extension of a polynomial matrix to a unimodular one is a matter of future research. In particular, one is led to wonder if the $\ell_2\left(\mathbb{N}, \mathbb{R}^{\mathtt{w}}\right)$-orthogonal complement of $\mathcal{B} \cap \ell_2\left(\mathbb{N}, \mathbb{R}^{\mathtt{w}}\right)$ can play a role in obtaining a complement of a $\mathcal{B}$. Or if the singular value decomposition of the truncated Hankel matrix

$$\begin{bmatrix} \tilde{w}(1) & \tilde{w}(2) & \cdots & \tilde{w}(t) & \cdots \\ \tilde{w}(2) & \tilde{w}(3) & \cdots & \tilde{w}(t+1) & \cdots \\ \vdots & \vdots & \vdots & \vdots & \vdots \\ \tilde{w}(L) & \tilde{w}(L+1) & \cdots & \tilde{w}(t+L-1) & \cdots \end{bmatrix}$$

can be used for complementing n with N to obtain a unimodular polynomial matrix. The left singular vector corresponding to the smallest singular value should serve to identify the element n in the left kernel, and the others somehow to find the complement N.

Acknowledgments

This research is supported by the Research Council KUL project CoE EF/05/006 (OPTEC), Optimization in Engineering, and by the Belgian Federal Science Policy Office: IUAP P6/04 (Dynamical systems, Control and Optimization, 2007-2011).

References

1. H. Akaike, Markovian representation of stochastic processes by canonical variables, *SIAM Journal on Control*, volume 13, pages 162–173, 1975.
2. A. Chiuso and G. Picci, Asymptotic variance of subspace estimates, *Journal of Econometrics*, volume 118, pages 257–291, 2004.
3. A. Chiuso and G. Picci, Consistency analysis of certain closed-loop subspace identification methods, *Automatica*, volume 41, pages 377–391, 2005.
4. D. Bauer, Comparing the CCA subspace method to pseudo maximum likelihood methods in the case of no exogenous inputs, *Journal of Time Series Analysis*, volume 26, pages 631–668, 2005.
5. M. Deistler, K. Peternell, and W. Scherrer, Consistency and relative efficiency of subspace methods, *Automatica*, volume 31, pages 185–1875, 1995.
6. W.E. Larimore, System identification, reduced order filters and modeling via canonical variate analysis, *Proceedings of the American Control Conference*, pages 445-451, 1983.
7. P. Van Overschee and B. L. M. De Moor, N4SID: Subspace algorithms for the identification of combined deterministic-stochastic systems, *Automatica*, volume 30, pages 75-93, 1994.
8. P. Van Overschee and B. L. M. De Moor, *Subspace Identification for Linear Systems: Theory, Implementation, Applications*, Kluwer Academic Press, 1996.
9. P. Rapisarda and J.C. Willems, State maps for linear systems, *SIAM Journal on Control and Optimization*, volume 35, pages 1053-1091, 1997.
10. M. Verhaegen, Identification of the deterministic part of MIMO state space models given in innovations form from input-output data, *Automatica*, volume 30, pages 61–74, 1994.
11. J. C. Willems, From time series to linear system, Part I. Finite dimensional linear time invariant systems, Part II. Exact modelling, Part III. Approximate modelling, *Automatica*, volume 22, pages 561-580 and 675-694, 1986, volume 23, pages 87-115, 1987.
12. J.C. Willems and P. Rapisarda, Balanced state representations with polynomial algebra, in *Directions in Mathematical Systems Theory and Optimization*, (edited by A. Rantzer and C.I. Byrnes), Springer Lecture Notes in Control and Information Sciences, volume 286, pages 345-357, 2002.
13. J.C. Willems, P. Rapisarda, I. Markovsky, and B. De Moor, A note on persistency of excitation, *Systems & Control Letters*, volume 54, pages 325-329, 2005.
14. J.C. Willems, Thoughts on system identification, in *Control of Uncertain Systems: Modelling, Approximation and Design* (edited by B.A. Francis, M.C. Smith, and J.C. Willems), Springer Verlag Lecture Notes on Control and Information Systems, volume 329, pages 389–416, 2006.

New Development of Digital Signal Processing Via Sampled-Data Control Theory

Yutaka Yamamoto

Department of Applied Analysis and Complex Dynamical Systems
Graduate School of Informatics
Kyoto University
Kyoto 606-8501, Japan
yy@i.kyoto-u.ac.jp
www-ics.acs.i.kyoto-u.ac.jp/~yy/

1 Foreword

It is a great pleasure to contribute this article to the special issue in honor of Giorgio Picci on the occasion of his 65th birthday. Throughout his career, stochastic methods in modeling and filtering have been central to Giorgio's research. This article intends to describe a new idea in digital filter design, but from a deterministic point of view. I hope that it can provide a contrasting viewpoint on noise and signals in some specific contexts.

2 Introduction

Digital signals are all around us: Jpeg and other format still images, MPEG2 format in moving images, CD, MP3 music sources, etc., just to name a few.

What is the advantage of digital? Why are they so prevailing?

First, they are inexpensive and very portable, due to the uniform quality that they guarantee. They are easy to copy, distribute, without worrying much about deterioration. This is in marked contrast to analog formats. One can easily see this by iterating analog photocopying three consecutive times. The quality will be noticeably deteriorated. In digital copying, this is hardly of concern. Highest quality analog signal processing, e.g., analog audio records, may still outperform digital processing, but it is more expensive, delicate, and often vulnerable.

What guarantees this high-quality reproduction? This is precisely due to the nature of digital. One represents the original analog signals via digital data which are

1. sampled, often in uniformly separated time and space,
2. quantized, and
3. saturated.

Such characteristics allow us to represent the original data via discrete set of numbers (digits), and this assures more flexibility in processing and high-quality reproduction precision.

A. Chiuso et al. (Eds.): Modeling, Estimation and Control, LNCIS 364, pp. 345–355, 2007.
springerlink.com

Thus, while digital processing often yields the impression of high-precision, we immediately see that quality deterioration is in a sense inevitable in the three steps above. To state it differently, it is this deterioration (and simplification) that guarantees the high reproduction performance and versatility in processing.

Thus digital signal processing is not an all-win game, as opposed to the common understanding of otherwise. We certainly lose some information contents in the digitization process above. One should therefore make a judicious choice in sampling and digitizing.

This article attempts to analyze some pertinent problems mainly in relation to sampling. This has been predominantly settled through the sampling theorem—the paradigm proposed by Shannon [6, 7, 13]. But this is not necessarily the best solution when signals are not nearly band-limited. We start with a somewhat critical overview of the current signal processing paradigm, and show how control/system theory can contribute to this situation.

3   The Shannon Paradigm

Suppose we are given a continuous-time signal $f(t)$, $t \in \mathbb{R}$. Sampling reads out its values with a discrete timing, mostly with a uniformly spaced time sequence $t = nh$, $n = \ldots, -1, 0, 1, \ldots$, etc. The period h is called the *sampling period*, and $\{f(nh)\}_{n=-\infty}^{\infty}$ *sampled values*.

Given the sampled values $\{f(nh)\}_{n=-\infty}^{\infty}$, we wish to reconstruct the original continuous-time signal $f(t)$ as much as possible, with high precision. But what do we mean by *high precision?* We obviously wish to mean that it is close to the *original signal* with respect to a certain performance measure. But then where is the *original signal* here?

In tracking systems in control, reference signals are measured and the error with the system output is available through measurement. In signal processing, such signals (to be tracked) are never available. Recovering such signals is precisely the objective of signal processing. This is a fundamental difference between control and signal processing. Then how can we apply control methodology to signal processing? Is it possible at all?

Before answering this question, let us review the common framework in the current theory of digital signal processing. We start with the sampling theorem.

The sampling theorem, usually attributed to Shannon[1], answers this question under the hypothesis of band-limited signals. That is, we assume that the frequency contents are limited to the frequency range lower than π/h:

Theorem 1 (Whittaker-Shannon). *Suppose that $f \in L^2$ is fully band-limited, i.e., there exists $\omega_0 \leq \pi/h$ such that the Fourier transform $\hat{f}$ of f satisfies*

$$\hat{f}(\omega) = 0, \quad |\omega| \geq \omega_0,. \tag{1}$$

---

[1] The history of repeated discoveries of the sampling theorem is quite involved. Shannon himself did not claim originality in this fact itself [6], although his name is very popularly attached to the theorem. The reader is referred to [13].

Then the following formula uniquely determines f:

$$f(t) = \sum_{n=-\infty}^{\infty} f(nh) \frac{\sin \pi(t/h - n)}{\pi(t/h - n)}. \tag{2}$$

This theorem says that if the original signal f contains no frequency components beyond π/h [rad/sec] (known as the Nyquist frequency), then f can be uniquely recovered from its sampled values.

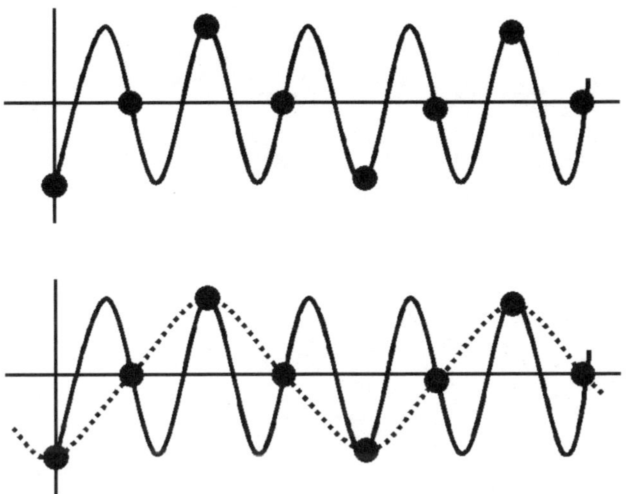

Fig. 1. Aliasing effect

Fig. 1 shows a typical situation. Consider the sine wave on the upper figure; its sampled values are indicated by black dots. The lower figure shows that these sampled values are compatible with a lower frequency sinusoid shown by the dotted curve. That is, the sampling period here does not have sufficient resolution that enables us to distinguish these two sinusoids. This phenomenon is called the *aliasing effect*, and the distortion induced by this lack of resolution is called an *aliasing distortion*. In other words, if the original signal does not contain high frequency components inducing such a behavior, it is recoverable from sampled values. The sampling theorem states that the Nyquist frequency π/h, which is the half of the sampling frequency, gives the limit of this faithful recovery.

This is essentially the content of the sampling theorem and is accepted as virtually the only guiding principle in digital signal processing for modeling analog characteristics.

3.1   Problems in the Shannon Paradigm

The principle above can however involve many problems. First, real signals are not necessarily band-limited as in Theorem 1. One may then wish this to hold approximately. However, in practice, the margin is mostly too small; for example, in the case of CD, the Nyquist frequency is 22.05kHz while the alleged audible limit is 20kHz: the margin is only 10%. Whether this is enough or not has been a long-lasting issue, but there are quite a few people who are against this format [5].

In any event, the spectra of musical sounds generally distribute very widely, way over 20kHz. If we sample them as they are, it will induce the aliasing distortion as the sampling theorem says, and hence one usually inserts an analog low-pass filter before sampling (called an *anti-aliasing filter*), with a very sharp cut-off characteristic, to avoid aliasing from high-frequency components.[2]

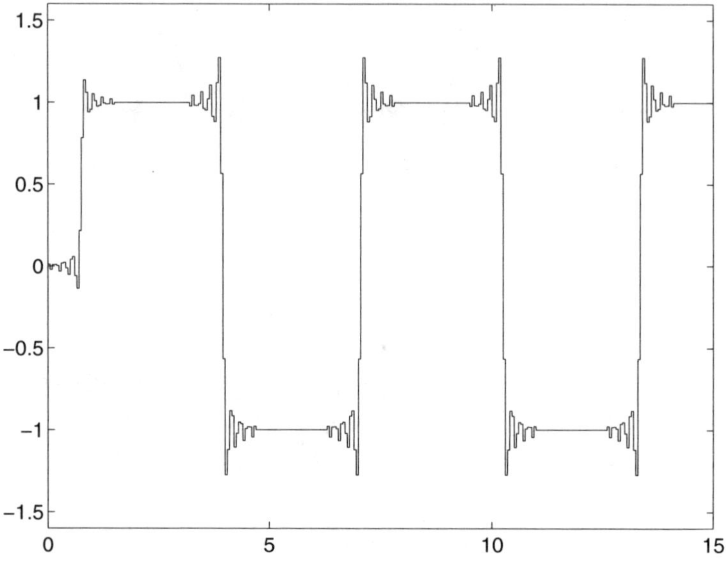

Fig. 2. Ringing due to the Gibbs phenomenon

This sharp roll-off characteristic however induces an unpleasant side-effect. For example, it induces a large amount of overshoot against square waves. This is called ringing, and is a result of the well-known Gibbs phenomenon in Fourier analysis. See Fig. 2 which shows a response of such a low-pass filter against a square wave. The common criticism against CD that it is often too metallic and harsh is probably due to this effect. To circumvent this effect, one needs more high-frequency components introducing a slower decay curve. The problem here is how we can sensibly accomplish this. To the best of author's knowledge, this *time-domain* Gibbs phenomenon[3] does not seem to have been an issue in signal processing.

---

[2] Ideally, it is desired that this filter cuts off from 1 to 0 at the cut-off frequency. This is called the (Shannon) *ideal filter*, but it is known not to be physically realizable.

[3] Oddly enough, the Gibbs phenomenon in the *frequency domain* has been a major issue in signal processing, because it was regarded harmful to realize an ideal low-pass characteristic, and various window-functions have been proposed. While the true performance should be measured in the signal domain, this preposterous attitude occurred from the objective of blindly pursuing the precision of the ideal low-pass filter.

Another problem here is that the reconstruction formula (2) is not causal. It requires infinitely many future samples to recover the current signal value. To circumvent this defect, one usually introduces a certain amount of delay in reconstruction, and allows to use finitely many future samples to reconstruct the present value. However, the convergence of this filter coefficients is slow, and it requires a large number of delays. This amounts to approximating the ideal low-pass filter that cuts off at or before the Nyquist frequency from 1 to 0. This ideal filter is not physically realizable, and approximation in the current signal processing techniques requires many, often ad hoc, techniques.

4 Control Theoretic Formulation

The Shannon paradigm may be summarized as follows:

- Given a sampling period h, we confine ourselves to the class of ideally band-limited functions whose spectrum is zero beyond the Nyquist frequency π/h.
- To force real signals into this class, one introduces a low-pass filter with a sharp cut-off characteristic.
- Once this class is fixed, one attempts to approximate the Shannon ideal low-pass filter that
    1. passes all frequency components below the Nyquist frequency, and
    2. stops all components beyond this frequency.

As we discussed already, the perfect band-limiting hypothesis is not a realistic assumption, and furthermore, the approximation problem as above leads to a filter of a long tail. Also, it introduces distortions as shown in Fig. 2. In addition, in approximating the ideal low-pass filter, one inevitably introduces a phase distortion around the cut-off frequency. The error due to this effect is often not evaluated.

To repeat, due to this unrealistic hypothesis, one introduces some artifacts, which in practice result in undesirable consequences not conceived in the ideal situation. In other words, we need to guarantee the performance when we are disturbed to a non-ideal situation. This is the question of robust control theory. *Can control theory help?*

First of all, we should find a way to reasonably confine our signal class to a more realistic class than perfectly band-limited functions.

Second, we should be able to set up a design block diagram that allows us to handle error signals. This will raise a problem, since we do not have the original signal to be compared with.

It is certainly not possible to *measure* a particular incoming signal. However, it is possible to conceive that a class of signals are fed into our digital filter to be designed, and consider the class of fictitious errors against all such signals. Suppose we have taken a class of functions $\mathcal{L}$, and let f be a signal in this class.

As a first step, consider the following abstract problem:

Consider $f \in \mathcal{L}$ and sample it with sampling period h to obtain $\mathcal{S}f$. Design a digital filter K such that, with a suitable hold device $\mathcal{H}$, the error $f - \mathcal{H}K\mathcal{S}f$ is as small as possible.

This is meaningless if we do not know f, but we can still discuss the *family of errors* corresponding to all such f. That is, we consider the *worst case error gain*:

$$\sup_{f\in\mathcal{L},f\neq 0} \frac{\|f - \mathcal{H}K\mathcal{S}f\|}{\|f\|} \tag{3}$$

and require that it be minimized. We are here not concerned with an individual f, but rather the worst case amongst all such $f \in \mathcal{L}$. Hence this is not a problem of minimizing the error for each specific (but unknown) signal but rather a problem formulated for the class $\mathcal{L}$. The disadvantage of unavailability of the error now disappears.

Of course, this abstract meta-problem is not guaranteed to be feasible. A remarkable accomplishment of modern sampled-data control theory is that this type of problem is solvable for a suitably defined class $\mathcal{L}$, in particular for L^2.

Let us now present a more concrete formulation. Consider, for example, the FFT (fast Fourier transform) readout of an orchestral music piece shown in Fig. 3. As this graph shows, the lower frequency range has more energy while the high frequency slowly decays. Suppose for the moment that this decay is governed by a proper rational function $F(s)$, and we take as $\mathcal{L}$ the set of all L^2 functions that are semi-band-limited by the decay curve governed by $F(s)$.

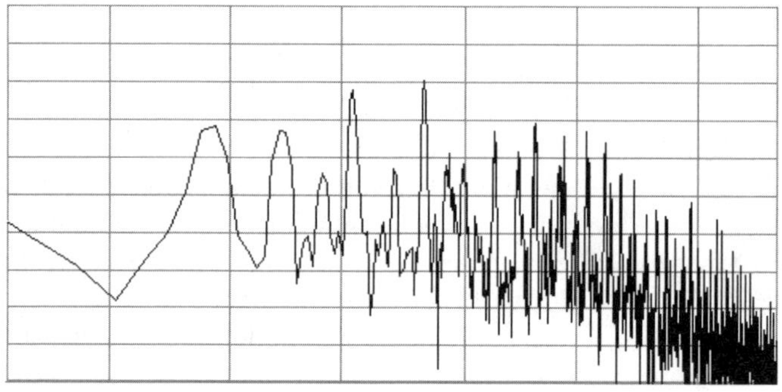

Fig. 3. FFT Bode plot of an orchestral music piece

Depending on each input signal, the output of $F(s)$ varies, but its high frequency decay is governed by $F(s)$. This analog signal is then sampled, and stored or transmitted as digital data. In CD this will be the recorded digital signal. The objective is to recover the original analog signal.

Let us express the signal reconstruction process in this process, allowing some amount of delay in the form of Fig. 4.

The block diagram 4 says the following: the external signal $w_c \in L^2$ is band-limited by going through the analog low-pass filter $F(s)$. As already noted, $F(s)$ is not an ideal filter and hence its bandwidth distributes beyond the Nyquist frequency. One can interpret $F(s)$ as a musical instrument, and w_c a driving signal. The obtained signal is sampled by the sampler $\mathcal{S}_h$ and becomes a digital signal. The objective here is how we can recover the original analog signal y_c as closely as possible. To make this possible, one needs to take a faster sampling period: the upsampler $\uparrow L$ makes the sampling period h/L by inserting $L - 1$ zeros between the original sampled points. The digital filter $K(z)$ is the one to be designed. The processed discrete-time signal then goes

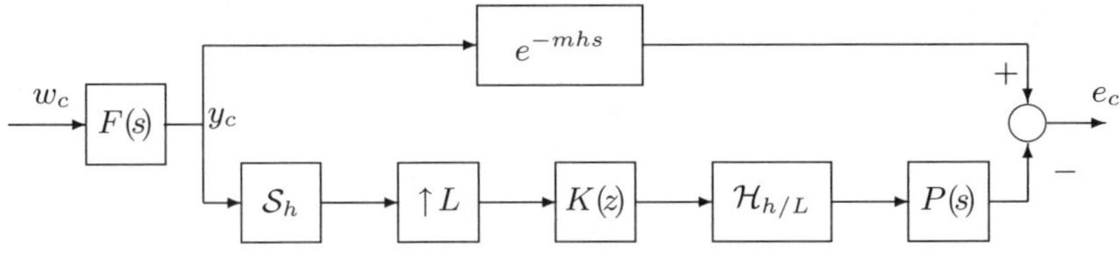

Fig. 4. Error system of a sampled-data design filter

through the hold device $\mathcal{H}_{h/L}$ and becomes a continuous-time signal. It is then further processed by an analog low-pass filter $P(s)$ to become the final analog output. This is the process of the lower part of the diagram. The design objective here is to make the difference between this output and the original analog signal y_c as small as possible. Since the processing via this digital filter inevitably induces a certain amount of delay, it is reasonable to compare the processed signal with *delayed* original signal $e^{-mhs}y_c$, rather than y_c itself. This is the idea of Fig. 4.

As a performance index, we take the L^2 induced norm from w_c to e_c (or the sampled-data H^∞ norm):

$$J := \sup_{w_c \in L^2, w_c \neq 0} \frac{\|e_c\|_2}{\|w_c\|_2} \tag{4}$$

Then this problem becomes a sampled-data H^∞ control problem. A problem here is that the delay here makes the problem infinite-dimensional, but it can be suitably approximated by the fast-sample/fast-hold approximation method [11,10].) Characteristic features here are

1. we can explicitly deal with the error signal e_c by setting up *a class of input signals*, and
2. the formulated problem is a sampled-data control problem, which is already known to be solvable.

For the solution of this problem, see [1, 3, 10, 11, 12], etc.

As we have noted already, the first point was not explicitly discussed previously, partly obscured by the perfect band-limiting assumption. Only by the procedure above, one can explicitly discuss the error and its performance level. It is exactly the second feature that enables us to solve the filtering problem that optimizes analog characteristics. The advantage of this feature cannot be more emphasized, because the current digital signal processing techniques can deal only with discrete-time problems, supported mainly by the fiction of the perfect band-limiting assumption.

Fig. 5 now shows the response to a square wave of a filter designed by the method here. We see that the Gibbs phenomenon is reduced to the minimum.

This can be applied to varied areas such as sound compression, sample-rate conversion etc. In sound compression, the bandwidth is often limited to a rather narrow range (e.g., only up to 12kHz), and this technique makes it possible to expand this to the original range by upsampling and filtering. This is patented [8,9] and already being marketed as sound processing LSI chips by Sanyo Semiconductors, and used in mobile phones and MP3 players; their cumulative total exceeds 2 million chips so far.

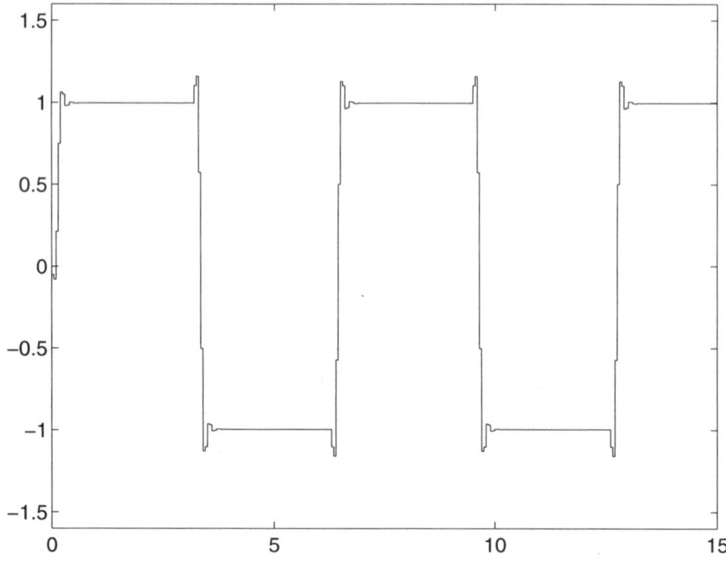

Fig. 5. Response of a sampled-data design filter against a square wave

5   Application to Images

The same idea can be applied to images. However, since images are two-dimensional, we should be careful about how our (essentially) one-dimensional method can be applied. There is no universal recipe for this, and the simplest is to apply this in two steps: first in the horizontal direction, save the temporary data in buffer memories, and then process in the vertical direction.

Fig. 6. The center of the Lena image

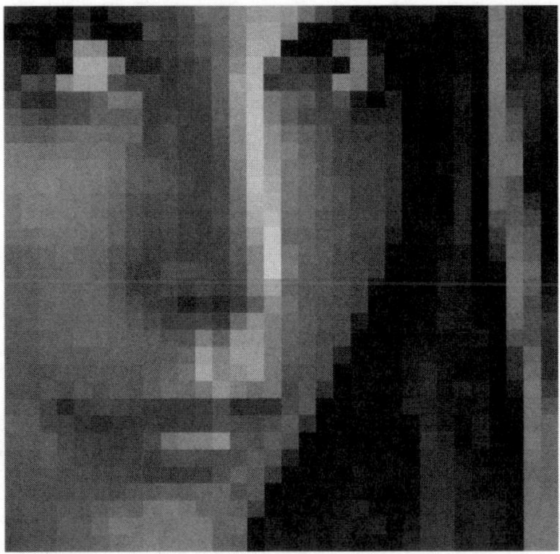

Fig. 7. Lena twice downsampled

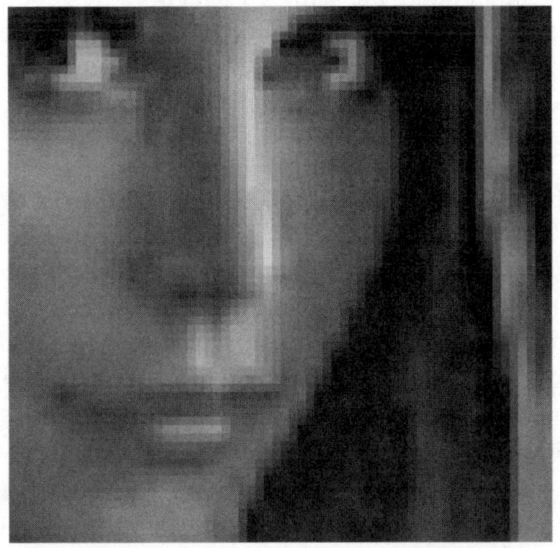

Fig. 8. Midpoint interpolation of Fig. 7

We can interpolate lost intersample data by the present framework. For example, take the well-known sample picture of Lena, Fig. 6. The next picture Fig. 7 shows its twice downsampled[4] (and degraded) image.

To recover the original resolution, we first upsample by the factor of 2. An obvious idea is to take the midpoint interpolation. This leads to the result of Fig. 8. However, the processed image still has jaggy boundaries, and the pupils look totally different from the

---

[4] I.e., decimated once every two points.

Fig. 9. Lena (via sampled-data filter)

original image. On the other hand, the sampled-data filter designed with a second order $F(s)$ resulted in Fig. 9. It is interesting to observe that some high frequency components are well interpolated around the pupils and the nose, and show a much smoother image. This is precisely the effect of interpolation beyond the Nyquist frequency.

6   Concluding Remarks and Related Work

We have shown a basic idea of applying sampled-data control theory to digital signal processing. While this is a very natural idea in that it enables us to interpolate the intersample behavior optimally, it also raises a fundamental difficulty in that reference (target) signals are not available. The key here is to set up a class of signals we want to track or reconstruct, and sampled-data H^∞ control theory provides an ideal framework for this purpose. That is, $F(s)$ in Fig. 4 models the high frequency roll-off, and considering all L^2 inputs to $F(s)$, one can discuss the possible error signals e_c derived out of this framework. This point has been quite implicit, or not even considered at all, in the digital signal processing literature.

Let us make a few remarks on related work. Chen and Francis [2] gave a first effort on applying sampled-data theory to signal processing, but as a discrete-time problem. The present author and co-workers have pursued the idea presented in this article in a number of papers; see, e.g., [3, 4, 8, 9, 12].

What does this theory suggest on signals and noise? As we see from the FFT Bode plot Fig. 3, the fluctuations are quite large in signals we process. We did not assume a noise model in our problem setting Fig. 4; instead, we have assumed an a priori energy decay curve governed by $F(s)$ there. In a sense, fluctuations from this assumption may be regarded as "noise." I hope that this would provide an auxiliary viewpoint on noise and signals, from a deterministic point of view.

References

1. Chen T., Francis B. A. (1995) Optimal Sampled-Data Control Systems. Springer, Berlin Heidelberg New York
2. Chen T., Francis B. A. (1995) Design of multirate filter banks by $\mathcal{H}_\infty$ opimization. IEEE Trans. Signal ProcessingSP-43: 2822–2830
3. Khargonekar P. P., Yamamoto Y. (1996) Delayed signal reconstruction using sampled-data control. Proc. 35th IEEE CDC: 1259–1263
4. Nagahara M., Yamamoto Y. (2000) A new design for sample-rate converters. Proc. 39th IEEE CDC: 4296–4301
5. Oohashi T., et al. (2000) Inaudible high-frequency sounds affect brain activity: hypersonic effect. Proc. Amer. Phisiological Soc.: 3548–3558
6. Shannon C. E. (1949) Communication in the presence of noise. Proc. IRE 37-1: 10-21; also reprinted in (1998) Proc. IEEE 447–457
7. Unser M. (2000) Sampling—50 years after Shannon. Proc. IEEE 88-4: 569–587
8. Yamamoto Y. (2006) Digital/analog converters and design method for pertinent filters. Japanese patent No. 3820331
9. Yamamoto Y. (2006) Sample-rate converters. Japanese patent No. 3851757
10. Yamamoto Y., Anderson B. D. O., Nagahara M. (2002) Approximating sampled-data systems with applications to digital redesign. Proc. 41st IEEE CDC: 3724–3729
11. Yamamoto Y., Madievski A. G., Anderson B. D. O. (1999) Approximation of frequency response for sampled-data control systems. Automatica: 35-4: 729-734
12. Yamamoto Y., Nagahara M., Fujioka H. (2000) Multirate signal reconstruction and filter design via sampled-data H^∞ control. Proc. MTNS 2000, Perpgnan, France
13. Zayed A. I. (1996) Advances in Shannon's Sampling Theory. CRC Press, Boca Raton

Printing: Mercedes-Druck, Berlin
Binding: Stein+Lehmann, Berlin

Lecture Notes in Control and Information Sciences

Edited by M. Thoma, M. Morari

Further volumes of this series can be found on our homepage:
springer.com